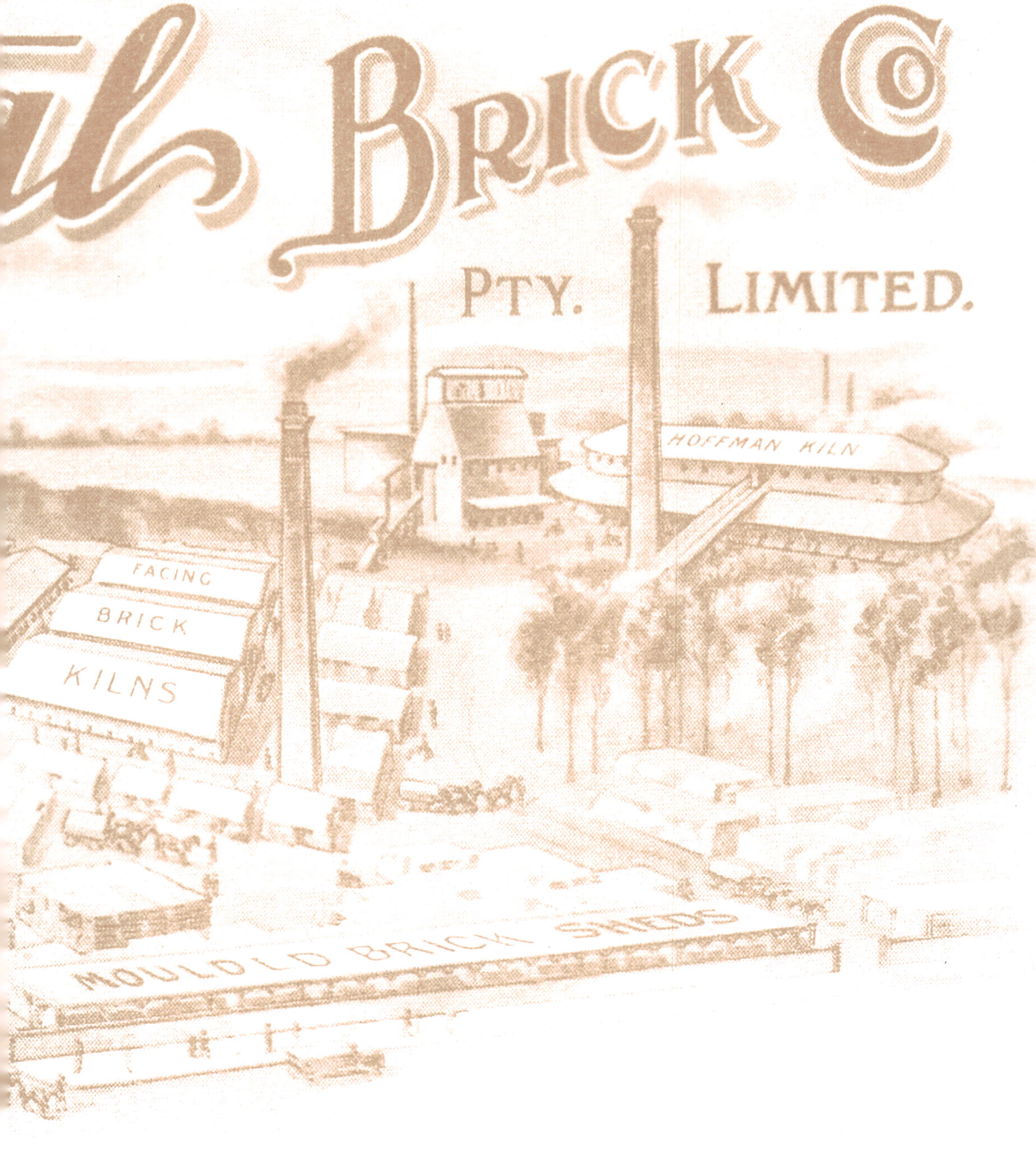
Brick Co
PTY. LIMITED.
HOFFMAN KILN
FACING
BRICK
KILNS
MOULDED BRICK SHEDS

THE BRICKMASTERS

1788 - 2008

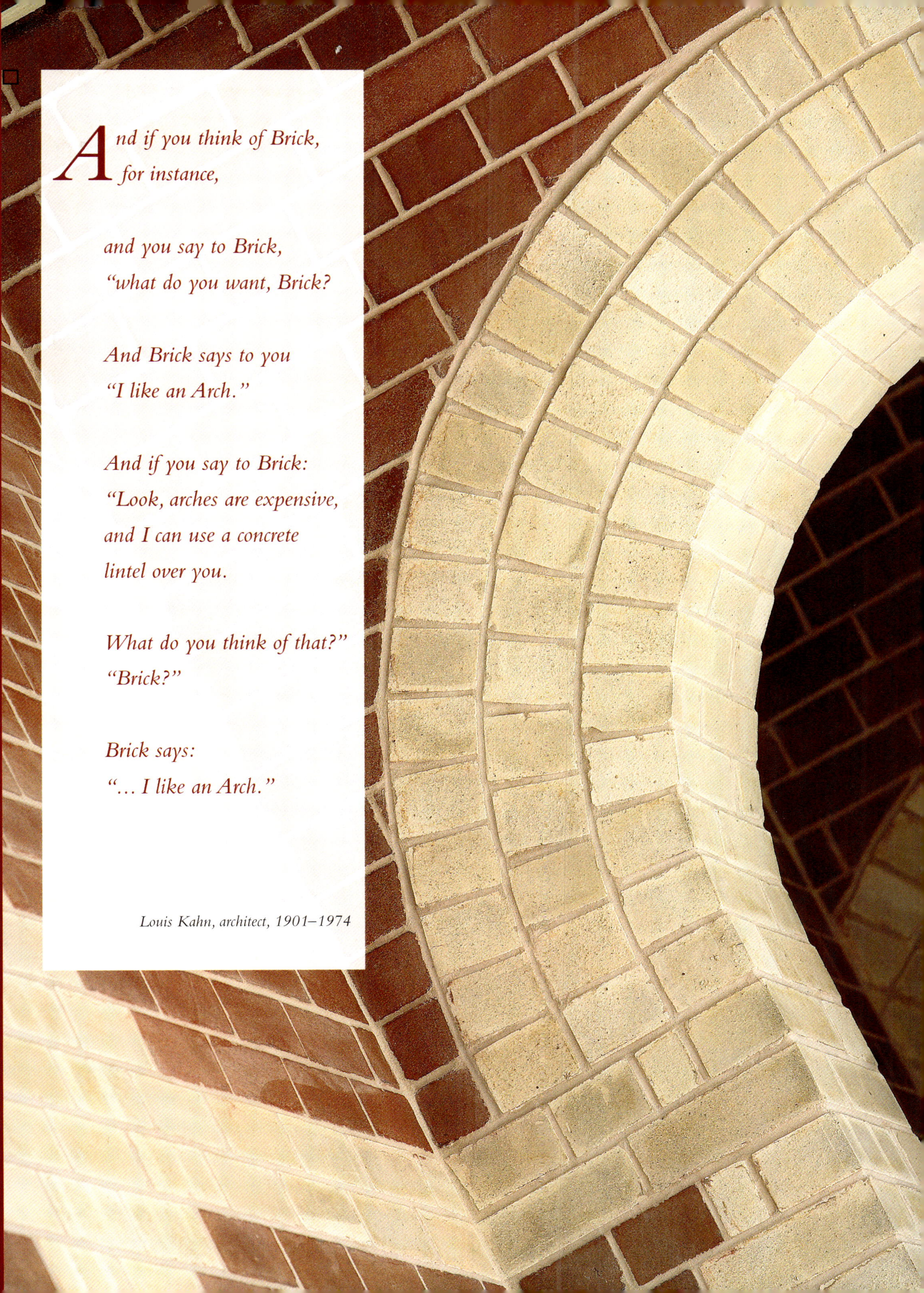

And if you think of Brick,
for instance,

and you say to Brick,
"what do you want, Brick?

And Brick says to you
"I like an Arch."

And if you say to Brick:
"Look, arches are expensive,
and I can use a concrete
lintel over you.

What do you think of that?"
"Brick?"

Brick says:
"... I like an Arch."

Louis Kahn, architect, 1901–1974

RON RINGER

THE BRICKMASTERS

1788 - 2008

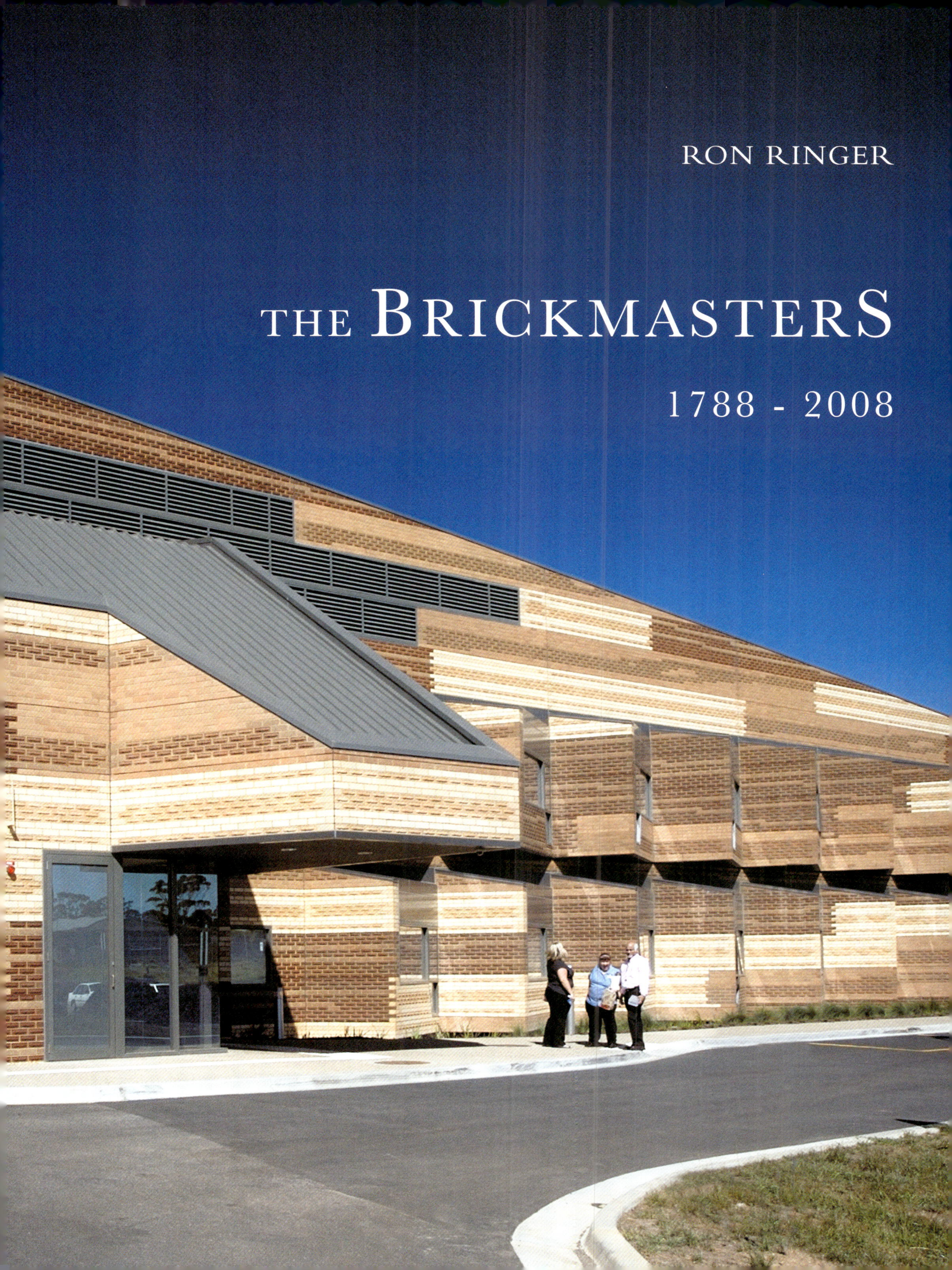

On the preceding pages:

Frontispiece:

Hyde Park Barracks, Sydney

Brick arch, Holy Name Chapel, St Ignatius College, Riverview, Sydney*

Brick arch, residential house, Centennial Park, Sydney

Hyde Park Barracks, Sydney

Sydney Technical College, Sydney

Mornington Centre, Victoria, showing polychromatic brickwork in contemporary style*

Following pages:

Pymble Ladies College, Sydney *

An engraved portrait of Arther Phillip Esq., Captain General and Commander in Chief in and over the territory of New South Wales. Frontispiece to: The Voyage of Governor Arthur Phillip to Botany Bay: With an Account of the Establishment of the Colonies of Port Jackson and Norfolk Island (1789). National Library of Australia

***** dry press face bricks, produced at Bowral Bricks, NSW***

Endpapers:

Front – Letterhead of Austral Brick Company Pty Ltd, circa 1933

Back – Letterhead of Bedford Brick Works, 1929

First published in Australia by Dry Press Publishing Pty Ltd,
Brickworks Limited,
Wallgrove Road, Horsley Park, NSW 2164, Australia

+61 2 9830 7700

ISBN 978-0-646-48825-7

Book design:
Kathy Meharry

Research and text:
Ron Ringer, email: ron@syntaxwriting.com.au

Printed by:
Green Giant Press Pte Ltd, Singapore

Contents

Acknowledgements

Lindsay Partridge, managing director of Austral Bricks, stood by this project from its inception, and for this I am truly grateful. To write a company history is one thing, but to set it within the context of Sydney, and Australia's unfolding story, is a more difficult task. Without his support, enthusiasm and patience nothing would have come of the endeavour.

As a record of the past I have kept to the known facts. As a chronicle of the development of Austral Bricks and its parent, Brickworks Ltd, no embellishment is needed to convey the story of change and continuity. At all times I have let the story speak for itself in a way that dignifies the life and times of the key players.

Along the way scores of people offered their advice and assistance. Technical advice and feedback on early drafts was freely given by a chorus of brickmasters, engineers, kiln builders, brick carters, industry spokespeople, trade unionists, members of the Dawes family, Austral management and staff and many more. Among their number were: Ross Bell, Alan Bentley, Brian Binskin, Pat and Norman Binskin, Rod Clark, Bill Dennis, Bob Edwards, Bill Emmett, Patrick Filmer-Sankey, David Fitzharris, Michael Flack, Richard Gentle, Ron Hagstrom, Vince Higgins, Tim Hirshman, Graham Holland, Adrian Howie, Cathy Inglis, Maralyn Keelty, Bill Larfield, Miles Lewis, Jim Liepins, Peter Mahony, Trevor Melksham, Jean Millner, Eddie Monson, Geoff Noble, Alex Payne, Jamie Pearson, Sandy Pearson, Jack Peterson, Bill Prest, Bob Rushton, Alex Sawtschuk, Peter Scott, John Shergold, Arthur Sherman, Rosemary Szente, Iain Thompson, John Tindall, Geoff Travis, Steve Wall, Frank Wallbank, Ric West and Doug Wilmot. Also Arthur Easton at the Mitchell Library for his kind assistance. The passing of three fine men among this number, Albert Burgis, Jim Millner and Bill Wheatley, was a sad loss. They were a mine of information and a pleasure to work with.

Another small band of friends and collaborators deserves special mention. Ray Dawes, the only surviving son of William King Dawes, and a former director of Brickworks Ltd, provided insight into critical events. Warwick Gemmell, the author of an earlier book that excited popular interest in brick pointed me in the right direction during the first few chapters. Peter Graham, a former student and long-time friend brought a critical eye and editorial skills to bear on the entire manuscript. Kathy Meharry, the designer, has produced an outstanding piece of work based on a concept by John Tuffin. Thanks to Josh Hill, photographer, and David Dickson, illustrator, who brought so much to life.

Last, but by no means least, I am grateful to my wonderful wife and friend, Robyn, and our children, Ben and Maddy for their patience at having to live with the brick story for so long, and who kept the home fires burning. Robyn, who read the manuscript and corrected errors of syntax and grammar, held me to my contract with you, the reader.

Preface

Brick has shaped the built environment more than any other building material used throughout recorded history. Wood is ephemeral and subject to rapid decay if left exposed to the elements. Stone is monumental, solid and part of the natural environment, and as a material whose use stretches back to antiquity it is the only possible rival to brick. Moreover, prior to the Industrial Revolution that swept Europe and North America during the late eighteenth and nineteenth centuries, buildings in steel and glass were unknown. The case for brick is difficult to fault. With its relative ease of manufacture and installation, cheapness and versatility as both a supporting structure or façade, brick is robust, practical and can be strikingly beautiful.

For our purposes the story begins with the discovery in 1788 of workable clays at Cockle Creek on what is now Darling Harbour in Sydney. Brickmaking in Australia not only enjoys a reputation for being the first extractive industry, but also the first manufacturing enterprise to take root in the fledgling convict colony. This work explores the development of brickmaking during early colonial times, noting the industrial transformations that carried the trade forward into the machine age of mass production. Dry press bricks, whether face or commons, used in thousands of engineering projects and countless homes, became the staple of building and construction until late into the twentieth century. From the outset a host of factors drove – and continue to drive – the industry including fashion, immigration, commerce and the business cycle, war, depression, government policy and, of course, quantum leaps in technology and its application to brickmaking. Like streams that coalesce into something broader and deeper, the development of the industry forms a river on which the history of both Sydney and European settlement in Australia is carried along.

Into this narrative The Austral Brick Company Limited made its entry in 1908 founded by a group of investors and builders led by Edwin Spackman, and establishing itself as a major player by the outbreak of war in 1914. Riding the boom of the 1920s and steering a path through the Depression years, the name Austral came to be associated with the remarkable figure of William King Dawes whose hold over Austral Bricks and Brickworks Limited extended for over a half a century. Without Dawes Austral Bricks may not have survived the devastating economic collapse. Certainly, the history of Brickworks Limited, established in 1934 by Dawes and other private brick manufacturers would have followed a different course altogether. Moreover, without the foresight of Jim Millner, the chairman of Washington H. Soul Pattinson (Soul Patts), who was the joint architect in 1969 with Dawes of the cross-holding with Brickworks, both companies may have succumbed to hostile takeover. Notwithstanding the contributions of these men, the achievements of an organisation cannot rest on two people alone. Dawes' three successors, Albert Burgis, Alan Bentley and Lindsay Partridge, the current managing director, took Brickworks and its flagbearer, Austral Bricks, to new heights. Under Partridge the company has

been transformed from a state-based operation to a national organisation that has no peer.

Outside the scope of this book lies the story of brickmaking's historic origins in the Fertile Crescent of modern day Iraq, Turkey, Iran and Jordan, which has been well documented by those who have gone before. Suffice to say, the earliest bricks can be traced back to Neolithic times (c.8300–7600BC) to Jericho where sun-dried mud bricks were laid in courses held together by more mud to make thick walls. Bricks fired in kilns only made their appearance in the third millennium BC, borrowing from the art and craft of potters whose handiwork clearly demonstrated the effect of high temperatures on clay. Fired bricks, however, remained a luxury article for thousands of years and were available only to those who commanded a supply of scarce and expensive fuels such as charcoal. As history demonstrates, without the permanent cultivation of grain crops and the construction of mud-brick dwellings human society would have remained nomadic and relatively primitive. Indeed, brick can explain a great deal about the past and the here and now as much as it can point to the future.

In recent times face brick in particular has undergone a renaissance among designers and architects. Dry-press bricks, when offset against stone, slate, steel and glass, confirm brick as quintessentially modern. Whether glazed, tinted, solid black, ebony, zinc, red earth – brick's enduring visual qualities borrow from an earlier age. Consider Berlin's Pergamon Museum and the original glazed ceramic bricks used to reconstruct the Gates of Ishtar, a remnant of the walls of Babylon. An echo of this grandeur can be seen in the work of the eighteenth century English architect, Sir Christopher Wren, whose brick buildings are still admired for their elegance and functionality. 'If you seek my monument, look around you', reads the inscription on his tombstone in St Paul's Cathedral, London. The legacy and continuing impact of brick is profound.

This book is a history that can be read on a number of levels and from different perspectives, any of which explore the achievements, the disappointments and the aspirations of a company that has provided, quite literally, the building blocks of inner city and suburban Sydney as we know it today. At another level the book will appeal to present and former brickmakers and those with an interest in the story of brickmaking in Australia. It can be also read as an alternative history of this great city. For business analysts and investors there is a wealth of detail to explore and appreciate.

Whatever the disposition of the reader *The Brickmasters, 1788–2008* is dedicated to an industry that will continue to influence the way we live and work.

Ron Ringer, Sydney, March 2008

Arthur Phillip Esq.

Captain General & Commander in Chief, in & over the Territory of New South Wales.

Published May 1. 1789 by I. Stockdale Piccadilly

Beginnings 1

Terra Australis – To build a house – An early setback – Clay is discovered – Brickfield Hill – Hard labour – The first brick carters – New farming settlements – Government House – A singular force – Private enterprise – Bloodworth's later career

First interview with the native women at Port Jackson, 29 January 1788. The pinnace stands offshore while those in the longboat meet the Aboriginal women for the first time. Captain Hunter wrote: 'They laughed immoderately, although trembling at the same time, through an idea of danger.' Watercolour by William Bradley from Bradley's 'A Voyage to New South Wales'. State Library of New South Wales

TERRA AUSTRALIS

Sydney was just ten days old on 6 February 1788 when the female convicts on the First Fleet set foot ashore. Contingents of marines and the 537 male prisoners were already hard at work, with tasks assigned and gangs organised by their convict overseers. There was much to be done in the fledgling colony of New South Wales.

Any relief those 180 women may have had at standing on solid ground after a voyage of 12 months was short-lived, for in the sweltering heat of that summer evening a spectacular thunderstorm erupted. Lightning knocked a sentry to the ground, temporarily blinding him. Valuable livestock were lost including a pig belonging to the lieutenant governor, two of the colony's lambs, and six sheep, one of which was killed by a falling tree. The storm no doubt provided an unexpected supplement of fresh meat. It may have been that from such beginnings grew a worthy Australian tradition, the barbecue.

Terrified, many of the newly arrived Europeans took refuge in their tents or prayed by their bunks as the storm raged. The historian and writer Tim Flannery offers some interesting and amusing speculative insight. 'With authority blinded or cowering under cover, the lower orders seized the moment. The sailors of the *Lady Penrhyn* obtained a double ration of rum to celebrate the offloading of the women convicts, and fortified with the ardent spirit they soon found amusement ...'[1] What followed turned into a free-for-all in which the consumption of liquor, fighting and fornicating seem to have been the general rule.

The shenanigans of this drunken bunch of seafarers and their willing accomplices must have vexed Governor Arthur Phillip. Yet as a naval officer familiar with the riotous behaviour of men allowed shore leave after months cooped up on ships, he would have taken such things in his stride.

Just how much of this would have been apparent to the Aboriginal people of the Eora clan is anyone's guess. Phillip was anxious to establish good relations with the native inhabitants and avoid incidents that might affect the 1,500 Eora who lived in the area between Botany Bay and Broken Bay. There were also the inhabitants of Sydney's northern shores to consider, people who called themselves Cadigaleans, Cadi being their name for Sydney harbour, and galean meaning 'the people of'.[2]

TO BUILD A HOUSE

The task of establishing what was, in reality, a vast open prison was immense. Tents were brought ashore to accommodate convicts, the soldiery, officers, stores, and even latrines. Offering little protection against the heavy downpours, canvas was no substitute for more permanent structures. Timber – abundant and readily available – was the first choice of building material. A sawpit had been commenced some weeks earlier at Botany Bay, the initial landing site some miles further down the coast. This location was soon abandoned in favour of Port Jackson with its plentiful supplies of fresh water. Dismantled and re-erected at Sydney Cove it allowed the carpenters (16 recruited from the ships and 12 from the convicts) to start work on timber milling. In a short space of time they had built officers' quarters, convict barracks and storehouses for the colony's meagre supplies.

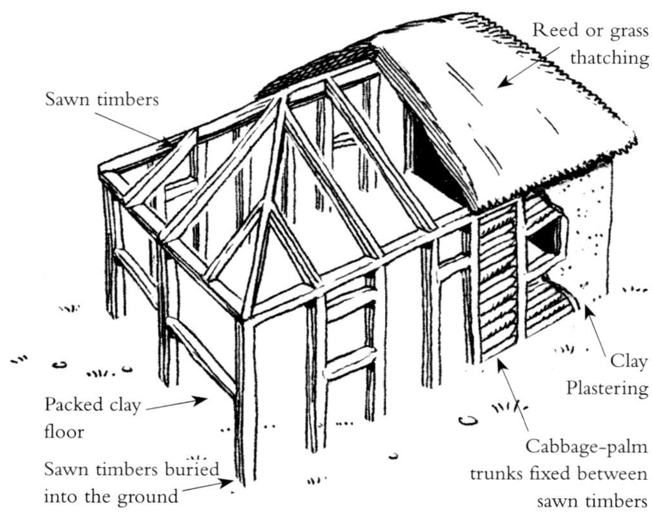

Cabbage palm hut, typical of those built in the early months of 1788. After Morton Herman

Not much to look at, the assortment of shacks dotted around the cove would have to suffice. Within weeks, however, most of the buildings were coming apart at the seams due to the warping, twisting, and rotting of the local wood used to build the crude wattle and daub hovels.[3] A pathetic sight they must have been. Remarked David Collins, judge-advocate and secretary to the colony: 'slight and temporary; every shower of rain washed a portion of the clay from between the interstices of the cabbage-tree .... their covering was never tight; their size was small and inconvenient.'[4]

Governor Phillip, a man of progressive views, envisaged durable buildings of quality, and was anxious that neither time nor labour should be further wasted upon such temporary measures.[5] Mindful of the tottering condition of so many

1. T. Flannery, *The Birth of Sydney*, Text Publishing, Melbourne, 1999, pp. 1–2.
2. Ibid. p. 8.
3. This building practice, widespread throughout England since medieval times, involved the application of an external skin of clay over a wooden frame.
4. D. Collins, *An Account of the English Colony in New South Wales [Volume 1]: With Remarks on the Dispositions, Customs, Manners etc of the Native Inhabitants of that Country. To Which are Added, Some Particulars of New Zealand: Compiled by Permission, From the Mss. of Lieutenant-Governor King*, (T. Cadell Jun. and W. Davies 1798) vol. 1, pp. 19, 71.
5. O.S. Pavlou, *The History of Bricks and Brick Making in New South Wales, 1788–1975*, Thesis, Faculty of Architecture, UNSW, Sydney, Ch. 2, p. 57.

Above: Cockle Bay circa 1819, the location of the present Darling Harbour. Watercolour by J. Taylor. Mitchell Library, State Library of New South Wales

temporary buildings, he never put aside the belief that in time the place would become a civilised town built in brick.

AN EARLY SETBACK

Phillip had anticipated the adoption of brick as a construction material in the planning of the first settlement. Records show that the cargo of the transport, *Scarborough*, included 5,000 bricks, relatively few in number, but presumably enough to construct solid foundations for a select number of government buildings.[6] Wooden boxes needed to hand mould wet clay to make bricks had also been stowed along with 5,448 squares of crown glass. Surprisingly, every form of building tool was lacking. Phillip's first request to London for stores listed house-axes, carpenter's axes, pitsaws, set-saws and crosscut saws, files, augers, nails, paint, lead and more besides. Indeed the expedition had been so badly equipped that spades were used for frying pans in those first few years of scarcity.[7]

There were more disappointments. Despite having familiarised himself with the trade and occupation of each of the fleet's 700 convicts, Phillip could find only one man skilled in the craft of making bricks. This was James Bloodworth, a London bricklayer and builder whose career had come to an abrupt end almost three years earlier, following his conviction in October 1785 on a charge of forgery. Sentenced to a seven-year term Bloodworth left England aboard the *Charlotte*, one of five convict transports contracted by the British Admiralty. Fortunately, several other men were identified as having a brick-related trade, including labourers Anthony Rope, Thomas Eccles and Walter Ruse, a farmer.

CLAY IS DISCOVERED

Still to be unearthed were deposits of workable clay necessary for the manufacture of bricks. A chance discovery was made when Phillip, anxious to supplement the colony's dwindling rations, set convicts to work clearing land for cultivation. One such group reported the presence of plentiful deposits of fine white clay about two kilometres from the settlement. Bloodworth himself may have been present at the time of the discovery. With evident satisfaction, George Worgan, 30-year-old surgeon of the *Sirius*, recorded in his diary on 28 February:

6. W. Gemmell, *And So We Graft From Six to Six – The Brick Makers of NSW*, Angus and Robertson, 1986, pp. 1–4.
7. Historical Records of Australia, Series 1, Vol. 1, p. 308.

Obverse of Sydney Cove medallion, 1789 (original issue) Josiah Wedgwood. Mitchell Library, State Library of New South Wales

Our other acquisition is the lighting on a soil which is seemingly fitted for making bricks, and eight or ten convicts of the trade are now employed in the business.[8]

Fortunately the clay field was located between two fresh water streams making the brickmakers' task a great deal easier. The two streams ran into a small inlet given the name Cockle Bay, which is now marked by Hay, Pier and Harbour Streets. Cockle Bay lay at the head of Long Cove, or Farm Cove, as it was first known. Set amid hilly country the 'Brickfields' was the focus of brickmaking until 1841. Thus, the month of March 1788, witnessed the establishment of brickmaking and mining in Australia, the earliest recorded industries in the colony.

BRICKFIELD HILL

Pressed into the service of the Crown, James Bloodworth was quick to grasp the significance of the finds along Cockle Creek. An intelligent man, his fortunes were about to change dramatically. Excavation of the clay and production of hand moulded bricks started immediately under Bloodworth's supervision. David Collins noted:

> *A gang of convicts was employed, under the direction of a person who understood the business, in making bricks at a spot about a mile from the settlement, at the head of Long Cove; at which place also two acres of ground were marked out for such officers as were willing to cultivate them and raise a little grain for their stock.*[9]

Soon the more easily won clay deposits gave out, forcing Bloodworth and his band of brickmakers to move further up the slope. Their path always followed the streams, one of which flowed across the present day George Street, just below Goulburn Street, then on into the bay. Before long an entire village – Brickfield Hill – had sprung up around the site, the area eventually comprising the upper part of what used to be known as Paddy's Market bounded by George, Campbell, Elizabeth and Goulburn Streets. Brickfield Hill remained a Sydney postal address until the introduction of postcodes in 1967.

Some idea of these early clay workings can be gained from George Worgan who, on the morning of 13 May 1788,

> *walked out today as far as the brick-grounds. It is a pleasant road through the wood about a mile or two from the village, for from the number of little huts and cots that appear now, just above ground, it has a villatick [village] appearance. I see they have made between 20 and 30,000 bricks and they were employed in digging out a kiln for the burning of them.*[10]

Bricks produced at Brickfield Hill were made from clays that formed part of a geological band referred to as Ashfield Shale, which gave them a distinctive reddish-brown colour.[11] There is no doubt that Phillip respected the judgement of Bloodworth in the matter of determining which clays were suitable for brickmaking, but he took the precaution of obtaining a second opinion from the greatest expert in the field of clay chemistry, Josiah Wedgwood. On 28 September 1788 samples of the clay were despatched to Sir Joseph Banks aboard one of the ships returning to England. Banks sent them on to Wedgwood's Staffordshire pottery at Etruria, where they were tested and pronounced excellent in quality by craftsmen who produced two medallions depicting two female figures, which symbolised labour and science.

8. George Worgan, *Journal of a First Fleet Surgeon*, cited in Jack Egan, *Buried Alive, Sydney 1788–92: Eyewitness accounts of the making of a nation*, Allen & Unwin, Sydney, 1999, p. 37.
9. Collins, op.cit p. 55.
10. Worgan, op.cit., p. 40.
11. J. Woodforde, *Bricks to Build a House*, Routledge, Kegan Paul, 1976, p. 10.

An inscription was added, *Sic fortis etruria crevit* (thus mighty Etruria grew) – as a design for Sydney.[12] Wedgwood, or his craftsman, may have had in mind the ancient land of Etruria and the Etruscans who gave succour to the emerging Greek and Roman civilisations.[13] Three years later further consignments of clay were despatched to Banks, including samples drawn from Sydney, a box of Norfolk Island clay and samples of white clays that had been discovered. Wedgwood's tests were doubly valuable for they confirmed the suitability of Sydney's clay for pottery, an equally important clay industry. In fact, pottery and brickworks, both of which were wood fired, were often combined in one enterprise at Brickfield Hill with the result that by 1804 several brickworks, combined with potteries – private and government-run – were operating in the area. The manufacture of clay pipes was also well established by this time.

HARD LABOUR

Worgan's mention of the pleasant road leading to Brickfield Hill with its village-like appearance must have been a comforting reminder of home. Yet within 15 months of its establishment, the area harboured many of the colony's most intractable and hardened convicts, sent there as a punishment for various misdemeanours: a prison within a prison. On short rations, and unused to hard manual labour, these men nevertheless provided the building materials for the establishment and growth of the colony.[14] The basic wage of just one shilling a year did little to discourage criminal activity.

> *The convicts who were employed in making bricks, living in huts by themselves on the spot where their work was performed, were suspected of being the perpetrators of most of the offences committed at Sydney; an order had been given, forbidding, under pain of punishment, their being seen in town after sunset.*[15]

In fact, burglary and assault in Sydney Cove grew to such an extent that night watchmen were recruited. On the night of Saturday 8 August 1789, the first 'police' reported for duty: ironically convicts who had property and were intent on its defence. The circumstances surrounding many of these burglaries were almost comic. According to one early colonist, mud mortar in walls 'was so soft that burglars will pick a hole through one of such in a very few minutes, no part of the house being safe, back, front and gable being equally inviting.'[16] With food in short supply and the plundering of the government stores, theft became a hanging matter.

Compounding the nuisance of petty crime was the escalation in the number of clashes with the native people due to the encroachment of white settlement on Aboriginal hunting lands. Labourers at Brickfield Hill were some of the first to be caught up in the violence.

> *A convict belonging to the brickmaker's gang had strayed into the woods for the purpose of collecting sweet tea [a herb] ... It was supposed that the convict in his search after this article had fallen in with a party of natives, who had killed him. A few days after this accident, a party of the convicts, sixteen in number, chiefly belonging to the brick-maker's gang, quitted the place of their employment, and providing themselves with stakes, set off toward Botany Bay, with a determination to revenge, upon whatever natives they should meet, the treatment which one of their brethren had received at the close of the last month. Near Botany Bay they fell in with the natives, but in a larger body than they expected or desired. According to their report, they were fifty in numbers ... it was certain, however that they were driven in by the natives, who killed one man and wounded six others.*[17]

THE FIRST BRICK CARTERS

Despite clashes with Aboriginal people the business of brickmaking continued to expand and was barely able to keep up with demand. On 4 June 1789 Collins reported:

12. *The Clay Products Journal of Australia*, October 1933, p. 9.
13. This motto can be seen inscribed on a coat of arms affixed to the exterior of Westpac bank on George Street, Sydney, facing directly opposite Martin Place.
14. Gemmell, op.cit., pp. 1–4.
15. Collins, op.cit., p. 97.
16. P. Cunningham, *Two Years in New South Wales*, Henry Colburn, London, 1827, Vol.1, pp. 61–2.
17. Ibid. p. 82.

View of Brickfield Hill or High Road to Parramatta on 11 August 1796. A team of bullocks attached to a primitive dray, struggling up the Brickfield Hill Road (now George Street), between Liverpool and Bathurst Streets, which was the only surveyed road south of the Burial Ground. Watercolour by Edward Dayes. National Library of Australia

> *The gang under the direction of the overseer employed at the brick fields had hitherto only made ten thousand bricks in a month. A kiln was now constructed in which thirty thousand might be burnt off in the same time, which number the overseer engaged to deliver.*[18]

Trees were felled and a road was made to transport the bricks from the flats (presumably the area adjacent to Cockle Creek) to Sydney.[19] In the absence of horses or bullocks, gangs of convicts were given the job of dragging heavy brick carts over the sharp rise to the main settlement more than a kilometre away towards Farm Cove. Each cart, laden with 750 kilograms of bricks, was hauled by a 12-man team which made nine trips a day. Carting was considered the most extreme punishment and highly dangerous on account of the primitive roads of the early colony. In 1796 bullocks, and later horses, replaced the convict chain gangs working the brick carts, but the deeply rutted tracks still posed a serious danger to vehicles and pedestrians alike. Eventually, during Macquarie's administration in the 1810s, the various brickyards employed 180 convicts who used 90 horses and several bullock teams to cart bricks and tiles to building sites in Sydney town.[20] An edict of 1806 instructed all empty public and private vehicles passing Brickfield Hill to load up with brickbats – broken bricks – and deposit them at various points along the road to Parramatta.[21] This early attempt to encourage road building seemed to have worked. During a visit to Sydney in 1803 by French explorer, Nicolas Baudin, the official artist, Charles-Alexandre Lesueur, observed that the road from Brickfield Hill to Botany Bay was paved for a distance of three miles.[22]

18. Ibid. p. 93.
19. From about the middle of the nineteenth century the area extending from Alexandria to Tempe was also referred to as 'The Flat'. Brian Binskin, an employee of Austral Bricks at the Eastwood yard, recalls this expression being used during his boyhood years in the early 1950s. According to Binskin the comment would have gone something like: 'I knew if I couldn't get a start anywhere else, I could always go back to The Flat and find work' (Interview: 8 August 2005). This may have been a general expression used to describe brick workings, which by their nature, required level ground, and may not have referred in this instance to the area of Brickfield Hill.
20. Malcolm J. Kennedy, *Hauling the Loads: A History of Australia's Working Horses and Bullocks*, Melbourne University Press, 1992, p. 53.
21. Gemmell, op.cit., p. 4.
22. Charles-Alexandre Lesueur joined an expedition led by Captain Nicolas Baudin which left Le Havre in October 1800 for New Holland or Terra Australis as it was also known. Lesueur took over as expedition artist when the official artists jumped ship in Mauritius. The two expedition ships, *Le Naturaliste* and *Le Géographe*, reached the west coast of Australia in 1801 and arrived in Sydney in May 1802. Baudin and his men were hospitably received in the British colony by Governor King and remained there for five months. They sailed in November 1802.

NEW FARMING SETTLEMENTS

Bloodworth continued to train brickmakers and bricklayers who figured prominently in Collins' diary which also referred to the establishment of a new farming settlement further up the Parramatta River at Rose Hill in September 1789.

> *The Governor is purposing to erect a capacious storehouse and a range of barracks at Rose Hill, a convict who understood the business of brick-making was sent up for the purpose of manufacturing a quantity sufficient for those buildings, a vein of clay having been found which it was supposed would burn into good bricks.* [23]

Unfortunately the bricks produced at Rose Hill, later renamed Parramatta, were not as good as those made at Sydney. Their colour was an attractive deep red when burned but they were not durable, which may have been due to the inexperience of Bloodworth's protégés. Meanwhile, exploration along the Hawkesbury River continued to push the boundaries of settlement. In 1791 a third convict farm was established at Toongabbie, several miles beyond Rose Hill. Not only did these farms help to alleviate critical food shortages in Sydney Town, they became major centres of population in themselves. Further afield, Van Diemen's Land – modern Tasmania – was settled relatively late, in 1803. Here, brick buildings were common by the 1820s due largely to a plentiful lime supply for mortar.

GOVERNMENT HOUSE

More than just a brickmaker and bricklayer, Bloodworth quickly established a reputation as a builder. Several months after having arrived in the colony he was asked to construct an official residence for Governor Phillip, whose original dwelling was a framed oilcloth house which he had purchased for £125 from Smith & Baber, the London manufacturer. It seems likely that Bloodworth had a hand in the design of the building, although it was attributed to Lieutenant Dawes.

23. Collins, op.cit., pp. 140,144.

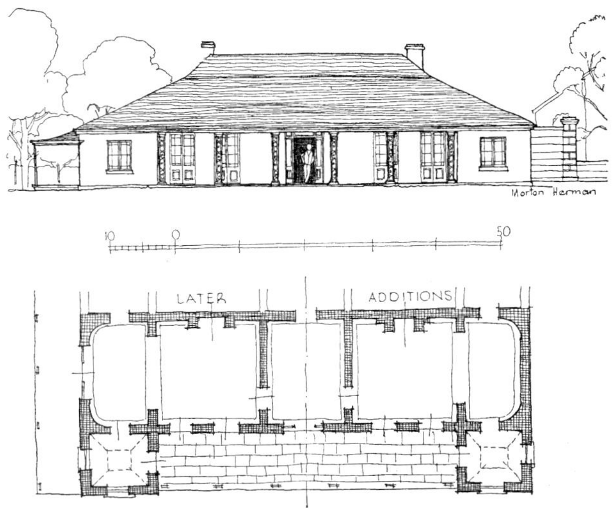

Elizabeth Farm, near Parramatta. Morton Herman

Phillip set Lieutenant Dawes to work locating a suitable site and within a short time he selected a hill on the west side of Farm Cove where Grosvenor Street now crosses George Street. Major Ross, the lieutenant governor, who had already started work on his own residence and parade grounds, thwarted Dawes in his plans. To avoid unnecessary friction with his second-in-charge and commander of marines, Phillip asked Dawes to find another piece of land on what is now the southwest corner of Phillip and Bridge Streets, currently occupied by the Museum of Sydney.

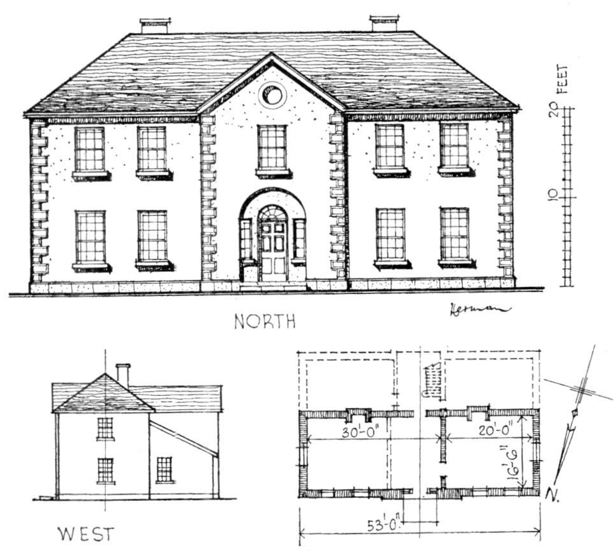

Government House. Formerly in Bridge Street, Sydney. Morton Herman

View of the Governor's house at Sydney in Port Jackson, New South Wales, January 1791. William Bradley. National Library of Australia

Work commenced towards the end of April 1788 and progressed steadily so that Governor Phillip was able to lay the foundation stone one month later. Soon after, Bloodworth's plans for a three-roomed house were amended, the footings now deemed sufficiently robust to support a two-storey building. Archaeological excavations during the 1980s revealed that they were built using mortar composed of mud and clay.[24] Brick and stone were used from the foundation to floor level, then solid brick walling with stone quoins to roof level. The internal walls were of single brick construction, the thickness of the external walls being one and a half bricks, which were rendered with pipe clay to protect against wind and rain. The only plasterer in the colony, James Thady, was quickly assigned to Bloodworth's gang. Thady, incidentally, is recorded as having been involved in a dispute between Ross and Phillip.[25]

The house was a typical example of Georgian architecture. A wooden staircase gave access to the first floor, which comprised bedrooms for Phillip and his party. The Aborigines were fascinated with the staircase, which allowed men to walk around above the heads of people standing on the ground floor. Arabanoo, an Aborigine captured at Manly and brought to Sydney, marvelled at the Governors' House with its staircase. He could not conceal his surprise at a *gunyah* – dwelling – in which some men could be inside and at the same time be walking over other men's heads.[26] In fact, the staircase was to be the only one in the colony for a long time to come. Tiles made at the brickyards covered the hipped roof, and wooden pegs made by the convict women secured them.

So it was that on 4 June 1789, just 16 months after the first landing at Sydney Cove, the denizens of Sydney gathered to celebrate the birthday of King George III and the opening of Government House.

24. H. Proudfoot, *The First Government House, Sydney;* March 1983, Report for the Department of Environment and Planning.

25. G. Mackaness, *Admiral Arthur Phillip, Founder of New South Wales, 1738–1814*, Angus and Robertson, 1937, p. 220.

26. Historical Records of New South Wales (HRNSW), 2, p. 794; Watkin Tench, pp. 9–11.

It was a 'bring your own bread' affair due to the severe food shortage in the colony.

Despite recurrent problems with dampness and rotting timbers, Government House remained the official residence in Sydney and the most imposing structure until 1845. Demolished two years later, its foundations disappeared beneath the pavement. The Mitchell Library holds some of the original bricks in addition to a copper plate commemorating the landing of the First Fleet. The latter was a chance find made in March 1899 by a Sydney workman, Robert McCann, in the course of excavation work on a telephone tunnel. McCann's pick unearthed the foundations of the old Government House about a metre below the footpath.[27]

A SINGULAR FORCE

Bloodworth continued his work, designing and building with brick and stone, mostly for the government and important officials. In November 1789, a brick house was commenced for Judge-Advocate Collins who was well pleased 'to occupy a brick house put together with mortar formed of the clay of the country, and covered with tiles ... of comparative comfort and convenience.'[28] Collins' house was the envy of his friends. Within a matter of months the Reverend Richard Johnson, Surveyor-General Augustus Alt and Surgeon White were housed in brick homes. Eager to secure the goodwill of the Aboriginal people, Phillip ordered a small brick building 12 feet square to be erected for his friend, Bennelong, on the eastern point adjoining the cove where the Sydney Opera House now stands.

In 1790 Bloodworth completed a large dry store on the east side of Sydney Cove. Unusually tall for a building of the day, the store possessed several striking features, including a hipped roof, an attic and the first dormer windows in Australia. In the same year Bloodworth was granted a pardon by Governor Phillip for good conduct, and won

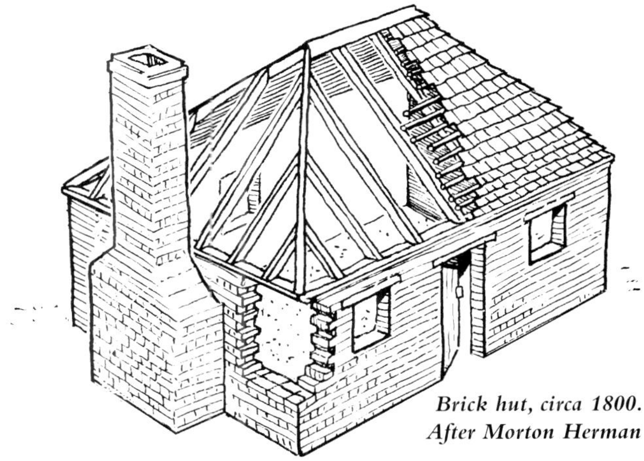

Brick hut, circa 1800. After Morton Herman

further recognition in 1791 when he was appointed supervisor of all brickmakers and bricklayers in the settlement. In 1792 his debt to society was cancelled in full.

> *James Bloodworth ... received the most distinguishing mark of approbation which the governor had in his power to give him being declared free, and at liberty to return to England whenever he should choose to quit the colony. Bloodworth had approved himself a most useful member of the settlement, in which there was not a house or building that did not owe something to him; and as his loss would be severely felt should he quit it while in its infancy, he bound himself by an agreement with the governor to work for two years longer in the colony, stipulating only to be fed and clothed during that time.*[29]

Captain Watkin Tench of the marines wrote an amusing account of early settlement in which he mentions Bloodworth with great affection. Despite being offered repatriation to England he chose to stay in Australia. Under Bloodworth's guidance building work continued at a rapid pace in the settlements at Sydney Cove, Parramatta and Toongabbie. In the course of 1793, one year after Phillip's departure as Lieutenant Governor, 160 houses were built in Sydney, each allowed 1,400 bricks for a chimney and a floor. Most of these were located along High Street and Pitt's Row (renamed George Street and Pitt Street by Governor Macquarie after 1810) and extended in a continuous line to the brickfields.[30]

27. O.S. Pavlou, *The History of Bricks and Brick Making in New South Wales, 1788–1975*, Thesis, Faculty of Architecture, UNSW, Sydney, 1975, p. 85.

28. Collins, op.cit., Vol. 1, p. 104.

29. Collins, op.cit., Vol. 1, p. 142.

30. Collins, op.cit., Vol. 1, p. 275.

Five years later Bloodworth's skills were put to use in the building of a 45-metre clock tower on Church Hill, near the present Anglican church of St Philip. For some years it remained the tallest structure in the colony, but was destroyed during a violent storm in June 1806. Government House at Parramatta (1799) is also attributed to Bloodworth and was constructed on the site of the first house, the foundations of which can still be seen.

Within 20 years or so of the construction of Government House at Sydney, the skyline came to be dominated by impressive houses commissioned by successful emancipists including Simeon Lord and William Kent.[31] Both of these men had risen to become highly successful merchants and entrepreneurs. While others (including officers of the New South Wales Corps) had made fortunes from trading, these merchants were the first to 'demonstrate their success materially in the size and appearance of their establishments'.[32]

Gentlemen, merchants and traders representing Sydney's expanding industrial base were eager to establish their credentials. For them a brick residence not only provided social cachet, but was also the ideal business premises from which they could maintain strict control over their affairs. Military men, on the other hand, appear to have had little interest in bricks and mortar, housed as they were in the town barracks – their accommodation ensured by military rank.

PRIVATE ENTERPRISE

One keen observer of the activities of Bloodworth and his protégés would have been John Palmer, purser of the *Sirius*. Appointed Commissary General to the First Fleet, Palmer completed his term in office in 1793 and received a land grant in Woolloomooloo, in recognition of his services.

31. Simeon Lord was transported in 1791 and by 1798 was recorded as acting as agent for some of the captains visiting Sydney. His commercial activities encompassed shipbuilding, ship owning, pearling, and the export of cedar, iron smelting and cloth manufacturing. Lord died at Botany in 1840.

32. J. Broadbent, *The Australian Colonial House: Architecture and Society in New South Wales, 1788–1842*, Hordern House, Sydney, 1997, p. 14.

Some time after, he opened a brick business and boat yard situated between what is now Palmer Street and Bourke Street. The brickyard serviced builders in the immediate area, making and supplying bricks to Sydney Town, although the yard was small by comparison to those at Brickfield Hill.

The arrival of further convict transports as well as free settlers had an immediate impact on the developing brick industry. Demand for brick housing rose, as did the quantity of bricks being turned out by the yards, both public and private. Skilled labour, especially hand moulders, was in short supply, and enterprising convicts were quick to take advantage of Governor Phillip's system of task work, which allowed work on private contracts after the day's official finishing time of three o'clock. Business was booming for convict brickmakers and freemen alike.

Tench wrote in 1790 of the work assigned to Samuel Wheeler, who was transported at the same time as Bloodworth:

> *Wheeler, with two tile stools, and one brick stool, was tasked to make and burn ready for use thirty thousand tiles and bricks per month; he had twenty-one hands to assist him.*[33]

It may be that under Bloodworth's supervision men such as Wheeler entered the craft of brickmaking. The reference to Wheeler indicates this was one of the biggest gangs on the field, although smaller gangs were producing proportional amounts. One belonged to John King, also on the First Fleet, about whom Tench observed: 'last year, with the assistance of sixteen men and two boys King made eleven thousand bricks weekly with two stools.'[34] He was assigned another four men on account of the increased distance needed to carry wood to fire the kilns. Indeed, the scarcity of bricks prompted King to claim that in England his bricks would have fetched one guinea per thousand at prices then being charged. Another brickmaker by the name of Becket is known to have been working at Rose Hill.

33. Watkin Tench, *Sydney's First Four Years*, Angus & Robertson, Sydney, 1961, p. 162.

34. Ibid., p. 193.

Private individuals in the market for bricks took what they could, usually the poorer, underfired bricks. The better quality fired bricks were requisitioned for official buildings. Needless to say, a thriving black market existed for those who could afford to pay. Whether used for official or private purposes, overall quality was extremely poor due to inadequate firing, which left bricks highly porous. Water could often be seen trickling down the inside walls of houses during heavy rain. Indeed, porous bricks and damp penetration remained a serious problem until the general introduction of dry press bricks and cavity walls in the nineteenth century.

BLOODWORTH'S LATER CAREER

Acceptance back into civil society must have given Bloodworth particular cause for satisfaction. In 1802 he was made a sergeant in Sydney's Loyal Association, a volunteer militia numbering about 200 men. Its counterpart in England was formed to act as a home guard in the event of an invasion by Napoleon's armies. For the authorities in Sydney the association's main purpose seems to have been to counter outbreaks of rebellion by disgruntled convicts and men who had gained their freedom through a ticket-of-leave.[35] The New South Wales Corps and the Loyal Association shared the same uniform, the only distinguishing feature being the cropped hair of the volunteers. It was also about this time that Bloodworth was farming a grant of 50 acres (20 hectares) at Petersham, which he later increased to 245 acres (99 hectares).

David Collins remained a champion of Bloodworth, writing in 1803 to Governor King on the eve of the establishment of a settlement at Port Phillip Bay:

> *I need not point out, how extremely useful he may be to me, so conversant as he is, not only in Constructing but in Planning the several Buildings that I must very shortly have in hand. I have not any one that I could place with so much satisfaction, to myself in all directions of my Carpenters, Sawyers, Bricklayers &. as this Person.*[36]

Shortly before his death from pneumonia on 21 March 1804 Bloodworth was appointed Superintendent of Building on a yearly salary of £100. This may have gone some way towards mitigating his reduced financial circumstances. Despite having the asset of his farm and a government salary, Bloodworth was insolvent when he died at his house, which was located in what was later called O'Connell Street, Sydney. He was survived by five of seven children from his marriage to Sarah Bellamy, who was herself a convict, having arrived on board the *Lady Penrhyn*.

Governor King ordered Bloodworth's passing to be marked by the equivalent of a state funeral, an honour never before given to a former convict. Bloodworth's death was of sufficient importance for the *Sydney Gazette* to carry this announcement on 25 March 1804:

> *On Wednesday last died, generally lamented, Mr. James Bloodworth for many years Superintendent of Buildings in the employ of the Government.*
>
> *He came to the Colony among its first inhabitants in the year 1788, and obtained the Appointment shortly after his arrival for his exemplary conduct. The first house in this part of the Southern Hemisphere was by him erected, as most of the public buildings since have been under his direction.*
>
> *To lament his loss he has left a Widow and five children, the youngest an infant now only one week old, and the complaint that terminated in his dissolution was supposed to proceed from a severe cold contracted about two months since. The attention and concern which prevailed at the interment of the deceased were sufficient testimonies of the respect with which he filled, and the integrity with which he uninterruptedly discharged the duties of a Public Trust*

35. The ticket-of-leave was a document issued to a convict enabling him or her to work for wages and choose a master. In 1790 Governor Phillip was given power to shorten the sentence of convicts, resulting in the development of this system.

36. Collins to King, 5 November 1803, *'Governor King's Letterbook'*, ML A2015, p. 345. 1797–1806.

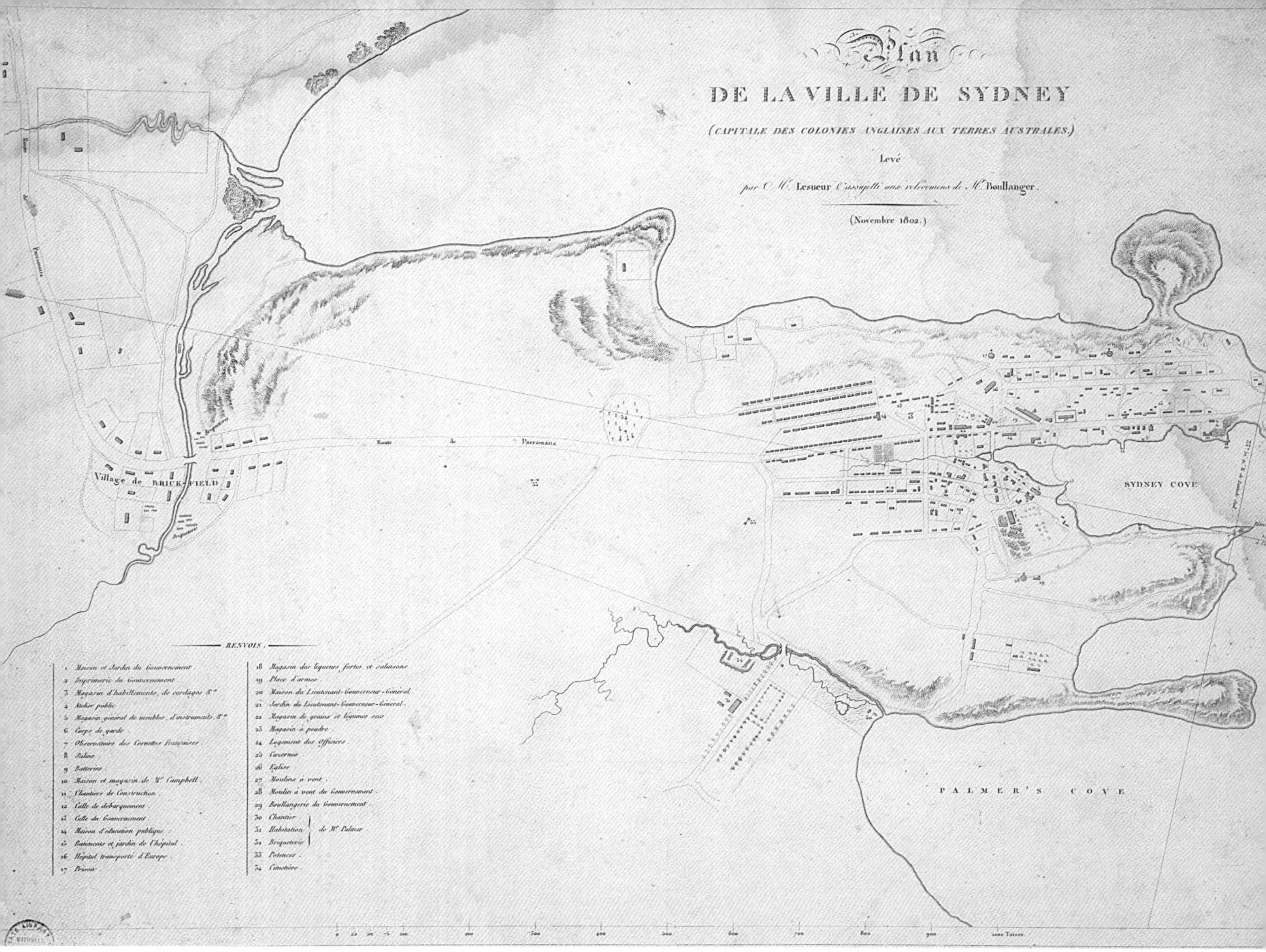

Lesueur was a gifted illustrator who commenced as assistant gunner aboard Le Géographe. He was appointed artist to the expedition following the death of zoologist René Maugé in February 1802 and confirmed as official artist when the leader of the expedition, Captain Nicholas Baudin, died in February 1803. Mitchell Library, State Library of New South Wales

> *during so long a period. His Excellency was pleased to order that a Funeral should be provided for at the Public Expense, and to show other marks of attention, to so old a servant of the Crown.*[37]

The article went on to describe the atmosphere of the funeral, mentioning the mourners in order of importance. Finally:

> *When near the Burial ground the Association were obliged to file off, for the accommodation of the friends of the deceased, and the populace, who were become very numerous, and when the remains were deposited approached the grave and performed Military Honours.*[38]

This last sign of respect was performed by his companions in the Loyal Association at what is now the site of the Sydney Town Hall. Thus ended the first chapter in the beginnings of brickmaking in Australia. By this time, the industry was established, and as the number of bricks and clay roofing tiles made at Brickfield Hill and Woolloomooloo increased, so did the number of brick buildings. With good quality building timbers being supplied from the Hawkesbury and the Hunter Valley, and glass being imported from England, the architecture of Sydney began to take shape.[39]

37. *Sydney Gazette*, 25 March 1804.
38. Ibid., 25 March 1804.
39. J. M. Freeland, *Architecture in Australia: A History*, F.W. Cheshire, Melbourne, 1968, pp. 34–36.

Early Manufacture 2

Along Cockle Creek – Pugging – Moulding – Drying – Clamp kilns – Scotch kilns – Building crisis – Plight of the limeburners – A social place

ALONG COCKLE CREEK

In the area of the Haymarket valley where swamps intersected by tidal creeks emptied into Cockle Creek, brickmakers worked full stretch to keep up with the burgeoning demand. At every stage the task was arduous and labour intensive with little by way of machinery to speed production. Let us, however, retreat a few steps here to consider the nature of bricks and brickmaking at the turn of the nineteenth century. In the absence of records it is reasonable to assume that brick manufacturing techniques in Australia were similar to those used in England, which had continued virtually unchanged for centuries.

A brickmaking team usually comprised at least three people – a clot moulder, a brickmaker and a take-off boy who removed the clot from the mould. Extra hands engaged in a variety of tasks, including hauling sand from Long Cove, carting the prepared clay from the pits to where bricks were made, and taking them away ready for drying. A production line it certainly was, albeit primitive by modern standards. From a pit adjacent to Cockle Creek, clay was excavated using shovels, transferred to shallow depressions in nearby rock outcrops, and mixed with fresh water.

The sodden mass of raw clay was then crushed and kneaded into a lumpy texture using the ends of small tree trunks or bare feet. Settling pits caught any slurry. Stones and vegetable matter were removed to avoid fractures in the bricks during drying and burning. The resulting 'puddled' or 'pugged' clay was left to stand or 'prove' for up to five days. Hand pugging was still being used until the late 1840s in country districts, although in the main settlements pug mills were quickly adopted, becoming the first step in the mechanisation of the brickmaking industry in Australia.[1]

Family at work moulding sandstock bricks and delivering them to the hack for drying

PUGGING

Introduced early from England where it had been in use for centuries, the pug mill was a circular barrel-shaped wooden tub with an upright revolving shaft through its centre. A beam measuring about five to six metres in length was attached to the shaft that could be driven either by manual labour or by horsepower. Revolving blades attached to the shaft cut and kneaded the clay into a plastic state. Masticated clay was forced through a hole in the base of the barrel, where it was cut off in pieces with a shovel and laid to one side. To achieve the right consistency for moulding, water was added during the pugging operation. Sand was thrown over the pugged clay and covered by sacking to reduce evaporation in the hot sun.

These early milling machines were a great improvement on hand pugging methods. Clay of a more homogeneous consistency could be mixed using less water, which made for stronger bricks of higher density. With manual labour at a premium the mechanical pug mill required just one operator to produce larger quantities of prepared clay. Being

1. J. Nangle, *Australian Building Practice: A Treatise for Australian Students of Building Construction*, Melbourne, G. Robertson & Co., 1900, p. 46.

Horse–operated pug mill at the brickyard of Fred and Harry Baker, brickmakers, Raymond Terrace. Watercolour by W. Green. Courtesy Pat Baker

portable it could easily be set up by the brickmaker wherever he chose and allowed a great deal of flexibility in deciding where to erect the kiln.

MOULDING

When the pugged clay was ready for moulding, the heavy, water-saturated mass was hand carted a short distance to a type of bench made from wood or metal. Measuring approximately two metres long by about one metre wide, and standing at a height of one metre, the bench was ideal for the task. Wheels were sometimes attached to two of the four legs which made it easier to move the table to the source of the clay.

Brick moulding usually took place under a hovel, a rough open-sided timber shelter about five metres square, with a roof supported on wooden posts, and thatched with reeds or covered with bark. Safely under cover, and protected by matting hung from the eaves, the brickmaker was able to work through rainy periods. Matting also prevented the hot winds from drying out the clay during the summer months. Small heaps of sand were placed at opposite ends of the moulding table, next to which stood a barrel of water in which a strip of timber, the 'strike', was kept.

On the table rested the brick mould, a bottomless rectangular box of timber, slightly deeper than the height of a brick. The extra depth allowed for a removable base about 25 millimetres thick, called the 'stock board', to be fitted snugly into the bottom of the mould. Slightly larger than the base of a burnt brick, the stock board was fitted with a 'kick' which was used to form a depression or 'frog' in the brick.

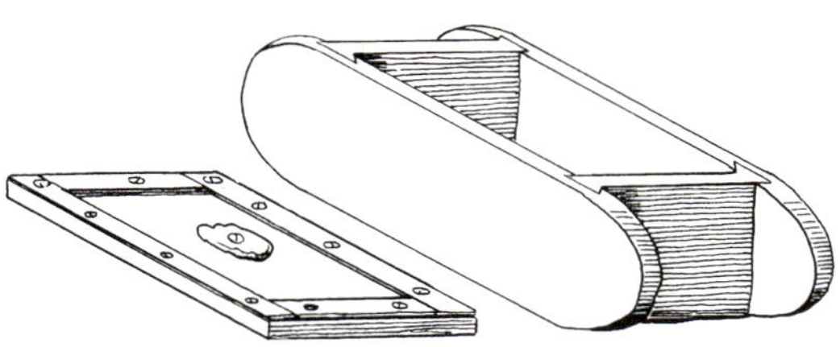

Wooden brick mould of the type used by hand moulders in Australia into the twentieth century

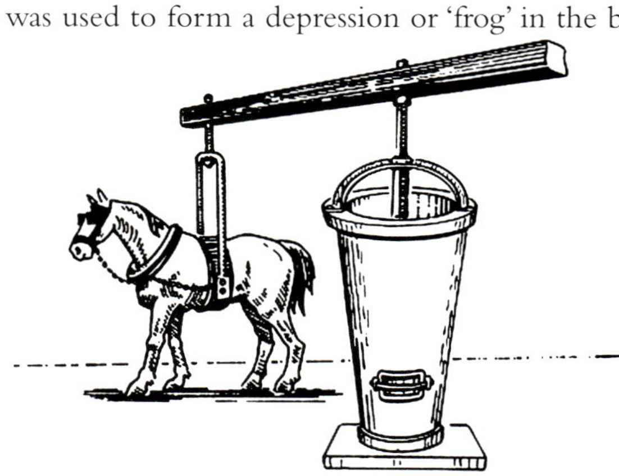

Horse–drawn pug mill. Such a mill is known to have been operating at Maitland in the mid 1830s

The frog served to identify the brickmaker and also enabled the mortar to bind more effectively with the brick. The top edges of the mould were lined with metal strips to prevent wear when the strike was scraped across them during moulding.

Slop moulding and sand moulding

Two methods of moulding – slop moulding and sand moulding – were practised at this time. Each used a different lubricant to slip the formed clay from the mould, either water or sand, whence the expression 'sandstock bricks'. Sand moulding was the more popular. Using this technique, the brickmaker sprinkled the stock board and mould with sand before positioning the mould over the stock board. The clot moulder then dusted the table with sand and picked up a 'clot', a piece of clay slightly larger than the finished brick. Kneading the clay into the rough shape of a brick before rolling it in sand, the moulder passed the doughy lump to the brickmaker who threw it forcefully into the mould. The clot was pressed deep into each corner to ensure a straight edge, later referred to as the 'arriss' with the advent of dry pressed bricks.

Any superfluous clay was removed by drawing the strike across the top of the mould, the residue being thrown back onto the heap of clay ready for re-use. The mould and its contents – the brick – were taken from the stock and placed on a pallet with the frog facing downward. A sharp tap was sufficient to detach the formed brick, leaving the mould free to receive another clot of pugged clay. 'Green' bricks, as they were called, were left to dry until they could be handled without loss of shape. A simple test was performed by pressing upon them with the finger which, in theory, left no marks. It was said that in England a slop moulder could produce up to 10,000 bricks a week and a sand moulder about 36,000 in the same period, an impressive rate of production.[2]

Occasionally a brick would fail to slide out, requiring the brickmaker to push it free, leaving thumb, finger

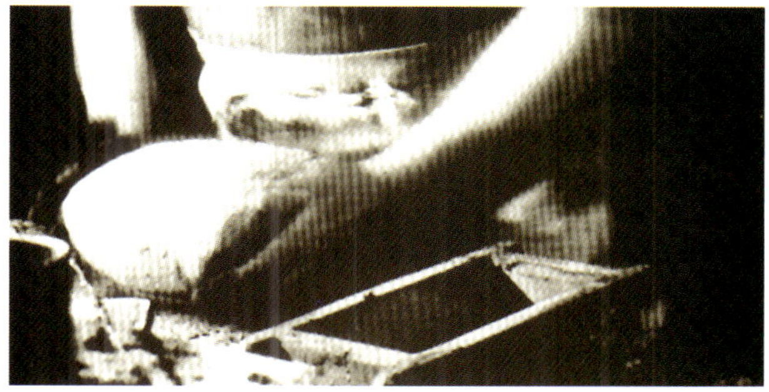

Sandstock brickmaker, England, circa 1935. Forming the clot by rolling it in dry sand on the bench

Throwing the clot into the mould. Standard-sized brick moulds were of hardwood. Later moulds were lined with steel

Cutting off the excess clay with a wire bow. A wooden scraper was used by Australian brickmakers

The full mould is lifted off the stock board and the brick turned out on to a pallet. Redland Bricks Ltd Collection

2. Bruce Shying mentions this figure in his booklet, *The Beginning of Brickwork in Australia*, Department of Technical and Further Education, 1976, but does not acknowledge a source, thus making it difficult to verify.

Thumbprints and other marks. Warwick Gemmell Collection

1. *Thumbprints made when forcing clay into mould*
2. *Handprint*
3. *Thumbprints and hand drawn arrow (Tasmanian brick)*
4. *Pipe impression and other marks*
5. *Thumbprints made when forcing clay from mould (Queanbeyan brick)*

"The Pack". Warwick Gemmell Collection

Objects fixed to stock to impress frog, from left. Nail, Military medal, jelly mould. Warwick Gemmell Collection

Tally marks. Warwick Gemmell Collection

and hand impressions in the bed and head of the brick. Thumbprints in particular may also have been made deliberately as tally marks by the overseer to record the number of bricks produced by a convict worker in the course of the day. Colonial builders using these hand moulded bricks were free of the restrictive building regulations of eighteenth century London, but still conformed to standard details and dimensions. For example, common stock bricks were roughly the size of a London brick which had been decreed by statute in 1739 to be eight and three quarter inches by four and one eighth inches by two and a half inches. This size varied across New South Wales and Victoria during these early days. Bricks used to build the original foundations of Old Government House, Parramatta, measured nine and a quarter inches by four inches by two and three quarter inches. Additions to the building during the time of Governor Macquarie used slightly smaller bricks. The point to note here is that the variation in brick size was influenced as much by the size of the mould as it was by popular demand, tradition or the type of brickmaking machinery employed. It seems that after about 1900 the dominance of the elegantly proportioned bricks of England came to an end.

Sandstock bricks, whether convict-made or those manufactured in private yards, improved in quality over time. In recent years sandstock has enjoyed a revival, with modern technology producing a brick superior in every way to its early nineteenth century counterpart. Needless to say, today's brickmakers – and their customers – have rediscovered the

Paw prints. Warwick Gemmell Collection

nostalgic appeal of the original. Impressed markings at this time were commonplace, but reveal little of these colonial brickmakers. From 1819 convict brickmakers left arrow marks, whereas private brick makers after about 1830 used a curious variety of marks affixed to the kick and pressed into the soft clay. These included animal designs, stars, heel prints, hearts, diamonds, spades and clubs, and even military insignia.

DRYING

It was now the turn of the take-off boy to load the pallets onto a cart and sprinkle the bricks with sand to further absorb moisture. Alternatively the green bricks might be loaded directly onto a barrow by the wheeler and taken to the stacking area, or hack. A moulding table was the scene of intense activity and kept two wheelers fully occupied. Brick barrows fitted with a wheel made of iron or steel were in continuous use from the earliest days of the colony to the 1930s, by which time solid or pneumatic rubber tyres had replaced all-metal wheels. At the hack bricks were unloaded and placed on the ground using two wooden paddles – 'clappers' – intended to avoid damage to the arrisses of the soft brick. Bricks warped through frequent handling were re-shaped using the

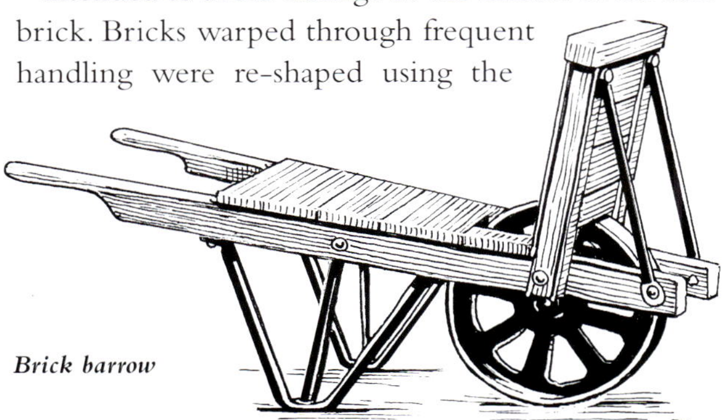

Brick barrow

paddles. Each brick was then laid at an angle ('skintled') to the run of the hack, with a small space left between each brick to aid the drying process.[3] The first row of bricks was skintled in one direction and the second in another to stabilise the structure and to ensure that the bricks had equal distribution

Drying sheds at the brickyard of Fred and Harry Baker, Raymond Terrace Road, Raymond Terrace, circa 1877

of weight. Once the hack had reached eight courses it was covered by reeds to protect against rain and the sun. Depending on the season, bricks were left to dry for anything up to a fortnight, or until they were ready for burning. In the summer heat two or three days were sufficient, but with demand constantly outpacing supply this was sometimes reduced to one day. During their time in the hack bricks were often marked by the paw prints of dogs, cats, possums and other small animals which roamed around the yards in search of food. Hack drying was, however, extremely slow. Damage to bricks was frequent and the wheeling, both to and from the hacks, skintling and general attention needed to expedite the drying process, added considerably to the costs of producing bricks.

CLAMP KILNS

In these early days no attempt was made to construct permanent kilns. Instead, the traditional Flemish method of building one-off kilns – clamps – was used, although it was greatly modified by individual brickmakers. Clamps are the oldest and most basic method of firing bricks and are still in use today around the world, especially in third world countries.[4] They were also cheap and easy to construct and maintain, although it took great skill to build an efficient clamp kiln. At Port Jackson they were located next to the clay pits with timber

3. Early in the nineteenth century, particularly in the Midlands and the north of England, hot-floor dryers were introduced, heated by underfloor flues piping hot air from a furnace or steam from the works engine. Further research might confirm if this was adopted in the colonial brick industry.

4. Nick Pezet, manager of The Austral Brick Company's plant at Golden Grove in South Australia, has first-hand experience of clamp kilns. For many years Pezet was a brickmaker in the south of England where he managed a plant that burned bricks in clamp kilns. Demand for the particular colour and quality of bricks fired in traditional clamps is responsible for the survival of these kilns in England and elsewhere in Europe.

Early twentieth century brickmakers at Burnt House Lane near Shide, Isle of Wight, UK. Clamp can be seen at upper left. Courtesy Isle of Wight Industrial Archaeology Society

hewn from the surrounding woodlands, and close to the main areas of settlement, which helped to keep transport costs to a minimum. Clamps were built on slightly higher ground to allow for adequate drainage. The base was compacted into the shape of a dish, so that when burning the clamp would have a tendency to lean inwards, which maintained structural integrity.

Generally, a solid foundation of fired bricks was laid, placed diagonally and on edge. The next two courses were made up entirely of green bricks on edge, laid parallel to the sides to form a heat seal. In the lower part of the kiln stoking tunnels (flues) were created at regular intervals from one to three metres apart. Packed with timber fuel, charcoal and cinders, the flues were used to light the fires in the clamp as well as replenish the fuel supply. Air for combustion also entered through these holes at the base of the clamp. As the clamp evolved, combustible material was scattered between the individual brick courses.

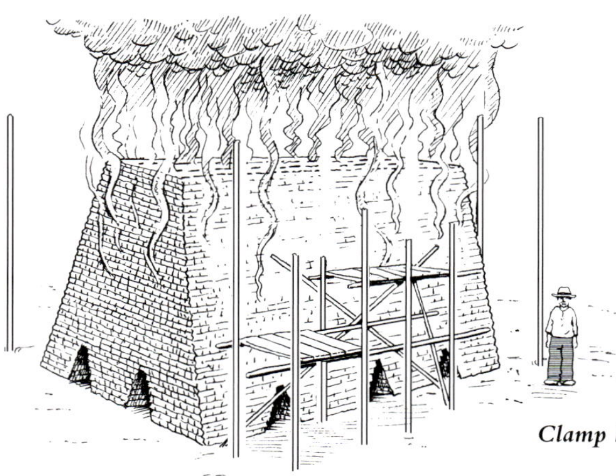

Clamp kiln

Soon, the clamp began to assume its familiar shape as rows of green bricks took shape, alternating as headers and stretchers in regular intersecting rows, and sealing the flues in the interior. At the base, a clamp could measure about 8 metres in length by 6 metres wide. At the top, allowing for the tapered effect, the clamp was slightly smaller at 7 by 5 metres, the finished clamp standing three to four metres high. Sometimes an external covering of daubed clay was applied to conserve heat, in which case they were referred to as 'scove' clamps or kilns.

FIRING THE CLAMP

Wood or charcoal was the basic fuel in the clamp, although coal was preferred since it was more efficient. Coalfields were being mined in the Newcastle district as early as 1797, with coal generally replacing wood in the brick kilns at work in the area. Escaped convicts discovered the first-known coal deposits near present-day Newcastle in 1791. It was during a search in 1797 for the escapees that a British soldier, Lieutenant John Shortland, found a coal seam close to a river he called, Coal River, later renamed the Hunter River. By 1800 shipments of coal from Newcastle had reached 4,000 tonnes, much of it to fuel industrial growth in Sydney. However, due to its high cost coal did not became the standard fuel until the introduction of continuous kilns during the 1870s.

Initially fire in the clamp was kept low by careful control of the draught, which brought the mass slowly up to firing temperature. When well alight the entrances were sealed with bricks and plastered with clay. Fissures would form in the clay plaster as it dried, permitting the entry of air necessary for combustion. Should the fire appear to dampen in part of the clamp, then flues could be re-opened and additional fuel poked into the heart of the clamp.

A pall of dense white vapour would often settle over the brickyard at this stage of the burn, due to the evaporation of moisture from the green bricks via the top of the clamp. After about three or four

5. New South Wales, Parliamentary Legislative Assembly, Newcastle Coal Heritage, speech by Kerry Hickey, MP, Minister for Mineral Resources, 30 March 2004

days clear air would issue, indicating the bricks were dry and that firing at full heat could proceed. After about a fortnight, or when the brickmaker judged the bricks to be properly burned, the fire was quenched and the clamp allowed to cool for a week or more.

LIMITATIONS

The limitations of the clamp kiln were numerous and eventually led to their demise. Each clamp was a one-off kiln, which had to be demolished to enable the bricks to be removed. Furthermore, clamp burning tended to produce bricks of extremely variable quality and colour, so that with the exception of bricks fired at the centre of the clamp, few bricks were uniform. There are several explanations for this. Firstly, consistent temperatures were difficult to achieve, so that up to 20 per cent of bricks produced by a basic clamp were either overfired or underfired. Overfiring fused the clay and sand sprinkled between the wet bricks to prevent them from sticking to each other. Put simply, the outside of the bricks attained a glass-like appearance, while the insides were rendered brittle.

In addition, bricks stacked nearer the flues and fuel supply were often distorted, which produced 'clinkers'. On the other hand, bricks from the outer face of the clamp were soft and underburnt creating the pleasant light-brown or salmon-coloured bricks, termed 'callows' or 'doughboys'. Since brickmakers could exercise no control over the rate of burn during the actual firing, clamps were extremely inefficient users of energy, and heat quickly dissipated into the atmosphere during firing and cooling.

Technology rather than habit and ignorance was the key determinant in the use of clamps and their immediate successor, scotch kilns. Until these updraught kilns were firmly established by about 1850 there was no viable alternative.

SCOTCH KILNS

Rapid population growth in the colony after 1820 spurred brickyards to produce ever-greater volumes, which made clamps increasingly unsatisfactory. As buildings became more sophisticated clients demanded a better quality brick, providing impetus for the introduction of more permanent brick kilns.

It was in England during the later seventeenth century that scotch kilns began to replace clamps when brickmakers discovered that the firing of bricks could be better controlled if they were loaded inside permanent walls. The basic object of these kilns was to generate an intense heat, to equalise it as much as possible over the enclosed space, and to maintain it in that space for a certain length of time. By the 1830s scotch kilns (also known as updraught or open kilns) were in general use throughout England.

GENERAL APPEARANCE

Scotch kilns made their appearance in Australia in the late 1820s, although bricks continued to be burned in traditional clamps until the 1860s or later.[6] In appearance the scotch kiln was a series of rectangular open-topped chambers about 3.5

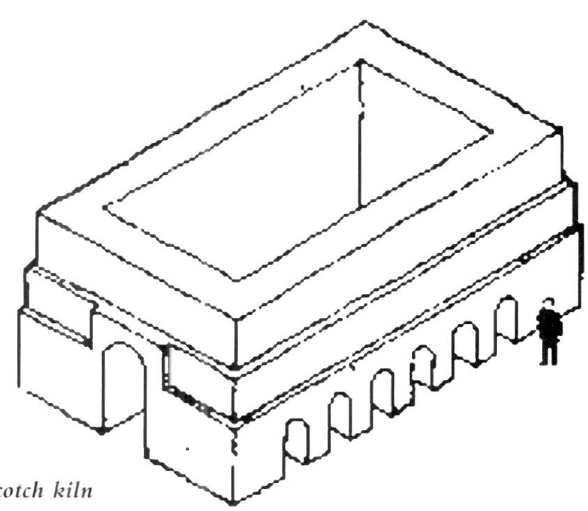

Scotch kiln

6. In England scotch kilns evolved from clamps in the later decades of the seventeenth century. A number of oil fired scotch kilns are still in use in yards at Aldeburgh, Suffolk, and near Chesham in Buckinghamshire. The newest dates from 1980. They survive despite inefficiency and fuel costs because they can produce beautifully coloured bricks.

Scotch or open kiln at Fred Baker's Metford yard, circa 1894 at East Maitland, NSW. State Library of New South Wales

metres wide by 3.6 metres high. It had a wide doorway at each end and six or more fireholes (also termed 'eyes') arranged along the sides and opposite each other. Fire channels about 23 centimetres wide by one metre high and corbelled over were formed between opposite fireholes.[7] A solid wall of bricks divided the kiln lengthwise to prevent cross draughts. A kiln of this size had a capacity of about 20,000 bricks.

To improve insulation the kiln could be built on the ground surface or slightly recessed into the ground using half-burnt bricks from earlier clamp burnings set in a clay or pug mortar. Much like its predecessor, the sidewalls were inclined, creating a tapered effect towards the top of the kiln, which was left open. Fuel to fire the kiln was packed on the floor of the structure. The green bricks were stacked on an open grate placed at such a height above the fire that the flames and heat could diffuse uniformly through the openings. In Australia it was usual to lay bricks on the compacted earth floor of the kiln, the system of open grates only appearing at the end of the 1840s, and possibly later.

The process of stacking the bricks was similar to that used in the clamps. Starting from the centre, bricks were arranged so that the tunnels or flues were formed by overlapping the bricks in the upper layers. Having built the centre of the stack, the men worked out towards the doors at both ends of the kiln. The bricks at each doorway were then protected by a casing of half burnt bricks smeared with pugged clay. Bricks forming the top layer were coated with clay to prevent heat escaping through the open top of the kiln. Judy Birmingham, an authority on colonial archaeology, comments that 'of such activities only the irregular pits where the clay was dug, or the circular horse walks, might survive, together with the reddened earth and broken brickbats of the early clamp where they were burnt'.[8]

BUILDING CRISIS

Whatever the inherent difficulties of using clamps or scotch kilns, they were nothing when compared to the shortage of lime used in mortar necessary for brickwork, and whitewash, an early form of exterior paint.

Lime, sand and fresh water were the chief constituents of mortar, with sand being plentiful around the bays and inlets of the colony. The same could not be said for lime. In June 1788 John White, Surgeon-General to the First Fleet, wrote in his journal that 'the stone of this country is well-calculated for building, but

7. A corbel was an overlapping arrangement of bricks in which each course extends farther in from the vertical of the wall than the course below.

8. J. Birmingham, *Industrial Archaeology in Australia*, Heinemann, Sydney, 1983, p. 54.

could any kind of cement be found to keep them? As for limestone we have yet to discover any in this country'. Indeed, Governor Phillip constantly appealed for limestone to be sent out as ballast in the ships from home. His successor, John Hunter, was equally hard-pressed to solve the lime shortage, as can be seen by this despatch to the Duke of Portland in 1796:

> *The brick buildings, barracks, store houses and officers dwelling houses have been for some time past in a state of rapid decay and crumbling ruins; to build such another set would be attended with great expense. The decay has been occasioned by the want of lime or proper cement in the beginning ... We ... are repairing and completely covering all the brick buildings so as to ensure their lasting at least 20 years to come.*[9]

Bricklayers bore the brunt of the shortage and were obliged to use puddled clay as a substitute for mortar, a frustrating practice in view of the heavy rainfalls that left buildings in a weakened state. The scarcity of lime mortar severely restricted building development for the first 20 years of the colony. Government House was the only two-storey structure built in that time, and the only one to use lime mortar.

SOURCES OF LIME

The main source of lime was sea shells, chiefly oyster. Live oysters made better lime than discarded shells and gangs of female convicts were set to work combing the foreshore of Sydney Cove. The supply of oysters was regarded as almost inexhaustible and they were scoured from the bays and estuaries with dredges and taken ashore in boatloads to be burned. Exploration and settlement during the early 1820s of the Clyde River on the south coast opened up a huge, new source. From one bed alone between 800 and 1000 bags of oysters were gathered weekly to be made into lime.[10] Further exploration around Sydney Cove would have found more shells, but at the time it was considered too dangerous to forage beyond the settlement for fear of attacks by Aborigines.

By 1804 supplies of lime were being shipped from the Hunter River and from settlements along the Hawkesbury River. Lieutenant-Colonel Paterson described the shell lime of the Hunter in a letter to Governor King:

> *I have not yet discovered anything like limestone, but the quantity of oyster shells on the beaches and inland is beyond conception; they are in some place for miles. These are four feet deep, without either sand or earth. Vessels might lay within a few yards of where they are found.* [11]

Eventually the lime shortage became so dire that it forced the authorities to intervene. In 1818 Governor Macquarie issued an edict forbidding the gathering of shells – once considered inexhaustible – from around the coves of Sydney Harbour. Informers were encouraged to report breaches of the law. A fine 'not exceeding five pounds sterling, one half of which is to go to the Informer and the other half to the police fund' was imposed.[12] The fact that Sydney rock oysters (Kentish oysters as they were referred to at the time) were considered a great delicacy might have had some bearing on Macquarie's edict.

To augment supplies Aboriginal middens along the coastal saltwater inlets of Cook's River and Botany Bay were excavated. In time the mudflats of Pittwater and Brisbane Waters yielded greater quantities of shells than Sydney or Botany. Exploited to the edge of extinction, and with the effects on the beds so noticeable, the government felt obliged to outlaw the burning of the live oysters in 1868. Well before this date, however, limestone of excellent quality was discovered in 1824, about 25 kilometres north of Bathurst and to the south at Argyle.

9. C. H. Bertie, *Story of Sydney*, Shakespeare Head Press, Sydney, 1933, p. 49.
10. V. Serventy & R. Raymond, (eds) *Australia's Wildlife Heritage, Our Plants and Animals*, Paul Hamlyn Pty Ltd, 1973, pp. 397–8.
11. HHRNSW, vol. IV, p. 415.
12. Colonial Secretary, In-Letters, State Records of New South Wales (SRNSW), 4/1741, Bun.12, No. 70–97.

MANUFACTURE OF LIME

Burning of shells for lime was a tedious process so that lime remained a scarce commodity. Once gathered the shells were washed, crushed and fired in specially constructed rectangular trenches. The shells were deposited in layers interspersed with wood to form a mound. The fire could burn for 24 hours or more causing the material to collapse into the pit where the heat was retained for several days. The cooled lime was then picked over to separate the unburned and over burned lumps from the good quicklime. Specially built kilns were also used to burn shells and these were dotted along the coastal inlets, especially around the mouth of Cook's River and in Botany Bay.

Fresh water was added and the slurry left for several days, by which time its chemical composition had changed, transforming it into slaked lime. When mixed with sand as well as hair from sheep, goats or cows, lime became a binding agent and was ready for use. Records tell of a dire shortage of animal hair for use in mortar at Norfolk Island in 1832. The governor of the penitentiary even ordered the shearing of the 400 Norfolk Island convicts to make good the shortfall.

The lime burners.
Zhenya Gay Collection

Technical Note – Limeburning

When oyster shells, containing calcium carbonate, $CaCO_3$ are burned, the end products are carbon dioxide CO_2 and calcium oxide CaO. The calcium oxide remains in a white powder or lump form known as lime, roach lime or quicklime. Water is added to the quicklime to make hydrated or slaked lime – calcium hydroxide $Ca(OH)_2$. This process is known as slaking. Slaking the quicklime gives off heat, causing rapid expansion of the lime and water mixture. Because of the vigour of this reaction with water, quicklime requires careful handling. By adding further water the slaked lime converts to a putty-like soft, smooth material that can be stored indefinitely if it is kept wet. The slaked lime is then mixed with sand and water to make mortar. On drying, the mortar reverts to calcium oxide, and absorbs carbon dioxide from the atmosphere to be reconstituted as calcium carbonate.[13]

PLIGHT OF THE LIMEBURNERS

Lime burning operations were hazardous and limeburners were afflicted by a combination of dust and heat. The dust inflamed the eyes causing temporary blindness.[14]

Years after receiving his convict pardon, James Tucker, who had been transported in 1826 for sending a threatening letter to his cousin, gave an account of his miseries under the pseudonym 'Ralph Rashleigh'. According to AustLit, the online resource for Australian literature, Tucker worked at the colonial architect's office in Sydney during the 1830s, winning a ticket of leave in 1835. He lost and regained his ticket several times for various offences, moving to Maitland, Port Macquarie, Goulburn and Moreton Bay before disappearing from convict records.[15] Tucker described the callous treatment of himself and other convict limeburners at the Limeburner's Bay settlement, north of Stockton on

13. J. Harrington, *An archaeological and historical overview of limeburning in Victoria*, Heritage Council Victoria, 2000.
14. Ralph Rashleigh, *Adventures of an Outlaw: The Memoirs of Ralph Rashleigh. A Penal Exile in Australia 1825–1844*, Jonathan Cape and Harrison Smith, New York, 1929, pp. 247–8.
15. Tucker may have authored several important works, including the early convict novel Ralph Rashleigh, which may have been written in the 1840s. It remained unpublished until 1929 when a heavily edited version was released. Some scholars believe that Tucker died in Sydney in 1888, but others have linked him to the earlier death of a James Tucker at the Liverpool Asylum in 1866.

the Hunter River. During the early morning hours half-starved convicts would be roused and put to work loading sacks of quicklime following the arrival of boats on the night tide.

Loading was especially hazardous, the quicklime being carried by the men who waded through water to the lighters riding at anchor. Serrated oyster shells underfoot would slice through even the most calloused feet which became infected. Inevitably, the quicklime became wet as it was manhandled through the shallows, and where it made contact with exposed skin would cause deep burns leaving the backs and shoulders of the convicts in a raw state and scarred with suppurating wounds.

Newcastle Brickyards

In 1822 the Government decided to start a Government brickyard and lime kiln at Newcastle to supply bricks for public works and private settlers. Newcastle was chosen because by this time Sydney was being adequately catered for by private enterprise. Prices from the state yard were 12s per thousand for bricks, 9d per bushel for lime, and bricklayers could be employed for 1s per day. These prices were comparable with the English rates at that time.

A SOCIAL PLACE

The increasing demand for bricks, tiles and pottery ensured that Brickfield Hill retained its position as the main supplier to government storehouses, the hospital, gaol and barracks. New brickworks and potteries appeared, and not just in Sydney Town; settlements up and down the coast, especially those in Van Diemen's Land, were voracious in their appetite for building materials, giving impetus to the continued growth of the industry.

The actual village at Brickfield Hill was described by French zoologist François Péron, a pioneer of anthropology and oceanography, as containing 'over forty houses, and numerous manufactories for tiles and earthenware crockery as well as bricks'.[16] As both a manufacturing centre and the last major settlement before Parramatta and the interior, Brickfield Hill was something of a magnet for travellers who frequented the brick built inns that served as coach stopovers, decades before the advent of the railway. No doubt, Péron would have heard that Brickfield Hill was a safe haven for pickpockets, footpads and those who fenced stolen goods. It also doubled as one of Sydney's earliest centres of entertainment. Governor Macquarie is said to have witnessed a play in a hayloft on the Hill, but most of the diversion was less highbrow. Grog shops and brothels, cock fighting, dog fighting and rat killing attracted large audiences on Sunday afternoons, in defiance of magistrates and constables. In August 1805, at the height of Britain's war with Napoleonic France, the *Sydney Gazette* reported a cockfighting contest in which 'the champion was 'Bone-a-part', a powerful party-coloured pile, and his antagonist, 'Sir Sidney', a ginger of true English breed'.[17] Following one round of fighting, the battered chooks had scarcely left the enclosure when a party of military and constables arrived on the scene.

> *In an instant the cocks in reserve were left to settle their own disputes; the setter-to set off; the heelers took to their heels; the owners disclaimed their property; and the better [sic] thought proper to decamp without adjusting their accounts.*[18]

Spectators cut and ran as far afield as the swamps along what is now Blackwattle Bay, a ten-minute sprint from Cockle Creek. Names were taken and arrests made and the six unfortunate cockerels found on the field of battle were ordered to the Orphan School, where, presumably, they made it no further than the dinner table.

In physical appearance, the Brickfield district was exceedingly unpleasant, the extensive diggings and

16. F. Péron, 'Description of Sydney Town, the capital of the English Colonies in New South Wales', *The Monthly Magazine*, 1 September 1809, p.127. Péron arrived in Sydney in May 1802 aboard Baudin's expedition that also carried Charles-Alexandre Lesueur mentioned in Chapter 1.

17. *Sydney Gazette*, Sunday 11 August 1805.

18. Ibid.

Blown away – A Sydney brickfielder

earthworks creating thick dust clouds which swept through the houses and down the broad streets of Sydney Town. On hot summer days the brown haze created by brick burning was exacerbated by 'brickfielders' – blinding storms of red brick dust whipped up by on-shore southerly winds.[19]

All was not well, however. As settlement encroached on the brickyards, Governor Macquarie's plan for Sydney called for the creation of an open space to be named Hyde Park. In his General Order of October 1810:

> *The Governor being desirous to prevent any Encroachment from being made on the Park by Brickmakers, and the Acting Surveyor having been directed to mark out for this purpose a Boundary line, dividing Hyde Park from the Brickfield … His Excellency commands and directs that none of those persons who have obtained permission to make Bricks, shall in future, on any pretence whatever, presume to cut up any ground for that purpose beyond the Line fixed upon as the Boundary for the Brickfields.*[20]

The demise of the government brick workings was still too far in the future to excite interest, especially from brickmakers who cared for nothing but the immediate business of mining clays and shale, and the cutting of timber to feed their hungry kilns. Yet, as the city's boundaries continued to expand, the continuing nuisance of smoke-belching brick kilns and the growing problem of criminality in the area were impossible to ignore. It is surprising that Brickfield Hill continued to operate as late as it did – up till about the year 1840.

19. Brickfielders were created by cold fronts moving up the coast of New South Wales. Winds closer to the coast moved faster and tended to produce an occasional burst of cold air known as a 'southerly buster' (also known as 'bursters'). Such a cold change is often preceded by strengthening north-westerly winds which blew into Sydney over the brick pits, picking up lots of red dust, hence the term 'brickfielder' which eventually gave way to the 'southerly buster'.

20. *Sydney Gazette*, 6 October 1810, p. 1.

Please note: According to the Geographic Names Register of New South Wales the name Cooks River (without an apostrophe) was not formally assigned till 9 March 1973.

Fuel to feed the fire 3

Supply and demand – A growing metropolis – Early colonial architecture – Cook's River – Demise of Brickfield Hill – 1843 Depression – Gold – Exodus

Sitting down to dinner – Rough weather during the voyage of the All-England Cricket Eleven to Canada in the year 1859. The scene might equally apply to the immigrant ships that sailed to Australia during the 1820s and 1830s. – H. Golding, The Wonder Book of Then and Now, Ward, Lock & Co. Limited, London, 1925

SUPPLY AND DEMAND

The fuel that fed the developing industry at Brickfield Hill and beyond was none other than people, for without new settlers – either free or in chains – building work in the colony would cease. And it was people that New South Wales began to attract in ever-increasing numbers.

From about 1815 the character of the colony underwent a profound change prompted by events overseas. With the defeat of Napoleon Bonaparte, Britain's wartime boom came to an abrupt end. Government supply contracts were terminated and the country was plunged into economic recession. Starting a new life in one of Britain's overseas colonies was an attractive prospect for all workers, regardless of their skills. Over the next two decades the flow of free settlers from the British Isles to the Australian colonies increased, albeit slowly at first. Publications expounding the attributes of such a move were circulating in England, with one describing the situation in 1819 thus:

> *For the artisan and mechanic, who are skilled in the works of utility ... there is ... no part of the world ... which affords an equal chance of success. To any, therefore, who have means of transporting themselves, and families to this colony, the removal would be in the highest degree advantageous. They could not fail to find immediate employment and receive a more liberal return for their labour, than they would be able to procure elsewhere. The blacksmith, carpenter ... stone mason, bricklayer, brickmaker ... sawyer, etc, etc.*

Easterly view of Sydney in 1821 taken from The Rocks where the southern approach to the Harbour Bridge lies today. Left of picture: men quarrying sandstone that gave The Rocks its name. The obelisk at the centre of the cross in what is now Macquarie Place (right of picture) still stands today. The house of emancipist Simeon Lord stands to the right. J. Taylor, circa 1819. Mitchell Library, State Library of New South Wales

> *would soon become independent ... For the master artisan and mechanic, the prospect of course is still more cheering.*[1]

And cheered they were, for New South Wales was desperately in need of skilled workers, especially brickmakers. Despite the availability of cheap convict labour, much of it was unskilled, and free settlers with a trade were able to command exorbitant wages from the moment they arrived. In 1824 the daily rate for a bricklayer was eight shillings, a handsome sum, while in 1835 a brickmaker was receiving up to ten shillings per thousand for his bricks.[2]

High levels of wages were hardly surprising in view of the handful of immigrants, a meagre 8,000, who arrived during the period 1821 to 1830. Their number was overshadowed by the 33,000 convicts transported during the same period. It was only in 1840, when the New South Wales Legislative Council abolished transportation, that free settlers exceeded the combined numbers of convicts and emancipists – more than 150,000 – living in the colony. Even then, however, the results were less than spectacular. The British government actively promoted free settlement so that between 1831 and 1842 there were 50,000 new arrivals. More than 3,000 migrants landed in Australia between January 1832 and July 1833,[3] with a further 76,000 making the journey during the decade of the 1840s.[4] Initially, government assistance was limited, posing a barrier to the immigration of mechanics and labourers. The cheapest passage to Sydney (steerage) still cost £40. On the other hand, Canada or the United States could be reached for just £5. The 58,000 immigrants who arrived from the British Isles between 1815 and 1840, were a tiny proportion of the one million who emigrated in that period. The vast majority – those classes who could have made the most significant contribution to the growth of industries in New South Wales – emigrated to Canada or the USA at a fraction of the cost.

It was also during this period that settlements were being established in the present states of Queensland and Western Australia. In Queensland the Moreton Bay District was established in 1824, and two years later the first bricks were being made from clay obtained in what is now Tank Street, Brisbane. The

1. W.C. Wentworth, *Statistical, Historical and Political Description of the Colony of New South Wales and its dependent Settlements in Van Diemen's Land, with a particular enumeration of the advantages which these colonies offer for emigration, and their superiority in many respects over those possessed by the United States of America*, London, Whittaker, 1819, p. 408.
2. *Sydney Gazette*, 10 February 1835, p.3, col. 3.
3. J.D. Lang, *An Historical and Statistical Account of New South Wales both as a Penal settlement and as a British Colony*. Vol. 1, London, Cochranes and McCrone, 1834, p. 280.
4. G. Sherington, *Australia's Immigrants*, Allen & Unwin, Sydney, 1990, p. 37.

brick towers of the Windmill, later the Observatory, were completed in 1827. On the other side of the continent, the Swan River colony in Western Australia was settled in 1829, and brickmaking commenced in 1831.

A GROWING METROPOLIS

Forty years after the arrival of the First Fleet, Sydney had travelled far, with a population of 11,000 scattered from Cockle Bay to the edge of Hyde Park, and south to Brickfield Hill. The decade following Macquarie's departure in 1822 witnessed a steady growth in Sydney in every respect. Development of the growing metropolis was marked by consolidation rather than expansion, with the boundaries of the town resting at the foot of Brickfield Hill to the south and Hyde Park to the east. It was only in the late 1830s that industry and workers' housing spread southwards to Surry Hills and Chippendale. The well-to-do gravitated to the élite new areas of Darlinghurst and Glebe Point.[5]

Most densely settled were the bustling dockland precincts of The Rocks and Millers Point with their cosmopolitan and colourful mix of visiting mariners, thieves, labourers, and men of business. Wharves, warehouses, bond stores and business premises made Sydney Cove a hive of activity, although in the absence of city planning, certain areas within The Rocks became notorious for their squalor. Here, a criminal underworld flourished, its reach extending far beyond the Cove itself.

Having developed sufficient infrastructure the town was now a large and diverse hub and trading centre in which the developing brick industry was beginning to play a vital part. One need look no further than the boom in speculative building which ran from the mid-1830s until 1842 when Sydney's population had reached more than 30,000. For most convicts and free settlers, however, Sydney was merely a staging post to the coastal plains and the Hunter Valley. In 1828 nearly all newly arrived convicts were assigned to landholders living beyond the township. And it was not as if the colony was unattractive or lacking in opportunity. Peter Cunningham, a surgeon who worked on board a number of convict transports, praised Sydney's physical charms and the peaceful nature of its inhabitants. In his two-volume work, *Two Years in New South Wales*, Cunningham noted:

> *On the site of Sydney alone what change has been effected! Where, thirty-eight years ago, not a human hut was to be counted, nor the slightest hum of commerce heard, we now have a city occupying a mile square, crowded with industrious citizens, and teeming with vehicles wheeling along the varied productions of the soil...*[6]

Charles Darwin, the naturalist, who visited Sydney in 1836 during his expedition aboard *H.M.S. Beagle*, was equally impressed. A diary entry for 12 January 1836 records:

> *Early in the morning a light air carried us towards the entrance of Port Jackson. A solitary lighthouse, built of white stone, alone told us that we were near a great and populous city. Having entered the harbour, it appears fine and spacious, with cliff-formed shores of horizontally stratified sandstone. The nearly level country is covered with thin scrubby trees, bespeaking the curse of sterility. Proceeding farther inland, the country improves: beautiful villas and nice cottages are here and there scattered along the beach. In the distance stone houses, two and three stories high, and windmills standing on the edge of a bank, pointed out to us the neighbourhood of the capital of Australia.*
>
> *At last we anchored within Sydney Cove. We found the little basin occupied by many large ships, and surrounded by warehouses. In the evening I walked through the town, and returned full of admiration at the whole scene. It is a most magnificent testimony to the power of the British nation. Here, in a less*

5. P. Bridges, *Foundations of Identity: Building Early Sydney 1788–1822*, Hale & Iremonger, Sydney, 1995, p. 151.

6. P. Cunningham, *Two Years in New South Wales*, facs. edn, p. 74, Vol. 2, Libraries Board of South Australia, Adelaide, 1996.

promising country, scores of years have done many more times more than an equal number of centuries have effected in South America. My first feeling was to congratulate myself that I was born an Englishman. Upon seeing more of the town afterwards, perhaps my admiration fell a little; but yet it is a fine town. The streets are regular, broad, clean, and kept in excellent order; the houses are of a good size, and the shops well furnished. It may be faithfully compared to the large suburbs which stretch out from London and a few other great towns in England; but not even near London or Birmingham is there an appearance of such rapid growth. The number of large houses and other buildings just finished was truly surprising; nevertheless, every one complained of the high rents and difficulty in procuring a house.[7]

EARLY COLONIAL ARCHITECTURE

It was not long before local design and building practices began to diverge from those of the mother country. This was largely in response to climatic differences and the availability of building materials. Within the vicinity of Sydney Cove and across town, the better houses were either single or double-storey detached houses, built of sandstone or brick, which was plastered and painted white, with front verandas.[8] Domestic architecture during the 1830s remained single-storey brick buildings with window and door openings edged by rubbed, flat arches in brick that were in contrast to the main walls. Windowsills were of stone which was more robust than brick and could withstand erosion by rainwater. The internal walls of brick buildings were rendered by the direct application of a 25 millimetre layer of hair mortar, finished with a thin coat of lime and sand. While shell lime could be used in either mortar or whitewash, the discovery of gypsum at Bathurst, and the importation of casks of plaster of Paris from London allowed a better quality finish to the settlement's buildings. When combined with lime, Roman cement, also imported from England during the 1830s, resulted in better quality mortars and stronger brickwork.[9]

For many years colonial architects preferred to use sandstone in official and public buildings. In fact, few other cities have been as profoundly influenced by its rocky foundation as Sydney, for the many workable outcrops of sandstone have given form and colour to its finest buildings, shaped its economy, guided its spread and protected its natural jewels – the rainforest gullies, coves and beaches made inaccessible to builders by its steep bluffs.[10]

The preference for sandstone was due mainly to the poor quality of locally produced bricks, but an element of snobbery may have been at work. As the Industrial Revolution in England gathered pace, brick was increasingly associated with textile mills, factories and warehouses, and also with cheap dwellings for the newly emerging working class. Brick was out of fashion, including the distinctive red and yellow bricks hitherto prized for their appearance. Unless it was faced in stone, a man of position did not care to live in a brick house. George III may have had something to do with the sweeping distaste. Architectural historian Arthur Oswald, writing in the 1930s, relates the story of a visit by the king to:

a fine brick house near Weymouth which its owner, Mr Morton Pitt, was accustomed to hear praised for its beauty. All the king said was, "Brick, Mr Pitt? Brick?" Accordingly, Mr Pitt took the hint and went to great expense of having his house faced with stone.[11]

Thirty years later in 1962 Alec Clifton-Taylor makes the same point, namely that 'throughout the whole of the eighteenth century, and well into the nineteenth, stone carried with it a social cachet which attached to no other material. The Georgians, although excellent builders of brick, had a yearning for stone beyond anything that had been felt in this

7. Journal of Researches into the Natural History and Geology of the Countries Visited During the Voyage of H.M.S. Beagle Round the World, under the command of Capt. Fitz Roy. R.N., John Murray, London, 1845, Chapter 19.
8. L. Turnbull, Sydney: *Biography of a City*, Random House Australia, Sydney, 1999, p. 84.
9. *The Sydney Gazette*, 10 February 1835, p. 2, col.1.
10. T. Flannery, *The Birth of Sydney*, Text Publishing, Melbourne, 1999, p. 8.
11. A. Oswald, *Country Houses of Dorset*, Country Life Ltd, London, 1935, cited in Woodforde, pp. 86–9.

Georgian Primitive, circa 1790. Simple brick-built house, typical of the pioneering days of all the colonies. The absence of sun shading was more suited to the cooler climate of Tasmania

Colonial Georgian, circa 1815. Note the addition of the veranda with its flat ceiling, forming part of the main roof of the house. Both illustrations after Morton Herman

country [Great Britain] hitherto. For monumental public buildings, stone was considered well-nigh essential; for houses, although not essential, it was certainly highly desirable.' [12]

The English fascination for stone soon made the leap to New South Wales where its superior appearance established stone as the preferred building material. Even when exposed brickwork became fashionable once again in the 1840s, the result was visually uninspiring due to the generally poor quality of the brick.[13] This may help to explain the widespread use of render and stucco on the façades of the homes of well-to-do people. By 1850 brickwork of many upper-class houses was covered with smooth-trowelled plaster on all surfaces visible to guests entering the front door. Outside only the back was left in naked brick, while inside the kitchen and rear utility rooms were left unplastered. According to social historian Robin Boyd, this attitude towards brick continued long after machine-produced bricks with regular arrisses were available. Until the late 1880s, bricks remained hidden beneath cement rendering, neatly ruled to simulate stone.[14] In the final outcome, brick triumphed for economic reasons. Even before the introduction of machinery bricks were inexpensive to make, provided the right clay was available. In fact, the cost was dramatically lower than those incurred in the quarrying, shaping and transporting of heavy blocks of stone, even for short distances.

Francis Greenway

Both private and public architecture practised in Australia during this period was largely an extension of the English late-Georgian style, and was referred to as the Macquarie style. Whether built of brick or faced in sandstone, early colonial architecture is best observed in the public buildings designed by Francis Howard Greenway. Convicted of forging a signature on a solicitor's document following his bankruptcy, Greenway was transported for 14 years in 1814. For a short while his career as an architect appeared to be at an end, although soon after his arrival Greenway's talents were discovered and he was granted a ticket of leave.

After setting up an architectural practice in George Street, he came to the attention of Governor Macquarie in 1816 who appointed him Sydney's first civil architect. The success of Greenway's lighthouse at South Head led to a number of prestigious commissions for government works, all of them noted for their elegant design and use of brick. One architectural fashion at this time, popularised by Greenway, was the use of a different coloured brick to define the brickwork of arches. To form arches, soft-burnt bricks were rubbed into the required taper.

12. A. Clifton-Taylor, *The Pattern of English Building*. Batsford, London, 1962, p. 62.
13. A. Plumridge & W. Meulenkamp, *Brickwork: Architecture & Design*, Harry N. Abrams, New York, 1993, p. 43.
14. R. Boyd, *Australia's Home: Its Origins, Builders and Occupiers*, Melbourne University Press, Melbourne, 1952, p. 107.

Francis Greenway's Hyde Park Convict Barracks, 1820. Drawing by William Hardy Wilson. At the left is the southern wing of the Rum Hospital which later became the Mint. National Library of Australia

Greenway's buildings included the new Government House in the Domain, the Female Factory at Parramatta, Hyde Park Barracks and a convict compound, St Luke's Church at Liverpool, St Matthew's at Windsor, St James's Church in King Street and a hospital at Liverpool. He also designed the 'Rum' Hospital, so called because Macquarie gave the builder 200,000 litres of rum as payment. The hospital's former north wing is now the central part of the New South Wales Parliament House in Macquarie Street. Another significant work was the stables for Government House, now the home of the Conservatorium of Music.[15]

From about the middle 1830s architecture in New South Wales began to undergo subtle changes. Until this time three broad styles had been evident. The first of these from 1788 to 1813 was characterised by a basic design comprising twin windows about a

15. Unfortunately, Greenway made enemies and fell out of favour when Macquarie accused him of overcharging. His public career ended in 1822 when Governor Brisbane dismissed him for having presented a large bill claiming commission for several years' work. In private practice, Greenway was unsuccessful and he retired to his rural property near Raymond Terrace in the lower Hunter Valley, where he lived in virtual obscurity until his death in September 1837.

Gothic revival, circa 1845. The drawing room now projects beyond the front of the house, creating an asymmetrical front. After Morton Herman

door, the absence of verandas, and possessing a hipped roof. The Macquarie style of public buildings from 1813 to 1821, mostly the work of Greenway, was austerely Georgian. Homesteads made up the third style in this period and were craftsmen-built and designed, 'their spreading plans and encompassing verandahs unconsciously speaking the first words in an Australian building vernacular'.[16]

A fourth style was now emerging, that of a refined pattern evoking Britain's Regency period. Houses for the well-to-do, it must be said, were magnificent in scale with huge ballrooms and drawing rooms, including semi-circular projecting bays ringed by verandas, high ceilings and sweeping staircases. This style of home, reliant on fine craftsmanship, could never keep pace with a popular building programme that catered for mass tastes and cheap construction. Notwithstanding, it lasted until the 1860s until overtaken by the growing challenge of Victorian tastes and economics.

A fifth style, Gothic Revival, championed by Edmund Thomas Blacket, seems to have had wider public appeal and established itself during the 1840s. 'The importance of Gothic was not to be measured in quantity; it was the first conscious exotic style, the first deliberate atmospheric scene-making, with high slate gables, carved bargeboards, stucco and a wooden porch.'[17]

16. R. Boyd, op.cit., p. 15.
17. R. Boyd, op.cit., p. 30.

Stretcher

English

Header

Flemish

Brick bonds – A variety of brick bonds are still in use, the most popular examples can be seen in these illustrations

The influence of architectural styles upon brick manufacturers was inevitable, especially with the growing popularity of polychromatic brickwork. This was the simple juxtaposition of different coloured bricks to achieve a decorative effect in buildings. Yet, this required good quality bricks and obliged the manufacturer to consider the clay types in relation to the final colour required. Little wonder that the fashion for polychromatic brick and paintwork was short lived; it was not until the early 1890s that they returned to favour.

BRICK BONDS

Bricklaying techniques are worthy of mention, and the art of bricklaying, like any other trade, demands both skill and an understanding of the materials. Handmade bricks were slightly wider at the top of the mould than at the bottom. The rough texture and uneven surface of the topside meant that when laying the bricks the bricklayer would place the wide surface uppermost. All bricks, therefore, had slightly overhanging faces and were set fractionally inside the line of the top edge of the course below, although the amount of overhang was slight. In a well constructed wall of handmade bricks, nothing but mortar could be seen from the top, whereas only brick was visible from below. The arrangement of bricks in interlocking patterns is known as bonding. Bonding brickwork helps to ensure a wall has strength and that the load is distributed evenly enabling the wall to resist sideways or lateral pressure. It is also decorative and adds aesthetic appeal. The strength in a brick wall is derived from overlapping individual bricks, either by a quarter brick (quarter bond) or a half brick (half bond). If the lap is less than a quarter brick then the strength of the wall will be greatly reduced. By the beginning of the eighteenth century a number of distinct regional styles had developed throughout England and Europe, including Flemish, Dutch and English bond. Colonial bricklayers in Australia used all of these styles.

COOK'S RIVER

The aspiration of any migrant is surely to make a better life for himself and his family. Beyond Sydney there was no shortage of opportunities, although many stayed around just long enough to get their bearings before pressing on into the hinterland. Others preferred to remain close by, so that a

View of Tempe on Cook's River, near Sydney, New South Wales, 1845. Watercolour by Conrad Martens. National Library of Australia

patchwork of settlements took shape within a radius of five kilometres, a journey of an hour or two by horseback or carriage.

One of these settlements was the southern side of Newtown, along the escarpment, and leading through the hamlet of St Peters down to Cook's River. Descriptions of the country along Cook's River by the early explorers were unflattering about the land's potential for food production. Captain John Hunter and Lieutenant Bradley both mentioned the shallowness of the water and large swamps, thickly wooded with mangrove trees. Captain Cook is known to have journeyed up the river as far as Tempe. On Empire Day, 1935, a tree was planted in Tempe Public School grounds to commemorate this event.

Despite its infertility the alluvial and deep shale soil of Cook's River and its surrounds offered more encouragement to new settlers than the harsh and unproductive sandstone of Sydney. Before the advent of farming these would have supported tall, open forests of Blackbutt and Sydney Blue Gum, the remains of which can still be found in the Sydney region and on the ridgeline of the North Shore. Other important species may have included turpentine, red mahogany, ironbark, white stringybark and the Sydney red gum.

Huge estates, initially in the hands of serving or retired military officers, were alienated over a period of decades. Many of those in receipt of the original grant either leased or sold their properties to people prepared to try farming in this previously unused region. Orchards were typical in the area. Records describe a Thomas Smyth who arrived as a marine sergeant in the First Fleet, later appointed Superintendent of Stores. For eight years he cultivated land on the heath at St Peters. According to the 'Return of Officers Holding Land' dated 14 August 1804, Smyth was a single man who held a total of 660 acres. Of this about one third was under wheat cultivation, 20 acres in maize, four

A view of Botany Bay, Cook's River, etc. from Norwood on the Parramatta River, 1857. Artist: Walter G. Mason. National Library of Australia

acres in barley and oats, and 121 acres as pasture. Smyth also possessed one male horse, one bull and a collection of livestock; he also employed nine convict labourers.

By 1830 the passage to St Peters and the Cook's River was well trodden with settlement along the river beginning to spread, and roads and crossings being made in several places. From where the railway bridge at Newtown now stands, the track turned due south and continued out to St Peters and Cook's River. The district also attracted many of Sydney's early élite. Dr Wardell had a large estate at Cook's River which, after his murder by a convict, was divided into three properties. One of these was owned by a Mr Mace, and was later purchased by Thomas Holt, an exceedingly wealthy businessmen. Other properties including the Unwin estate were the setting for stately homes, with the result that a number of striking Georgian villas could be found in the district of St Peters and Cook's River.

Leslie Duguid built a home in the area, as did A.B. Spark whose impressive colonial mansion, *Tempe*[18] was outstanding. *Tivoli*, a large colonial-style Georgian villa designed by John Verge, was built about 1840 for the wealthy ironmonger, Richard Reilly, and demolished in 1918. In the intervening

St Peters Anglican Church, Cook's River, one of Sydney's oldest churches, was consecrated in 1839. The church served an area extending from the coast to Bankstown, and from the harbour to Port Hacking. The village named after the church was the work of Alexander Brodie Spark, who built nearby Tempe House in 1836

years much of the surrounding land had long been relegated to a large block on the edge of suburbia. Yet grand residences of this kind in the colony of New South Wales were modest by comparison with those in England.

For many years, however, St Peters and Tempe were considered distant enough from Sydney to deter ordinary workmen and their families from settling in the district. This situation prevailed until about 1839 when the larger estates were subdivided and much of the land was thrown open to the public and

18. L.G. Norman, *Historical Notes on Newtown*, Council of the City of Sydney, 1963, p. 7.

sold. The change was accelerated after 1838 when stonemasons and labourers were brought to Tempe to build a dam across Cook's River. Governor Bourke, fearing the possibility of a breakdown in Sydney's water supply from ponds in Botany and Randwick (in what is now Centennial Park), had a stone dam built in order to conserve the fresh water flowing down the river. Unfortunately the dam was not a success, the water being tainted by saltwater. Indeed, the flats on either side of the river had been so impregnated with saltwater that the scheme was eventually abandoned.

Adding to the appeal of this rural idyll was the view across Shea's Creek to distant Botany Bay which would have impressed early settlers with its tranquillity. Sadly, it was not to last, due to the fact that the southern part of Marrickville, and particularly the area of St Peters and Tempe, was ideal for brickmaking. With its supplies of clay soil, water, and timber to provide fuel for brick firing, the area soon became a magnet for industry which jostled for space alongside other activities. From the 1830s it presented a chequered scene with gardens, grazing cattle, and cultivated farms competing with the chimneystacks of brick pits, the excavations of the quarries and the pungent smells of tanneries.

Elsewhere, private brickmakers continued to set up, providing keen competition to the government grounds at Brickfield Hill. By 1837 many small yards were operating in Pitt, Sussex, Castlereagh, George, Goulburn, Phillip, Elizabeth and Clarence Streets. According to the General Post Office Directory in 1839, there were 26 bricklayers and brickmakers in the whole of New South Wales, a figure far exceeded by the number of unregistered businesses.

DEMISE OF BRICKFIELD HILL

Unfortunately, the good fortune that accompanied the building boom and the growth in private brick manufacturies were dealt a blow with the closure of the government yards at Brickfield Hill. For more than 40 years the centre of Sydney's brickmaking industry had provided the major supply of bricks, yet by 1840 its presence was a hindrance to the rapidly growing metropolis. There was at this time not a single tree in Hyde Park, and the bare ground increased the intensity of violent windstorms which swept up grit and dust in enormous volume – the 'brickfielders'. Quite apart from brickfielders and polluting kilns the area was noted for its crime, which was just as bad in 1840 as it had been earlier in the century. Historian Geoffrey Scott refers to statements made by a colonist soon after the turn of the twentieth century. The old man recalled that when he arrived in Sydney in 1838 barbaric sports still flourished around the pubs in Brickfield Hill.

> *"Horse racing, dog fighting, cock fighting, rat killing and prize fighting were all the rage", he related with a sort of nostalgic gusto. "Scarcely anything else was thought of, and it was very hard for a young fellow to settle down and become steady". In one Brickfield Hill tavern he saw two bulldogs fight, and when they had sunk their teeth into each other "the owner of one dog chopped off two of his feet to show the breed of the dog that still kept on fighting!" In another pub, the unsettled youth saw a dog kill sixty rats in one minute. The Willow Tree in Pitt Street was also the site of a well-known rat pit.*[19]

Towards the end of 1841 brickmaking operations ceased on the Brickfield Hill. The following newspaper article appeared soon after:

> *BRICKFIELDS: Government having taken these fields into their own hands, prior to their being disposed of in small allotments, several attempts have been made to discover eligible ground for the manufacture of bricks in the neighbourhood of Windsor, but hitherto without success. Many industrious hardworking people have consequently been thrown out of employ, while others who have to supply contracts may find themselves in rather an awkward predicament. It is much to be hoped that by perseverance, the brickmaker may yet be more fortunate in his search than he has been and that his diligence will be rewarded by the accomplishment of his wishes.*[20]

19. G. Scott, Sydney's *Highways of History*, Georgian House, Melbourne, 1958, pp. 49–52.
20. *The Australian*, 21 October 1841, p. 4, col. 1.

The closure of Brickfield Hill may have served a number of purposes including the elimination of pollution, as well as clearing out an area known for its criminal activity. Improving vehicular access would also have been a major consideration in view of the hilly topography of Brickfield Hill. James Maclehose, a visitor to the colony, remarked that 'up to the middle of last year, [1837] the ascent of Brickfield Hill was not only steep and difficult but actually dangerous ….'[21] He goes on to say that excavation and levelling work on the Hill towards the end of the 1830s had already removed about five metres of rock from this massive outcrop.

> *Upwards of one million cubic feet of rubbish changed their situation by manual labour, a great proportion of which was solid freestone rock.*[22]

The rubble was used to infill the Haymarket valley, with excavation continuing over the next few decades. Indeed at the beginning of the last century, when the foundations of the Haymarket branch of the Savings Bank at the corner of George and Hay Streets were being prepared, the remains of a baker's oven was discovered about two metres below the pavement.[23]

The removal of brickworks and potteries from Brickfield Hill was also an opportunity for the government to conduct further house cleaning. During 1838 a comprehensive set of building regulations had been introduced, ostensibly for safety reasons. Aimed at fire control, they specified the thickness of external and party walls, as well as the manner of construction of fireplaces, chimneys and flues. The new regulations not only marked the victory of brick construction over timber in Sydney, but also paved the way for a boom in terrace housing after 1845.[24]

21. J. Maclehose, *Picture of Sydney and Stranger's Guide*, Sydney, 1838.
22. Ibid.
23. *Truth*, 22 August 1926, Vol D, p. 210.
24. J. M. Freeland, *Architecture in Australia: a History*, F. W. Cheshire, Victoria, 1968, pp. 84–89.

For the people, by the people: Vision of the Times
Agitation for self-government commenced in the colony of New South Wales towards the end of the convict period. In 1823 the authorities established an unelected Legislative Council to advise the Governor, followed soon after by an Executive Council. William Charles Wentworth, a prominent reformist, pressed for self-government in the newspaper he founded,* The Australian. *New South Wales was granted a part-elected Legislative Council in 1843. Monthly Almanac and Illustrated Commentator, 1856

Yet, closure of the government brickyard spelled the end of private manufacture within the city of Sydney. In the absence of alternative sites and with continued demand for bricks the industry found itself in crisis which took several years to resolve. Slowly the evicted brickmakers moved to the adjacent areas of Balmain, Glebe, Newtown, Redfern, Alexandria, Marrickville, St Peters, Camperdown and Waterloo where new clay deposits and timber for firing were present. Many of the yards in these emerging suburbs had been worked by several small yards for some time, although the firms were modest enterprises of five or six employees, with a proportionately small output. Unfortunately, brickmaking was an environmental disaster for Cook's River and Shea's Creek; within a generation they were little better than open sewers. In the early 1890s the latter was straightened, channelled and thenceforth known as the Alexandra Canal.

1843 DEPRESSION

The demise of Brickfield Hill and its association with the historic figure of Bloodworth was a turning point for the industry. From this point forward, the rapidly evolving industrial conurbation of southeast Sydney became a driving force behind the economic development of Sydney.

The move to new locations, however, came in the midst of a devastating drought that gripped New South Wales from 1837 to 1841. Hardest hit was the wool industry, the staple export and biggest single contributor to the economy of the Australian colonies. Prices held up until 1840, then collapsed, creating an ever-widening ripple effect. Moreover, the surge in building that had followed in the wake of immigration was mostly speculative, over-extending the industry. As the depression took hold, most construction work came to an end apart from some improvements by the Government of New South Wales to public buildings along Bridge and Macquarie Streets. Where there had been 26 brickmakers in Sydney in 1839, five remained in 1855.[25] British investors withdrew, banks failed and with capital flowing back to London confidence evaporated.

The depression was the first real setback to the economic well being of the industry as well as the first in a long period of consolidation within the Sydney brick industry. Recovery was some years away and only towards the end of the 1840s did the depression begin to lift. Immigration picked up although the imbalance created by an oversupply of skilled mechanics prompted the Colonization Commissioners to adjust the charges for the passage from England according to the emigrant's trade. Henceforth agricultural labourers were charged £2, mechanics (including bricklayers and brickmakers) £5, and labourers £7.[26] Wages were similarly scaled so that

25. W. Gemmell, *And So We Graft From Six to Six – The Brick Makers of NSW*, Angus and Robertson, Sydney, 1986, pp. 1–5.

Town Wages per diem, 1854

	s. d		s. d.
Painters	13 0	to	16 8
Carpenters	15 0		20 0
Joiners	17 0		20 0
Plasterers	25 0		30 0
labourers	12 00		15 0
Bricklayers	25 0		30 0
labourers	16 0		18 0
Masons	25 0		30 0
labourers	10 0		12 0
Carters	8 0		10 0
Quarrymen	18 0		20 0
Blacksmiths	14 0		16 0
Female servants	8 0 per week		12 0 per week
Male servants	15 0 per week		25 0 per week

Country Wages per year, with rations, 1854

Farm labourers	£40 – £50
Man and wife	£60 – £70
Watchmen	£25 – £30
Shepherds	£30 – £40

W.B. Haigh, Commission and Labour Agent, 481 George Street, Sydney, 9 June 1854. Source: The Shipping Gazette and Sydney General Trade List, Vol 11, Monday 12 June 1854, No 539, p. 108

bricklayers received £37 per annum, or 5s per day in 1849.[27] Five years later – at the height of gold rush – a skilled bricklayer was fetching between 25–30s per day, which was an enormous sum of money. By comparison to all other trades represented at the time, bricklayers, stonemasons and plasterers were at a premium, as the above table indicates.

By the time Victoria was separated from New South Wales with the enactment in the British Parliament of the Australian Colonies Government Act (1850), brickmaking in New South Wales had been transformed. From a state-owned industry worked by convict gangs and administered by the government, it was now largely in the hands of private enterprise, giving the general public access to good quality building bricks. For the most part, however, the trade continued to be a comparatively small-scale operation confined to the making of sandstock bricks. Yet the coming gold rush was about to change the colonies irrevocably, and with it the brick manufacturing industry.

26. S. Sidney (ed.), *Sydney's Emigrant's Journal and Traveller's Magazine, Second Series*, London, Orr & Co., 1850, p. 286.

27. J. H. Burton, *Emigrant's Manual: Australia, New Zealand and America and South Africa: Emigration in its practical application to individuals and communities*, William and Robert Chambers, Edinburgh, 1851, p. 134.

GOLD

English-born Edward Hargraves was already an experienced gold prospector when he returned to his sheep run outside Bathurst in early 1851. Convinced that the gold he had come across in small amounts in California could be found in abundance in Australia, he taught four young Australians how to use the Californian techniques of panning and cradling. Hargraves made no secret of his find and claimed the reward of £10,000 offered by the government to the first person to strike gold. By mid-May 1851, 300 diggers were living in a canvas town that Hargraves named Ophir. Soon, people were pouring across the Blue Mountains in hundreds of carriages and coaches, drays and carts, on horseback, but mainly on foot carrying their belongings or trundling them in wheelbarrows. Even richer strikes were made at Sofala on the Turon River. Later that year gold was discovered at Clunes and Ballarat in Victoria, where the diggings were bigger and more bountiful.

The Prospect – Image of Gold digger – Melbourne Punch 1855

The discovery of gold transformed Australia. News spread quickly, and in a matter of months thousands of people began to arrive from around the world. Among their number were genuine settler-migrants who had been granted assisted passage. The majority, however, were opportunists who wanted their share of the gold bonanza.

Sydney and Melbourne were the obvious points of entry for the new arrivals. Most were from the British Isles and Scandinavia, although increasing numbers of Chinese made the long voyage. Anxious to escape poverty, and with families to support back home, Chinese gold diggers were prepared to work longer hours – and for lower wages. It was not long before policies adopted by the colonial legislatures of New South Wales and Victoria made things difficult for Chinese prospectors. Regarded with suspicion on account of their religion, language and culture, it seems they were condemned more for being hard workers than for any other reason.

The Aspect – Image of Gold digger – Melbourne Punch 1855

Eventually most prospectors gave up on the diggings and gravitated towards the larger towns. New migrants, whatever their place of origin, were especially welcome since they helped to make up for the drain of manpower to the gold fields during the early phases of the rushes. What proportion sailed back home is unclear, but many would have found it easier to start a new life in Australia.

Not all immigrants aspired to be gold diggers; a large proportion travelled no further than Australia's coastal fringe, choosing to settle in the thriving commercial and industrial centres including Sydney, Newcastle, Brisbane and Melbourne.[28] Between 1851 and 1861 Sydney's population rose from 54,000 to 96,000. Small suburbs had already established themselves in Balmain, Glebe, Newtown, Paddington, Redfern and St Leonards. In the course of the next 20 years they were joined by others, and collectively referred to as inner city suburbs. The long-term expansion of population to these outlying areas was noticeable; while more than 80 per cent of Sydney's population lived in central Sydney in 1851, by 1901 it was only 23 per cent.

Visitors were impressed, and commented on the prosperous appearance of the town, noting that it was well ordered. Indeed, Sydney was far from isolated. A thriving seaport, the city boasted regular steamship runs between Europe and Australia, was connected by railway to Parramatta in 1855, and linked by telegraph line to Adelaide and Melbourne from 1858. By 1881 Sydney's population had reached 225,000 and by 1901 was approaching half a million.

Much the same applied to Australia generally. From 1841 to 1850 the population of the Australian colonies increased by 20,000 people each year. The rate of increase jumped sharply in 1851 to over 32,000, rising to a phenomenal 86,000 in 1852. Thereafter, the number of people rose by nearly 90,000 per annum until 1858, after which time the rate of increase slowed. Overall 465,000 people arrived in Australia during this period, with just under half being offered passage as assisted migrants.[29]

28. C. H. Bertie, *The Story of Sydney*, Shakespeare Head Press, Australia, 1933, p. 93.
29. F. Crowley, *A Documentary History of Australia, vol. 2 (Colonial History)*, Thomas Nelson, Melbourne 1973–1980, p. 248.

HIGH COSTS OF LABOUR

As mentioned previously, the price of labour rose dramatically due to the gold rushes although they declined in later years. In 1857 bricklayers in Sydney were receiving 6s 6d to 7s 6d per day while brickmakers were charging 30s per thousand bricks at the kiln. By the end of the following year, wages for a bricklayer had reached 13s 10d per day, reflecting the extreme shortage of bricklayers due to the continued pull of the goldfields. At one point the price of bricks at the kilns reached an unprecedented £15 per thousand.[30] Workingmen were quick to take advantage of the labour crisis with stonemasons and carpenters in New South Wales successfully negotiating the world's first 48-hour working week.

Unfortunately, the high price of both labour and bricks was a serious deterrent to prospective homebuilders. A cottage of four rooms with 14 inch brick walls, kitchen and outhouse of broad paling cost upwards of £400 to erect, which was an impossible sum and far beyond the means of most immigrants.[31] Inferior houses could save about 15 per cent on costs by reducing the walls to nine inch thickness. It was a field day for unscrupulous developers and fly-by-night builders who were quick to set up in business but often failed in their duty of care.

EXODUS

Another significant impact was the general exodus of brickmakers from Sydney during the early years of the gold rushes. In 1858 only eight brickmakers were recorded as still working in the Sydney area, the majority operating in Newtown, Camperdown and Surry Hills, with just 30 brickmaking establishments in the whole of New South Wales.[32] Unlike the yards in country towns, which were almost completely abandoned, the larger works in settled towns such as Newcastle and Sydney survived this period of instability.

30. *Australian Gold Digger's Magazine*, 1 October 1852, p. 33.
31. *Colonial Magazine*, 1850, pp. 137–9.
32. Sir H. Parkes, *Freehold Homes in a Gold Country: Two Public Addresses on the present condition and natural resources of the Colony of New South Wales, delivered at Derby and Birmingham*, Hudson Ellis, Birmingham, 1861, p. 11.

The city and harbour of Sydney from near Vaucluse. 1852. Hand coloured lithograph by Thomas Shotter Boys. National Library of Australia

As gold fever subsided many returned to their trade, helping to stabilise the industry at a time when the larger brickworks established in the 1840s were being called upon to produce huge quantities of bricks. Before long the imbalance between town and country was corrected, and with the injection of wealth the fortunes of the brick industry revived in areas surrounding the gold fields. Mining towns were in desperate need of brick dwellings to replace timber and mud hovels, with the result that local brickworks thrived. Excavations by the gold prospectors presented an unexpected bonus for they uncovered deep deposits of white clay and pipe clay which were exploited by the local brickyards.

For the year 1852 prospects for brickmakers appeared excellent:

> *A good clay for bricks is found in many parts of Australia. In towns, as soon as the resident population recover from the first effects of the gold fever, they must prepare houses for the influx of emigrants.*[33]

With demand on the increase, the brick industry was about to enter a new era and with it a sweeping technological revolution.

33. S. Sidney, *The Three Colonies of Australia; New South Wales, Victoria, South Australia; their pastures, copper mines and Gold Fields*, Ingram Cooke, London, 1852, p. 242.

Sketch showing the water reserve and the country between Sydney and Botany by William Henry Wells, 1855. Mitchell Library, State Library of New South Wales

An Industry Transformed 4

Innovation – Crushing and mixing – Developments in hand moulding – Plastic moulding – Frederick Goodsell – Downdraught kilns – Friedrich Hoffman – Patent kilns – Fireclay and refractories – Dry press moulding – George Foster & Sons – Impact of dry press methods – The lime industry – Victoria – Taking stock

Arrival of the first railway train at Parramatta from Sydney. Wood engraving by Walter G. Mason. National Library of Australia

INNOVATION

The gold rush and rapid population growth that followed in its wake transformed the manufacturing of bricks and tiles in Australia. Transformation was part of a wider process of industrialisation taking hold in Sydney and Melbourne. Progress was slow when compared to changes sweeping Britain, continental Europe and the USA, but by 1850 industry was already at work on the western side of Sydney around Sussex, York and Kent Streets, and later at Pyrmont, near to which lay the wharves and dockland area serving the coal-shipping industry. Diggers returning from the goldfields in northern Victoria mixed with new arrivals to form the nucleus of labourers and factory workers who made their home in the area. Some also gravitated towards the brick pits of St Peters and Camperdown.

Sydney was on the move, a trend that gained further momentum in 1855 when the colony's first railway line was opened between Sydney and Parramatta. The building of a more extensive suburban railway network connecting the west, south and north of Sydney, and paving the way for new suburbs and industry, had to wait until after 1875.

Meanwhile the brick industry took advantage of the introduction of modern steam powered machinery. By the 1880s steam had transformed brickmaking from a cottage industry to a sophisticated and highly capitalised enterprise. It was arguably the biggest single factor in the industry's modernisation, although the demand for better quality bricks from a growing, and increasingly affluent, middle class played its part.

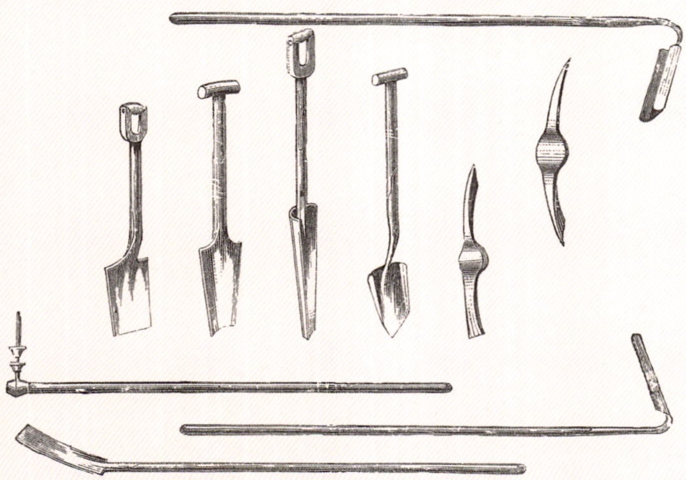

Top: *Filling the trucks with shale and clay*
Middle: *Spades and other implements used in digging clay and shale. A. B. Searle, circa 1850*
Above: *Sinking the holes in preparation for laying the charge of gunpowder*

SLOW START

Change is incremental, patchy and seldom follows pre-determined rules. Nowhere was this more evident than in the extraction of the dense clays and hardened shale. Such back-breaking work was performed by labourers using pick and shovel. Manual excavation at brick pits in Australia continued well into the twentieth century, unlike England and the USA where steam powered excavators were operating as early as the 1820s. The pace was intense, the work monotonous, and the men subject to the vagaries of the weather. Along the Waterloo flats the Sydney summer heat, often well over 30°C for days at a time, would have been stifling and made all the more unbearable when the southerlies failed to materialise. Late afternoon downpours, typical of Sydney's pattern of summer weather, may have provided some temporary relief, but turned claypits into quagmires. When combined with huge volumes of thick smoke and dust, conditions for the men, especially those working in the kilns, were appalling. Only the toughest – and driven – could achieve tally every day. The yard foreman of the time was hired for his ability to fight, for it was brawn rather than brains that kept men going in these harsh conditions.

As a matter of fact, beer – or the promise of it – was just as much an incentive to stay on the job. The end of day usually signalled a general exodus to hostelries in the St Peters district. James Richards' *White Horse Inn* (later known as the *White Horse Hotel*) was popular, and so too was *The United*, adjacent to the railway line at Mitchell Road, St Peters. The delightfully named *Cottage of Content* on Newtown Road was never short of customers. All of these public houses were known to be operating by 1863. Another favourite of those engaged in the brick trade was the *Cricketer's Arms Hotel*, one of Newtown's earliest watering holes. Originally known as the *Bricklayer's Arms*, the hotel at 285 King Street was kept by a Mr Eggleton who succeeded William Meekes, formerly keeper of the toll bar.

The brickmaking industry has always been associated with exploitation, hardship and drunkenness and the observation of the Englishman, Reverend James Dennett, in 1866 might just as well have applied to Australia.

> *Drunkenness is the curse of the working class in every trade in this country but it seems tenfold intensified in that of brickmaking.*[1]

No brickyard in England at this time was without its beer shop, and men and women were paid in the local hostelries. This practice transferred to the colony, and there may have been an expectation that the public house was the accepted venue for the settlement of wages. Brickyards were also the haunt of drunks and the down and out in search of a warm place on a cold night. In recalling the fate of one unfortunate, Charles Dickens might just as well have been writing about the brickyards of St Peters.

> *He begged his bread from door to door. Every halfpenny he could wring from the pity or credulity of those to whom he addressed himself, was spent in the old way. A year passed over his head; the roof of a jail was the only one that had sheltered him for many months. He slept under archways, and in brickfields — anywhere, where there was some warmth or shelter from the cold and rain. But in the last stage of poverty, disease, and houseless want, he was a drunkard still.*[2]

The standard garb for the workingman was a pair of heavy serge pants held up with belts and braces, a collarless flannel shirt[3] and hob-nailed boots. Hard wearing, practical, and comfortable – up to a point. The poorest would have gone barefoot. Infrequent rest breaks – 'smokos' – were taken under the watchful eyes of the overseer who rigidly enforced discipline. In the days prior to mass-produced, machine-rolled cigarettes, a clay pipe of rough tobacco was an enjoyable diversion, even among women. Certainly, it held no stigma, as it does today.

1. J. Woodforde, *Bricks to Build a House*, Routledge and Kegan Paul, London, 1976, p. 101.
2. C. Dickens, 'The Drunkard's Death', Ch. 12, *Sketches by Boz*, 1836.
3. As recently as the 1970s shirts could still be purchased with detachable collars, which could be fixed in place by means of a 'stud' pressed through the top buttonhole.

Pug mill for small works, circa 1900 – A century after settlement, steam-powered pug mills prepared clay for use in handmade bricks. This illustration shows a typical all metal, 2–6 horsepower grinding mill

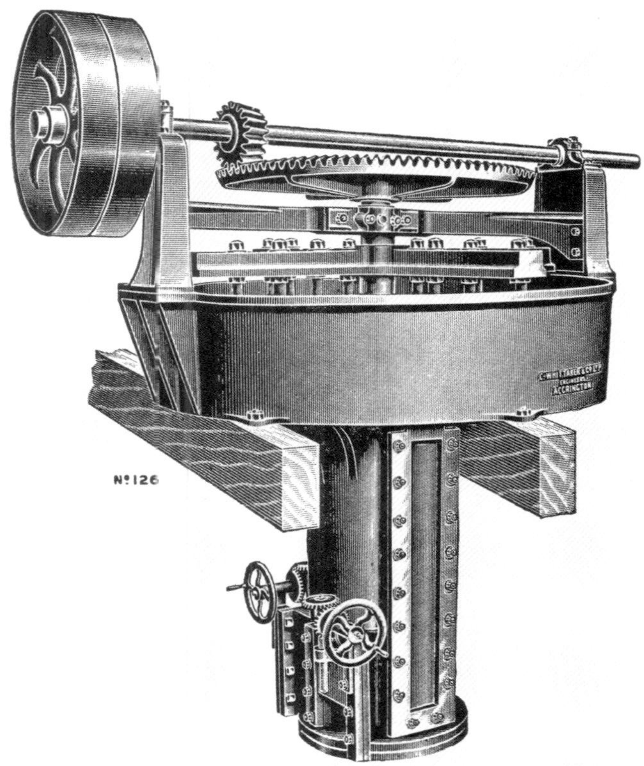

Pan-mixer with vertical pug mill, circa 1900 – This powerful mill had two sets of knives, one in the large pan and another in the barrel. The former was so arranged that at each revolution the clay is taken one step nearer to the centre of the mill. The mill was well suited to hard clays that did not mix readily with water, e.g. shales and fireclays

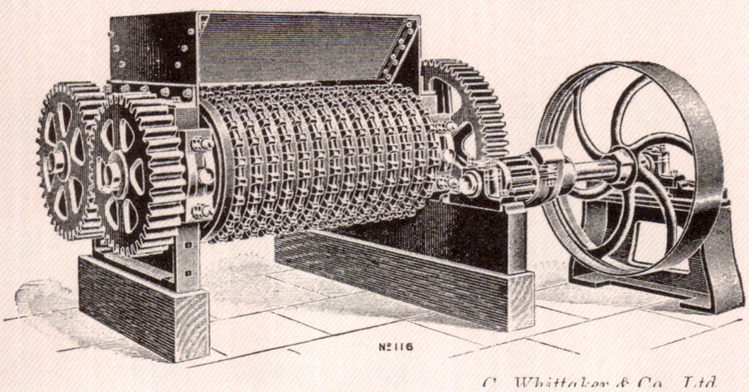

Crushing machinery – kibbling rollers and toothed crushing mills

Many designs of projections and grooves were used in crushing and rolling machines. Among the best were the hedgehog (toothed) rollers and kibbling rollers. The projections on one roller engage with those on another. The material caught between the rollers is broken up allowing a succeeding pair of smooth rollers to pulverise it

Wet-pan mill for grinding and mixing wet clays

A wet pan was another type of grinding machine. It consisted of a cast-iron pan approximately three metres in diameter, with portions of the bottom perforated with holes. Heavy cast iron rollers were carried around the pan, crushing the clay and shale, and forcing it through the holes

Blasting at the quarry was the first stage in the extraction of shale. Holes were bored in the shale and clay bed. Slabs were blown out with blasting powder, a dangerous activity and one that frequently showered the neighbourhood with dust and debris. The process was more or less continuous, with new boreholes being drilled using a sledgehammer and chisel, while other men broke the huge slabs into smaller pieces. The dirt was shovelled into hand or horse-drawn rail trucks and taken to a central area.

From quarry pit to the plant, clay was transported by wheelbarrows, a slow and tedious process, especially when the diggings were located at an inconvenient distance from the plant. Muddy conditions after wet weather created numerous difficulties and slowed down work as barrows became bogged in the worn, rutted paths. The bigger firms that consumed large quantities of clay installed portable tramways along which ran small wagons. Traction was achieved using a winch turned by hand or by horse. The wagons were made of iron with a central buffer and couplings. Emptying was performed by tipping the container over on one side, for which there was a pivot connection to the frame. Steam engines were well suited to this type of work.

CRUSHING AND MIXING

Before long the use of steam-powered machines began to transform the way clays and shales were prepared. Soon after the gold discoveries in New South Wales and Victoria, machines capable of breaking open the quartz and releasing the ore from the dirt and rock were being exported to the colony by mining entrepreneurs. Swan's Patent Crushing Mill with its set of heavy rollers connected to a central shaft and driven by a steam engine was typical. Such mills with their rotating grinding wheels that held the dirt were refinements of the original Chilean Mill in use from ancient times. In its basic form two vertical millstones on a watertight base of bed stone were turned by waterpower. The millstones were attached in such a way that they rolled on their edges, at the same time rotating clockwise on the shaft. As the mill crushed the rock, water was introduced to wash away the ore and allow the heavier gold to sink to the bottom.

The larger brick firms were quick to adopt this technology which was well suited to pulverise the harder Wianamatta shales found throughout the Sydney region. Grinding mills were efficient and could be worked by either horse or steam power. Indeed present day mills in the few surviving dry press brickmaking operations bear a close resemblance to those used in this period.[4] The simplicity of their operation made them versatile and efficient with little scope for improvement. Over the next decade, grinding and crushing machines of every description, most of them steam-driven, were imported to service the brickmaking industry. By 1900 machinery driven by belts connected to steam engines had been installed in all major brickyards to crush and mix the hard shales. Most of this heavy machinery was manufactured in Great Britain and was extremely robust. Until recently it was still being used in a handful of yards, including Bowral Bricks in New South Wales.

DEVELOPMENTS IN HAND MOULDING

Advances in grinding techniques worked in favour of brick moulding which also adopted labour saving machinery. By the middle decades of the 19th century the rapid pace of technological innovation in brick moulding was beginning to transform the industry in Great Britain and the Australian colonies. In fact, during the period 1820 to about 1850 British inventors patented 109 brickmaking machines and kilns. Some of these were displayed at the Great Exhibition of 1851 in London. Even the smallest of firms used the new mechanical devices. One type imported to the colony was the screw-operated John Whitehead Press, which achieved a pressure of 15–18 tonnes. Using this machine two men could produce about 7,000 bricks per day.[5] Whitehead's press with its daily rate of 15,000 bricks was also popular and clearly outmatched the tally of a traditional hand moulder who was hard-put to shape more than about 1,300 bricks.

4. Dry press bricks were formed on a mechanical, steam-driven press. A quantity of finely crushed shale and clay and about 10 per cent water was introduced into a box and reduced under pressure to the shape of a brick. See Ch.14 for a full description of the process of dry press brickmaking.

5. A.B. Searle, *Modern Brickmaking*, Ernest Benn, London, 1920, p. 78.

Adjustable lever press made by John Whitehead & Co. Ltd, England, circa 1890. The weight lever is adjusted so that the pressure can be adapted to bricks of varying thickness

Press for handmade bricks made by the Brightside Foundry and Engineering Co. Ltd, England, circa 1890. A single motion of the lever closes the box and presses the brick. The reverse motion of the lever opens the box and raises the brick. When operated by one man and a boy this machine could press 5,000 bricks per day

Loading the Hoffman kiln at Derrabassi in the Punjab 2002. By joining two medium size Hoffman kilns together, the plant manager of Pioneer Brick, Gurpal Singh Bajwa, created one magnificent Hoffman kiln with a brick capacity of 1.6 million. When loaded with wall tiles, pavers and bricks it holds almost two million pieces. Brian Binskin

In the case of the lever press, pugged clay was thrown into moulds which were recessed into the surface of the table. A lever was depressed forcing the clay into the shape of a brick. By repeating the action several times, bricks were re-pressed to ensure adequate compaction of the clay, producing a brick with fine, straight edges (arriss). A return mechanism forced the clay brick upwards and out of the mould. Such devices were popular with hand moulders who used them to re-press mechanically extruded green bricks, using the frog to help further compress the body of the clay.

Unfortunately brickmaking machines of all types were extremely expensive. According to *The Builder*, a trade journal, machines for moulding bricks cost approximately £1,400,[6] although the economies of scale were considerable and justified the expense. It was estimated that 20,000 bricks hand produced by four gangs over a 10 hour day cost the brickmaker £3 16s, but only £1 9s when produced by machine.[7] Progress towards mechanisation was still painfully slow. While moulding had been extensively mechanised in England and the USA, family-run yards in Australia, many of them short of capital, continued to use traditional methods. Handmade bricks continue to be manufactured, especially in developing countries such as India where family traditions have remained unchanged for generations.

PLASTIC MOULDING

There was never any question that traditional methods of hand moulding, even where hand or horse-powered machinery was used to speed the process, would eventually be replaced by newer techniques of mass production. Steam-powered machinery blazed the trail, enabling brickmakers to better exploit the range of clays found in different parts of Australia. From about the mid-1860s it was common to hear the expressions 'stiff plastic' and

6. *The Builder*, 1856, p. 22.

7. *The Builder*, 1852, p. 385.

'semi plastic' used to describe types of bricks being produced under steam power. The making of plastic bricks, either stiff or semi plastic, was a response to the gradual introduction of steam power to the brickmaking industry. In one sense, the process represents a transition between the hand operation of the mid-nineteenth century and the dry press process commonly in use by 1900.[8] But what exactly was a 'plastic brick'? In *Modern Brickmaking* (1911), A.B. Searle defined plastic bricks as those:

> *made from clay which has been converted into a highly plastic paste, or in which the plasticity has been developed as fully as possible. All hand-made bricks and tiles are of this kind, but the term is also used in connexion with machine-made goods, particularly with loamy clays. The main difference between this and the stiff plastic process is the greater quantity of water added to the clay, which necessitates thorough treatment and more careful drying.*[9]

SEMI PLASTIC BRICKS

One machine used in Australia to produce plastic bricks was a mixer-extruder which produced wire cut bricks using semi plastic clays, i.e. stiffer than that preferred by hand moulders. The process was ideal for the softer, alluvial clays of Victoria. The bricks produced were popular with architects because of their smooth face, although the brick had no frog. They could also be made with the same kind of plastic clay used for hand made bricks.

The earliest wire cut machines, known as 'stupids', consisted of a clay-filled box with its hinged lid clamped down. A hand-turned rack and pinion worked a piston inside, extruding the clay through a rectangular mouthpiece of similar section to the length and breadth of a brick. The box had to be refilled after each piston stroke, although continuous extrusion was adopted after 1860. A column of clay measuring about 0.9 metres long emerged from the machine and was pushed sideways against a frame of taut wires, which sliced the clay into about ten bricks with wastage at each end. When partly dry the bricks required finishing in hand-operated presses.[10] The Goodsell works mentioned in the next section made extruded wire-cut plastic shale bricks, and the plant's circular revolving table could produce 15,000 bricks compared to an output of 1,250 a day by earlier methods.

Wire cut machine patented by Clayton (England), circa 1860. Clay was fed in at the top and fell between a pair of crushing rolls into the extruder. Two clay columns were extruded sideways, cut by wires on a hand-operated frame and the bricks placed on barrows. Two small tanks over the mouthpieces held water for lubricating the clay column.

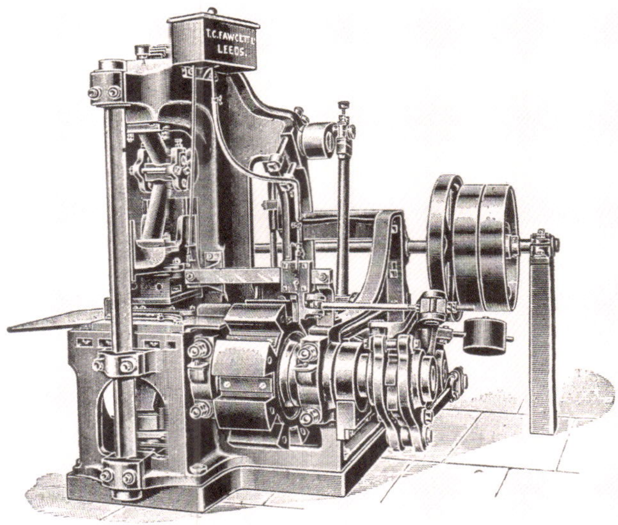

Stiff plastic brick machine with open clot moulds. Clay fell down a chute from the screens into a mixer where a little water was added, and thence into a pug mill. After being pugged it was forced into a mould in the cogwheel. At the same time, the mix in another mould was forced out and sent under a press where it was shaped.

8. N. Peek and C. Pratten, *Working the Clays: the Brickmakers of the Ashfield District*, Ashfield & District Historical Society, 1996, p. 137.
9. A.B. Searle, *Modern Brickmaking*, London, Ernest Benn, 1911, p. 120.
10. M. Hammond, *Bricks and Brickmaking*, Shire Publications Ltd, Buckinghamshire, 2001, p. 14.

Clay and shale trucks. Tip trucks running on iron rails were common in most brick pits by the turn of the century. Arthur Koppel

Despite the apparent success of the extrusion technique and its widespread use in Australia and overseas, this method had several disadvantages. Clay was insufficiently compressed during extrusion and the bricks produced were little harder than those moulded by hand. Further, because the clay was so wet the green bricks were dried in a shed prior to firing, resulting in shrinkage, a lack of uniformity, and blunt arrisses. Unsurprisingly, they were ill suited to the quality of building being erected in the boom period during the second half of the nineteenth century. To overcome some of these deficiencies extruded wire-cut bricks were usually re-pressed in a machine similar to the lever press, and ejected with a frog identifying the manufacturer. If re-pressing was to be performed, and to allow redistribution of the material after compression, the bricks were made shorter but deeper than the final required dimensions. Goodsell was just one of many firms that manufactured plastic shale bricks in the St Peters district. The Red Cross Steam Brickworks, which operated from 1886 to 1897 in Victoria Road, Marrickville, also specialised in white and moulded semi plastic bricks.

STIFF PLASTIC METHODS

On the other hand, bricks made using the stiff plastic method were superior in almost every respect. Here the object was to produce bricks that could be set direct into the kiln, obviating the need for drying. The process, however, could only be used with harder clays and shales, and was generally unsuitable for surface and the more plastic clays. This was an important advance since previously shale was considered to have had no commercial value. Brick of the stiff plastic type, made from good shale, was metallic in character, absorbed only a small percentage of water, resisted high crushing stresses, and consequently was often used for engineering work.

Powerful grinding mills, the new breed connected by belt drives to steam engines, were needed in order to reduce the shales to powder. During the grinding process, water was added to make the material sufficiently plastic to be pressed, but not so soft as for hand moulding. The bricks were then shaped in power-driven clot moulds or presses and the resulting common bricks were often of such consistency that they could be taken direct from a machine and set 20 courses high without being deformed. Conversely, facing bricks produced by the stiff plastic process were of necessity dried prior to burning. Where large quantities of facing bricks were required they were immediately re-pressed in a nearby machine to achieve a greater density and strength. Most of the failures in the stiff plastic process appear to have been due to attempts to shorten the process of manufacture by omitting tempering, pugging or drying.

By the 1880s moulding and cutting machines had developed to a high degree of sophistication suited to particular types of plant and production. Yet within a decade a quantum leap in brickmaking technology arrived in the form of the dry press, pioneered by English firms Bradley and Craven and the Platt Brothers of Oldham. Within a decade dry press had established itself as the preferred method of brick moulding.

FREDERICK GOODSELL

One pioneer producer of shale plastic bricks was the Newtown brickmaker, Frederick Goodsell, whose firm commenced manufacture in the early 1870s. Determined to capitalise on the burgeoning demand for bricks, Goodsell and his business partners established a steam brickworks on a large acreage at the junction of Edgeware Road and May Street, Newtown. In fact, he claimed to be the first manufacturer in Australia to produce bricks of the plastic type. *The Sydney Morning Herald* of 17 June 1870 gives an account of the opening of the plant. The article declares that 'Messrs. Goodsell, Tye, and Hughes, who have for many years engaged in the work of brickmaking at Newtown, have recently erected a very large and complete establishment for the manufacture of common bricks, fire-bricks, drain pipes, and flooring-tiles by steam machinery upon a greatly improved principle.'

The newspaper article goes on to say that:

> *the brick-making machine is constructed on the principle of forming the bricks in moulds. The clay in the first instance is conveyed by a tramway, worked by steam, to two iron rollers, weighing about four tons, which have a small space between them, and are made to revolve in contrary directions. Consequently they draw in the clay with which they are fed, and it comes highly compressed. It then passes into a powerful pug-mill, where it is thoroughly mixed and worked up with hot water, previously to being conveyed into the machine, where the bricks are moulded, undergoing a pressure of thirty tons. The bricks, when delivered by the machine, are carried to fine commodious drying sheds, instead of being exposed to the weather. The newly formed bricks are much stiffer than those made by hand, and require only half the time to dry. From there being so little water in the clay and from its undergoing so severe a compression, the bricks produced are remarkably compact, strong, and heavy, each brick weighing upwards of 10 lbs... The machine is capable of turning out 15,000 bricks daily.*[11]

STEAM BRICKS

Steam-power was the motive force behind the new-fangled belt-driven brick machinery, yet steam was also introduced during production. The expression 'steam bricks' requires further explanation. During the steaming process, dry clay and shale stamped from moulds was subjected to a bath of pressurised steam which penetrated through to the core. The procedure eliminated the long drying times prior to firing and also saved a great deal of space around the yard. Less water also meant correspondingly less distortion during burning, with the end result being an extremely hard brick. Moreover, by adjusting the firing time, bricks could be produced that ranged in colour from cream through reds to dark brown.

There were, however, distinct limitations, not the least being that highly pressed bricks were impervious to steam. Without sufficient moisture the inside of the brick emerged from the kiln with a friable and crumbly core. This left the exterior over burnt – vitrified – with an exceptionally tough outer skin of two or three centimetres. Conversely, excessive steaming simply turned the outer skin to slush before it could burn. With the introduction of the Hoffman kiln and dry press machinery, however, steam processes gradually disappeared.

DOWNDRAUGHT KILNS

Developments in moulding and pressing were also accompanied by advances in kiln technology and firing. By the mid 1850s clamp kilns were all but obsolete in the main settlements, having been replaced by scotch kilns. Wood firing had also given way to coal or coke firing, with the exception of country areas where clamps persisted and were still able to use plentiful supplies of local forest timber.

11. *Sydney Morning Herald*, 17 June 1870, p. 5.

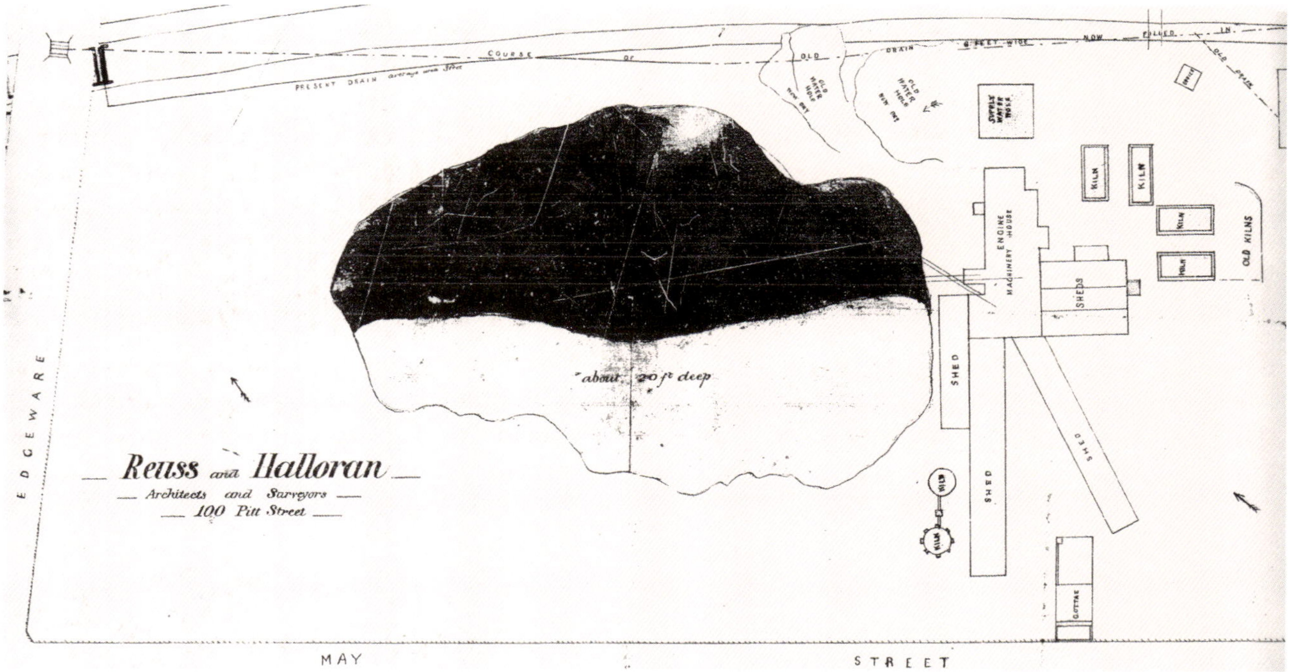

At some point during 1883, prior to the construction of the railway, Reuss and Halloran, a firm of architects and surveyors, was commissioned to survey the brick pits along the boundaries of May Street and Egdeware Road, Newtown. The extent of Goodsell's brick operations can be seen in this map. Goodsell sold his brickmaking interests to Speare in the 1890s. The pit was eventually sold to Marrickvillle Council, filled and renamed Camdenville Park. Mitchell Library

By the early 1860s a new type of kiln was well established – downdraught kilns, or dome kilns, characterised by a permanent arched brick roof, fire holes and wickets in the walls.[12] The shape and design of these kilns strictly controlled the distribution of heat. The fire and its powerful draft of hot gases circulated upwards within the chamber to the roof through flues built against the walls, then downwards through the stacked bricks. A perforated floor and underground flues were regulated by a series of dampers which controlled the rate and intensity of burn within the kiln, as well as drawing exhaust fumes to a detached chimney. Corrugated iron roofing protected some downdraught kilns. This type of kiln was also referred to as an intermittent kiln because it needed to be cooled after each burning to allow the bricks to be removed. The principle was also applied to the firing of pottery in beehive kilns.

12. This reference to a wicket refers to the entrance to the kiln chambers, through which the barrowman entered with his load of green bricks, ready for stacking (setting) in the kiln. The wicket was closed off and daubed with mud to seal the chamber, ready for burning. The size and shape of the aperture was reminiscent of a cricket wicket, and the game was frequently played against this small arch.

Unlike clamps and open kilns, downdraughts were comparatively fuel-efficient, burning anything between 500 and 1,000 kilograms of coal per 1,000 bricks. Moreover it was easier to control the process of oxidation or reduction, which produced an excess

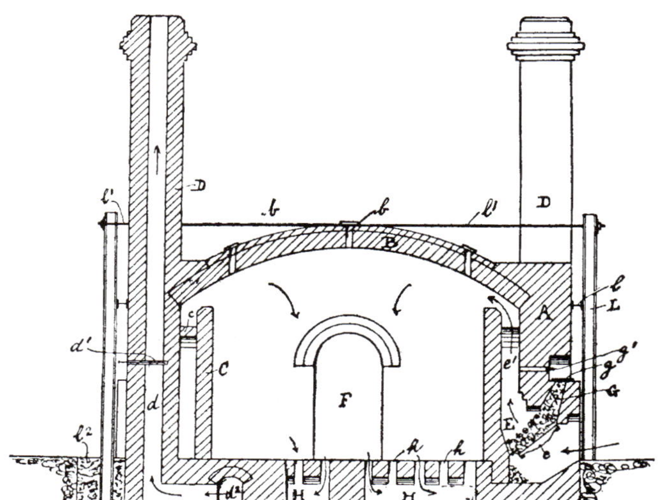

Cross section of downdraught kiln – the flame, fire gases and air rise through vertical flues built into the kiln wall. After deflection from the roof, the heat is distributed among the bricks.

or deficiency of oxygen in the kiln atmosphere, thereby determining the colour of the burnt brick. Because heat was more evenly distributed through the bricks it was easier to obtain particular colours and densities. The capacity of such kilns varied from 13,000 to 100,000 bricks, with the largest used to manufacture blue bricks, which were favoured by engineers.

Downdraught kilns continued in use long after technology rendered them obsolete, because of their ability to make bricks with a distinctive colour. This applied especially to those bricks from the heart of the kiln stack where conditions produced a brick that was hard, well-shaped and finely coloured. However, underburning and vitrification were common. Moreover, only one batch of bricks could be fired at a time. After each firing the kiln was shut down and emptied. Despite being more fuel efficient than its predecessor, each new firing of a downdraught kiln was a laborious process and costly in terms of fuel. Fuel was the biggest single cost in the manufacturing, which tended to inflate the price of bricks at the kiln, although brick prices trended downwards due to the introduction of new technology. Despite claims to the contrary, most of the kilns patented by the mid-nineteenth century were either too slow or too costly to operate. In all types of kilns the stacking (setting) processes were identical. Bricks were laid apart and each layer was sprinkled with sand to prevent them sticking together.

WATER RESISTANCE

Weather-proofing qualities were improved by the simple expedient of throwing salt over bricks in the early stages of firing. Salt glazing vitrified the outer crust, but most people considered it unsightly, which helps to explain the continuing practice of painting or rendering brickwork with stucco plaster. Firing of the bricks, however, required careful supervision, and in the absence of skilled or watchful burners, poor quality bricks continued to be manufactured. According to Smythe, a contemporary observer, bricks produced in Melbourne in 1855 'are light and porous, and readily yield in any direction to a slight blow. They absorb water very rapidly. When it had absorbed its maximum quantity it was easily crumbled between the fingers. Subjected to the ordinary tests the common red brick very rapidly disintegrates. Such a material is certainly not suited for building three storeys high; and yet a considerable number of houses in this city are so constructed.'[13]

FRIEDRICH HOFFMAN

Clearly more needed to be done to improve the strength of bricks and the rate of manufacture. The developments that transformed industry starting in the 1870s were in tempo with rapid growth in Australia. English inventors, James Gibbs and Thomas Ainslie, who founded the highly innovative Ainslie Brick and Tile Machine Company, were the first to patent multi-chamber kilns, in 1841 and 1843 respectively.

However, the main architect of change was the German inventor Friedrich Hoffman who patented his design for a radically different type of kiln in 1858. Hoffman's 'continuous' kiln was circular in plan and worked on the principle that the fire, unlike in earlier kilns, was never extinguished. The kiln comprised 12 or more firing chambers partitioned by means of a brick wall and connected to the outside by an arched opening in the side of the kiln. Above was a corrugated iron structure which protected the kiln and also the workers who were able to move about during the firing process. Once lit the kiln was not allowed go out and the fire moved progressively from one chamber to another with the bricks undergoing drying, pre-heating, firing and cooling in turn. At any one time a proportion of the chambers was left empty, although for a short time only. The entire process was controlled manually by the manipulation of dampers and sliders and required great skill and experience.

Hoffman went on to develop a rectangular form of his kiln in 1870, solving the earlier problem of brick

13. R.B. Smythe, 'On the Comparative Value and Durability of Building Materials in Use in Melbourne', Transactions of the Philosophical Society of Victoria, Vol. 1, 1855, James Blundell, Melbourne, 1855, pp. 30, 32.

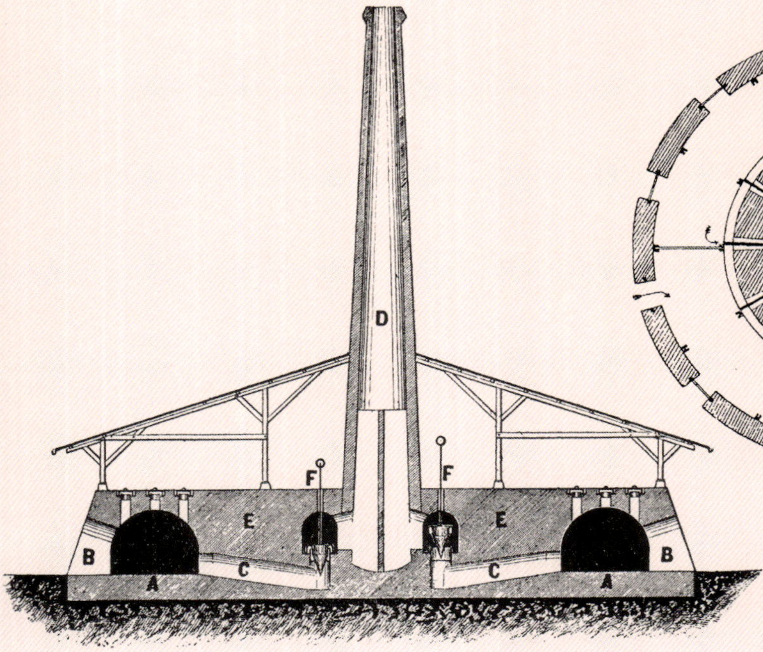

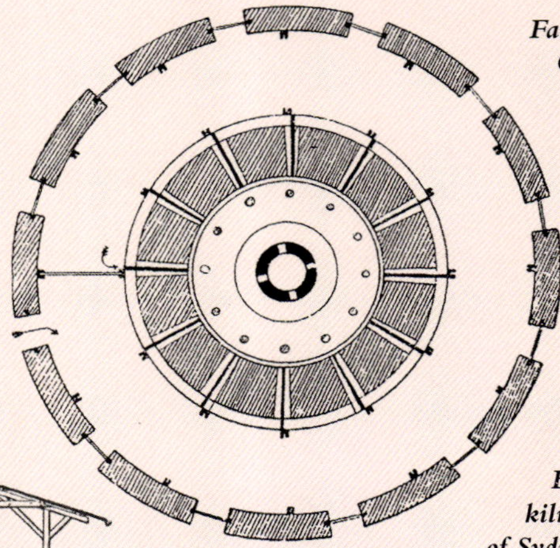

Far left: Section of the Hoffman kiln. Green bricks are set in the chambers (A) and loaded and unloaded through wickets on the side of the kiln (B). Powdered coal is fed through the fire holes from the roof. Exhaust gases are drawn through the flues (C) which can be regulated by dampers or valves (F). Gases vent to the chimney (D)

Left: The basic process of burning, as originally conceived by Hoffman

Below: Model of Hoffman's circular kiln, circa 1880. Powerhouse Museum of Sydney

discolouration caused by the inadequate venting of hot gases. Hoffman kilns had a total capacity of about 40,000 bricks per firing and were fuel efficient, burning between 120 to 220 kilograms of fuel for every thousand bricks. In fact, they were second-to-none in being able to fire large quantities of bricks quickly and efficiently. Hoffman kilns were also constructed to burn lime used in the manufacture of mortar. With improvements this type of kiln became the standard for brick manufacture in the Western world and remained so until after the end of the Second World War.[14]

The eventual triumph of the Hoffman process caused a fall in the price of bricks. In 1880 hand-made bricks cost 8s 6d a thousand which compared unfavourably to Hoffman bricks at 2s a thousand. The second important result was the beginnings of standardisation in brick size. In Australia this appears to have come about as a result of the establishment of the Hoffman Patent Steam Brick Company Ltd in 1870. This company not only burned bricks in Hoffman kilns, but also produced bricks by a wholly mechanised and continuous process which made German sized bricks measuring nine inches by four and a half inches by three inches.[15] The few surviving Hoffman kilns can be seen at the former works of the Hoffman Steam Brick Company in Brunswick and Standard Brickworks, Box Hill, Victoria. A fully equipped Hoffman brick manufacturing operation was extremely capital intensive. For example apart from the three large Hoffman kilns and engineering workshops the Brunswick works consisted of three dry press machines used to press bricks and two powerful steam engines – one to haul clay out of the pit, and another to ensure adequate drainage. The cost in the early 1870s was something in the order of £50,000, which probably amounts to several million dollars at today's prices.

14. Austral Bricks' Eastwood yard used this type of kiln until the plant's closure in 2000. Dome kilns were used at the company's Punchbowl factory until the 1990s and remain in use at the yards of its Bowral plant, Bowral Bricks.

15. J.M. Freeland, *Architecture in Australia: A History*, Penguin Books, Ringwood, 1972, p. 146.

PATENT KILNS

In Australia there were many local variations to the Hoffman, and these were often referred to as 'patent kilns'. The main difference between the simple Hoffman kiln, originally circular in shape, and a patent kiln was in configuration. Patent kilns were made with two semicircular sections connected together by two straight portions so as to form an ellipse with flattened sides. James Nangle, architect and author, writing early last century in *Australian Building Practice*, commented that:

> *a kiln much in use about Sydney is rectangular instead of circular in plan, with rounded ends, and the continuous chamber consists of two straight tunnels built side by side with the flue leading to the smoke stack between them.*[16]

CENTENNIAL BRICK KILN

One variation on the traditional Hoffman kiln was the Centennial brick kiln. According to its designers, Isaac Button, John Goodsell and Edward Peters, it represented a great advance.[17] The main advantage of this type of patent kiln was its ability to carry out – simultaneously – the processes of setting, burning, cooling, drying and discharging of products using less fuel and labour than the standard Hoffman. The Centennial must have been truly enormous; the wickets were large enough to allow a six-horse dray to load from within the walls of the kiln. Button, Peters and Goodsell installed a Centennial kiln at Goodsell's Newtown plant and also at the yard belonging to the Croydon Steam Brick Company. The kiln consisted of 18 chambers, each of which held 35,000 bricks. The company reckoned to turn out 200,000 bricks a fortnight, the cost of fuel being 1s 6d per thousand as against 3s to 10s with other kilns. The yard manager advised Burwood Council in 1888 that the two new patent kilns to be erected at his works in the next month were expected to eliminate the smoke problem that had been causing complaints among local residents. Only one patent kiln was subsequently erected at this brickyard. The type of continuous kiln erected at Croydon was a feature of many brickyards in Sydney's inner-western suburbs by 1890, including the Excelsior, Burwood, Ashfield and South Ashfield works. Indeed, by the end of the century every major brickyard was using either a Hoffman or Centennial continuous kiln.

SERCOMBE CHAMBER PATENT BARREL KILN

The Sercombe Chamber Patent Barrel kiln, registered by F. J. Sercombe and Mr Wolstonecroft, was yet another refinement on the Hoffman. The first of these kilns was built at their Bendigo brickyard in Victoria. In fact, by 1900 there were over 350 Sercombe kilns in operation in various parts of the world, superseding the earlier, less efficient Scotch kilns. The principle feature of the kiln was the creation of a perfect draught from, or to, the top, and throughout all the chambers. This ensured a steady and continuous draught in all the chambers burning simultaneously. Heated air could be transferred from the cooling chambers to those areas containing the raw products, and also allowed residual moisture in the green bricks to be driven off as steam and drawn out at the top, eliminating the problem of staining.[18]

AN EARLY TUNNEL KILN

Another development – an early tunnel kiln – deserves special mention. In 1880, the Junction Brickworks at Granville closed after only two years' production due to a combination of high initial capital costs, a misalignment between the raw material and brickmaking processes, and the difficulty of establishing mass production of bricks.

Henry Gurney, a Sydney engineer, purchased the failed company and then onsold it to John Hay Goodlet who is mentioned in the next chapter. Gurney was granted a patent for the invention of a kiln which allowed the burning of bricks loaded on a moving trolley.

16. J. Nangle, *Australian Building Practice*, 3rd edition, William Brooks, Sydney, 1925, p. 74.
17. Victoria Legislative Assembly, Patents, 1891, p. 51.
18. O. S. Pavlou, *The History of Bricks and Brick Making in New South Wales, 1788–1975*, Thesis, Faculty of Architecture, UNSW, Sydney, p. 435.

The 'Centennial' Patent Brick Kiln, circa 1889. Builder & Contractor's News, 9 February 1889

Fuel was fed through the bottom of the kiln. This is quite probably the earliest attempt to develop tunnel kiln firing methods in Australia, although the idea was not widely adopted. We do know, however, that after 1889 Gurney installed one of these kilns at his Steam Brickworks in Botany Road, Alexandria and another at the Oakes Brickworks in Neutral Bay.

FIRECLAY AND REFRACTORIES

Increasing demand for clay products throughout Australia propelled the industry, especially kiln technology, forward into modernity. Yet without refractory brick (also referred to as fireclay), fire-resistant bricks used to line the chambers, kilns would have been functionally useless, and it is to refractory that we now turn our attention.

Refractory bricks were often manufactured by the producers of ordinary building brick because only a low level of technical knowledge was required. Raw fireclay is usually found in conjunction with coal deposits and has a different chemical composition to other brick clays and shales. Refractory bricks were used extensively in iron and steel foundries, and for many decades, until a domestic industry established itself, they were imported from England or the USA.

Some small-scale refractory brickworks were established in Sydney, one of them an English firm which manufactured heat resistant magnesite bricks for use in blast furnaces.[19] The company that claimed to be the first producer of firebricks in Australia was Hughes and Drury's Newcastle and Burwood Fireclay and Brickworks located in the Junction area of Newcastle in about 1868.[20] The success of this firm, and another established by Joseph Pendlebury, is evidence of the growing market for refractory products in Australia's developing economy. Hughes and Drury commenced production in 1866 and came to dominate the colonial market in the second half of the nineteenth century. The firm supplied many industries in Sydney and regions north of Newcastle. In 1888 it was producing 20,000 firebricks per week, and 5,000 feet of drainpipe.[21] Some time later the Newcastle factory ceased production of firebricks, concentrating instead on drainpipes until its closure in the early 1980s.

In 1901 Joseph Pendlebury established a refractory brickworks near Wollongong on the south coast of New South Wales. Pendlebury had over 40 years' experience in brick production, first in England,

19. Today, alumina magnesite carbon brick is made from high quality bauxite, corundum,and fused magnesia. It has low thermal conductivity, good isolation ability and appropriate residual expansion. It is mainly used in the fabrication of ladle side walls and bottoms in iron and metal smelting processes.

20. *Newcastle Pilot*, 5 September 1868.

21. *Aldine Centennial History*, Vol. II, Sydney: The Aldine Publishing Co, 1888.

then as foreman at the local Bulli Brick and Tile Company. He was a well-respected and expert burner, one of the most important skills within refractory production.

Imported refractory bricks continued to provide the bulk of supplies, whether from Victoria and South Australia or the UK and the USA.[22] From the middle of the century between 100,000 and one million firebricks are estimated to have been imported into New South Wales alone every year.

DRY PRESS MOULDING

Innovation in brick moulding machinery continued with the most significant – and far-reaching – advances being made during the late 1880s. About this time heavy, belt-driven machines powered by steam engines were imported from England in increasing numbers. Designed to compress finely ground dirt mixed with about 10 per cent water, the so-called 'dry press' machines were particularly suited to the processing of the harder clays and shales found among the claypits of St Peters in Sydney.

The new technology enabled dry clays and shales to be delivered by hand barrows or rail trucks direct to the grinding pans where a small quantity of water was added to produce clay with more uniform particles. From there it was fed into a hopper at the top of the dry press machine which loosely deposited a quantity of clay into moulds of the same width and length as a brick, but deeper than the required thickness. A plunger, which exactly fitted the mould, was forced downward under heavy pressure (about 900 tonnes per square inch) squeezing the clay to the size required.

Patents registers and contemporary journals of the early 1870s and 1880s can help us to determine the most widely used machines at this time. The first of these brick machines were imported from England and manufactured by Thomas Fawcett Ltd of Leeds. The single-brick Fawcett machine proved ideal in the colonial situation and could convert the ground dry clay into dry press bricks of good quality, and were especially suited to the raw materials found at St Peters. The first steam-operated dry press brickworks was started in the St Peters district by W.G. Collins and Company in 1878, with other yards throughout the state quick to follow suit. Six years later there were a dozen such brickyards in Sydney and seven operating in Melbourne.

Dry press machines produced by the English engineering firm of Bradley & Craven also became popular among Sydney brick manufacturers. James Horsley Craven, son of the founder, even visited Australia during the late 1870s to exhibit his father's wares. Following the introduction of the first dry press in 1878, Bradley & Craven almost exclusively supplied the colonial market in the 20-year period before the advent of Foster and Sons Pty Ltd.

If Bradley & Craven's machines were not the first dry press machines to be developed, they were a significant improvement on previous models. The patent holder asserted that it was difficult to make perfectly sound and solid bricks from completely dry clay. The air lodged in the interstices of the clay dust was not always completely expelled by a single compression of the plunger. By reapplying (re-pressing) the plunger, the Bradley & Craven machine forced air out of the mass and consolidated the brick to an extent not achievable by a single pressure. In this machine two and even three distinct pressings of increasing magnitude were given to each brick. Imported dry press machines cost between £4,000 and £5,000, a significant amount of money in those days, although for any manufacturer wishing to attract investment the production advantages over traditional methods were clear.

PLATT BROTHERS, OLDHAM

One manufacturer who competed vigorously with Bradley & Craven deserves special mention. Platt Brothers & Company was established in 1821 by two brothers who lived in the mill town of Oldham, which lies in the foothills of the Pennines, 12 kilometres north-east of Manchester in England.

22. The New South Wales Statistical Register 1870–1901 categorised all refractory bricks as firebricks, even though useage differed according to the type of kiln and material being heated.

The town's wealth was built on cotton spinning and the manufacture of textile machinery. In its heyday Oldham was the greatest cotton-spinning town in the world, boasting more spindles than the whole of the United States. In fact, by the end of the nineteenth century Platt Brothers employed over 14,000 workers, a remarkable number of people by today's standards. The firm made a great variety of industrial machinery in addition to spinning and weaving machines. Among them were brick presses. Soon the aptly named 'Platts' became indispensable to the mass production of commons and face bricks (used in the construction of a building's façade) by the dry press method.[23]

Like its competitors the Platt press was a steam-powered hydraulic stamp that could turn out bricks at the rate of 20,000 per day or 2,800 per hour. The Platt was driven by a line shaft connected directly to a steam engine which was later replaced by an electric motor with a capacity of about 300 horse power. The press exerted a powerful hydraulic force that stamped bricks from a pre-set steel mould packed with a mixture of crushed clay and shale. Bricks were ejected from the mould which was then refilled with a fresh batch of clay. The working of a Platt was accompanied by deafening thuds as the press forced the clay into its familiar brick shape. Platts, by all accounts, were also extremely dangerous, with a tendency to pulverise the hands of slow-witted operators.

Operating the press was a machine driver, his offsider, and as many 'wheelers' as it took to load the green bricks onto hand barrows in the days prior to the introduction of forklifts. To achieve greater compaction of the clay and shale mix, the operator could slow the press using a pulley connected to the line shaft. This was especially important when producing higher quality face bricks. Since face bricks required a greater density, the slower the machine pressed, the more effectively it squeezed the air out of the clay mix.

Name plate from dry press machine manufactured by the firm of Platt Brothers, Oldham, 1911. Ron Ringer Collection

From the 1870s onwards Platt machines were exported in large numbers to countries throughout the British Empire including Australia, where they found a ready market in the major brick-producing areas of Sydney and Melbourne. Having the larger population, Sydney was the biggest market in Australia for Platt machines, which overtook Bradley & Craven. The precise number of machines exported to Australia is difficult to gauge, but may have been in the low hundreds.

GEORGE FOSTER & SONS

Over the course of time, however, it became clear that traditional Platts, engineered to produce bricks in the wetter English conditions, were not as well suited to handling local clays and shales. The Sydney engineering firm of George Foster & Sons was quick to see an opportunity. With local conditions in mind Foster's made everything from pressure tanks, pit trucks, crushers, roll crushers, grinding pans, to a local version of the Platt, and in due course it became nationally famous.

George Foster, a young immigrant from Nottingham, set up on his own in 1898 at King Street, St Peters, having gained experience working at Hodkinson's, a Sydney manufacturer of brickmaking machinery.[24] According to brickmaker Brian Binskin, a former long-serving employee of Austral Bricks' parent

23. The term commons is used to describe cheaper bricks used to build the body of the house or structure. Face bricks were used to 'face' buildings and were superior in terms of their overall quality – the straightness of the arriss, and smoothness of the surface. Face bricks were burned in a beehive kiln using a different process and fewer quantities were produced, making them more expensive.

24. *The Clay Products Journal of Australia (CPJA)*, October 1934, p. 11.

company, Brickworks Limited, Platt Brothers may have appointed Fosters as a licensed repairer. Certainly, compliance plates salvaged from discarded machinery suggest that production of brick presses and a range of brick manufacturing machinery had already commenced at Foster's Newtown factory around the year 1911.

The main point of difference between the British-manufactured Platt dry press and Foster's version was the pressure, or lifting plates, which in the case of the latter were more robustly engineered. These were located on both the top shaft where the charger holding the prepared shale mix was cleared, and on the bottom shaft. Both shafts on the Foster's press had a working life of up to nine years, which was a great improvement on the English-made machines. Ongoing maintenance was a necessary requirement and the operator was obliged to regularly tighten the die bolts holding the plates onto the bottom head to prevent serious damage to the machine and personal injury.

Soon after the outbreak of World War One in 1914, and with it the German U-boat threat to Allied shipping, Platt Brothers withdrew from supplying dry press machines to the Australian market. Foster stepped in to fill the gap and subsequently diversified into tile-making machinery. In 1920 the firm had grown to such an extent that it relocated to larger premises on the opposite side of King Street. At about this time, George Foster died and was succeeded by his son, also named George, who along with his three other brothers continued in the business.[25] At its peak during the 1950s the company employed nearly 300 people. During the 1970s, the Sydney engineering firm of William Wallbank purchased Foster's. Today, Frank Wallbank continues to manufacture a modern version of the dry press brick plant, sold as the Wallbank Press. Modern extrusion and tunnel kiln production methods mean that the industry is again reliant on foreign technology, much of it imported from France and Italy. Sadly the former home of Foster's operations in St Peters fell into disrepair and was recently demolished to make way for an apartment block.

ENGINEERING FIRMS

Foster was not the only engineering firm to manufacture brickmaking machinery in Sydney. Blanks and Lefebvre at their Invicta foundry in Mitchell Street, Glebe, patented the Invicta Brickmaking Machine in 1890 which proved a great success for the company. By 1892 five of their machines were in use at the Excelsior Steam Brickmaking Company in Marrickville, with the Gore Hill Brickworks operating an Invicta to the great satisfaction of the manager, Mr G.H. Collings. Richard Davies made numerous improvements to dry press machines in 1889 at his Croydon foundry and was highly regarded by Sydney brickmasters and yard managers alike. Marrickville Engineering Works was another firm that serviced Platt machines.

In Victoria, Naylor and Williams patented the Semi Plastic Automatic Brick Moulding and Pressing Machine in 1888, which was manufactured by Jacques Brothers, engineers in the town of Richmond. Also in Victoria was Edward Parry, the manager of the Blackburn Brickworks who patented the Improved Brick-Making Machine for Plastic Moulding in 1890 with the assistance of Edward Seitz, an engineer. Inventions and improvements were common during this period with Samuel Whitely of Paddington, Sydney, patenting a machine in 1886 which combined the plastic and extrusion methods of manufacture.

IMPACT OF DRY PRESS METHODS

Dry press machines won the day, their great advantage being that they dispensed with the need to pug clay in the traditional way. This rendered hand and horse-driven pug mills obsolete within a matter of years.

25. *The Clay Products Journal of Australia (CPJA)*, October 1934, p. 11.

The dry press also produced a denser brick that suffered less shrinkage in drying and baking. Bricks still needed to be dried in the hack before being fired, but developments in kiln technology shortened the drying period. Moreover, steam-driven moulding machinery increased the production capacity enormously, so that brickmaking became a continuous process with substantial profits to be made, rather than a seasonal occupation. It was not until the late 1890s that colonial brickmakers refined the dry press method sufficiently for it to become the major brickmaking technique employed in New South Wales.

There were certain disadvantages worth mentioning. If full compaction in the press was not achieved, fired bricks tended to lack strength and solidity, and were often porous in texture. On the other hand, extreme pressure produced a dense brick with smooth faces that not only prevented the evaporation of moisture within the brick, but was also insufficiently porous to allow bonding with the mortar. For technology that was designed in the second half of the nineteenth century, it is surprising that dry press machines remain largely unchanged and in use today, with no substitute having emerged.

Finally the establishment of firms such as George Foster & Sons illustrates a local trend towards self-sufficiency. This had been apparent since the late 1870s with the creation of several small firms in Sydney and Melbourne specialising in the manufacture of parts and accessories for the imported machines. In fact, homegrown manufacturing ultimately displaced many of the English machines, one notable example being the engineering firm of William Wallbank in Auburn, Sydney, mentioned previously.

THE LIME INDUSTRY

Innovations in methods of brick manufacturing were not matched in the production of lime where techniques remained inefficient and based on antiquated practices. Nonetheless sufficient quantities of lime were generally available and shortages no longer hampered the progress of the colonial building industry.

Botany Bay and Cook's River continued to supply most of the lime used in Sydney from oyster and cockleshells obtained from the extensive mud flats. William and John Buckridge erected open kilns near Cook's River.[26] Edward Blaxland's kilns were operating near Ermington to supply Parramatta.[27] Sydney Harbour was also an important shell source. By the 1860s the lime trade was a lucrative business and some 20 vessels, mostly small ketches, plied the harbour. Ketches from Brisbane Water and Pittwater brought cargoes of shells and landed them at Lime Burner's Wharf.[28]

By the mid-1860s, lime kilns were in operation on the waterside at Millers Point,[29] and at a lime wharf on Sussex Street. Numerous kilns dotted the uninhabited shores of Middle Harbour. In fact, wherever large supplies of shells could be obtained, lime kilns proliferated. Newcastle was keenly competitive, which tended to lower the price of lime. When the Newcastle supplies could not reach Sydney due to bad weather, prices quickly rose.

VICTORIA

The profound changes overtaking the brick industry in New South Wales were mirrored in Victoria where mechanisation proceeded in response to the growing population. The finer alluvial clays of Victoria were also more amenable to mechanical moulding due to their highly plastic nature. Although major improvements were made in the 1850s to brick moulding and firing, Victoria lagged behind, mostly, it seems, on account of transporting heavy loads on the State's poor roads.

Victoria's first bricks were made in 1827 at Corinella on Western Port, eight years before the foundation of Melbourne. Melbourne's first brick structure is said to be a chimney built in 1835 for John Batman by William Buckley, the newly rehabilitated 'wild

26. *Truth*, 21 May 1916, Vol. B., p. 165.

27. S.T. Leigh, *The Handbook to Sydney and suburbs with a plan of the city and map of the roads of the Colony, Sydney*, S.T. Leigh, 1867, p. 65.

28. Journal and Proceedings, Royal Australian Historical Society (R.A.H.S.), Vol. 48, pp. 307–9.

29. George Alderton's huge beehive kiln stood on the rocky harbour shores and ceased operations in 1890 when the land was resumed for the construction of the railway at Miller's Point. It was finally demolished when work on the Sydney Harbour Bridge pylons was commenced.

white man', who may have abandoned white society many years before, but not his bricklaying skills. Bricks were also imported from Van Diemen's Land for use in Melbourne. Melbourne's oldest existing brick building is believed to be Tavistock House in Flinders Lane, dating from the 1840s.

Industrial development in the Victorian colony followed much the same pattern as elsewhere; proximity to raw materials and markets were decisive factors. In Melbourne, rich seams of alluvial clay, surfacing at many points, underlay much of the developing city. The towns of Kew, Hawthorn, Malvern, and Oakleigh established their own brickyards, as did regional towns such as those at Echuca, Ballarat, Bendigo and Portland.

During the 1860s several brick firms were established in Northcote, Thornbury, Preston and Reservoir and to the southwest at Burwood and Box Hill. One of these was Cornwell's works (established in 1861) at Brunswick which later changed its name to the Brunswick Potteries and Brickworks. The firm produced bricks of all types at the rate of 120,000 per week. As the market grew Cornwell expanded his product range to meet the needs of the large local population and became the first manufacturer in Victoria to produce roofing tiles of the Marseilles pattern. Another large pottery established in 1858 was the Bendigo Pottery at Epsom in Victoria, which was to become one of the colony's most successful manufactories in the following years. Its founder was G.D. Guthrie.

The 1870s were a turning point for the Victorian industry with the development of a railway network linking Melbourne to outlying regions. Many of the smaller, regional brickworks closed as the more efficient mechanised brick press and Hoffman kilns were adopted. One such firm, founded by three railway construction contractors[30] who purchased land in Dawson Street, Brunswick, began making bricks using the traditional methods. In 1870, however, the partners formed the Hoffman Patent Steam Brick Company, installed mechanised plant and Australia's first Hoffman kiln at their Brunswick yard and commenced production in 1872. By 1879 over 120 million bricks had been made, and their quality was said to be:

> *superior to anything of the kind in the colonies, and had carried off the prizes at all the Colonial Exhibitions and had been honourably mentioned at the Philadelphia Exhibition.* [31]

TAKING STOCK

The decades of the '70s and '80s were prosperous years for the brick industry in Australia as it approached its centenary. There is no doubt that the introduction of the Hoffman kiln and dry press brick moulding machines was vitally important. The process of modernisation through mechanisation was no longer possible to ignore. The few progressives who had installed modern machinery were now joined by an increasing number of manufacturers anxious not to be left behind by the competition. In the 30-year period before the turn of the century, mechanisation in the brick industry effectively severed the link with Australia's first brickmakers. In that time most of the old hand moulders who had clung to traditional techniques dating back thousands of years had disappeared, to be replaced by larger firms with mechanical presses and continuous kilns. The revolution in brickmaking was now virtually complete, although the old ways were yet to disappear completely. Apart from minor refinements the craft of the brickmaker continued in much the same way until the 1950s when the next wave of change engulfed the industry through the widespread adoption of tunnel kiln technology.

30. Railway engineers used bricks in great quantities for the construction of tunnels, cuttings, platforms, and bridges.

31. *The Clay Products Journal Australia*, March 1955, p. 11.

Above: Daley's Brickworks. Marrickville Local Studies Library

Right: The Surry Hills works of Goodlet and Smith

Below: The Standsure Brickworks – Established by Thomas Daley, the Standsure Brick Works was a prominent landmark in Marrickville before its closure in the early 1920s. Marrickville Local Studies Library

The Brick Masters 5

Transition – Child labour – Protective associations and benevolent societies – Waterloo Flats – New arrivals – Family ties – Brick cartage – Josiah Gentle – William Judd – Westward expansion – Goodlet and Smith – North of Sydney Harbour

TRANSITION

For Sydney's brick industry the 1870s and 1880s were years of transition and expansion. Until that time brickmakers were mostly family concerns or small partnerships confined to the burning of sandstock bricks. Skills were passed down from father to son and those who excelled in the craft of brickmaking were often referred to as brick masters. Equipped with basic mechanical equipment new firms established themselves and continued to use hand-moulding techniques. Relatively few made the transition to a large-scale industrial enterprise which required large amounts of capital.

However, it soon became evident that hand-moulding yards were no longer profitable when competing with the highly mechanised plant of the larger works. Even some of the larger firms vanished from the metropolitan area, having exhausted the clays in their pits. Daley's brickworks and the Toyer Brothers' pit at Dibble Avenue, Marrickville, are just two examples. Thomas Daley established Standsure Brick Company in 1886 at Sydenham Road on what is now Henson Park. Two years later his company was turning out an average of 100,000 dry press bricks per week. Standsure's luck ran for nearly 40 years before the pit closed in 1924 some years after Daley's death. James Toyer was a small operator and went on to become a nurseryman, founding a successful business known as Toyer and Sons.[1] In addition, the brickmaker Walter Halverson became a grocer, and was joined in this occupation by Charles Burling and Francis Russell. The two brothers, Jack and Abe Cox, well known among brickmakers at St Peters, closed their small yards and became jockeys, Abe being one of the foremost in Western Australia.

Yet for every defeat there were success stories. Some of the old sandstock brickmakers, pug boys and brick masters left for country districts where they could still practise their craft using traditional methods. A greater number still were absorbed in the steam yards, with some fortunate enough to become yard owners. Charles Cooper began the steam brickworks at Ryde and Bankstown. William Edwards, James Squires, William Judd, and Henry Woodley were long associated with St Peters Steam Brickworks which they established. Judd went on to found the Hurstville Steam Brick Company at Mortdale in 1884, and Woodley the Federal and Carlton Brickworks. In most cases the major shareholders in these companies were brick masters (or their sons) who had seen the writing on the wall for sole traders and were sufficiently entrepreneurial to seize an opportunity.

BOOM CONDITIONS

High levels of immigration sustained the Sydney building industry. Add to this a sharp decline in brick prices in the 1870s due to more efficient firing methods using coal instead of wood, and it is little wonder that boom conditions prevailed in the east coast colonies. An idea of the importance of the industry in 1890 can be gleaned from an article in the *Sydney Morning Herald*, stating that as many as one and a half million machine made bricks were being manufactured each day in Alexandria alone. Using records taken from Sands' Directory for the years 1860 to 1900 we can chart the expansion of brickmakers and the location of their yards. *See list of appendices.*

1. James Toyer's father, George Toyer, gave evidence before the Select Committee on the Employment of Children, 1876. James Toyer and his five brothers worked in the brick pits from an early age.

***On the Vag* – Unusual brick marking, quite possibly a practical joke, that suggests collusion between the brickmaker (not necessarily the owner of the yard), the burner who fired the bricks, the brick carter, and bricklayer or building contractor. 'On the Vag' (after the Vagrancy Act) was a term used to describe a vagrant. The inclusion of the brick was certainly deliberate and would have gone unnoticed by the church elders until it was too late for it to be easily removed.**

Bricks for the church at Lochinvar may have been sourced from any of the four large brickyards in East Maitland. These were Turton (Waratah Works), Auckett, Pont and Baker, the latter being the most likely supplier. William Henry Baker arrived in Australia from England in 1847. His brickmaking firm was the first in the Maitland district to sell sandstock bricks. Jack Krajewski

CHILD LABOUR

The concentration of the industry into fewer hands may have been one of the catalysts that prompted closer government scrutiny. The Shops and Factories Act, which encompassed the brick industry in broad terms, provided a general set of regulations, but its effectiveness was questionable. Increasingly there were calls for greater regulation of working conditions and shorter working hours. Public concern was such that in 1875 a Select Committee on the Employment of Children was established by the New South Wales Legislative Assembly. Members of the committee took evidence from individuals and brick masters at a variety of brickyards in the St Peters and Waterloo districts.

The committee established that a typical working day was 10 hours and the use of children as puggers, widespread. In fact it was commonplace for a child to remove eight or nine tonnes of clay dirt in a single shift. During prolonged heavy rain work stopped altogether, although in what the brickmakers termed 'drabbly weather' the men went into the pit to dig clays, assisted by the boys. A child's pay was anything from 10s to 14s a week.[2]

2. Select Committee of the NSW Legislative Assembly on the Employment of Children, 1875. Evidence was taken by the Committee about the employment of children in the St Peters brickyards in January 1876.

***Above:* St Peters Church, Cook's River, circa 1855–1864, on what is now Princes Highway. The Sunday School building with its gable roof can be seen to the right of the picture. Unknown artist – St Peters Church, Cook's River, Sydney**

***Right:* Children's cricket team, St Peters, 1920s – Children pose for the camera. They would have differed little in appearance from their counterparts mentioned in the 1875 report into child labour. Many of them were the sons of local brickmakers. St Peter's Church, Cook's River, Sydney**

James Cook, a working brickmaker, told the enquiry that not one of the six children of a brickmaker working the yard next door could 'read or write so that I can understand him,' although quite a number were extremely good at arithmetic. This was not unusual in view of the fact that puggers and moulders would need to keep a mental note of how many bricks they made. School education, however, was not a high priority for the poor families of the district. According to the evidence of Mr Guille, the schoolmaster at St Peters Certified Church of England School at Newtown, children were regularly taken out of the Infants' Department to work on the brickfields.

There is scarcely a trade where there are so many illiterates as are to be found among the brickmakers of St Peters, and I impute it to the early age at which they have to go pugging up.[3]

Another respondent was Mr Alfred Dawes, uncle of the future general manager of Austral Bricks and founder of Brickworks Limited, William King Dawes. Alfred Dawes was more sympathetic to the beneficial effects of hard work:

Mr Dawes says that many a time as a lad he fell asleep, tired out, when sitting at his evening meal. It

3. Ibid.

was a case of survival of the fittest; those who stuck to the job suffered no ill effects physically, and a great many of them were rewarded by becoming captains of industry in later life.

Resort to fisticuffs was common, and Mr Dawes was an eyewitness of many stand-up fights at which the combatants pummelled each other while fellow employees formed a ring. Quite a number bred game fowls and fighting dogs.[4]

One committee member, the brickmaker Frederick J. Goodsell, agreed with Dawes that such intensive labour was a healthy improver of men:

As regards hurting them, I could produce as fine a man as you could see in New South Wales who are brickmakers, and who have been puggers-up. The younger they take to it the better brickmakers they are, and the easier they go through their work.[5]

Members of the committee looked for instances of abuse, but were hard put to find any evidence. In one exchange, brickmaker James Cook admitted to there being rare cases of drunkenness among the adults. Cricket seems to have been played on Sundays with 'no great indulgence in any vicious amusement'.

PROTECTIVE ASSOCIATIONS AND BENEVOLENT SOCIETIES

Discontent over pay and conditions became an increasing source of friction between brickworkers and their employers, especially with mechanisation and the growth of large-scale production facilities. As early as 1856 New South Wales stonemasons and carpenters had negotiated the world's first 48-hour working week, but this was by no means applied to other workers.[6] Following this, brickworkers were quick to organise themselves with the formation in Sydney of the Operative Bricklayers' Union in 1858.

4. A. Dawes, 'Five Generations of Brickmakers: Remarkable Record of Dawes Family in Australia', *The Clay Products Journal of Australia*, 1 October 1935, p. 15.
5. Select Committee, op.cit.
6. The Australian Operative Stonemason Society was central to the first Eight Hour Day campaign in Sydney in 1858, and called its members 'masons'. The society met in 'lodges' established by a 'Grand Lodge'.

By the early 1870s workers in Australia were agitating for an eight-hour day and forming themselves into benevolent societies, one example being the Bricklayers' Society. Friendly societies – mutual benefit associations – were particularly active from the 1830s, providing medical assistance for members and their families and payments when members were unable to work through illness, as well as funeral benefits. Work stoppages in the brickmaking industry were rare, the only known general strike taking place on 23 June 1886 in a bid to gain a 48-hour week, down from the expected 6-day working week of 56 hours. Evidently the strike failed for they were still working the 10-hour day in 1892.

Employers also made common cause. A rash of friendly societies were established in the mid 1840s, one of them being the Brickmakers' Friendly Society. Petersham Municipal Council minutes for 30 November 1875 record that the society rented the Petersham Town Hall for its meetings in 1875. In July 1879 the Brickmasters' Protective Association of New South Wales – referred to by its critics as the 'Brick Tryst' – was established as a mutually beneficial combination of employers in the brick industry, and to protect against bad debts. For instance if a customer, usually a builder, was behind in his payments for bricks supplied, the job would be marked as 'posted'. A notice to this effect was entered on a list and notification posted by the secretary of the association so that any customer who remained on the list received no further supplies. If the account was settled a further notice was sent to all members informing them of the fact.[7]

The original membership roll comprised about 120 members, and included important names such as: Bakewell brothers, Robert Fowler, Frederick Goodsell, Josiah Gentle, the Tye brothers, Thomas McGill and many others. Alfred Dawes, whose memoirs were published in *The Clay Products Journal of Australia* during 1948, claimed that in 1887 the brickmakers of the Parramatta district may have formed a local branch for the protection of their

7. Alfred Dawes' reminiscences, in *The Clay Products Journal of Australia*, 2 December 1935, p. 22.

H. Waddington's Vulcan Brick Works, set back from Barwon Park Road, is known to have been operating during the 1870s. Evidence of Waddington's ownership can been seen on the map of Needham's Grant, Waterloo Estate, prepared by Reuss & Halloran in 1881. The yard was acquired by E. Vickery & Sons in 1891 and continued to operate as the Vulcan Brick Works. Rows of bricks, drying under reed-covered hacks, can be seen in the foreground. The kiln next to the machine shop was probably of the open type. Robert Hutchinson Collection

trade against the systematic swindling by bogus contractors. The Dawes family was almost unique in having had a continuous involvement with the brick industry for nearly a century and a half.

The Brickmasters' Protective Association ceased to function from about 1893. Exactly who its successor was is difficult to ascertain, but some of its role appears to have been taken over several years later by the Metropolitan Brick Company Ltd which acted as a selling and distribution agency for the brick companies.

Not all brickmakers were against shorter working hours. The opening of the Eight Hours Co-operative Brick Tile and Pottery Company Ltd on Government Road at Merrylands in 1887 marked a move unprecedented by any brick company. The co-operative's stated object was to 'establish a co-operative brickworks on purely mutual principles'. Employees were encouraged to take up shares in the new venture, with preference given to workmen. Machinery of the latest design was installed and brickmaking sheds and kilns erected. The capacity of the works was between 10,000 and 15,000 bricks per day and enamelled bricks and tiles were also manufactured.[8] The experiment was not a success as the company went into liquidation several years later.

8. *Australasian Builder and Contractors' News*, 13 August 1887, p. 221.

WATERLOO FLATS

Following their ejection from Brickfield Hill in 1839 many brickmakers had settled in an area known as the Waterloo Flats. Indeed the clay soils of St Peters, reputed to be the best brick land in Sydney, encouraged a high concentration of brickworks on the 'Flats' which extended from Waterloo through Marrickville and south-westward to nearby Mortdale. Retired brickmakers interviewed by the author believe that local brickyards produced most of the bricks used to build the city of Sydney over a period spanning 125 years. Anecdotal evidence such as this is always open to question, but for construction work on the south side of Sydney Harbour, it is reasonable to suppose that most bricks were sourced from the brickyards at St Peters since they were the most numerous. Waterloo Flats was renamed as part of Alexandria in 1863, five years prior to the incorporation of Alexandria as a municipal entity in 1868.

Over time the brickpits changed the social and economic complexion of the St Peters–Tempe area. The expansion of industry and improved transport links created a more industrial suburb populated mostly by working class people. Development came at a price, and some of the grand homes of the area were demolished to make way for the brick pits, while other estates were subdivided to accommodate the growing population. By the early 1880s open countryside was more or less a thing of the past, and air pollution had become a serious issue, making the district less desirable as a place to live. An investigation was launched in 1882, although the findings of the Noxious and Offensive Trades Inquiry Commission merely acknowledged brickworks to be a Third Class trade. In effect this meant that manufacturers did not require the consent of residents to operate, although permission from local authorities was needed. Police surveillance was mandatory, presumably on account of the use of explosives to blast shale from the deepening quarries.[9]

NEW ARRIVALS

Who were the men working the clays in the brickpits of colonial Sydney? A comprehensive survey of all brickmakers operating in the St Peters-Newtown district is beyond the scope of this book, although some deserve special mention. Prominent among them were the Goodsell family, Rupert Cook, the Harber brothers, and the Burling and Tye families. Once household names, they are now mostly forgotten. All of them were masters of their craft, and more than a few connected by marriage.

Unnoticed during the year 1839 was the arrival in Sydney of two migrant families – Dawes and Goodsell – who were destined to leave their mark on the brickmaking industry. John 'George' Dawes and his wife, Louisa, sailed aboard the immigrant ship, *SS Orient*, from Plymouth in October 1838, landing near Circular Quay in April 1839.[10] Originally from Peasmarsh in Sussex, Dawes and his new wife were people of their time, both from large families. Even prior to their departure there were six mouths to feed, with many more to follow in due course. They were also strict Wesleyans, devout and hardworking folk who had been sponsored by an immigration scheme for the sum of £18. George Dawes was not the only family member to emigrate. At or about

George and Louisa Dawes, circa 1880, also a younger Louisa, left.

George Dawes was born on 28 September 1802 and died in May 1886 at his Newtown home after having been taken ill whilst attending church. His wife Louisa was born on 6 May 1809 and died in September 1886.

Source: Unpublished manuscript and family tree compiled by Alfred George Naasson Dawes before his death in 1949. Rosemary Szente Collection

the same time a number of his siblings migrated to Australia, with the remainder taking passage for the United States.[11] Of the two families the name Dawes carries more weight across history's long reach, for it was William King Dawes, great grandson of George and Louisa, who shaped the brick industry in Australia over a period of 50 years from 1930 until his death in 1981. And it is Dawes who dominates the story of the Austral Brick Company and Brickworks Limited.

Towards the end of 1839 John Goodsell and his family, also from Peasmarsh, arrived in Sydney on the *Hastings*. There was more to Goodsell than met the eye. Years before, in 1825, George Dawes had married Goodsell's sister, Eliza, which made the two men brothers-in-law. Eleven years later in 1836 Eliza died leaving her four young children motherless. Still a young man, George remarried, this time to 30-year-old widow, Louisa Payne, who had two children of her own by a previous marriage.[12]

9. New South Wales Commissions: *Noxious and Offensive Trades Enquiry Commission*. Report of the Royal Commission appointed on 20 November 1882, Sydney Government Printer, 1883, p. 14.

10. *Shipping Arrivals Australian B Genealogical Research 1826–1842*, ed Carol Baxter, pending publication. SRNSW, Ref 4/4784.

11. Dawes was from a family of 23 boys and one girl.

12. George Dawes' marriage to Eliza Goodsell had produced four children – Mary Anne (12), Ellen (8), George Jnr (6), Catherine 'Kitty' (4). Louisa Payne's (nee Rattan) children by her first marriage were Nicholas Payne (died at birth), Anne (10) and Dennis (4). Her marriage to George Dawes produced Orientia (born en route in 1838, and named after the vessel), Winifred (1840), Naasson (1843), Eliza (1844), Emma (1847), William Aaron (1849), Naomi Louisa (1852).

George Dawes' connection to John Goodsell was typical of the development of the brick industry during the nineteenth century – the intermarriage of brickmaking families, many of them Non-Conformists. Finding little work in Sydney, he spent several years in the bush as a labourer. Returning to the metropolis, the family settled in the fast-growing suburb of Newtown where, some time after 1845, John took over the brickworks of his brother. Sands,[13] an early residential and commercial directory, lists one of his sons, Frederick, operating as a brickmaker at St Peters in 1865, almost certainly at the same yard. [14]

THE GREAT AND THE GOOD

Like many of their contemporaries, the Goodsells were active in the local community and also prominent philanthropists. John, the family patriarch, found time to serve on the committee of the Sydney Mechanics' School of Arts in 1856. Frederick was on the Cook's River Road Trust in 1858, and also performed light duties as treasurer of the Wesleyan Sabbath School's Committee. Much to the delight of local children, his mighty six-horse teams would be pressed into service for the annual Sunday School picnic. Piled aboard the dusty brick carts the precious cargo would make its way from the Wesleyan School to either Bondi or Coogee where sand, surf and a feed lay in store.[15] Frederick Goodsell also cultivated a network of trusted contract suppliers, and in early 1863 we find George Dawes, previously mentioned as a newly arrived immigrant, making bricks under contract at 15s per thousand. A decade or so later, Dawes' grandson also appears to have worked for the firm of Goodsell and Tye.[16] At some point during the 1860s, possibly after the death of his father in 1863, Frederick and three of his sons, Alfred, John and the confusingly named Henry Wesley Goodsell, established a partnership with fellow Wesleyan and brickmaker, Alfred Tye. Frederick's eldest son, Alfred Goodsell, was one among many interesting characters of these early days and was the principal if not the only brick mould and barrow maker in the St Peters district. He made everything for the trade from the 'strike', a piece of wood used to smooth the face of the brick, to the brick barrow. Alfred Goodsell used the staves of well-seasoned beer casks in the manufacture of his brick moulds, and all eight of his sons assisted him in the business. One grandson of note was the late Sir John William Goodsell, Secretary to the Treasurer and Chairman of the Public Service Board of New South Wales. The Goodsell Building on Chifley Square in Sydney which houses government offices is a reminder of the work of this outstanding public servant.

The Methodists held regular services at R. Bentley's Brick Yards in 1859 and also other brickyards at this time. Mr Garner of the brickyards is noted as a circuit steward and preacher in the pamphlet, 'Preachers Plans of the Primitive Methodist Church' held at the Mitchell Library.

CLAYS, SHALES AND RAILWAY LINES

Meanwhile, bricks manufactured at the Goodsell's Newtown yard – also known as the Newtown Brickworks – were stamped in the frog as 'G. and T. Newtown', the first lot being used in Farmers' building (now Myers) at the corner of George and Market Street, Sydney. The Jubilee Souvenir of Newtown Municipal Council mentions that Goodsell's machine-produced red bricks were much admired.[17] Certainly, they were extremely popular, and other brickyards were quick to adopt the new technology. Unable to keep up with demand, Goodsell purchased additional machines. A variety of shapes were produced at the yard, among them bull-nose, splays, and coping moulds for window sills, door jambs, string courses and arches.

13. Sands Directory was a residential and business directory for the city and suburbs of Sydney, published between 1858 and 1932.
14. Frederick John Goodsell was born in 1814 in England and arrived in Sydney with his parents in 1838. He died in Newtown in 1888. His three sons, John Wesley (1834–1921), Alfred (1830–1922), and Henry Wesley (1836–1922) all worked in the family brickmaking firm. Alfred's two sons, Charles Wesley (1857–1942) and William Henry (1855–1942), were also brickmakers whose names are recorded in Sands Directory.
15. Frederick Goodsell was also a councillor and alderman on St Peters Municipal Council over a 20-year period from 1866–1886.
16. A. Dawes. 'Five Generations of Brickmakers: Remarkable Record of Dawes Family in Australia', *The Clay Products Journal of Australia*, 1 October 1935, p. 13.
17. Jubilee of the Municipal Council of St Peters, 1912.

In the early 1880s the firm installed a continuous tunnel kiln similar to the one patented by the engineer, Henry Gurney. Unfortunately, after a few years technical difficulties led to failure and its subsequent demolition. Soon after, in 1885, John and H.W. Goodsell took over management of the company from their elderly father, Frederick, who died three years later. The brothers also acquired Alfred Tye's share of the firm. Tye may have seen his chance to finally quit the partnership following the resumption by the state government some years earlier of a corridor of land running through the property. The land was directly in the path of the Illawarra Railway that was commenced in 1876 and finally opened in 1884. Like many other local businessmen who owned land in the vicinity, John and his brother invested money in the railway company and seem to have profited as a result. In fact, with the station at St Peters no more than a stone's throw away, local brickyards were well placed to attract new workers and settlers to the district.

The precise details remain unclear, but it seems that in the late 1880s the brothers established a successor company, the Goodsell Steam Brickworks Company, on the nine-acre site at St Peters. Until the demise of the company the Goodsells are known to have operated four brickworks, two in Newtown and one in St Peters and Macdonald Town respectively. Hand written notes left by Frederick's son, Henry Wesley (H.W.) Goodsell are worth noting for they speak not only of the importance of brickmaking to the district of St Peters, but also hint at the gradual decline of the family business.

> *I am a brickmaker, 33 years experienced. Since 1879 there has been great increase in the business. I knew few other lands about Sydney that contain the same variety of clay as our property. I have heard all the evidence with reference to improvement of clay on the Lord Street side. That is correct. Bricks could not be made from clay shale in consequence of the iron stone there which has the effect of bursting the bricks. I can make any coloured bricks that are asked for. Before resumption [of land for railway construction] our production of bricks was about 160,000, now 102,000.*
>
> *It costs more to process the clay now, that it cannot be got in such quantity – at the rate we are now going on we may continue 3 or 4 years longer, but only not make DP [dry pressed] for more than 2 years. Have examined the clay in two holes on Lord Street side of rail and consider it as good as that in pit, if not better. Patent kilns were placed on … ground after forming partnership in 1870. It cost without patent £5,000 or £6,000. The kiln was sold in consequence of being flooded out and not being able to get sufficient clay.*
>
> *One patent kiln produced about 100,000 bricks, 4 other kilns produced 50,000. I now have 7 kilns which only burn 103,000 bricks. Burnt all sorts of bricks in patent kiln. We sold the patent for £1,000. It is now being used at North Shore. Our plant was in very good order. Was not present when offer was made to float the Company.*[18]

Clearly, the depletion of clay reserves below the subsoil was having a marked effect on production. It may have been that the plant's crushing and grinding machinery was ill equipped to process the ironstone present in bands of hard shales. Ironstone is a fine-grained, heavy and compact sedimentary rock composed mainly of iron oxide and clay or sand, and mostly occurs in the form of nodules. Dynamite was often used to blast it out where it had been formed, usually at a depth of 80–100 feet below the surface. However, since it could not be processed through the grinding mills and crushers the men would leave it on the side of the pit.

The eventual fate of the Goodsell plant at John Street, Newtown and its sale to the brickmaker Peter Speare in 1891 is, however, shrouded in mystery. Financial commitments may have left the Goodsell family heavily exposed to the depression of the early 1890s. Correspondence between Goodsell and the State Government over the resumption of brickmaking land for the construction of the

18. Copy of handwritten manuscript lent to the author by Warwick Gemmell.

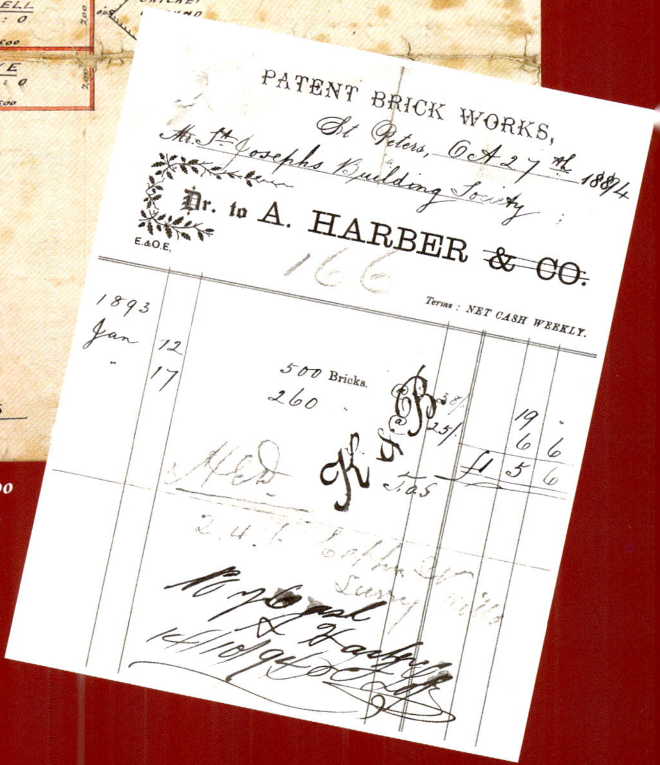

Above: Brick leases at Newtown and St Peters, on Needham's Grant and the Waterloo Estate (part of Sir Daniel Cooper's Estate), circa 1880 Reuss & Halloran, Surveyors, 1881. Mitchell Library

Right: A. Harber, Patent Brick Works, St Peters, 1894. The original date printed on this receipt has been overwritten with a new date, 27 October 1894. The status of Abel Harber's firm has changed with the deletion of '& Co'. By this stage the business may have been in liquidation and the remaining stock in the process of being sold. Society of Australian Genealogists.

Illawarra railway line reveals that Goodsell was left out of pocket. Whatever the cause of the decline in the family's fortunes, the yard continued to operate as the Newtown Steam Brickworks in the hands of its new owner, Speare.

Following the Second World War the land was resumed by Marrickville Council, put to use as a rubbish tip, then turned over to parkland and renamed Camdenville Park in 1957. In 1912 Ernest Speare formed Speare's Brick and Pipe Works Pty Ltd at Tempe and the yard survived into the 1970s when it was closed and the land purchased by the winemaking firm of Penfolds. Members of the Goodsell family lie buried in the graveyards of local churches, St Peter's and St Stephen's at Newtown.[19]

FAMILY TIES

The Goodsell family was just one of many that plied its trade in brickmaking across the district, although others figured prominently. Henry Burling operated at Lord Street, Newtown in 1863 and was later joined by two brothers. In 1886 Henry moved his business to the brickfields surrounding Alexandria, where his son was also active in the trade. Another local operator, Robert Porter, settled in Wardell Road, Marrickville in 1868 where he operated a brickworks from 1871 to 1876. The yard now lies under Livingstone Oval in Marrickville Park. The Playford brothers were also a successful family firm with works at New Canterbury Road, Marrickville, and at Windsor Road, Dulwich Hill, from the 1860s until 1882. Playford's bricks were used in the construction of the Royal Prince Alfred Hospital. Anthony Blamire was one of those small-scale brickmakers who moved from site to site after exhausting the supply of clay at each location. Blamire operated a number of brick pits over a 15-year period, in Canterbury, Petersham and Marrickville. Blamire Lane off Illawarra Road is named after him.

William, John and Thomas Bakewell were English migrants whose original trade was bricklaying. Their firm, Bakewell Bros Beulah Pottery and Brick Works, produced double pressed facing bricks and moulded bricks. Naasson Dawes was listed as manager of Bakewell Brothers from 1891–1904. William Edwards was originally a sandstock brickmaker and remained in business almost continuously from 1865 to 1877. His brother, James, started a brickyard with a Mr Gardner at Neutral Bay. Following the success of these works Gardner established a brickyard in Enfield. James went on to establish a yard at Mortdale and later the Thornton yard near Newcastle.

Another local businessman, Frederick C. Curlewis, set up the Warren Brickworks adjacent to the Tye Brothers' yard at Chapel Street, Marrickville. Curlewis was in trade from 1885 to 1902 and moved to the brickfields at Mitchell Road, Alexandria, in 1887 where a serious fire damaged his machinery and engine room.[20] Curlewis is thought to have been the first to introduce the open kiln fired facing brick, setting and burning them to obtain the colour of a face brick. Curlewis commanded the trade in facing bricks which were superior in quality and cheaper in price to the double pressed plastic facing bricks that were quickly losing market to the more popular dry pressed bricks.

Rupert Cook was another English migrant, and one of the more enterprising brickmakers who made the transition to a machine-moulding operation. He started working in the Marrickville district in 1863 but left Australia in 1873. Returning to the colony in 1879, Cook established a successful brickworks in Addison Road, where he manufactured mostly plastic bricks, wire cuts and double pressed bricks. Cook was one of the first brickmakers in the Sydney area to produce white-glazed or enamelled bricks and received first prize at the 1896 Royal Agricultural Society of New South Wales Exhibition.

19. In the course of research the author happened upon a reference to a legal dispute, Goodsell versus Goodsell that was heard in the Equity Division of the Supreme Court of New South Wales. The exhibits book lists correspondence relating to the court case which took place over a six-year period from 1928 to 1934 (Exhibits Book, 1927–63, Vol: 3/17304). Records indicate that there may have been a family disagreement over disposal of the estate.

20. *Australasian Builder and Contractors' News*, 13 August 1887, p. 236.

St Peters Brickworks, circa 1887–1901. This simple painting (possibly a gift to W. Mercer) clearly shows a clamp kiln to the left of the two cottages at a Newtown brickyard. To the right can be seen a brick carter and his horse and cart. Despite technological developments and the introduction of mass-production, brickmaking remained a small-scale operation until the turn of the century. D. Edwards. Mitchell Library, State Library of New South Wales

The three Harber brothers – Abel, Emmanuel and Isaac – were prominent in brickmaking and politics in St Peters over a 30-year period commencing in 1863 and are mentioned in the 1875 enquiry into the employment of children. During the 1870s they are known to have been producing semi plastic bricks at various sites, but were forced to close in 1893 due to the depression. It seems that they lost heavily due to their involvement in the construction of the Imperial Arcade, Sydney. Possibly, the builder defaulted on payment for the bricks supplied. A more likely scenario is the direct association with the building firm. We know that a great many brickmakers were shareholders in building companies, and formed close working relationships with builders, for whom they became regular suppliers. The collapse of a building firm would surely have had serious consequences for its investors.

BRICK CARTAGE

Despite being in competition with one another, brickmakers were bound together by a common thread – the need to ensure delivery of their bricks. From kiln to customer, taking the bricks to market was a matter for horse and cart. Horses were commonplace and played a vital part in the day-to-day activity of most people. Men, women and children rode horses, people arrived in town on horseback or in a horse-driven vehicle and the main streets had hitching posts at the front of shops, hotels and houses.[21] Commercial and manufacturing activity in Sydney and Melbourne created a huge demand for both light and heavy haulage. Building trades were especially reliant on heavy haulage, transporting bricks, stone, sand, lime, timber and shingles to sites around the cities.

By the 1820s the two-wheeled heavy pole or bullock dray was in general use, its specific function being to carry stores and produce on unmade roads over long distances. Clydesdale horses imported from Scotland were well-suited to the task of heavy haulage. Four-wheel box type wagons began to replace the dray from about the 1860s, and refinements in the form of a flat tabletop were being extensively used in the late 1880s. The impact on road surfaces was mitigated slightly by the use of wide tyres, but from the 1850s to the end of the century, colonial parliaments repeatedly debated the question of vehicle regulation, but to little effect.

In his memoirs, Alfred Dawes recalled the widespread existence of toll bars across Sydney and its principal thoroughfares until 1887, perhaps even later. Tolls for the brick carters would often have to be paid twice. For example a load delivered to the Randwick district from St Peters would encounter the tollgate on the Newtown Road (now King Street) near the Deaf, Dumb and Blind Institution.

21. M. J. Kennedy, *Hauling the Loads: A History of Australia's Working Horses and Bullocks*, Melbourne University Publishing, 1992, pp. 180–181.

Tollgate on Parramatta Road, Sydney. Government Printing Office Collection. State Library of New South Wales

Tolls increased the cost of cartage approximately by 1s per thousand bricks and were considered a nuisance.[22]

Of equal nuisance value to Sydney's bricklayers was the spoilage of bricks delivered to the building site. Many of them were broken completely, the result of rough handling during loading and unloading of the carts. More often than not, bricks were simply tipped in bulk onto the site, or casually tossed onto the ground. One attempt to remedy this problem – a new form of dumping wagon – met with success. At the point of delivery, the wagon was backed up, causing the rear wheels to remain stationary and the body of the wagon to slide over the rear axle and tilt rearward.[23]

The cost of cartage, however, made building with brick an expensive enterprise. In 1887 it cost 5s to cart 1,000 bricks less than one kilometre in any direction from the kilns at St Peters, 12s to King Street in the city, and 14s to Circular Quay.

22. A. Dawes, Early Brickmaking in Australia: Slump of the Nineties and After, *The Clay Products Journal of Australia*, 2 December 1935, p. 22.

23. *Town and Country Journal*, 16 January, 1875, p. 97.

Tip tray wagon – Town and Country Journal, January 1875

JOSIAH GENTLE

Charles and Josiah Gentle deserve special mention as brickmakers who made high quality bricks, and were also noted philanthropists. Charles, the elder of the two, was born in Bedfordshire, England, in 1845, and his brother Josiah in 1847. The death of their father, Edward Gentle at the age of 24, soon after Josiah's birth put an end to the family's piano-making business. In 1850 his widow married James Rawlings, a brickmaker who plied his trade in what was the principal industry of Bedfordshire. In early 1852 Rawlings and his extended family left England as assisted migrants, arriving in Sydney on the *John Gray* in November. Soon after they settled in New Town (Newtown) where he found work in the various brickyards of the district. Josiah began as a brickmaker in Waterloo Flats during the 1860s and is first listed in Sands' Directory in 1869 and is recorded as living in Maria Street (now Darley), Newtown until 1880. By 1871 at the relatively young age of 25 Gentle was already a successful brickmaker working in collaboration with his step-father. By the mid 1870s his brother, Charles, appears in the Directory as one of only two brickmakers at Alexandria.

Josiah may even have owned a yard in Petersham, which had been relocated from Waterloo in the late 1870s. From newspaper obituaries we know that by 1877 the Bedford Brickworks was already established on the corner of Cook's River Road and Waterloo Road (Mitchell Road). The yard was one of the best equipped in Sydney, comprising an elaborate two-storey office administration building and over a dozen buildings, a Hoffman kiln and four chimney stacks. Like Rupert Cook, Josiah Gentle successfully made the transition from hand moulding to a dry press operation. Charles quit the trade in 1888 and Josiah continued as proprietor until about 1895, at which point his son, Stanley 'Stan' Josiah, assumed control. Josiah senior died in 1912. An active Freemason, which was common among builders and the brickmaking fraternity, Gentle was remembered for his many acts of generosity, especially during the depression of the 1890s. A devout Wesleyan Methodist, he had made a donation of land in 1871 for the building of the Primitive Methodist Church in May Street. Following his death, Bedford Brickworks passed to his sons, Emanuel James Gentle, Stan and Josiah Jubilee Gentle. Stan went on to found the Ashfield Brick Co. Ltd and was also a director of The Austral Brick Company Limited. Josiah Jubilee did not continue in brickmaking.

WILLIAM JUDD

William George Judd, another brick pioneer, mentioned earlier, was born in 1847 in a house on Pitt Street, Sydney. As a young lad he commenced in the brickmaking trade among the 'sandstock men' of St Peters,[24] married the daughter of a Queensland brickyard proprietor, and in his adult years entered municipal life. Well acquainted with other public-minded industrialists such as the Goodsells, he served as mayor of St Peters during the 1880s, and was the driving force behind the construction of town halls at St Peters, West Botany and Sutherland. In 1885 in company with others he founded the brickworks at Mortdale and after moving to Arncliffe, again stood for council and was elected mayor of Rockdale. Judd's Hurstville Steam Brick Company was a vital part of the local economy, for its bricks were used in the Queen Victoria Building and other city constructions.[25] Like other manufacturers of substance, Judd built workers cottages some of which survive today in Judd Street in Banksia and at Mortdale.

Judd's son, William, continued the business following the death of his father early in the new century. According to the late Ernest Alfred Crome, a photographer and historian, William Judd became a weekend resident on Lot 47 in the early years of the new century. His wife, Elizabeth, leased Lot 46 next door, as well as Lot 39 just across Attunga Road on the southern corner of Wyralla Road. Having established his retreat at Yowie Bay, Judd entered into the local government scene in the shire and became president of the shire council for its first two years and vice president of the Local Government Association in 1906. He was later elected to parliament as the representative for Canterbury. During his period as shire president Judd was conscious of the intrusion of petty crime and juvenile delinquency. He spoke about the presence of the 'larrikin element at shire beaches and that a strong hand from the beginning would keep out the ratbag fringe.'[26]

24. While he was a young child, Judd's father sailed for the American goldfields, in the course of which he was shipwrecked and then rescued by a passing ship. He was dropped off on an island (whereabouts unknown) and remained there for four years before returning home. His wife began a laundry business to support herself and son William, but one day the young five year old was kidnapped. A laundry employee retrieved the boy and gave the kidnapper a thrashing.

25. The main contractor on the Queen Victoria Building was the firm of Howie and Moffitt, about which more will be said in later chapters. Archibald Howie was a significant shareholder in The Austral Brick Company Limited.

26. E.A. Crome, unpublished manuscript, from the collection of Ross Bell, retired brick carter at The Austral Brick Company Limited.

The Mortdale yard closed during the Great Depression of the 1930s, although Hurstville remained open until 1972. The following year there was nothing left after bulldozers flattened the site, which was later acquired by the local county council and the Department of Education. The site eventually became the St George Campus of the University of New South Wales, which is now a TAFE college.

NAASSON DAWES

Men such as Gentle and Judd were highly regarded in the trade and respected by the builders, contractors and merchants of Sydney. The story of one of their contemporaries, Naasson Dawes, is of particular interest for it provides a direct link to William King Dawes who shaped the brick industry like no other person during the twentieth century.

Naasson Dawes was born in March 1843 at Glebe, the first child of brickmaker George Dawes whom we encountered earlier. Still in his early 20s Dawes became the manager of Alfred Tye's Alexandria yard, quite possibly through the latter's business dealings with a relative, Frederick Goodsell. It was Naasson's father, George Dawes, who had married Frederick's sister, Eliza, before their departure for Sydney. Naasson carried on the tradition, when he married Alfred Tye's daughter in October 1865. The union produced ten children, one of whom was William Frederick Dawes, father of William King Dawes.

Naasson Dawes continued to work for Tye and his partners Goodsell and Hughes, although for how long is a matter for speculation. The 1884 Sands' Directory lists him as general manager of Bakewell Brothers at Erskineville. Dawes was an experienced and knowledgeable brickmaster and went on to manage various companies, notably the New South Wales Brick Company Limited (almost certainly from its creation in 1902 until 1906) and the Federal Brick Company Ltd. Naasson Dawes was struck down by a heart attack at the age of 74 in June 1917 while travelling aboard a tram near Central Railway Station. He is buried in Rookwood Cemetery in the family grave.

Naasson Dawes. Rosemary Szente Collection

WESTWARD EXPANSION

The district of St Peters may have had the largest concentration of brickmakers, but it was not unique. With the growth of suburban Sydney an astonishing number of yards were established across Sydney. The tendency to cluster in certain geographical areas was due to local clays and shales, as well as a dramatic surge in building activity within the surrounding area.

For many decades Annandale was a focus of brickmaking activity. In 1858 George Blackhall was still making bricks at Glebe, although local pits were unable to keep up with demand from building firms. Blackhall packed up his works and moved to Leichhardt where deposits of the ubiquitous Wianamatta shales were more easily exploited. Historian Warwick Gemmell records that by 1871 no fewer than ten brickmakers were established in Leichhardt, including Henry Hughes, Charles Linney, and John Chalkling. In 1882 the Sydney Land and Building Company was known to be operating in the area.

From Ashfield to Ryde scores of smaller operators worked the clays. Little is known about these individuals apart from a simple entry in local rate books. William Randle, a railway contractor who supplied bricks for works on the Sydney–Parramatta line, was probably the first major brickmaker in Ashfield. In nearby Petersham, John Fyle's brickyard formerly located on the present site of the Mungo Scott flourmill may have been established as early as the 1840s.

Another brickmaker, William Keen, learned his craft at St Peters then moved to Ashfield in 1867 where he set up a small brickworks near his wooden cottage in Lang Street. Keen's firm was eventually taken over by the Excelsior Brick Company Limited. In the same year, a William Bottle of Croydon is recorded as operating a small brickyard in Lucas Road, Croydon. William Bottle may have been in partnership with a relative, Lucas Bottle, after whom Lucas Road is named.

With good profits to be made, the industry attracted entrepreneurs. In 1879 prominent Sydney retailer and businessman Anthony Hordern leased his Ashfield property and brickmaking interests at Boundary Street in Croydon to a family business in the name of Israel, George, William and Eliza Cottle. By 1880 there was a sizeable pit on Hordern's land, indicating the success of the brickmaking trade. Two years later, Hordern established the Croydon Steam Brick Company and became principal shareholder.[27] A victim of the 1930s depression, the firm went into voluntary liquidation in September 1935, its successor being Croydon Brickworks Pty Ltd. Three years later the latter is recorded as having a substantial shareholding in Brickworks Limited, although it was finally wound up in September 1947 and its assets acquired by the aforementioned company.

Accompanying the activities of these brickmakers was the ever-present nuisance of pollution which gave rise to many complaints and enquiries, most of which went unanswered. At the Croydon Steam Brickworks, for example, the manager advised Burwood Council in June 1888 that the two new patent kilns to be erected at his works in the next month were expected to eliminate the smoke problem that had been causing complaints among local residents. Only one patent kiln was subsequently erected at this brickyard. Despite the crippling effects of the 1890s depression, Hordern's firm remained in production until the Great Depression of the 1930s.

Meanwhile Richard Brand's brickmaking venture established in 1881 was short lived. An engineer by training, Brand purchased land along Cheltenham Road but went broke, eventually selling his plant and land to the Excelsior Brick Company. Further afield Charles Cooper, once a sandstock brickmaker, began the steam operated brickworks at Ryde and Bankstown during the 1870s.

GOODLET AND SMITH

As early as the 1790s bricks and pottery were being produced across a wide area in western Sydney now encompassing the suburbs of Granville, Merrylands, Rose Hill and Parramatta. Historian Terry Kass notes that numerous, small brickyards in the Granville–Merrylands area were operating during the 1850s. Most of them were small, family concerns with a modest output of hand moulded sandstock bricks.

The brick industry, however, was expanding rapidly, a trend that underpinned the growing expansion of Sydney, especially along rail lines. To avoid the congestion on the Waterloo Flats many prospective brickyard owners preferred to locate their plant in outer suburbs where land was cheaper and suitable clay was available. One of these people was John Hay Goodlet (1835–1914), joint owner of the Sydney hardware and building supplies firm of Goodlet and Smith.[28] The firm operated one of the largest sawmills and joinery works in the colony at Darling

27. E. Dunlop, *Between Two Highways, The Story of Early Croydon*, published by Dunlop, 1969.

28. John Goodlet was born in March 1835 at Leith, Scotland, and migrated to Australia in 1852 when he was 17. Arriving in Melbourne he found employment with a firm of timber merchants, C. and J. Smith. In 1855, he came to Sydney where he continued working with the company and in 1862 established a partnership in Sydney with James Smith. Goodlet married twice but had no children. A noted philanthropist he was also a Commissioner for the Sydney City Railway, a director of the AMP Society, and a prominent member of the Presbyterian Church, to which he left most of his considerable estate upon his death in 1914. *Source: Australian Dictionary of National Biography, Online Edition.*

Above: Goodlet & Smith's 1904 catalogue showing the Ultimo steam sawmills and wharf (top) and the Granville brickworks (below). Goodlet and Smith Prices Current: October 1904. Mitchell Library

Right: John Hay Goodlet – Goodlet moved to Ashfield in 1873 and was a generous benefactor of the Ashfield Presbyterian Church until his death in 1914. The historic Goodlet Sunday School Fellowship building in Knox Street, Ashfield, part of Ashfield Presbyterian Church, owes its existence to Goodlet. Canterbury City Council Local History Photograph Collection

Harbour, timber mills on the coast, and a fleet of three ships to serve them. The acquisition in 1884 of the Junction Brick Company Limited at Granville suited the company's diversification into building materials.

Until the downturn of the 1890s Goodlet's fleet of ships was active in the transport of bricks. The depression of the 1890s obliged the company to sell them with the result that by 1905 the majority of their bricks were being transported by rail from Granville siding to country areas. Quite possibly the firm of Goodlet and Smith was an early user of the Hoffman kiln. What is certain is that following its incorporation as a limited company in September 1890, the company built a new rectangular Hoffman brick kiln. This occurred soon after Goodlet's return from a fact-finding visit to Europe. Goodlet's Granville site was a dry pressed yard, and there were additional plants at Waterloo, Granville and Riley Street at Surry Hills.[29] All of the kilns at Granville were coal fired and the exhaust fumes dispelled through the 150 feet high chimney, was a striking landmark. Isaac Paton, the bricklayer who erected the stack in 1878, is reputed to have downed a bottle of brandy every morning before scaling the chimney to commence work.[30]

The company prospered and in 1894, despite the onset of the 1890s depression, it became the first manufacturer of terracotta tiles of the Marseilles pattern in Australia. The success of its products lay in the preparation of the clay or the manufacturing process, and in the use of drying chambers to evaporate excess moisture before kilning. These chambers were heated by conduits linked to the kilns and contained iron-framed shelves upon which the tiles were set to dry.

Goodlet and Smith suffered a decline from 1907 which closely followed the increasing age and eventual death of Goodlet himself. Increasingly, the

29. The Waterloo plant was one of the most sophisticated in the colonies, containing a continuous kiln for firing the raw articles and a large installation of power-driven machinery.

30. T. Fowlie, *A History of Granville*, Mitchell Library, A.1492, Vol. 1, p. 288.

company became less significant over time. It was taken over by Newbold Silica Firebrick Company Ltd in 1955. The latter firm ran Goodlet and Smith and another acquisition, Clyde Brick Company in Auburn, as one division of the company dedicated to producing building bricks. Production on the Granville site ceased in 1985.

NORTH OF SYDNEY HARBOUR

Isolated by water from the central hub of the city, Sydney's north shore soon made up for lost time. Ferry traffic between Circular Quay and North Sydney increased six-fold between 1880 and 1913. Commuter villages around the harbour such as Hunters Hill, Mosman, Neutral Bay and Cremorne rapidly became commuter suburbs.[31] In North Sydney local brickworks sprang up, taking advantage of the presence of suitable clays. Where bricks had been carted across the harbour on the horse ferry they were now close at hand with yards in Gore Hill and Artarmon.

31. L.H. Turnbull, *Sydney: Biography of a City*, Random House Australia, Sydney, 1999, p. 109.

Buildings and people go together, and where people reside transport links develop. North Sydney's building boom was fuelled by a network of tramlines, as well as the opening of the North Shore rail line in 1893. This opened up the entire north shore as far as Hornsby, and brought city jobs within easy reach of the more remote parts of Sydney. Following the turn of the century the exodus from the inner city almost bordered on panic when a wide range of social classes fled the plague-infested older parts of the city.[32]

Edward Lanceley, who left England to settle in New Zealand in 1876, was one of a number of men who cashed in on the growing importance of the North Sydney district as a centre for brickmaking. Young and foot-loose, Lanceley moved to Sydney in 1880 where he worked at Newtown, but troubled by the swarms of mosquitoes that bred in the swamps along Shea's Creek, he decided to return to New Zealand. Unable to secure a passage for some weeks, and with

32. Arthur Paine, a carter working at Sydney central wharf, was the first person known to have died of bubonic plague on 19 January 1900. By 12 March a dozen people had succumbed to the disease which is thought to have been carried by rats from ships in port. Cartoonists had a field day, identifying all manner of culprits.

A Brickyard Song

Oh! We belong to the brick-yard "push"
And work ten hours a day;
We've a pug mill and the best machines,
And our "Boss" is a first rate pay.

Bill Bradbury takes the kilns in hand,
With Dick Tappin for his mate;
Jim Strange is engine-driver,
And keeps things up to date.

But, since he won the racing sweep,
He is often on a "bike"
And takes the shine out of the rest –
Wal, Harry, George and Ike.

Ike Perkins is machine-man now,
And so is Harry Brown,
While our "pressers" I can safely say,
Are the best in all the town.

Jack Miller and Wal Campbell
Are running neck and neck,
But when George Strange is on the job
He can always make a cheque.

Our "nippers" too, are very good
And always earn their screw,
And for a consultation
They keep a "bob" or two.

So up goes the trolley, lads,
With Bill Campbell in the pit –
The bricks are cut out quickly
And for the press are fit.

Then off they go on barrows,
They're pressed and branded too,
And laid out underneath the sheds
In rows so straight and true.

And when they're dry, away they go
For Bill and Dick to set –
The M C bricks have such a name
Large orders now we get.

And so we graft from six to six
And joke and laugh all day;
Of course we often have some fun
On the night we draw our pay.

Last month our "Boss" invited us
Down the river for a pull,
So in the traps we all took seats
Excepting young Fred Bull.

But Freddy drew the line at that
And wouldn't share our fun,
For a jolly pleasant day we had
With fishing line and gun.

Bill and Dick are going to buy
A spinning-jenny soon;
Bill will take the ready cash,
Dick keeps the folks in tune.

So, it's "Twenty to one the Bolter!"
You'll hear Dick singing out;
While Bill invites the public
To pick the winner out.

So plonk your tanner on the horse,
You'll find that all is fair,
And if you win you'll get your coin –
Both Bill and Dick are square.

Source. Liverpool Herald, 24 December 1898, p. 7.

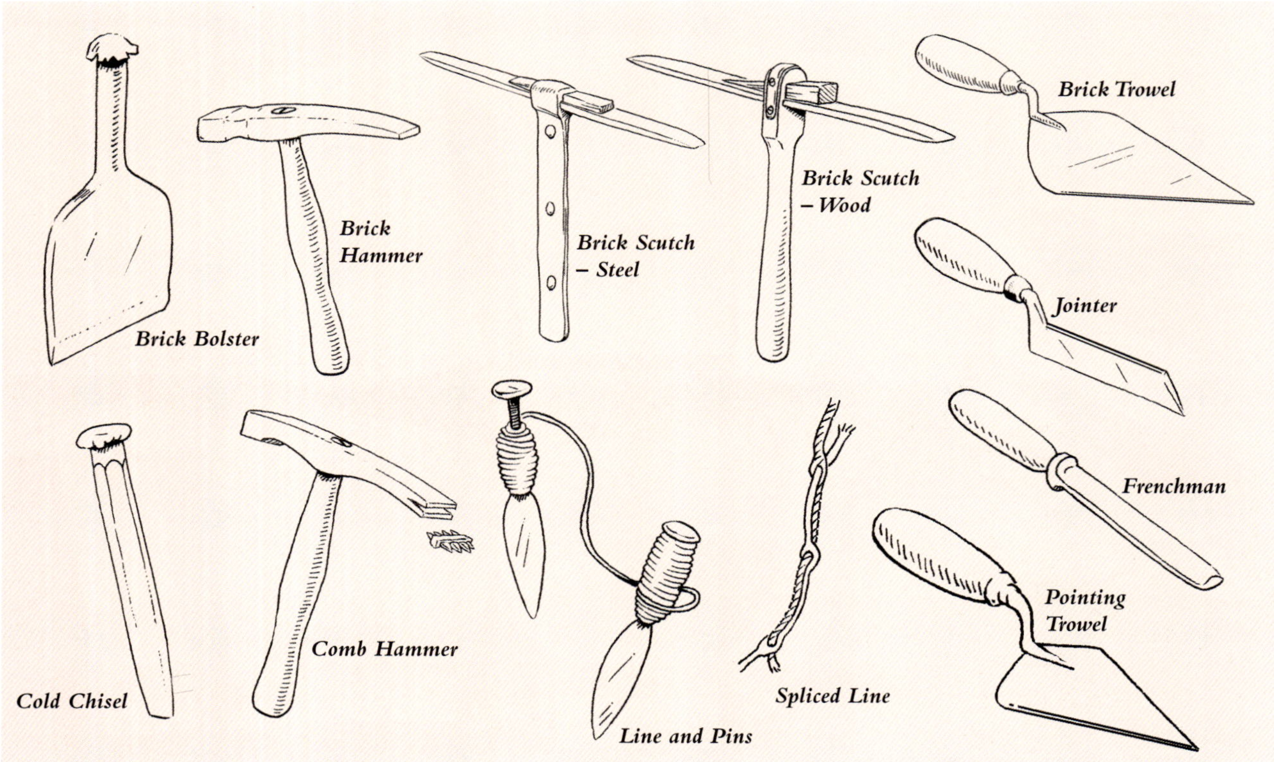

Tools of the bricklayer

little more than the clothes on his back, a tent, a bedroll, and sixpence in his pocket, Lanceley made for Gore Hill, then dense bushland, in search of employment at one of the many brickyards. Finding the work to his liking he decided to stay.

Putting his back into the work, Lanceley was good at his adoptive trade. Soon after he was asked to manage a yard owned by Messrs J.B. Magney and G.J. Weynton, two brickmasters who had previously worked the clays of St Peters. Magney and Weynton had purchased an existing brickyard at Gore Hill, setting up a small machine capable of turning out 10,000 bricks a day. Subsequently, they took Edward Lanceley into the partnership with the result that under his direction the firm prospered. In the absence of proper records, tracing the provenance of brick companies from this time is difficult, to say the least. Just how the partnership between Lanceley, Magney and Weynton metamorphosed into the North Sydney Brick and Tile Company requires some explanation, but can be understood by reference to the Land Company of Australasia Ltd. Established in 1885, this company owned and operated the Gore Hill Brick Works which was situated on part of the land now occupied by the Australian Broadcasting Corporation, North Sydney TAFE and the Royal North Shore Hospital. Underlying the whole estate on which the works was located was a bed of blue shale with pipeclay immediately above it. By 1890 the yard was said to have been the largest brickworks in New South Wales, and also contained the largest Hoffman kiln in the southern hemisphere. Members of Parliament and bankers attended the opening ceremony in 1888.

Unfortunately, the Land Company was yet another victim of the 1890s depression and failed in the 1893 bank crash. It was then taken over by the recently formed North Sydney Brick and Tile Company (first listed in Sands' Directory in 1894) and in due course became one of Sydney's most successful yards. In fact the original stamp in the brick of this company was 'MWL' – Magney, Weynton, and Lanceley – indicating that the partnership was still active. The Gore Hill plant, referred to as the No. 1

The Gore Hill Brickworks. Illustrated Sydney News, 5 September 1889. Inset: Receipt for bricks purchased from Gore Hill Brick Works

yard, was fully mechanised and the common bricks and tiles produced were burnt in the two Hoffman kilns, or in the case of face bricks, in downdraught kilns.

In 1903 Lanceley expanded the business, opening a new brick kiln (No. 2 yard) in Herbert Street near the St Leonards railway station. In 1914 he founded Ryde Brick & Tile Works. Still at Gore Hill, although a short distance away, two other yards were in business, including the Northern Suburbs Brick Company (1913–1970), Wilson's Brick and Tile Works (1897–1930), and the Oakes Steam Brick Company at St Leonards (1888–1892). Further up the road at Hotham Parade, Artarmon, the Butcher brothers owned and operated a successful brickyard from 1905 to 1950.

As the century drew to a close the clay products industry in Sydney was firmly in the ascendant. Burnt clay and shale was king in the construction industry, and for the builders of housing, railway sidings, tunnellers, drainers, commercial offices, government works, and bridges it stayed that way for decades to come. It was only during the 1960s that reinforced concrete and concrete blocks began to displace traditional building methods. Even then the use of concrete was favoured mainly in the construction of multi-storey buildings where bricks were used as a decorative feature, rather than as a structural form.

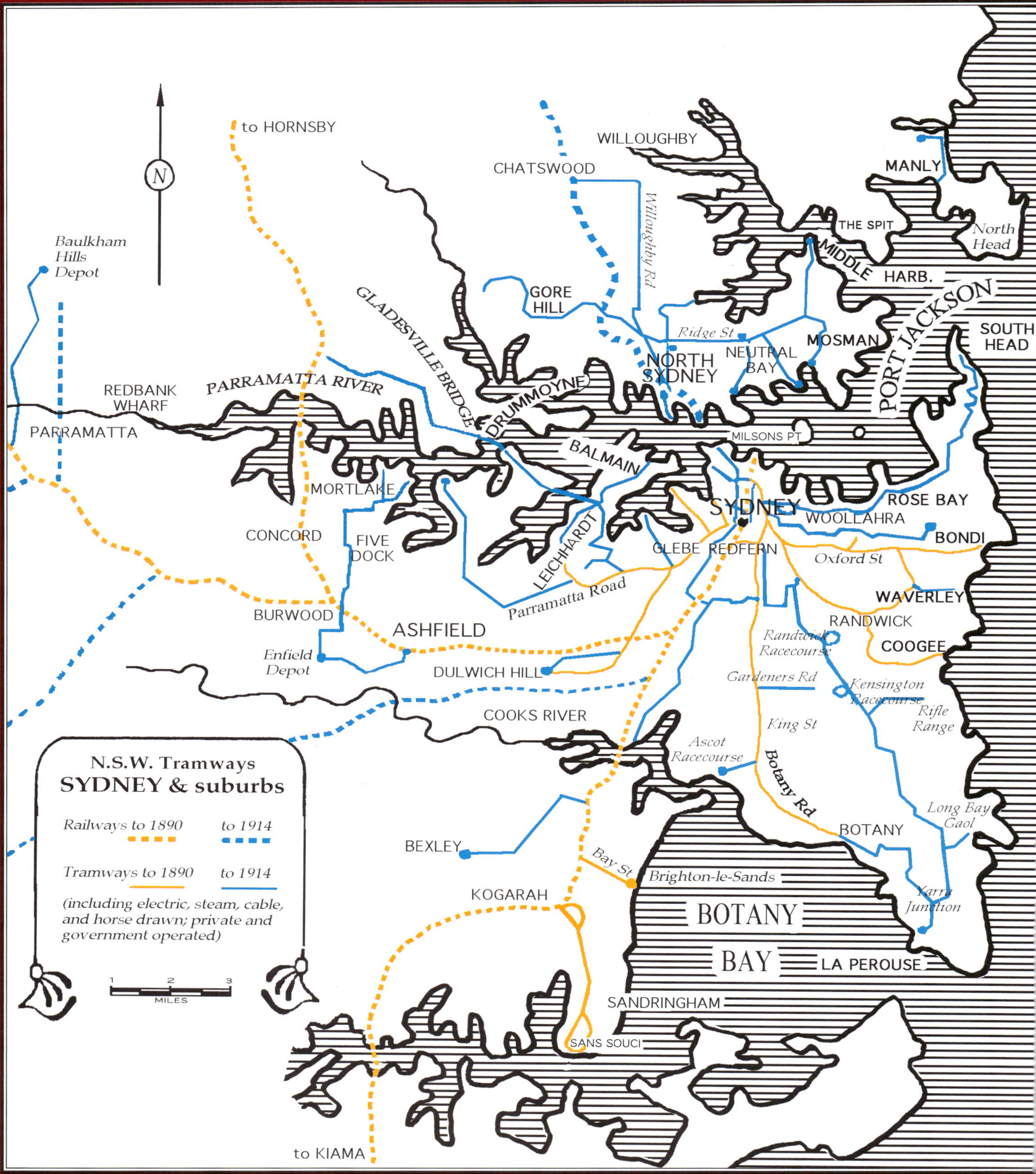

Tram and rail lines in Sydney, 1890 and 1914 – Maps based on annual reports of the Commissioner of Railways and NSW Government Tourist Bureau, Trips Around Sydney, 1914. State Rail Authority of New South Wales

A Place to Live 6

Bricks and mortar – Going places – Planning Sydney – Infrastructure – Middle age spread – Experimentation and innovation – Painting the town red – Public architecture – A life to lead – In the doldrums – Revival of the industry

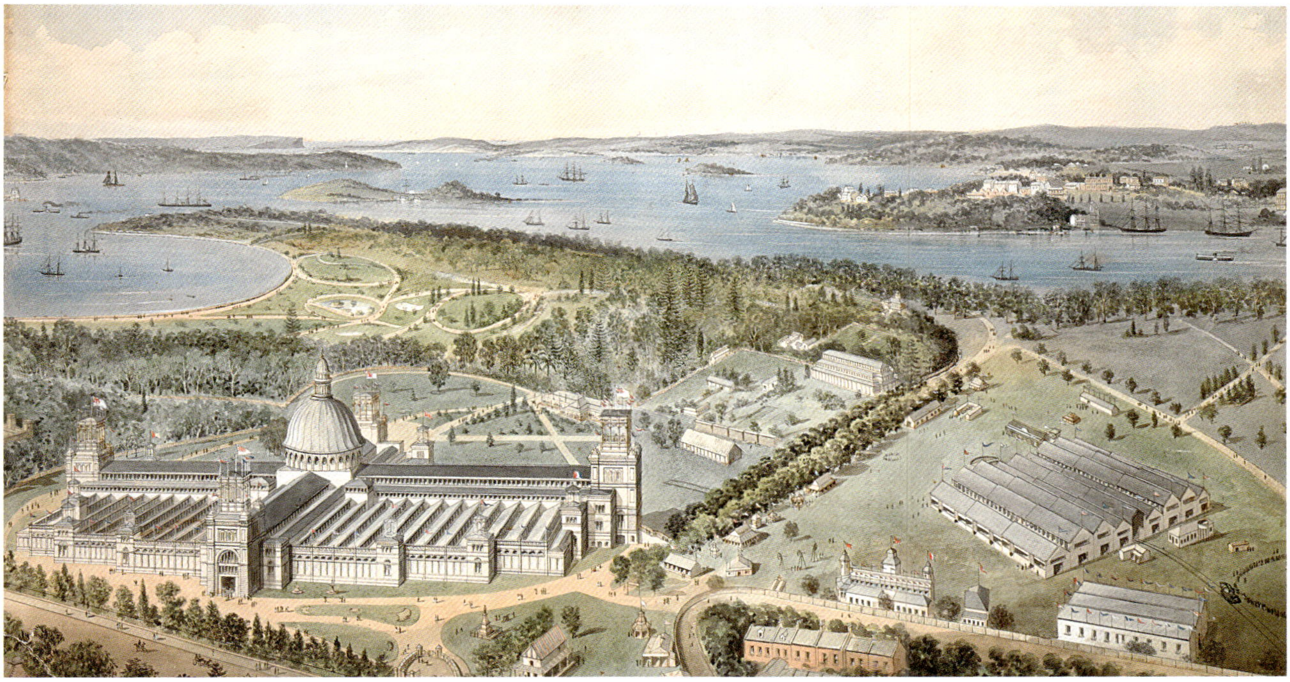

Garden Palace, Macquarie Street, Exhibition Building, Sydney,1879, burnt down 1882. Mitchell Library. State Library of New South Wales

BRICKS AND MORTAR

Apart from a brief downturn in the economy during the 1870s Australia enjoyed boom conditions until the depression years of the 1890s. From the early 1850s every major settlement grew in size, especially the capital cities of Sydney and Melbourne, a phenomenon that would have pleased brickmasters everywhere. Between 1851 and 1900 the population of Sydney rose from 54,000 to nearly 500,000, making it one of the largest cities in the world.

For journalist James Inglis, a recent arrival from New Zealand, who walked, trammed and trained his way around Sydney in 1879, the city was well and truly on the move.[1]

1. James Inglis was born in Scotland in 1845 and migrated to New Zealand at age 19 where he joined the west coast gold rushes. In 1866 he went to India and became a famine commissioner in Bhagalpur in 1875. Arriving in Australia in 1877 he became a journalist, businessman, investor and founder of the brand, 'Billy Tea'. Inglis settled in Strathfield and was elected to the Legislative Assembly of the New South Wales Parliament. He died in 1908.

The overflow of bricks and mortar has spread like a lava flood, over the adjacent slopes, heights and valleys, till the houses now lie, pile on pile, tier on tier, and succeed each other row after row, street after street, far into the surrounding country; and the eruption is still in active play ...

The sand drives are covered with cottages, the very marshes have a crop of dwellings that are springing up, like mushrooms; often, alas, like that very fragile and brittle and little calculated to withstand a lengthened wear and tear, nurtured in corruption and redolent of putridity and decay ... suburban extension is proceeding with wonderful speed.

Everywhere the sound of the workmen's tools is heard, all through the busy day. Brick-yards, iron foundries, saw-mills and joinery establishments are in

Above: Circular Quay, Sydney 1895. Sir Francis Boileau. National Library of Australia

Right: Circular Quay, westside, 1870. Small Picture File collection, Mitchell Library, State Library of New South Wales

full activity, and at present the building trades are in constant and vigorous employment.[2]

Inglis was clearly impressed with the sheer audacity of Sydney's building splurge, just one of many. He also notes the growing pains that accompany an expanding population – *houses … row after row, street after street, far into the countryside.* And so it goes, even today as has often been remarked: 'it'll be a great city when it's finished.'

No doubt, in the course of his visit in 1879 Inglis spent time in the recently completed and imposing steel and glass Garden Palace situated in the Botanic Gardens which was part of the Government Domain. Built to showcase an international exhibition of invention and industry from around the world, it was the pride of the city.[3] He would also have taken the newly opened horse-drawn (later steam) tram running from Hunter Street to the Devonshire Street railway terminal at Redfern. Such was the popularity of the exhibition that the government purchased many of the major exhibits and set up elsewhere the Technological, Industrial and Sanitary Museum, the grandparent of today's Powerhouse Museum. Unfortunately, fire destroyed the Garden Palace on the night of 22 September 1882 depriving Sydney of a magnificent landmark building.

2. J. Inglis, *Our Australian Cousins*, Macmillan & Co, London, 1880, pp. 144–6.

3. Colonial Architect James Barnet designed and supervised the building of the Garden Palace which was erected in eight months using electric lights imported from England so that work could be carried out at night.

View of Sydney from Paddington, 1870. Watercolour by George Penkivil Slade. National Library of Australia

GOING PLACES

Carrying the rapid pace of suburbanisation was an embryonic transport hub. From the 1850s, and again in the 1880s, a spate of railway construction encouraged the drift of middle class people away from the city. By 1898, about the time the decision was made to electrify parts of the system, a network of tramlines extending over 50 kilometres was in use.[4] By 1914 construction work at the main wharves of Circular Quay allowed ferry passengers to connect with the west to east tram route from North Sydney to the Spit, which had been electrified as early as 1893.

Additional tram routes were added over the years, but the railway system was still without a direct link to the city's commercial business district. Until 1906 when Central Railway Station was opened, both the western and southern rail lines terminated at Redfern, three kilometres south of the harbour's port facilities.

The network of tram and ferry services across the lower North Shore were in the vanguard of suburbanisation that began to reach into outlying Balmoral and Manly. In fact, the building of a military road to the garrison on Middle Head after 1880 had encouraged greater accessibility to this peninsula. As one historian observed: 'between the years 1890 and the outbreak of World War I in 1914, there was such a frenzy of land selling and building activity that Mosman seemed like a district in the throes of a mining boom'.[5] For the first time, the price of residential and suburban building lots across the lower North Shore catered for professional and business people, as well as clerical workers. Mosman, incorporated as a municipality in 1893, soon became a thriving, fashionable suburb. By 1906 the borough's population was slightly over the 10,000 mark with more of the well-to-do anxious to secure their place on the water.

Development of these outer suburbs gathered pace after the turn of the century as more and more families quit the overcrowded, and crime-ridden, inner city districts such as Glebe, Newtown, Redfern, Surry Hills, Paddington and Woolloomooloo. These poorer areas were increasingly prey to groups or 'pushes' of young male thugs, euphemistically

4. Sydney's tramways were placed under the control of the Railway Commissioners until 1930, when a separate government tramways department was set up.

5. Jack Carroll, cited in *The Settlement and Growth of Mosman*, Brookvale, 1953, p. 19.

Government House from the Domain, 1850. Hand coloured lithograph by Conrad Martens. National Library of Australia

referred to as larrikins, who waylaid people going about their everyday (or night) business. For example, the inner city working class suburb of Waterloo, renowned for its many brickworks, was a favourite haunt and the setting for Louis Stone's 1911 novel, *Jonah*, after the larrikin hero, Jonah. Many of the places frequented by Jonah – Chinatown and Paddy's Market, Botany Road, Sydney Harbour and its foreshores remain iconic of Sydney. Yet, developments in transport failed to keep pace with the growing suburbs. Getting about Sydney was both difficult and extremely time-consuming. Communications between the city and the north shore were limited and Sydney Harbour was severely congested.

In 1900 whole swathes of the city – east and west – were still without the railway. They were also increasingly devoid of much of the native flora and fauna. With the coming of the western railway lines, eucalypts and she-oaks were felled to make railway sleepers or used as firewood. Tanneries created an almost insatiable demand for bark, especially from local wattles – green wattle – used to make tanning solutions to cure leather. A huge swathe of land between Prospect and Penrith became a supplier of clay for bricks, sand and gravel for cement, and from the 1880s blue metal for roads. As early as 1856 the naturalist, the Rev. William Woolls of Parramatta pointed out that:

> *Owing to changes which are taking place amongst us the 'native botany' will soon become a matter of history. Perhaps some of your readers may smile at this, but nevertheless it is a fact, for owing to the destruction of native plants on the one side of Parramatta by the waterworks, and on the other side by the railway, I fear that several of the Myrtaceae [eucalypt family] and Leguminosae [pea family] have already disappeared.*[6]

PLANNING SYDNEY

The observations of Woolls and others did not go entirely unnoticed. Momentum was also building to put aside open space within the rapidly expanding metropolis, although Sydney's development went largely unplanned. Delays in civic works created serious deficiencies in the provision of basic services. Controls over building and housing standards were practically non-existent. The streets of Sydney were often dark and dingy, and the buildings noteworthy only for their extreme ugliness, with no scope for beautification. Slum clearance was highly problematic, but where were the inhabitants to go?

6. W. Woolls, quoted in L. Gilbert, *William Woolls, 1814–1893 'a most useful colonist'*, Mulini Press, Canberra, 1985.

Apart from anything else many of the new suburbs lacked imagination as places in which to live. In view of wholly inadequate regulatory powers and the tendency for government bodies to compete with each other, such problems appeared to be overwhelming.

In 1880 colonial politician and 'father of Federation', Sir Henry Parkes introduced legislation that had far reaching consequences for the future shape of the city. The Land for Public Purposes Acquisition Act was duly passed and the New South Wales colonial government put aside funds to obtain land for various public objects 'such as for the purposes of water supply, for fortifications, sidings, light houses, public schools and the formation of recreation grounds and public parks'.

Important influences at home and from overseas were at work. The push for urban planning and slum clearance was well underway in Europe and North America and lent some urgency to local debate about what must be done. The transformation of Paris under Baron Haussmann during the 1860s and 1870s (critics condemned him for having destroyed the medieval city), and the formation of the reformist London County Council in 1888, were closely followed by public figures and politicians in Australia. Some of them even embarked upon fact finding tours which laid bare the inadequacy of Sydney's planning provisions.

INFRASTRUCTURE

Influenced by politicians such as Parkes, vital infrastructure began to take shape over a period of decades, albeit in piecemeal fashion. The Crown Street Reservoir constructed in 1858–9 amidst some controversy is still in operation today and remains the oldest working reservoir in Australia. Brickmakers were called upon to manufacture large quantities of waterproof bricks fired with a vitreous enamel coating. Over a million bricks were used in building the telephone tunnels of Sydney in the 1890s, and from 1880 a start was made on sewage outfalls to improve standards of public health and hygiene.

The first of these drained high lying land and included the Bondi Ocean Outfall at Ben Buckler Headland, and the Southern Ocean Outfall at Bondi. The latter discharged waste water processed by the Botany Sewerage Farm which operated between 1886 and 1916. Lower lying areas simply dumped untreated sewage into Sydney Harbour. Imposing a sense of order on the chaos was Sydney's Metropolitan Water and Sewerage Board formed in 1888 to organise all water supply and sewerage services in the city, its ultimate objective being the reticulation of such facilities to every house in the city. This bold vision was progressively realised in the coming decades, although the house building boom after the Second World War left many recently established outlying suburbs reliant on private tanks. For the time being, however, the 'dunny man' was a regular visitor to large parts of inner city Sydney. Both the Bondi and Southern Ocean outfalls vented noxious gases through a series of impressive brick chimneys.[7] The stack at Hunter Street Lewisham with its inlaid brickwork is just one example. By the turn of the century engineers were experimenting with the use of reinforced concrete in major infrastructure projects. The Johnstons Creek and Whites Creek sewer aqueducts (circa 1897), part of the new Sydney Sewerage Scheme, were originally intended to be of brick construction, but this was abandoned in favour of reinforced concrete. Both of these spans still exist and serve their original purpose. Controversy was never far behind and the circumstances surrounding the use of concrete in the aqueducts were the subject of a Royal Commission.

PARKS AND GARDENS

Satisfying the public thirst for recreation space was an expensive business and there was never sufficient money to buy back land, so that the areas reserved were often of poor quality and the pattern of distribution uneven. Joseph Henry Maiden, director

7. Sewers built during a later period were often concrete with steel reinforcement constructions and tended to be less durable than brick arch sewers.

Population of Sydney 1788-1901

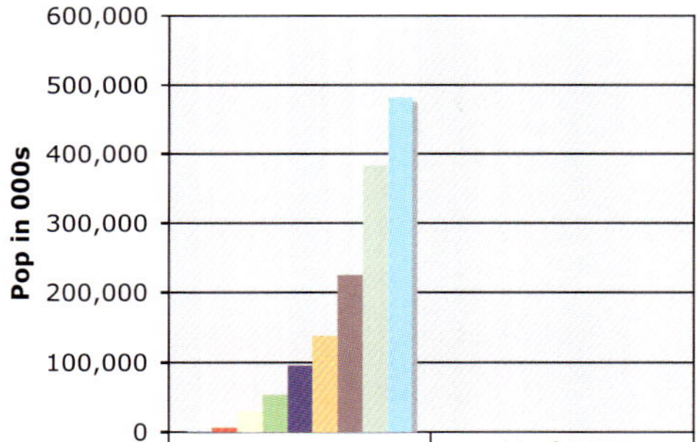

Years	Sydney	% of NSW population
1788	1,024	0
1810	6,156	0
1841	29,973	19.4
1851	53,924	29.7
1861	95,789	27.3
1871	138,000	27.4
1881	225,000	30
1891	383,000	34
1901	482,000	35.6

Years

Population of Melbourne 1788-1901

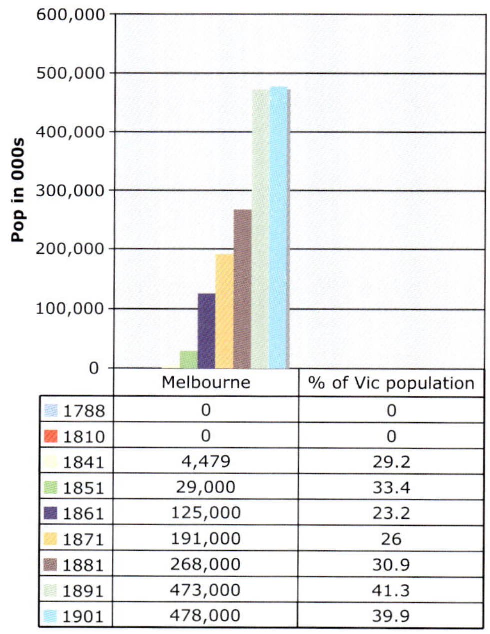

	Melbourne	% of Vic population
1788	0	0
1810	0	0
1841	4,479	29.2
1851	29,000	33.4
1861	125,000	23.2
1871	191,000	26
1881	268,000	30.9
1891	473,000	41.3
1901	478,000	39.9

Source: ed. J. Davidson, The Sydney – Melbourne Book, Allen & Unwin, Sydney, 1986, statistical appendix. The data used in this table are from censuses and information derived from Commonwealth, State and Colonial Statisticians, with the exception of 1788 and 1810. For a general discussion of problems related to determining accurate statistics see G.J.R. Linge, The Delimitation of Urban Boundaries, Canberra, 1965

of the Botanic Gardens in 1896, was an early and enthusiastic promoter of parks as recreation areas. Yet, with more people using public space there was also a sharp rise in criminal activity, although it is reasonable to assume that the Victorian criminal underworld had never really strayed from its old haunts in and around Brickfields Hill. In 1904 the *Sydney Morning Herald* criticised the authorities responsible for Hyde, Cook and Phillip parks for allowing them to become haunts for 'the garrotter, the sneak-thief and the black-mailing prowler'. The newspaper lamented the use of Hyde Park in particular as a:

> *resting place of hundreds of destitute men … there is something radically wrong when a reserve that is intended to promote the delight as well as the health of families resident in the city has to be avoided because the seats … are the abode of vermin, disease and germs.*[8]

Spurred into action the New South Wales government transferred control of three urban spaces to the City Council which installed electric lighting, pleasing crowds of 'orderly law-abiding people [who listened] to the music provided by … the City bands'.[9]

Population density in inner Sydney suburbs

	People per acre	
Municipality	**1871**	**1891**
Sydney (city)	25.9	37.4
Newtown	8.9	36.8
Paddington	10.2	44.1
Redfern	14.5	46.9
Darlington	25.0	61.9

Source: T.A. Coughlan, General Report on the Eleventh Census of New South Wales, Sydney, Government Printer, 1894, p. 121

8. *Sydney Morning Herald*, 15 June 1904.
9. Sydney City Council Archives. *Proceedings of the Municipal Council*, 1906, p. 192.

MIDDLE AGE SPREAD

Far away from alleyways and lairs of the inner city were the new developments of Strathfield, Concord, Lane Cove, Hunters Hill and Waverley. All of these were deemed to be the outer suburbs of Sydney in 1901, although the trend towards the fringes of the city was already discernible. Whereas in 1851 more than 80 per cent of Sydney's population lived in central Sydney, 50 years later the figure had fallen dramatically to a meagre 23 per cent. Of course, this was simply a response by those who could afford to settle in more desirable locations. Overall, the process of urbanisation throughout Australia quickened towards the end of the century, with 13 per cent of the country's population living in Sydney by 1901. Rural poverty, industrialisation and high immigration in the decade before the First World War were largely responsible for this trend. In fact, one of the consequences of urbanisation and the growth in large cities such as Sydney and Melbourne was the shift in the balance of political power away from farmers and pastoralists to the industrial worker. Many of them, including employers, demanded protection for their business, wages and employment through tariffs.

EXPERIMENTATION AND INNOVATION

The booming towns and cities, especially Sydney and Melbourne, during the late nineteenth century gave brick architecture an unparalleled opportunity for experimentation and innovation. Huge quantities of brick and clay products were required for the new factories, commercial premises, houses, railway stations, viaducts and pump houses as the Australian economy developed. In the new subdivisions architectural styles favoured brick, so that for brickmaking firms the future looked bright. There was also the added assurance that capital invested in machinery to increase production had been well spent.

During the 1860s the imaginative use of different coloured bricks in public buildings – polychrome brickwork – began to make its appearance, although more so in Melbourne than Sydney. An early

'Grotesque' dragon – 'Grotesque' dragons and griffins were imported from France along with the Marseilles tile from about 1890. These items were mostly popular in Melbourne, although examples can be found in other capital cities. Photograph by Josh Hill

example was the Independent Church (1867–8) by Melbourne architect Joseph Reed who had designed the State Library of Victoria in 1854. Reed is believed to have popularised the use of polychrome brickwork in Melbourne following a visit to the Italian kingdom of Lombardy where multi-coloured brickwork could be seen in many of the buildings. He was also a major exponent of the use of brick in public buildings.

The two-tone effect of polychromatic brickwork was often achieved by creating a subtle colour contrast between the red brick walls and the dark brown or liver-brick surrounds of window and door openings. Despite its appeal, polychrome brickwork made little impact on building methods and architectural tastes until the quality of brick improved. This had to wait until later in the century with the introduction of new dry-pressed processing techniques which allowed for greater diversity in colour.

The decoration of parapets through patterned brickwork was especially popular among architects, although two problems had to be overcome: the unsightly appearance of the ragged edge of the cut brickwork around curves, and the lack of colour contrast between the two slightly different shades of brick used in construction. Both were solved with the use of tuck-pointing which reached the height of its popularity in the 1880s and 1890s. Tuck-pointing emphasised the regularity of the brickwork and consisted of crisp lines of a special type of mortar, plasterer's putty, which was either white or coloured and made from a slurry of fine, well slaked lime.

Those who practised the craft of tuck-pointing raked out the joints of the brickwork about a centimetre from the surface. A new mortar mix, coloured to match the bricks, was carefully inserted in the joints so that the whole wall took on a uniform appearance. Water was allowed to evaporate from the mortar before the plasterer's putty was applied in thin grooves previously cut into fresh brick mortar using a special trowel. The result provided an appearance of precise and highly contrasting brickwork. Tuck-pointing at sixpence an hour was a time-consuming activity but relatively inexpensive.[10]

Widdicombe brick at bottom. Warwick Gemmell Collection

10. H. Fraser, *The Federation House: Australia's Own Style*, New Holland Publishers, 2002, p. 60.

RISING DAMP

If polychromatic brickwork was an aesthetic matter, not so was the problem of damp which plagued householders and builders alike. What was to be done? According to Miles Lewis, William Atkinson's 1805 publication *Views of Picturesque Cottages with Plans*[11] contains a reference to the cavity – 'hollow' – wall that was often used in the building of ice houses. Certainly, their use was recorded in various parts of Britain and the United States as early as the 1850s. C. B. Mayes in the Australian Builders' Price-Book, published in Melbourne in 1862, mentions use of the cavity wall which appears to have established itself in Bendigo in 1869 and throughout Victoria by 1885.

The cavity wall consisted of two parallel brick skins separated by a gap of 50 millimetres (2 inches) held together with metal ties, usually iron bars. The outer leaf of brickwork bore the brunt of driving rain, whereas the cavity prevented the entry of damp and insulated the wall during hot weather. The two walls were bonded using a variety of methods, including iron ties, and cross laid bricks, an interesting example being the so-called Widdicombe Brick manufactured by the firm of T. H. Widdicombe of Portarlington, Victoria, after the type patented in England by George Jennings. This new product was reported in the *Australian Mechanic* in November 1872.[12] The cavity wall had many advantages. It provided excellent insulation against heat or cold, used no more bricks than a nine inch wall, and was watertight. Furthermore, it was strong enough to carry two storeys and was neatly presented without headers (brick ends) showing on either face. With the brick and terracotta styles beginning to come into vogue, it was ideal and remained the standard brick construction in Australia for the next 50 years. After 1900 internal partitions were built in four and a half inch brickwork, except in better quality two-storey homes where they remained 9 inches thick.

11. W. Atkinson, *Views of picturesque cottages, with plans: selected from a collection of drawings taken in different parts of England, and intended as hints for the improvement of village scenery*, Gregg, Farnborough, England, 1971 edition.
12. *The Australian Mechanic*, 15 November 1872, p. 134.

With the appearance of the cavity wall in the early 1890s, houses threw off their mantle of grey stucco drabness, to emerge as red and buff-coloured brickwork.[13] In 1903 John Barlow testified to the importance of the cavity wall:

> *A few years back it was absolutely necessary (if our houses were to be kept dry) that the walls should be covered with paint or stucco, but the general use of cavity walls, and the comparatively low price at which excellent face bricks can now be obtained, have rendered it possible for architects to impart a new and more lively character to their work.*[14]

At the same time, the traditional Flemish, English and Colonial bond of solid, interlocking brickwork gradually lost ground to stretcher bond which found its expression in the Federation house. There was even a proposal in 1888 by architect, John Ellis, who advocated glass bricks; 'if made hollow … they would be about twelve times the strength of an ordinary clay brick and about half the weight… Buildings would be healthier, lighter and more durable.'[15] Ellis never lived to see his idea take shape, and it was only during the 1940s that glass bricks made their appearance in the more avant-garde experiments thrown up, from time to time, by architectural fashion.

BRICK VENEER

Cavity wall construction was not the only innovation in house building during the period. Spacious houses and double brick construction appealed to the more affluent individuals, but they remained prohibitively expensive for working people. This may have prompted the introduction of brick veneer construction, another attempt to eliminate the dampness found in weatherboard houses. Early examples were crude, with bricks built around wattle and daub. Architectural historian J. M. Freeland notes that Swan Hill was the location of the first brick veneer building in Australia, and possibly the world. Peter Beveridge, a settler, is said to have burnt a kiln of bricks and used them to build an impermeable skin around his pine cabin in 1850. After 1851, a standard system evolved whereby a brick skin was tied with wire to an inner structural timber frame. It was cheap, effective, easy to maintain and became a status symbol in Australia's post gold rush society.[16]

HOME FINANCE

Innovation was also at work in another less obvious way. The construction of brick houses and commercial premises was entirely dependent on the means of people to pay. As early as 1843 a novel system to fund the purchasing of a house was introduced to Australia from England. The Starr-Bowkett system, originally drawn up in 1843 by a London doctor, T. E. Bowkett, permitted a number of people to pay a small amount each week into the funds of the Starr-Bowkett Society. After about one year a draw was held and the lucky winner was lent sufficient money to enable him to purchase a house. The person would continue to pay the weekly subscription and also start to pay off the loan on which no interest was charged. This meant that further draws could be held at ever decreasing intervals. When every member had paid off his draw the society was wound up and its surplus distributed to members.

Richard Benjamin Starr partnered Bowkett in the operation of the parent organisation, but after a disagreement around 1870 Starr continued on his own. He modified the scheme to some extent and charged each new society for the use of his rules. Starr-Bowkett societies were supplemented by the growing number of building societies that were founded to pay for the construction of well-built cottages sold to working men and their families on reasonable terms. The St Joseph's Permanent Investment and Benefit Building Society was established in 1868

13. H. Fraser op.cit., p. 60.
14. H. Fraser, op. cit., p. 60.
15. R. Boyd, op. cit., p. 39.

16. J.M. Freeland, *Architecture in Australia*, F.W. Cheshire, Victoria, 1968, p. 111

Above: Fowler's pottery, Camperdown, 1864. The expansion of Sydney from 1840 to 1914: A pictorial contrast, J. Feldheim, Melbourne 1914. Robert Hutchison Collection

Right: Fowler's Pottery, Camperdown, 1914. This photograph was taken in the same year that the pottery moved from Camperdown to Marrickville. Robert Hutchison Collection

by the Australasian Holy Catholic Guild of St Mary and St Joseph. Throughout its existence, the society maintained strong ties with the Catholic Church.[17] The society's properties were mainly terraces or cottages in numerous inner city suburbs.[18] Most properties were in clusters near the local parish church or school. Unfortunately, for every building society established with honourable intentions, there were many that simply overreached themselves. Some lent up to 90 per cent of construction costs on terms which were extended over 50 years. Most of them collapsed along with the banks as soon as the depression hit in the early nineties.

17. The society went through two name changes before it became Lisgar Investment and Building Society. In 1979 the society was acquired by National Mutual, and later sold to the State Bank, which eventually wound it up.

18. These included Redfern, Surry Hills, Ultimo, Erskineville, Glebe, Paddington, Waverley, Balmain, Leichhardt, Annandale, Waterloo, Marrickville, Petersham, Newtown, Camperdown, Woolloomooloo, with a few properties in Manly, Hunter's Hill and North Sydney.

PAINTING THE TOWN RED

Unlike innovations in bricks, buildings and architecture, the manufacture of roofing tiles was a great deal slower to develop. Terracotta products were difficult to produce, and despite its expense and tendency to conduct heat in summer weather, slate remained in favour until the First World War.

One exception was the pottery works of Enoch Fowler which commenced operations before 1850. For many years Fowler's business enjoyed a near monopoly. In 1850 he relocated the works from Broadway (near the former department store of Grace Brothers) to Parramatta Road at Camperdown, on what was still the fringes of Sydney, a distance of about three kilometres.[19] The new site, bounded by

19. A limited company was formed in 1919 and the works transferred to Marrickville where the business was extended to include the production of floor and wall tiles and white enamelled fireclay ware. A branch factory was established in Victoria at Thomaston, in 1927, for the manufacture of sanitary earthenware and crockery.

Australia and Denison Streets, covered 3.5 acres. By the late 1860s continued expansion had doubled the acreage and the firm was renamed Fowler's Pottery, Fire Brick and Drain Pipe Manufactory to reflect the range of products manufactured. While in this locality, Fowler's manufactured glazed stoneware included vinegar jars, ginger beer bottles and general household stoneware. Success of the venture led to the production of earthenware sanitary ware and household crockery. Fowler was quick to apply steam driven machinery and modern production techniques.

MARSEILLES TILES

To end the dependency of the colony on slate, which was brought to Australia in ships as ballast, steps were taken to import from terracotta roofing tile manufacturers in Marseilles, France. Such firms laid claim to their tiles being weather tight and non-conductors of heat, and tests in Sydney and Melbourne soon proved this to be the case. The Marseilles, or French Pattern tile, was developed from a prototype designed by the Gilardoni Brothers in Alsace in the mid 1850s. These machine-made interlocking roofing tiles were maintenance free, durable and waterproof, and designed to be laid with interrupted vertical joins which aided ventilation. Small quantities of Marseilles tiles were being imported into the colony from about 1860 by the Melbourne firm of Messrs Henry Lange and Company.[20]

The expiry of Gilardoni Brothers' patent rights enabled a number of tile manufacturers in and around Marseilles to adopt and improve the original concept. These manufacturers were collectively known as the Société Anonyme des Tuileries de Marseilles and, after about 1900, as the Société Generale des Tuileries de Marseilles et Cie. The syndicate marketed this new interlocking tile in large numbers and in two colours – grey for Europe and red for tropical climates such as were found in South America and Australia. In 1888 the Syndicate won a medal for roofing tiles at the Melbourne Exhibition. The Melbourne furniture firm of W.H. Rocke and Company was the first to import them to Australia in quantity after 1886 although isolated shipments may have arrived earlier. However, when Rocke's subsidiary operating in Sydney became insolvent in 1892 and was unable to handle the sale of the tile shipment, the Wunderlich brothers took responsibility for the transaction. Indeed it was the Belgian Consul in Sydney who urged the art metalwork firm to take them over, thus launching the largest terracotta firm in Australia. By 1893 it was the sole agent for the importation of Marseilles tile and by 1914 had landed 75 million in Sydney.

In 1908 the company went public to form Wunderlich Limited so that an increase in its capital base allowed the company to expand at a time when the building industry was enjoying boom conditions. At its experimental tilery at Rosehill near Parramatta, Wunderlich began by reproducing almost exactly the French Pattern tile, ie the Marseilles tile. By 1910 no other roofing material was able to compete in Melbourne, Sydney or Adelaide. In fact, Wunderlich's foresight paid off; when shipments from France ceased after the outbreak of war their factories were already producing one and a half million tiles per year.

Affluent society in Victorian Australia was quick to embrace terracotta because it enabled expressive architecture by means of decorative sculptural forms, and also because it produced a surface unaffected by atmospheric pollution. Wunderlich was right in claiming to have 'painted the town red' through the importation of red terracotta roofing tiles. The drab grey slates and galvanised iron and the dun colour of timber shingles were gradually replaced by the distinctive orange-red of the imported tiles that characterised so many Australian suburbs. Their growing popularity was also due to the adoption of English Queen Anne, a style of architecture originating from the reign of Queen Anne during the first decade of the 18th century. With its emphasis on high-pitched roofs and red accents, grey slates

20. *Colonial Mining Journal*, September 1860, p. 8.

Entrance to the Domain with three unidentified people and view of St Mary's Roman Catholic Cathedral. National Library of Australia

gave way to red tiles. Bricks were tuck pointed. The boom in housing construction before and after the depression of the early 1890s was another factor. Apart from anything else the Marseilles terracotta roof tile was regarded as superior to any other form of roofing material.

PUBLIC ARCHITECTURE

Meanwhile, the architects Walter Liberty Vernon[21] and William Wilkinson Wardell[22] were creating a distinctly Australian style in public buildings. Vernon became Government Architect in New South Wales, and his work brought together Victorian Classical Revival and early twentieth century forms of architecture. Vernon's office introduced a new approach to public buildings with an Arts and Craft style as shown through fire stations at Darlinghurst and Pyrmont, post offices and country courthouses. He also built a number of major public buildings, such as the Mitchell wing at the State Library, the Art Gallery of New South Wales, Fisher Library at the University of Sydney and Central Railway Station.

Wardell's interest in Classical Academic and Classical Gothic styles produced the imposing St Mary's Cathedral, Sydney, and St John's College at the University of Sydney. Under their influence, highly decorated office buildings in the city were built, including those in the Pitt Street Mall between King and Market Streets. Sydney's Town Hall is another example of the classical style embraced by Wardell.

JOHN HORBURY HUNT

Perhaps the most celebrated – and talented – architect of the time was John Horbury Hunt (1838–1904), renowned for the radical architectural style that he

21. Walter Liberty Vernon emigrated to Australia from England in 1883 and was appointed Government Architect in 1890, taking over from James Barnet who had occupied the position for 30 years.

22. William Wilkinson Wardell (1823–1899) emigrated to Australia from England, settling in Melbourne in 1858. Wardell had enjoyed a successful career as a church architect in England, and was appointed chief architect for the colonial government of Victoria soon after his arrival. He later became inspector-general of public works, overseeing many of Melbourne's early major projects. Wardell was an exponent of the Gothic Revival movement and his churches include St Patrick's Cathedral and St Ignatius Church in Richmond in Melbourne, and St Mary's Cathedral in Sydney. He also designed Melbourne's Government House and Venetian styled banking chambers for the English Scottish and Australian Chartered Bank.

John Horbury Hunt (left) and Lloyd Taylor, 1889. Mitchell Library. State Library of New South Wales

introduced to Sydney and rural New South Wales. Canadian-born Hunt spent his formative years in Boston, Massachusetts, and in 1863 at the age of 24 sailed for Australia to escape the American Civil War. His commissioned work, almost 100 identified buildings, can still be found throughout the state and includes churches, cathedrals, stables, schools and private houses.

Hunt was eccentric and cut a somewhat comical figure pedalling through the streets of Sydney in the 1860s. A short man, he disguised the fact by wearing an outfit of knee-length frockcoat, brightly coloured waistcoat, high-waisted striped trousers, built-up shoes and spats, and a flared bell-topper hat with ventilation holes concealed by the hatband. His suits contained numerous pockets to hold drawing instruments and his hat had a compartment for drawing paper.

Hunt's 1881 design for the Rose Bay convent housing the Sisters of the Society of the Sacred Heart was a tour de force. The chapel is widely recognised as a masterpiece and its interior is structurally breathtaking. The stone vaulted roof was the first of its kind in Australia and has never been surpassed.[23] Hunt criticised the stylistic pretence of other architects, which did little to endear him to establishment figures who rejected his designs for the Art Gallery of New South Wales as being too 'eclectic'. This was exacerbated by his conceit, lack of tact and quick temper, which led to him frequently parting company with colleagues and collaborators.[24] Yet, as President of the Institute of Architects of New South Wales (1889–95), Hunt tried to rid the industry of 'jerry builders' and 'quack architects'.

Hunt's contemporaries, at home and overseas, admired his designs for their siting, asymmetrical

23. P. Reynolds, L. Muir, J. Hughes, *John Horbury Hunt. Radical Architect, 1838–1904*, Historic Houses Trust, 2002.

24. The building and architectural community in New South Wales continues to honour Hunt through an annual competition, the John Horbury Hunt Award for Excellence in Brickwork. In 2002 the Historic Houses Trust of New South Wales also published the work *John Horbury Hunt: Radical Architect 1838–1904*.

PA Federation Poem – Hop, The Bulletin 1898 AUSTRALIA'S GLORIOUS FOURTH

balance, excellent brickwork and craftsmanship. Indeed, he was a vocal advocate for the return to brickwork and was at the forefront of a worldwide movement where every brick served structural, rather than ornamental purposes. Sadly, Hunt died the death of a pauper in 1904. A lonely man during the final years of his life, Hunt suffered the ignominious fate of having his work largely ignored. Today, however, John Horbury Hunt is considered to have been one of those who laid the foundations of modern Australian architecture.

A LIFE TO LEAD

Regrettably, beautiful architecture – public and private – befitting a major capital city remained in short supply. Despite flashes of genius, Sydney's built environment never matched the grandeur of colonial Melbourne and other prominent Victorian towns. This might appear surprising in view of increasing prosperity brought about by a booming export trade in grain, wool and refrigerated meat, as well as an influx of new settlers. Most of the newcomers flocked to the large towns and cities and by the end of the nineteenth century Australia was the second most urbanised society in the world after Britain. In fact, by the time the recession of the 1890s took hold, Melbourne covered an area of 164,000 acres with an average population density of three people per acre. Sydney averaged four people per acre in its older and marginally more compact form over 96,000 acres.[25]

Urbanisation and the inexorable spread of brick buildings may have been the catalyst for expression of another sort. Starting in the 1880s and continuing well into the next century, Australia experienced great advances in the arts, painting, and especially literature. *The Bulletin,* founded in 1880, quickly became a centre for nationalistic discourse on politics and literature. Prominent among its writers were J.F. Archibald and later A.G. Stephens both of whom encouraged young talent such as 'Banjo' Paterson, Henry Lawson, Joseph Furphy and Victor Daley. Paterson mythologised Australian bush culture through his many ballads, drawing on a long tradition that embraced convicts, bushmen and bushrangers. Interestingly, pastoral images of nature and the bush evidenced in the work of paintings of late nineteenth and early-twentieth century artists, especially those of the Heidelberg School, contrasted sharply with the suburban reality of most people.

25. A. Davison, *'Australian Suburban Imaginaries of Nature: Towards a Prospective History'*, Australian Humanities Review, Issue 37, 2005.

Envy, An Everyday Street Scene – Hop, The Bulletin, 1894. Artist J. King

The poems and writings of Henry Lawson express the profound unease and ambivalence felt by many, especially young men, towards the city. Sydney academic, Brian Kiernan, explores the paradox that despite Australia being one of the most highly urbanised countries in the world from the 1890s, its literature was preoccupied with the countryside. 'For Lawson and others of that generation there was a choice, or a tension, between the two realities of 'Sydney or the Bush'.[26] A sense of rural nostalgia – the search for a lost Eden – pervades Lawson's writing, finding expression in the somewhat bleak imagery he uses to depict city life around the turn of the century – street faces, larrikins and their molls, doss houses and life in Sydney's docklands, made all the more threatening by gaslight and the shadows. Furthermore, Lawson was a socialist and champion of the under-dog who believed in that quintessential Australian notion of 'mateship'. Lawson was not alone in his bleak depiction of city life in Sydney and his quest for a more equitable society. William Lane (who led the failed New Australia venture to Paraguay) and Christopher Brennan were two other notables who explored Sydney in prose and verse, although few among the working masses shared their somewhat pessimistic view of the world. Meanwhile, popular forms of theatre such as vaudeville with its bawdy shows guaranteed to scandalise 'wowsers', continued to gain in popularity. The Tivoli Theatre was a popular venue for such music hall entertainment.

Indeed, this busy engagement of the citizenry in defining itself through singing, writing and drawing the rustic figures of the bushman, the drover, the prospector and the farmer enabled the Australian continent to acquire new meaning in the imaginations of newcomers and the previously settled. The inclusion of the kangaroo and emu on the new nation's crest recognised Australia's unique native fauna and its association with its people.

IN THE DOLDRUMS

Boom times are just that – ephemeral, and generally followed by a bust. According to Sydney historian, Lucy Turnbull, capital inflows peaked between 1886 and 1890, with huge amounts ploughed into frenzied land speculation and building, especially in Victoria. The appetite for foreign goods – and the export of homegrown produce – was equally insatiable. The tonnage of shipping in New South Wales almost doubled between 1871 and 1881, and then doubled again in the following ten years to 1891.[27] Compared to Sydney only London, Liverpool and Hull in the British Empire had goods of a greater value passing through their ports.

All of this was about to change. There were early warning signs that a depression was on the way; financial institutions were particularly vulnerable and heavily exposed to a property bubble. Sooner or later the bubble would burst. Indebtedness, both public and private, was rampant with government railways operating at a loss, and pastoralists falling

26. B. Kiernan, 'Sydney or the Bush. Some literary images', in J. Roe (ed), *Twentieth Century Sydney. Studies in Urban and Social History*, Hale & Iremonger, 1980, p. 148.

27. T.A. Coghlan, *Results of the Census of the Seven Colonies of Australasia*, General Post Office, Sydney, 1894, p. 87.

heavily into debt to purchase their properties. To make matters worse a severe drought bit deeply into the rural economy from the mid-80s. In itself the drought was devastating, but it also had the effect of accelerating the long-term drift of the rural population towards the towns and cities. Relief schemes were introduced, with the New South Wales Government putting the unemployed to work on public projects. Just as they had done in the 1840s, the price of wool and metals and other export commodities fell. In 1891, for the first time in many years, colonial governments failed to raise the money they wanted on the London market. Capital inflows from Great Britain dried up. The embryonic global economy was plunged into depression with Britain, its colonies and many European countries sharing the same fate.

In Australia, credit restrictions were introduced and governments drastically reduced spending. Highly mortgaged companies and financial institutions, some of them fronts for criminal directors intent on fleecing money from unsuspecting investors, were hit by a wave of defaults and quickly collapsed. By the end of 1891 two banks had failed and within months there was a run on the Savings Bank of New South Wales. A string of bank collapses followed over the next two years, taking the economy with it. Although the financial crisis had largely passed by the end of 1894 a further drought set in until 1903, blighting all efforts to revive the economy. In fact, it was only after this point that recovery became a reality. The crisis was most severe in Victoria where land speculation had been at its most intense. High unemployment in Melbourne led to emigration into country districts and to other colonies.

IMPACT ON THE BRICK INDUSTRY

The impact on the brick industry was immediate for it brought decades of growth to an abrupt halt. Everyone connected with the trade during the 1890s had a very lean time, both workers and employers. This disastrous state of affairs affected everyone dependent upon the industry. Not only did the shareholders suffer through the inability of the companies to pay dividends, but also the shares became practically unsaleable.

Country districts were particularly hard-hit as were the small family businesses, and many yards were forced to close. In 1890 Sydney had 113 yards, by 1896 it had 42 and in 1900 the number was 32. Sydney brick production collapsed with the annual output of bricks falling from 276 million in 1889 to 40 million in 1894. In an effort to keep the industry afloat, wages were cut and many adult employees received less than 10d per hour. Brick carters received from 30s to 36s per week. For bricklayers, the average wages in 1886 had been 12s to 13s for an 8-hour day, whereas a decade later they were receiving 9s 8d in Sydney and 7s 6d in Melbourne.

The collapse of brick prices was even more spectacular. On the eve of the depression prices had been at an all time high. At a meeting of the Brickmasters' Protective Association of New South Wales in 1889 brickmakers fixed a uniform price of 47s per thousand bricks. Thereafter prices collapsed, forced even lower by fierce price-cutting on the part of the manufacturers. During the lean years following 1892, prices fell heavily, sometimes as low as 17s per thousand bricks at the kiln. In September 1894 some brickmakers even accused the association of causing bankruptcy by being selective in the work handed out to its members.

THE IMPACT ON MELBOURNE BRICKYARDS

The depression devastated the brickyards of Melbourne with the result that brick manufacturing companies established in the inner suburbs and the environs of the metropolis found themselves in a parlous condition. The smaller yards were beyond hope of financial redemption, and the larger yards actively competed against each other to such an extent that bricks were being sold in 1894 from 14s to 16s per thousand at the kiln, which was below the cost of production.

A sudden rash of strikes before the depression, disrupted the brick and building trades and were an ill-omen. At a meeting of the Bricklayers Society in August 1888, financial assistance was voted for the Ballarat bricklayers who were striking to secure a 45-hour week. Interruptions to production sent many yards to the wall. It was a time of great upheaval and uncertainty. Work was suspended at several Melbourne brickworks for more than four months in 1890, owing to the scarcity and increased price of coal used in the kilns to burn the bricks. The Hoffman–Northcote and Tunstall brick companies came to a complete standstill for five weeks, which was a catastrophe in that it required the extinguishing of the fires of the continuous kilns. Restarting a kiln from cold could take anything up to a fortnight. Meanwhile, the Northcote Brick Company was distressed to learn that a rival brick manufacturer was intending to set up on its doorstep. The cheekily named New Northcote Brick Company began production in 1896 and appears to have prospered.[28] Those firms still under contract to produce bricks for builders, slashed workforce numbers, cut costs, and were able to negotiate higher prices.

Following the lead set by the brickmakers of New South Wales, a group of Melbourne manufacturers comprising Hoffman, Northcote, Butler's, Fritsch-Holzer, and New Northcote formed themselves into an association some time during 1896. This was not for the purpose of exploiting the public but to regulate the output and price of bricks in order to prevent outright collapse. Arrangements were made to spread brick orders, to close brickyards, and to lease certain companies outside the association for periods of three or five years. Conspiracy theorists cried foul, although with very little evidence to support their claims. Given that there was always going to be ill-will between the advocates of state intervention and free marketers, it is not surprising that many in the labour movement regarded private brick manufacturers with great suspicion. Indeed, the formation of protective associations during this period seems to imply that the industry was being run by a cartel. Little has been written about this practice in Australia for the simple reason that few are prepared to take the witness stand. Generally speaking the motivation for cartels (especially those in the United States) was to restrain trade to the advantage of the members. There is no doubt that brickmasters, whether in New South Wales or Victoria, acted in concert with one another to set prices, collect debts, and market their products, yet this was a far cry from running a cartel to the detriment of the customer. Essentially, it was a defensive measure intended to shore up the industry at a time of crisis, not as some believe an example of monopoly capitalism. The behaviour of brick manufacturers during the depressions of the 1890s and early 1930s would seem to bear this out.

State Of Victoria, Brick Production

	Brick output 000s	Brick & stone houses as a proportion of all houses	Price per 1,000 bricks	Number of Melbourne yards	Victorian employees engaged in brick & pottery industries
1889	277,897			88	3,250
1890	276,651				3,240
1891	207,270	44.91%			2,340
1892	80,172			72	1,340
1893	48,467		22s to 17s		890
1894	40,109	(decade low)	16s to 14s	51	960
1895	64,205		16s		1,010
1896	55,682		22s to 30s	29	1,120
1897	51,048		26s to 32s		940
1898	64,511		32s to 27s to 34s		1,100
1899	82,751		34s to 30s to 34s		1,300
1900	85,387		34s to 40s		1,320
1901	84,898	46.46%	40s		1,580
1902	90,545		40s		1,453
1903	77,827		40s		1,583
1904	80,027		40s		1,433
1905	90,990		40s to 38s		1,513
1906	112,966		38s		1,825
1909	129,303		38s		1,695
1910	145,810		38s		1,847
1911	153,945	45.91%	38s		1,968
1912	180,724		38s		2,149
1913	175,645		38s to 33s to 35s		2,060

Source: Statistics compiled by Jack Petersen, 22 November 1991

28. The Hoffman and New Northcote companies were fierce rivals and continued to grow into the first half of the twentieth century, principally by acquisition. Ultimately, the Hoffman Patent Steam Brick Company was itself taken over in 1960 by the Clifton Brick Company, founded in 1891, and a name widely respected in Melbourne until its demise in the 1990s. The assets of Clifton were eventually acquired by Brickworks Limited.

Bricklayers Society banner. Unions NSW

Tilelayers Society banner. Unions NSW

In testimony given to the Royal Commission on the Brick Manufacturing Industry of Victoria (May 1913) into the state of the industry discussed in Chapter 7, Sir Charles Langford, a director of Excelsior Brick Company in Sydney, stated that he did not think it had been to the disadvantage of the building trade of Melbourne that a number of brickyards were kept closed. 'I think,' he said, 'they (the builders) should be very thankful they (the yards) were kept closed, because if they had all kept on they would have been all closed and they would have got nothing out of it.' [29]

REVIVAL OF THE INDUSTRY

The effects of the depression on the industry were not all negative. At a time when competition between firms was high there was now an incentive to improve matters, especially the mediocre quality and limited range of bricks. A meeting of the newly formed Sydney Brick Company in 1893 discussed the state of the depressed market:

> *(Brickmakers) were making efforts to bring their bricks to one uniform size. And if they succeeded in getting a uniform size, they would give their attention to obtaining a uniformity of colour; and uniformity of colour would be selected out of the number of kilns to select from.* [30]

We know very little about the Sydney Brick Company, but it was quite probably a selling agency that worked on behalf of brickmakers, having been established soon after the winding up of the Brickmasters' Protective Association of New South Wales in 1893. The Metropolitan Brick Company Ltd may have taken over from the Sydney Brick Company after 1900 and was active until 1938.

29. Royal Commission on the Brick Manufacturing Industry in Victoria, 20 May 1913.

30. *Building and Engineering Journal*, 25 March 1893, p. 112.

Comments such as those made at the aforementioned meeting were timely for the simple reason that during times of slack trade builders and architects could afford to be choosy. During the depths of the depression George Fischer 'deplored the fact that brickmakers still showed no interest in turning out better bricks or taking outsiders into their confidence'.[31] Nangle stated in 1894 that handmade brickyards were 'turning out much better bricks than those from the patent kilns'.[32] Yet despite machine-made bricks having a more densely packed face, they were crumbly in the centre and could neither be rubbed nor cut by bricklayers. Poor quality products only served to make matters worse, and until building activity finally picked up, Sydney and Melbourne yards were stacked out with millions of underburnt, poorly manufactured bricks. A few brickmakers had reason to be optimistic with the establishment of a small export trade carried by ships of Burns-Philp to Fiji and the South Sea Islands. There was even an import trade when the building justified the additional cost. High-quality bricks from Britain, Germany and the USA were still being landed at the wharves up to the mid-1890s.[33] Several attempts were made to remedy the situation, one being a comprehensive list of characteristics to classify bricks (see table below).

With the worst of the depression behind them, several of the larger companies during this post-depression period were quick to address such problems. By 1896 a range of high quality, shaped bricks was available. The Greengate Hotel in Killara, Sydney, established in the 1830s, was extensively remodelled in 1902, the work involving the use of 150 different types of moulded and ornamental bricks produced by Goodlet and Smith to the architect's specification.[34]

The industry was also more receptive to the various academic disciplines on how to improve their

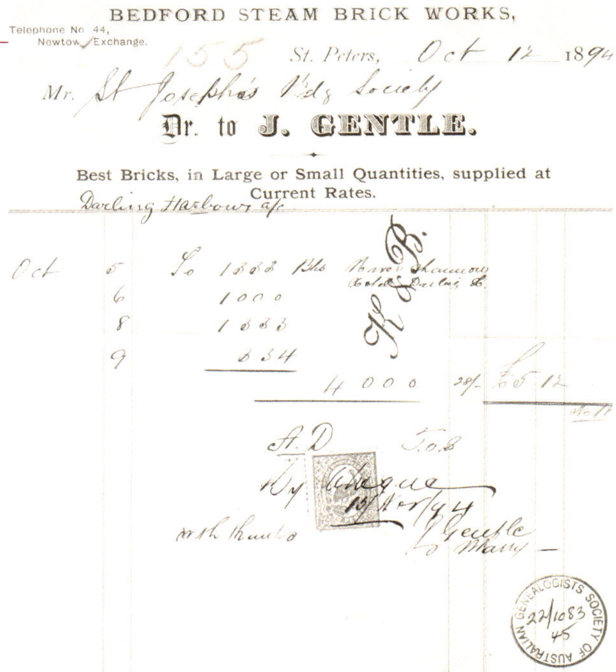
BEDFORD STEAM BRICK WORKS,
Telephone No. 44,
Newtown Exchange.
St. Peters, Oct 12 1894
Mr. St Joseph's Bldg Society
Dr. to J. GENTLE.
Best Bricks, in Large or Small Quantities, supplied at Current Rates.

Brick receipt, Bedford Steam Brick Works, 1894. A receipt signed by Josiah Gentle showing payment of 28s for a total of 4,000 bricks, a clear indication of the effect of the depression on brick prices. The bricks may possibly have been seconds, judging by the price. Society of Australian Genealogists

product. By 1900 the Department of Mines in New South Wales was conducting tests on clays and shale to determine their chemical and physical properties. The testing of bricks for their crushing strength, while well established in Europe and North America, was not common practice in Australia until the late 1890s. As early as 1878 regional testing laboratories were being set up in Germany at the urging of

Brick type	Description
Callows or underburnt bricks	Low-priced and used for internal partition walls.
Clinkers or overburnt vitrified bricks	Low-priced and used for foundations.
Ordinary common bricks	Handmade or machine-made by the wire cut or dry-press process. Burnt in a continuous kiln.
OK or open kiln face bricks	Superior common brick in quality and colour, made using the dry-press method and burnt in old-type open kilns. Usually applied to facings on buildings of lesser quality and grandeur. Burnt in a downdraught kiln and used for buildings of superior quality.
Double-pressed face bricks made by the plastic process Enamelled bricks	Limited application, e.g. lavatories, washrooms and wet areas.

31. J.M. Freeland, *Architecture in Australia: A History*, Penguin Books, Ringwood, 1972, p. 199.
32. *Building and Engineering Journal*, 4 February 1893, p. 41.
33. Bricks from overseas were usually shipped to Australia as ballast on trading vessels. Most were firebricks used for refractory purposes.
34. O.S. Pavlou, *The History of Bricks and Brick Making in New South Wales, 1788–1975*, Thesis, Faculty of Architecture, UNSW, Sydney, p. 479.

Bridge Street, 1900. National Library of Australia

the Union of German Architects' and Engineers' Societies. Conferences were even held on the testing of building materials in the cities of Munich, Dresden, Berlin and Vienna. In France the École Nationale des Ponts et Chaussées began testing bricks in 1881. In 1884 the USA introduced brick testing when the United States Army was required to test the bricks of the New Pension House in Washington, reporting that they had withstood a crushing pressure of 10,000 pounds per square inch. Interest in brick testing may have been prompted by the collapse of a house in London's Haymarket in January 1878. The death of an oyster seller prompted authorities to tighten building regulations, although the cause of the accident was found to be a builder's negligence in allowing a great weight to rest on a small area of brickwork. In Australia compression tests were first conducted at a number of places in 1891, including the engineering laboratory of the University of Sydney, and the Technical College of New South Wales at Ultimo. The college was also the first educational establishment in Australia to introduce a school of bricklaying in the same year.[35]

Slowly, confidence returned to the market and the brick trade was able to re-establish its viability in the building industry. When Australia ceased to be a group of colonies in 1901 patrons could expect to have their buildings designed by competent architects and built with the bricks of the highest quality. On the other hand, working class housing – jerry built and dingy – was a different matter altogether. For the first time, however, dry press methods made possible the production of a wide variety of brick shapes that could be used as decorative features, edging for footpaths, or capping bricks for walls. Referred to as 'specials' by modern brickmakers, shaped bricks became increasingly popular from the 1920s onward.

Federation of the Australian colonies in 1901 not only marked the birth of a new nation but a resurgence in confidence. A new century was about to dawn.

35. *Illustrated Sydney News*, 26 September 1891, p. 10.

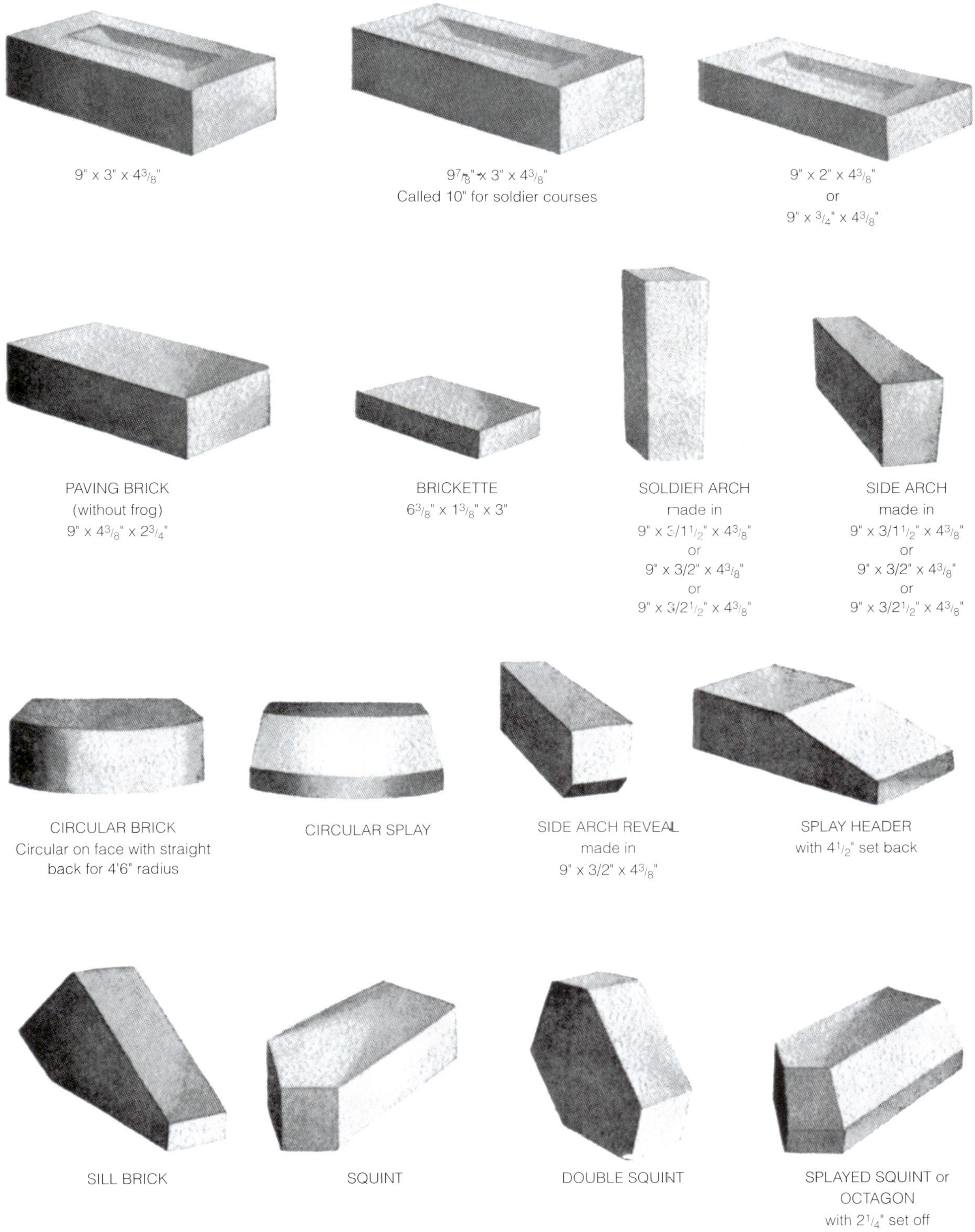

Brick shapes: A wide range of brick shapes was made possible by the introduction of dry press methods and the use of specially shaped brick moulds for limited runs. These plates are from the trade catalogue of the Central Brick & Tile Company Limited, circa 1938. Moulded bricks were popular with many architects and were used as features in walls, parapets, and architraves around doorways. Author's Collection

Anticlockwise: Office building, Macquarie Street, Sydney, insets showing brickwork in Nightingale Nurses Hostel, Sydney Hospital, Sydney Technical Institute

The Austral Brick Company Limited 7

Turning point – Federation House – Slack trade – Birth of Austral – A matter for speculation – Property acquisitions – A developing relationship – Out on strike – Housing boom – New century startups – A state-owned brickyard – An eventful year – A bigger pool – Royal Commission in Victoria – The coming of war

TURNING POINT

On 9 July 1900 Great Britain's 80-year-old monarch, Victoria, gave her assent to an Act of Parliament that would bring Australia's six colonies together as a federal state. Six months later on 1 January 1901 at a ceremony attended by 60,000 people in Centennial Park, Sydney, the Commonwealth of Australia was proclaimed and the first Governor-General, the Earl of Hopetoun, was sworn in.

The birth of a new nation and the death of Queen Victoria just weeks after the gathering at Centennial Park were symbolic, for these two events commenced the long process of loosening ties between parent and colony. The assertion of nationhood was made even more significant several months later in May 1901 when the Duke of York, later King George V, officially opened Parliament at Melbourne's Exhibition Hall. Australia's first government was formed by the leader of the Liberal Protectionists, an urbane Sydney lawyer called Edmund Barton. One of three main parliamentary groupings, Liberal Protectionists governed with the support of John Christian Watson and the Labor Party which held the balance of power in both Houses. A third group led by George Reid was made up of Conservative Free Traders who became the opposition. Common tariffs were enacted, a foreign policy outlined, and the High Court was established. Defence and telegraph services were placed under federal control and immigration laws unified, which led to the introduction of the White Australia policy whereby Chinese people and labourers from the South Sea Islands (Kanakas) in particular were returned to their home. A ban on any further immigration by non-whites remained in force for decades to come.

Longstanding rivalry between Sydney and Melbourne became even more pronounced in the aftermath of federation with arguments over the location of the nation's capital. Brisbane, Perth, Hobart and Adelaide, however, were never serious contenders for the position. On the other hand, neither Sydney nor Melbourne could express the entirety of what it meant to be Australian, and for this reason a separate site – Canberra – was chosen.

South Sea Labour Trade – This cartoon appeared in the Bulletin in 1896. It satirises the practice of using people from the Pacific islands as forced labour. Some Australians can trace their ancestry to islanders brought to far north Queensland during the late 19th century. Artist: Phil May, 1896

FEDERATION HOUSE

A new housing style referred to as 'Federation' quickly established itself in the decade following 1901. Its most notable feature was the introduction of red brickwork, tuck-pointed to achieve a pleasing symmetry, and complemented by a roof of distinctive Marseilles roof tiles.

Moulded terracotta façade, Macquarie Street, Sydney. ***Photograph by Josh Hill***

Federation architecture was at its most influential in Sydney and Melbourne with their expanding middle classes and garden estates. From fashionable suburbs such as Melbourne's Kew and Camberwell, and Sydney's Burwood and Mosman, the Federation style spread throughout rural areas and the nation generally. Country people adapted it to suit local building materials such as timber, corrugated iron and occasionally stone.

Architects who were influential in the development of the Federation style borrowed and modified elements of English Queen Anne, the American Shingle style, and the Arts and Craft Movement which, in the case of the latter, had its origins in England in the mid-1800s. The aforementioned was a reaction against the mass-produced goods of the Industrial Revolution and the intricate and over-elaborate forms of Victorian styles in art, architecture, furniture and design.

One expression of Arts and Crafts, Art Nouveau, was widely influential and can be found in the Paris Metro, as well as in jewellery and posters of the time. Art Nouveau first appeared in Brussels in 1893 and before the end of the century had translated to Australia where radical architects adapted the style in city office and professional buildings. By 1901 Art Nouveau was all the rage. Such European influences when fused with local interpretation resulted in a general movement towards simplification of form and detail, evident in Australian houses from 1910 onwards. Hugh Fraser has written extensively on the Federation style of architecture and notes that such houses were characterised by steeply pitched gabled and hipped roofs covering the entire upper floor. Tall, tapered chimneys often punctuated these terracotta-shingled roofs with their leadlight windows, and off-white roughcast walls with contrasting wooden trim.[1] Melbourne architect Kenneth Milne who moved to Adelaide in 1909 was a local proponent of the style, as was B.J. Waterhouse. Following a career in the New South Wales Government

1. H. Fraser and R. Joyce, *The Federation House: Australia's Own Style*, New Holland, Sydney, 2002, p. 32.

Architect's office, Waterhouse started a successful private practice and built numerous houses on the northern foreshore of Sydney Harbour.

SLACK TRADE

With the turn of the century and the gradual improvement in the economy, demand for building materials began to revive. Unfortunately the casual indifference shown by many brickmakers of the 1880s began to reappear. Brick carting continued to be an ad hoc affair. For example if a builder requested a supply of bricks he was obliged by the yard to send his own vehicle. It was not uncommon to see half a dozen drays queued at opening time waiting for a load of bricks.[2] Furthermore, despite signs of recovery the dead hand of the 1890s depression was still being felt as late as 1904. An article in the *Sydney Daily Telegraph* of that year illustrates the plight of the industry:

> *During the past few months reports have been current of an extraordinary dullness in the brick making industry and of the probable closing down of large establishments in the suburban area. A visit to the yards verified the rumours, for there are millions of bricks stacked for which there is little or no demand. Within a few weeks, unless things improve, a great number of the larger brickworks will have practically suspended business, throwing a large number of men out of employment.*
>
> *'Never during my fifty years of brick making', said the manager of one of the largest brickyards 'has there been such a slackness of trade.'*[3]

A severe drought compounded the problems, preventing the economy from making a full recovery. Brickmakers placed responsibility for the depressed state of the trade on the Commonwealth Government's Arbitration Act of 1904 which determined wages on a national scale. Rupert Cook, the owner of a large works at Enfield and Marrickville, claimed:

> *The present Government has so interfered with trade and labour and individual liberty that people will not invest their money here. The Arbitration Act has been the great curse of the land. One man I know, who used to pay thousands of pounds in wages, closed down directly the Arbitration Act came into force. All my life I have taken part in ameliorating the condition of my workmen. Many years ago I was instrumental in getting the working hours of brick makers reduced, but when I am interfered with and tyrannised over, and told what wages I shall pay, and whom I shall employ, rather than submit I will clear out of the State. I have been 60 years in the trade, here and in England, and never have I seen the industry in such a state as now.*[4]

This early piece of Commonwealth legislation established a framework for industrial and workplace relations that was to endure into the early years of the twenty-first century. Yet, for brickmakers, and no doubt many other employers, there was indignation at what they saw as unwarranted interference in the relationship between themselves and their workers. In fact, the issue of government intervention in matters of trade and industry became highly contentious with the opening of the State Brickworks in New South Wales in 1912.

BIRTH OF AUSTRAL

Had it not been for the arrival of a new player, The Austral Brick Company Limited, the year 1908 might otherwise have gone down in the history of brickmaking as being somewhat uneventful. In due course the firm established itself as an aggressive, dominant player in the Sydney brick industry, and later, at a national level, became one of the largest manufacturers of brick and clay products in Australia. But why is it that of all the new startup companies during this period, the name Austral Bricks stands out? To answer this question we must go back to its point of origin.

2. W. Gemmell, *And So We Graft from Six to Six: the Brickmakers of New South Wales*, Angus & Robertson, Sydney, 1986, p. 15.
3. *Sydney Daily Telegraph*, 1904, cited in *The Clay Products Journal of Australia*, 1 October 1935, p. 23.
4. A. Dawes, 'Early Brickmaking in Australia, 'Slump' of the 'Nineties, and After', *The Clay Products Journal of Australia*, 2 December 1935, p. 22.

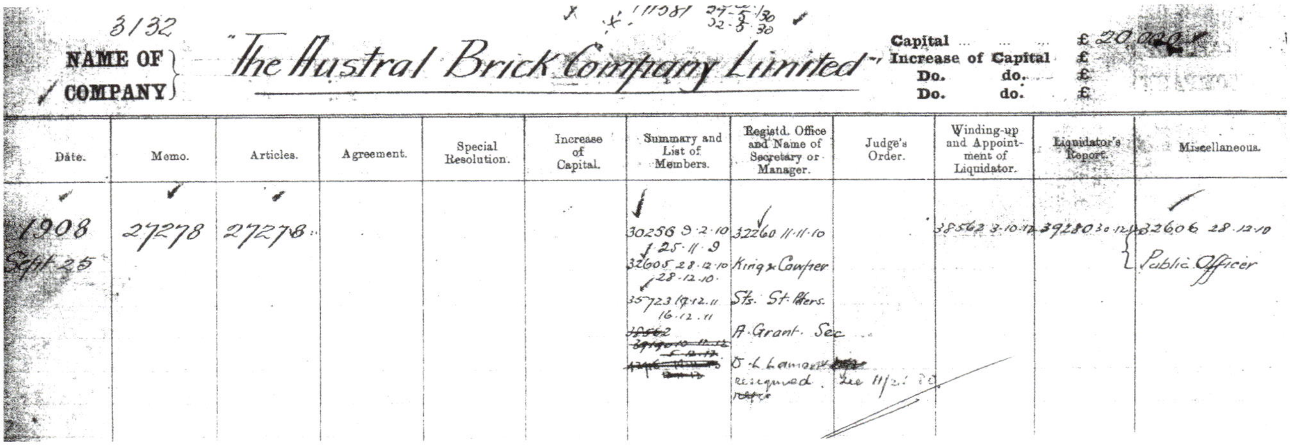

3132

NAME OF COMPANY } The Austral Brick Company Limited

Capital ... £ 20,000

Increase of Capital £

Do. do. £

Do. do. £

Date.	Memo.	Articles.	Agreement.	Special Resolution.	Increase of Capital.	Summary and List of Members.	Registd. Office and Name of Secretary or Manager.	Judge's Order.	Winding-up and Appointment of Liquidator.	Liquidator's Report.	Miscellaneous.
1908 Sept 25	27278	27278				30256 3.2.10 / 25.11.9 / 32605 28.12.10 / 28.12.10 / 35723 19.12.11 / 16.12.11	32260 11.11.10 King & Cowper Sts. St Peters. A. Grant. Sec		38562 3.10.12	39280 30.12	32606 28.12.10 Public Officer

The Austral Brick Company Limited, registered in September 1908. State Records of New South Wales

The story of Austral Bricks begins in the first decade of the twentieth century when the Commonwealth of Australia was still in its infancy. In its 1910 edition Sands' Directory features an advertisement for the Austral Brick Company Limited, although this tells us nothing about when it was established. As for the name, Austral derived from a straightforward anglicising of the Latin adjective 'australis' meaning 'southern'. Perhaps the name was simply a deft piece of marketing intended to position the company as a home-grown Australian company.

The Register of Companies for the period 1905-1910 records that the Austral Brick Company Limited was incorporated on 25 September 1908 with a capital of £20,000.[5] Its registered office, a small nondescript brick building not much bigger than a typical garage attached to a suburban home, was on the corner of King and Cowper Street, St Peters. The new arrival was a public company, although there is no record of it having been listed on the Sydney Stock Exchange. It might well have been some form of partnership comprising no more than three or four people. Business historian Diane Hutchinson notes that corporate regulation at this time resided with individual colonies and from 1901 with individual states.[6] Disclosure requirements in New South Wales were considerably weaker than those adopted by Victoria from 1896.[7] Moreover,

ANTHONY HORDERNS'—THE SHOP FOR THE CLASSES.

Bri TRADES AND PROFESSIONS. Bro 1495

Dean Henry and Son, Constitution rd, Petersham
Dive Samuel, senr., Long Bay rd, Botany
Flemington Brick, Tile and Pottery Proprietary Co., Ltd., 107 Pitt st
Gentle J., Bedford Brick Works, King st, St. Peters. Tel. 44 Newtown
Hart and Gallagher, 361-365 Canterbury New rd, Petersham
Hodkinson and Wright, 23 Norton st, Manly
Metropolitan Brick Co., Ltd.—W. H. Nicholls, J.P., manager, 12 Castlereagh st. Tel. 4009
Sims James, Brixton rd, Rookwood

BRICKWORKS.

See also Brick Manufacturers.

Auburn Brick, Tile, and Pottery Co., Prince's rd, Auburn

AUSTRAL BRICK CO. LIMITED,
Manufacturers of all classes of Bricks, King and Cowper sts, St. Peters. Tel. 609 Newtown

BAKEWELL BROS.,
COULSON ST., ERSKINEVILLE
(NEAR SYDNEY).
Manufacturers of Chemical and Sanitary Goods and Potteryware of all descriptions. Floor and Wall Tiles.
Telephone, 19 Newtown.

Bedford Brickworks—Josiah Gentle, proprietor, King st, St. Peters. Tel. 44 Newtown
Beulah Brickworks, Mitchell rd, Alexandria

BUTCHER BROS, & CO,
Manufacturers of every description of Bricks, Hotham Parade, Gore Hill, North Sydney. Tel. 198 Chatswood

CARRINGTON STEAM BRICK CO., LTD.—Arthur H. Penn, Secretary, Barwon Park rd, St. Peters. Telephone 125 Newtown
Cook Rupert, Burwood rd and Mitchell st, Enfield
Croydon Steam Brick Co., Limited—R. W. Davies, sec., Webb st, Croydon
Dive Samuel, Long Bay rd, Botany
EXCELSIOR BRICKWORKS—W. J. Downton, Proprietor; Thomas J. Joiner, manager, 7 Church st west, Croydon. Tel. 86 Ashfield
Federal Brick Co., Ltd., steam brickworks, Mitchell rd, Alexandria. Tel. 277 Newtown
Federal Brick Co. (Carlton branch)—H. Woodley, manager, Forest rd, H'ville
Fowler R., pottery and brickworks (estab 1837), George st, Camperdown. Tel. 22 Newtown. And Macquarie st, Parramatta. Tel. 310 Parramatta. And Darby st, Newcastle. Tel. 517 Newcastle. (see advt. opp. Potteries)

GENTLE J.
Manufacturer of Every Description of Bricks. Bedford Brickworks: King st, St. Peters. Tel. 44 Newtown

GOODLET AND SMITH LIMITED.
767 Harris st—(See Advt. on Front Pages of Directory)

Hurstville Steam Brick Works—Harold A. Judd, manager, Judd st, Mortdale
Illawarra Fireclay and Brick Co., Ltd., 2 Bridge st
Kuring-gai Brick and Pottery works, 23 Norton st, Manly
Liverpool Steam Brick Works Co., Ltd.—G. S. Travers, jun., sec., 12 Castlereagh street. Tel. 1778 City
Metropolitan Brick Co., Ltd.—W. H. Nicholls, J.P., manager, 12 Castlereagh st. Tel. 4009
Newtown Steam Brickworks—S. Speare, John and Goodsell sts, Newtown. Tel. 150 Newtown

NORTH SYDNEY BRICK AND TILE CO.
J. B. Magney, H. O. Weynton, R. R. Lancely, Proprietors, Gore Hill, North Sydney. Tel. 222 North Sydney

PETERSHAM BRICKWORKS (The)—Hy. Dean and Son, Constitution Road, Petersham. Tel. 182, Petersham
Standsure Brickworks—T. Daley, proprietor, Sydenham rd, Marrickville
Strathfield and Enfield Steam Brick Works, Water st, Enfield
Sydney Brickworks—Jas. H. Charlesworth, manager, Unwin's Bridge rd, M'ville

ST. PETERS BRICK CO., LTD. (THE)
Barwon Park rd, Alexandria. Tel. 86 Newtown. W. Edwards, Manager, 31 Pearl st, Newtown. Tel. 366 N'town

THE N.S.W. BRICK CO. LIMITED,
Huntley st, Alexandria, Sydney, Manufacturers of every description of Bricks. Tel. 412, Newtown

Warren Brick Co.—F. C. Curlewis, manager, Mitchell rd, Alexandria. Tel. No. 135 Newtown
Waterloo Brick Co., Ltd.—James Edwards, manager, Botany rd, Waterloo
"Wilson's," Herbert st, Gore Hill

BROKERS (GENERAL).

See also respective headings under which they trade

Alexander D. S., 44 Carrington st
Arnold G. and Co., 19b Sussex st
Bauer and Co., 108 Pitt st
Baumberg A. H., 117 Pitt st
Booker J., Vickery's chambers, 84 Pitt st
Bryant F. B., 108 Pitt st
Cadden Robert and Co., 64 Pitt st
Chard W. H. and Co., 8 Young st
Craig C. H. and Co, 161 Sussex st
Douglas Percy and Co., 18 Barrack st
Dowling Neville, 18 Bridge st
Fraser, Uther and Co., Ltd., 8 Spring st
Fuller E. T. A., Woolbrokers' chambers, Young st
Godwin G. W., 60 Castlereagh st
Hagen F., 182 Pitt st
Ham and Owen, 17 Bridge st
Hanson W. G., 5 Bond st
Harper W. J. and Co., 114a Pitt st
Houlder Brothers and Co., Ltd., Exchange Corner, 63 Pitt st
Houston William, 9 Bridge st
Johnson N. A., 16 Pitt st
Johnston Gordon G., 36 Moore st
Joseph H., 158 Pitt st
Lawler J. T. and Co., 141 Sussex st
McDonald Robert and Co., 18 Barrack st
McGregor P., 141 Sussex st
McNall Harold H., 109 Pitt st
Macnamara J. P., 157 Sussex st
Madden D. A. and Co., 19 Bridge st
Madden and Hixson, 19 Bridge st
Michael Dick, Norwich chambers, 56 Hunter st
Muir D. R., 34 Pitt st
Newell W. J. and Co., 170 Sussex st
Niccol C., 287 Clarence st
Oelrichs and Martin, 113 Sussex st
Paine Warren A., 82 Pitt st
Peisley G., 38 Elizabeth st
Phillips and Barker, 58 Pitt st
Powick and Co., 93 Pitt st
Roberts Wynn, 18 Bond st
Tournier Charles, 25a Pitt st
Triggs H. J., 158 Pitt st
Walker R. W., 20 Loftus st
William John S., 55 Pitt st
Wood Reuben and Wood, Ltd., E. S. and A Bank buildings, corner George and King sts
Wright and Bruce, Ltd., 20 Loftus st

BRONZERS.

WILLINGTON BROS.
587 George st.—(See advt. opp. Electroplaters)

BROOM AND BRUSHMAKERS.

Campbell and Kinsella, 367 Pitt st
City Brush Factory, 62 Regent st
Federal Broom Co., 12 Fred st, Leichhardt
Grubmeier H., 72 Foucart st, Rozelle
Hamilton and Co. (Lond.), 72 King st
Harriott Bros., Annesley st, Leichhardt
Hippit Charles T., Portland st, Waterloo
Industrial Blind Institution, Boomerang st
Jessop Charles, 3 Little Queen st
McMahon Terence, 37 Smith st
MOXHAM W. J. AND CO., Commission, General Merchants and Millet Broom Manufacturers, 85 Clarence st, Sydney. Tel. 139 Central and 255 Redfern
Palmer Bros., 21 Shepherd st
Piper William, 91 Campbell st, Newtown
Schofield Brush Co., 3 Prospect st, Pad'ton

SMITH JOHN AND CO.
City Broom Factory, 58 Marlborough st, Surry Hills

Stanford and Pilley, McClelland place
Sydney Broom Manufacturing Co., 85 Clarence st
Taylor Samuel, 5-7 Dalley st
Windeler H., 35 Castlereagh st, Redfern
Wyld Alfred, 2 Buckland st
Wyld H. J. and Co., 89 Francis st, Glebe and 2 Franklyn st, Glebe
Zan Bros., Moore st, Drummoyne

WOOD AND COMPANY CONDUCT HIGH-GRADE FUNERAL WORK 10% LOWER THAN OTHER FIRMS.

Sands' Directory. The first mention of Austral Brick Company on King Street at St Peters is made in the 1910 edition of Sands. No 485

5. Register of Companies, 1905–1910, State Records of New South Wales, 11/11784.

6. D. Hutchinson and P. Lee, 'Disclosure, Imperfect Information and the Development of an Equity Market', Paper presented at the Fourth Accounting History International Conference, Braga 2005.

7. New South Wales enacted its Companies Act in 1874. This Act was broadly modelled on the English joint stock acts which regulated the incorporation of limited liability companies. Hutchinson and Lee in 'Disclosure, Imperfect Information and the Development of an Equity Market' (2005) state that the Australian colonies departed from the English model. Companies registered in New South Wales were able to adopt modified articles with substantially fewer disclosure requirements, a departure from English custom where auditing and accounting standards were more stringent.

Edwin Spackman (far left) – Sydney historian and photographer, Ernest Alfred Crome (1902–1987), took this photograph during the late 1920s. Crome knew Spackman better than many others. He was a frequent visitor to Weeroona, Spackman's Attunga Road property on Yowie Bay. Spackman built the cottage as a holiday home and boatshed on land opposite Deer Park in 1911. A close relationship existed between the Dawes family and Edwin Spackman. The brickmaker, Naasson Dawes, is on the right of the picture holding a baby. Compare this image with his portrait inset. Ross Bell Collection

there may have been no distinction between a company registered as 'Proprietary Limited' and one registered as 'Limited'.

A MATTER FOR SPECULATION

Papers that can identify beyond doubt the men behind the new company no longer exist, but according to Ray Dawes whose father, William King Dawes, was managing director of the company for over 50 years, Austral Brick was the creation of a group of prominent Sydney builders. Clearly, the new venture brought together people who had money to invest, knew something about brickmaking, and were most probably involved in Sydney's building industry. Former employees who joined the company in the 1930s believe that Edwin 'Ted' Spackman was the driving force behind the new company, and also its first managing director. Born in 1873 Spackman started out as a surveyor and in due course became a successful builder. By the turn of the century he was already a wealthy man, living in Silver Street, St Peters, and the owner of a string of residential and commercial properties, including a retail emporium. Spackman came from a large family and one of his brothers, a builder whose fortunes were in a constant state of flux, lived opposite the Austral yard. Like other men engaged in the building trade Spackman was also an active freemason, and in March 1903 had joined Lodge Manoah 41 which met at the Masonic Centre, Castlereagh Street. Curiously, his occupation is listed as being 'Iron Manager'.[8] Spackman cut an impressive figure and had the appearance of an English gent – tall, pucker, impeccably dressed with a penchant for butterfly collars, and sporting a neatly clipped moustache. Fastidious in all his ways, Spackman was also a stalwart of the May Street Methodist church at St Peters and is remembered by those who knew him as a devout, although somewhat austere man of German ancestry. He was a member of the local council serving as a representative for the Brompton

8. Spackman was made a Master Mason on 21 January 1904. He left that lodge on 17 June 1935 and did not rejoin freemasonry. Information supplied by kind permission of the United Grand Lodge of New South Wales and Australian Capital Territory.

Edwin Spackman, circa 1939. Ross Bell Collection

Ward and as Mayor of St Peters from 1903 to 1907. In a letter to the Council dated 31 December 1911 Spackman resigned as alderman, probably due to his heavy work commitments in the brick company, or perhaps he felt there may have been a conflict of interest.

Spackman's knowledge of the local real estate market was second to none and he was instrumental in acquiring various properties on King Street, and later at Canal Road, St Peters.

Another player – and a close associate of Spackman – was the delightfully named Ernest Alfred Turtle who operated a real estate agency in King Street, Newtown. Over the years Turtle purchased a great deal of land for Austral.[9] Bruce Parks acquired the agency during the early 1930s, becoming deeply involved in land and property acquisitions on behalf of Austral Bricks and later, Brickworks Limited. Yet another investor was the builder, W.C. Turland, who is later named as company secretary in correspondence that passed between the firm and St Peters Council. His immediate predecessor in this position seems to have been Mr A. Grant according to records held by State Records of New South Wales. Yet, Austral was small fry and just another operation among many others at this time. Up against the Bedford Brick Works on the corner of Mitchell Road and King Street, it hardly rated a mention.

PROPERTY ACQUISITIONS

Over the next few months Spackman and his associates pushed ahead with plans to establish a business, purchasing two properties in December 1908. Both were situated on land that held plentiful reserves of excellent clays and shales. The first of these was a Georgian-style colonial house, *Petersleigh*, owned by Mrs Alice Jane Boyd whose husband Thomas had died earlier the same year. Located at 238 King Street the house, various buildings including stables and land occupied 12 acres and were bounded by the present Cowper Street and Canal Road. The entrance gates of *Petersleigh* were directly opposite Edith Street. Overlooking Botany Bay, views from the house must have been stunning. The St Peters Rate Book valued Alice Boyd's property at £5,000 with rates of £69 17s 5d paid for the half year to July 1908, a handsome sum of money indicating the extent of acreage and number of buildings.[10]

Petersleigh

Petersleigh was built in the 1830s by Henry Kerrison James who had emigrated from England along with Alexander Brodie Spark. James and Spark were friends and supporters of Sir Charles Cowper, the Premier of New South Wales from 1856. The merchant and customs agent, Michael Metcalfe, was the next to own *Petersleigh* and he added a wing to accommodate his large family of four boys and four girls. *Petersleigh* had extensive stabling, and a well on the property provided fresh water. Canal Road separated the estate from *Heathcote*, the residence of Thomas Chalder. Metcalfe died at *Petersleigh* in 1890 leaving a substantial estate to his wife, Agnes Georgianna, whom he had married in 1845. Thomas Boyd acquired the property on her death and lived there with his wife, Alice J. Boyd, until his death in 1908.

9. Minute Book of St Peters Municipality, A95/141.9, the County of Cumberland, Parish of St Peters, North Ward. Valuation Book for the Triennial Period ending 31 December 1913.

10. Minute Book of St Peters Municipality, A95/141.9, the County of Cumberland, Parish of St Peters, North Ward. Valuation Book for the Triennial Period commencing January 1st 1908.

Meanwhile, a short stroll away at 196 King Street lived Henry Stephen Bond whose property occupied three acres of prime land valued at £1,200.[11] Bond's home, *Finningham*, between Cowper Street and *Petersleigh*, had been erected in the late 1860s. Like *Petersleigh*, it was set back from the Princes Highway and the two houses were in alignment with each other. At some point before its construction a racecourse may even have existed on the site. According to E.G. Boden who wrote a short piece on the Tempe Wesleyan Chapel:

> *On the opposite side of Ricketty Street, and occupying the site of the Austral Brick Works, and extending down the flats to the canal (then a muddy creek), was the Barwon Park Racecourse. In its day this was a very popular course. Being rather small and an irregular egg-shape, it was only used for about 10 years.* [12]

Henry Bond must have received an offer he could not refuse, because he sold up about the same time as Alice Boyd and moved elsewhere. By January 1909 both properties had been formally transferred into the hands of their new owner, and in June they were combined as one. The Valuation Book for the three-year period commencing 1 January 1908 clearly shows the names of Boyd and Bond which had been entered when they were owner-occupiers. A neat line has been ruled through their record and a handwritten note appears, that of:

> *Austral Brick Company. Now combined as one assessment. Brickworks and Land. About 16 acres.*[13]

The purchase of these two properties was sufficiently important for the *Sydney Morning Herald* to report in October 1910 that *Petersleigh* and *Finningham* were now owned by the Austral Brick Company.[14]

The year 1908 that witnessed the establishment of the Austral Brick Company also marked the expansion

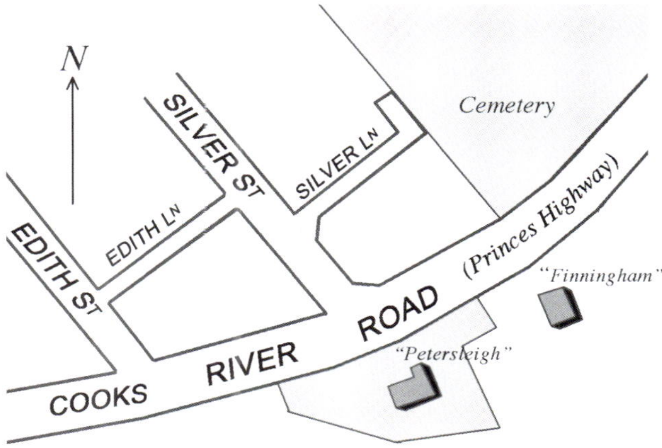

Location of Petersleigh and Finningham in relation to Cook's River Road. No photographs survive of either of these two mid-Victorian buildings.

of Wunderlich, the tile manufacturer mentioned in the previous chapter. Likewise, the engineer, George Foster, commenced operations in King Street, St Peters, in 1908. Situated directly opposite Gentle's yard, Foster's engineering business grew rapidly and was well patronised by local brickmakers. For many years Foster had been contracted by the English firm of Platt Brothers to maintain brickmaking equipment exported to the colonies. This arrangement appears to have ceased abruptly on the outbreak of war, at which point Foster developed grinding pans and dry press machines of his own design.

With the onset of the First World War in 1914 George Foster even went into the manufacture of tile-making machinery.[15] The Newbold Silica Fire-Brick Company also made its entrance in 1908. Founded by Fred and Arthur Newbold, their Lithgow plant became the first company in Australia, and its most significant, to produce silica firebricks.[16] In fact, the Lithgow factory was set up in close proximity to the only functioning iron and steel furnaces in the country at that time. There was a constant demand for firebricks (also referred to as refractory bricks) made from a special type of clay. Such bricks were used to line the interior of furnaces and kilns to retain the heat and also to prevent the disintegration of the kiln itself which was constructed using common bricks.

11. Ibid.
12. E.G. Boden, *Tempe Wesleyan Chapel 1855–1939: 80-Year Romance of a Small Suburban Mission*, The Australasian Methodist Historical Society, *Journal and Proceedings*, Vol. VIII – Part 2, January 1940, p. 376.
13. A reference in the remarks to 'Petersleigh' App 2100, Volume 80, Folio 21, January 29, 1869. Lands Title Office. Minute Book of St Peters Municipality, A95/141.9, the County of Cumberland, Parish of St Peters, North Ward.
14. *Sydney Morning Herald*, 20 October 1910, p. 7.
15. *The Clay Products Journal of Australia*, October 1934, p. 11.
16. The Newbold Silica Fire-Brick Co, Company Catalogue, 1918, p. 5.

St Peters Railway station, 1907. The chimneys of the Bedford Brickworks are visible in the background. Robert Hutchinson Collection

Foster dry brick press

A DEVELOPING RELATIONSHIP

A flurry of activity now took place at Austral's Cowper Street office. Having acquired the necessary land to erect a brickmaking operation, a letter from W.C. Turland, the company secretary, giving notice of Austral's intention to build a plant on the site, was read at the St Peters Council meeting held on 28 December 1908.[17] Thereafter more correspondence followed with a letter in January 1909 asking the company to consider ceding 20 feet along their frontage for the widening of the road.[18] A polite note in reply advised that every consideration would be given to the request, although the company would like to see the plans first.[19] Another round of correspondence in April 1909 follows with Austral seeking permission to 'use explosives at their new works in King Street.'[20] Of course, came the reply, there would be no objection provided they were used in reasonable quantities, all proper precautions be taken in their use, and the interests

17. Minute Book of St Peters Municipality, A95/141.9, the County of Cumberland, Parish of St Peters, North Ward. Rate Book for year ending 31 December 1908. W.C. Turland is also mentioned in the Rate Book for year ending 31 December 1910.
18. Ibid, the Rate Book for year ending 31 December 1909: 25 January 1909.
19. *Sydney Morning Herald*, 8 February 1909.
20. *Sydney Morning Herald*, 19 April 1909.

George Foster & Sons, St Peters, circa 1914. Mitchell Library, State Library of New South Wales

of the general public were not jeopardised in any way.[21] Unfortunately, the yard foreman may have been over-enthusiastic in the use of dynamite, because complaints soon began to reach the council over blasting activities at the yard. Damage to the public thoroughfare was noted, and an apologetic Austral Brick Company reassured the council that the employee responsible for the disruption would be swiftly dismissed at the conclusion of his present shift – and that it would not happen again. The culprit may have been John Aske, the plant manager, for his tenure appears to have been short-lived. In the same year he was replaced by David R. Rogers who is listed after 1912 as an investor and director, a position he retained for many years.[22] Rogers' appointment as manager, and the fact that he lived several doors away from Spackman in Silver Street, St Peters, was no coincidence. Clearly, the two men went back a long way.

Throughout 1909 building work continued without interruption, including the construction of kilns, workshops, a chimney and the outline of excavations in the brickpit itself. Whether acquired secondhand, purchased overseas or manufactured by local engineering firms, machinery and equipment were installed. These included dry press brick machines supplied by Platt Brothers in England, as well as grinding pans to crush the hard shales. Overseas shipments would have arrived either at Circular Quay or at one of the nearby wharves. It would then have been despatched by dray to Central Station where it was freighted to St Peters, not far from the Austral yard. Shipments arriving by rail freight at St Peters station would have been hauled into the yards by horse or bullock teams. Whatever their means of transport, the passage of heavy machinery down King Street to St Peters must have been an unusual sight, although this was not recorded in local newspapers.

In July a letter was sent to the council intimating that the company's efforts to overcome the smoke nuisance from the engine chimney had failed and the installation of a gas suction plant was being pursued with all possible speed. The letter also asked that any action involved with the matter be deferred until this work was complete.[23] This is a clear indication that brick production was in full swing by the winter of 1909, which is surprising in view of the fact that land had only been acquired in December of the previous year. Entry points to the site had become a pressing matter and over the next few months The Austral Brick Company made a number of urgent requests for improvements in road access.[24]

21. Ibid.
22. Sands' Directory, 1910.
23. *Sydney Morning Herald*, 26 July 1909.
24. Ibid.

Patent kiln under construction, circa 1910. Patent kiln (location unknown) showing wooden barrel vaulting in place prior to the construction of the firing chambers. This kiln would have been very similar to those built at the Austral yard, St Peters. Robert Hutchison Collection

OUT ON STRIKE

The year 1909 ended with a brick strike that started in late November, dragging on through December and into early January. The stoppage was extensively reported in the Sydney newspapers.

> *Sir, in the building column of your issue for November 30 there appears a statement that the supply of bricks for the additions to the Trades Hall is plentiful. I am under the impression that no common bricks had been delivered to the Trades Hall in the last 9 days and that until the strike ends there is little probability of any being delivered.*

In an obvious reference to the lack of progress being made on the Trades Hall job, the writer, clearly not a union sympathiser, then goes on to say:

> *Of course, it may be that under the system of work employed at the Trades Hall a comparatively small number of bricks will last quite a long time.*[25]

Six weeks later, on 11 January 1910, another letter in the *Sydney Morning Herald* laments a complete standstill in the industry.

> *Not a brick in any of the yards; empty scaffolds in the various jobs in the course of progress; bricklayers idle everywhere, while in the architects' offices are to be found unsigned contracts and plans for buildings not yet tendered for as well as orders for drawings which cannot be prepared, while there is so much uncertainty about the supply of bricks. That is the position today with respect to the building trades of the metropolis.*[26]

Things went from bad to worse and during April and May the brick workers' strike was exacerbated by a coal famine brought about by miners in the State coalmines on the south coast who had also walked off the job. Like any other brick manufacturer, Austral was affected by the disruption to production.

25. *Sydney Morning Herald*, 1 December 1909, p. 7, 'Bricks for Trades Hall', letter to the editor.

26. *Sydney Morning Herald*, 11 January 1910, p. 10, 'Mr. Henley MLA', letter to the editor.

By June the effects of the strike were beginning to bite. On 27 June, the *Sydney Morning Herald* published a letter from the parliamentarian, Mr Henley MLA, lamenting the sorry state of affairs. 'Yards were 50 million [bricks] short of orders now in hand and it would take three months to pull up back orders going at full pressure'. Henley may have been connected with the brick industry in some way, for he appears to be quite knowledgeable about the subject. 'With the return of coal supplies following the end of the strike', he claimed, 'it would take three weeks after the kilns are lighted before bricks can be sent out to the different jobs. The coal famine has put back the building boom of Sydney for many months.' [27]

HOUSING BOOM

Eventually the various strikes were settled and the men went back to work, although a strong suspicion took hold that a combine of brick manufacturers was operating to the detriment of the public. Conspiracy theories were given added weight when the NSW Brick Masters' Association increased the price of bricks from 38s to 42s per thousand at the kiln. Manufacturers in other states followed suit and raised their prices. The timing of this should not be overlooked. Immediately before, on 1 June 1910, striking workers had been granted a pay rise by the New South Wales Wages Board.[28] Producers covered their increased costs by passing them on to the customer, provoking an indignant response by builders and others in the trade. Public anger was vented in the press, which carried articles attacking the so-called 'brick combine'. So widespread was the feeling that brickmakers were holding the public to ransom, that those in government believed that the state must take action. In a published statement, Andrew Fisher, the Labor Party Prime Minister who had taken office in April 1910, announced that all trusts and combines would be dealt with by the federal government. 'We shall do our duty,' he declared.[29] And this is exactly what happened, for several months later the New South Wales government voted to establish a state-run brickworks at Homebush.

Fears of price fixing and anti-competitive activity echoed similar concerns held by many in the United States. As early as 1890, Americans were becoming increasingly alarmed at the spread of trusts and large industrial combinations which had come to exercise monopolistic control over the market. President Theodore Roosevelt used the Supreme Court to establish the right of the federal government to regulate industry, and introduced his so-called Square Deal which saw a number of Bills enforcing anti-trust laws pass through Congress. This riled many industrialists, especially the rail magnate, J.P. Morgan, who attacked Roosevelt's plans.

Despite the furore in the wake of the brick strike and price rises, the pre-war years in New South Wales, and Australia generally, witnessed growing prosperity and a boom in house building. Driving this were the offspring of migrants who had settled in Australia during the 1880s. In addition an increasing number of more recent immigrants (over 97,000 in 1912 alone) was feeding the property market.

PLANNING SYDNEY

As the boom gathered pace there was also a move by state governments to provide low-cost land and housing on terms that even the poorest worker could afford. This was a noble thought, although home ownership for the masses was still decades away. Planning for a 'Greater Sydney' was a vital requirement if the city was to accommodate its growing population. In 1900 two conferences were held, attended by local politicians and civil servants. Little came of the discussions but the subject was revived in 1906 when Premier Sir Joseph Carruthers, a champion of local government, introduced his Local Government Act (Town Planning). The successful passage of the Act through State Parliament rationalised the system of local government in New South Wales and established the major principles of a modern planning system. Underpinning this apparent victory for commonsense was a fierce, and protracted, debate pitting the centralists who

27. *Sydney Morning Herald*, 27 June 1910, p. 4.
28. A. Dawes, 'Early Brickmaking in Australia 'Slump' of the 'Nineties, and After', *The Clay Products Journal of Australia*, 2 December 1935, p. 22.
29. *The Clay Products Journal of Australia*, 1 January 1936, pp. 24–25.

Nielsen Park, 2007. Photograph by Josh Hill

believed in a Greater Sydney unified through central planning powers wielded by a City Council, against the federalists – municipal councils acting together. The importance of the debate should not be underestimated and was the subject of intense interest among newspapers, professional commentators, and politicians of the day. Journals such as *The Bulletin, Building* (the journal of the Master Builders), and *Art and Architecture* actively promoted the cause of town planning which resulted in the appointment of a Royal Commission into the Improvement of Sydney. Unfortunately, the general public failed to recognise the link between the need for civic reform and urban renewal. The centralists lost out.

Carruther's successor, Premier Wade, announced the commission on 27 February 1908, but it was some months before the terms of reference and personnel were agreed. The newly appointed Lord Mayor of Sydney, Thomas Hughes, presided over the enquiry and evidence was received from local experts, pressure groups and from overseas authorities. The findings were issued as a Final Report in June 1909 and numerous recommendations made including major improvements to the road, rail and tram systems. Slum clearance and the acquisition of land for workers' housing in the suburbs found particular favour and the commission endorsed a Building Act. The city council, architects and builders had pressed strongly for such an Act, hoping to rein in speculative builders, and improve housing standards and health conditions. The Federated Master Builders Association was particularly active on this front. Founded on 19 November 1890 the association was a national body with branches in each of the states. At its tenth convention which met in Sydney on 18 October 1909 and the following days, there was talk of 'rebuilding the City of Sydney'. Two prominent figures, Thomas Loveridge and H.J. Thompson, are recorded as national president and New South Wales president respectively. Apart from being master builders, both men were heavily involved in the brick industry. Thompson presented a paper in which he spoke of the registration of builders and their obligations.

> *In the first place the builder's business is one of the most intricate. He has to deal with materials and men, both of which are always changing ... Beyond all this he has to keep a wary eye on the City Building Act, or the building regulations of any of the forty-one or more municipalities, the Water and Sewerage Board, Gas Company, traffic regulations, and the Scaffolds and Lifts Acts; and if he manages to steer clear of all these, there is the old drag-net, the Police Offences Act, waiting to trap him at every turn.*[30]

30. Proceedings of the Federated Master Builders Association of Australasia, Vol. II, 1909, p. 99.

Practical men who did more than simply erect buildings or construct engineering works, builders resented the extent to which government red tape threatened to make their task almost impossible. They were also fiercely independent conservative-minded entrepreneurs who were suspicious of the growing influence of socialism within the ranks of government.

The Commission's most important contribution was its attempt to develop a consistent overall policy in development control and planning. Niels Nielsen, the Danish-born Minister for Lands in the first Labor government which came to power in New South Wales in 1910, announced plans to resume large tracts of foreshore. Nielsen created Ashton Park, an area encompassing Bradleys Head and later Taronga Park Zoo, the Hermitage Reserve, and Nielsen Park, a popular harbourside bathing and picnic spot that was on the ferry route between Circular Quay and Watsons Bay.

THE 'GARDEN SUBURB'

As the debate on town planning was being aired publicly, the concept of a garden city made its appearance. The idea emerged in Britain in the late 1890s and came to influence the development of a number of suburbs in Sydney after 1910. Campbell Carmichael and John Rowland Dacey, both members of the state government, introduced the design to New South Wales, although Dacey had been waging a decade-long campaign for low-cost government housing for working class people. Early examples in Sydney include Daceyville Garden Suburb in Randwick, a model development initiated in 1912, the Matraville Soldier's Garden Village, and Haberfield. Dacey died in 1912 and his vision was partly carried out by the Holman ministries from 1913–1920, under the supervision of J.D. Fitzgerald. William Holman had been voted into office with a vigorous socialist housing policy. Blaming housing troubles on landlords and contractors, including private brick manufacturers, he was determined to eliminate all of these groups. The building of the model suburb of Daceyville disbarred private companies from tendering for any government work. Instead, the Works Department organised construction using day labour, but as in most projects lacking a pre-determined level of expenditure, the Daceyville cottages grew in cost to the point that a frustrated government was obliged to revert to using contractors.

It was about this time that the first few examples of what became known as the 'Bungalow Style' began to appear. In England it was customary for a single storey dwelling to be referred to as a bungalow. Its Australian counterpart adopted a roof with a low-pitched roof, designed intentionally to provide greater protection against the sun through wider eaves. In time these were referred to as American Bungalow or Californian Bungalow types.

NEW CENTURY STARTUPS

Boom conditions from about the year 1910 until the outbreak of the First World War in August 1914 also witnessed the appearance of a number of brick companies. Line entries – often no more than that – in the register of companies attest to the creation of a great many brick enterprises. A swag of them can be traced even further back over a ten-year period commencing in 1900 and culminating in the creation of the State Brickworks in 1911.

Starting in 1900 Henry Woodley, already part owner of the St Peters Steam Brickworks with James Squires, William Judd[31] and William Edwards, established the Federal Brick Company Ltd on May Street at St Peters. Woodley, a founding director, was later joined by the Queensland-born, E.S. Phippard who was also appointed chairman. Phippard was a prominent builder with several major works to his credit in Brisbane, including the Colonial Treasury Building. His company built the Commonwealth Bank in Martin Place, but went bankrupt in the process.[32] Phippard's son, Frederick, an educated man with a Bachelor of Mining Engineering degree from Sydney University, joined the business

31. William Judd was a keen bowls player in his retirement and in 1948 we find reference to him in the *Clay Products Journal of Australia* as a stalwart of the St George No. 1 Pennant team.

32. E.S. Phippard was a keen player of lawn bowls and was President of the Randwick Bowling Club in 1917. His son, Frederick G. Phippard, occupied this position from 1926–1931.

E.S. Phippard, building contractor and director of The Austral Brick Company Limited, circa 1914. Courtesy Tony Phippard

after about 1920 and was proprietor of Federal's Alexandria yard.

A year or so later the *Sydney Morning Herald* published a major feature – 'Continuous Burning Brick Kiln – Lighting a new kiln' – on the opening of the new plant.[33] Naasson Dawes, the firm's general manager, was cheered for his speech which detailed how, despite the Federal having only one kiln and being just 12 months old, the company had been able to turn out over 8 million bricks and pay £10,500 in wages.[34]

> *After 5 years of depression the building trades had looked up wonderfully and he [Dawes] had no doubt that now Federation had been accomplished prosperity would rise to such a height that the firm in common with others would require to keep the machinery constantly in motion in order to compete with the demand.* [35]

Several years after it commenced operations the Federal was incorporated as a limited liability company on 20 July 1906, with a working capital of £90,000. This was increased to £200,000 on 28 August 1912, an indication that things were going well. The plant operated at Mitchell Road, Alexandria, between 1907 and 1947, and by 1918 a branch, referred as the Carlton yard, had been established at Hurstville. The Alexandria yard made bricks for Long Bay Gaol, shipping them out by tram from a siding within the works.

The year 1902 also provided a rich crop of startups. One of these was the Warren Brick Company established by F.C. Curlewis and located on Mitchell Road. The company temporarily ceased operations in 1911, reopened several years later then closed during the depression. The Jubilee Steam Brickworks on Unwins Bridge Road, St Peters, also made its appearance and was ably managed by Arthur Charlesworth from 1902 to 1919.

Charlesworth was a mover and shaker in the local brick industry and was involved in the New South Brick Company Limited at Huntley Street, Alexandria. The National Brickworks Company Limited opened at Thornleigh in 1902, adjacent to the northern railway line on a property covering seven acres fronting Pennant Hills Road.[36] The first annual meeting was held on 20 April 1903 on the floor of a new type of kiln, the 'Hardy', designed by the son of Edward Blackett, the celebrated church architect. The National produced a unique form of interlocking hollow brick as well as common open-kiln shale bricks, and later expanded its operations into land at nearby Dartford Road.

Over at Gore Hill, two young brick pioneers in their twenties, Frederick Thomas Kirby and his brother-in-law, Charles Butcher, set up a yard at Hotham Parade, Gore Hill. Both men turned the first sods with their picks. Charles Butcher had only one hand, the result of an accident with a horse, although he still managed to work as yard foreman

33. *Sydney Morning Herald*, 22 April 1901, p. 4, 'Continuous Burning Brick Kiln — Lighting a new kiln'.
34. Naasson Dawes may also have had some involvement with the New South Wales Brick Company Pty Ltd.
35. *Sydney Morning Herald*, 22 April 1901, p. 4.
36. Like many other yards adjacent to railway lines, a siding was built in the yard enabling the transport of bricks by flat-top rail cars.

Horse team from the yard of the Butcher Brothers, Artarmon, circa 1914. This early photograph shows Fred Kirby's father, Herbert Stow 'Stan' Kirby, who was a brick carter for the firm with his team of Suffolk Punch horses. The cart had a carrying capacity of about 1,000 bricks. 'Sam' (lead horse), 'Punch' and 'Rowdy' (centre), 'Ginger' (in the shaft). Warwick Gemmell Collection

and manager. Samuel Butcher, his brother, would sit on the edge of the pit watching his workers; anyone not pulling his weight was thrown sixpence and told not to come back. Eventually, the two Butcher brothers took over the business and formed a company that was incorporated on 18 March 1913 with a capital of £143,000.[37] Well respected by his industry peers, Sam Butcher was appointed to the Board of the National Brickworks in 1914. His Artarmon works was demolished in 1951.

NUMBER 12 CASTLEREAGH STREET

Also making its appearance in December 1906 was a company called Builders Exchange Limited, which, for many years, occupied chambers at 12 Castlereagh Street, Sydney. Numerous references to the Exchange can be found in documents relating to the activities of brick companies and building firms who used the premises as a registered office, meeting place, and convenient location from which payments were made to contractors. Certainly, the Exchange was extremely important in the dealings of brick manufacturers, and assumed greater significance over the course of the next two decades. Its company secretary was Bertram 'Bert' Moseley, a prominent player in the brick industry, and about whom we shall discover more in later chapters.

Closely associated with Builders Exchange was the Suburban Land and Investment Company, or SLIC, as it was known. SLIC started life on 8 September 1911 at a registered office at Goodlet Street, Ashfield, named after the brickmaker. Its directors included the builders Alexander Stuart, Edwin Samuel Phippard, Arthur Scott, William Mitchell, and William Stuart (brother of Alexander), as well as an accountant, William Henry Nicholls, and John Charles Waine, a land speculator. Alexander Stuart, managing director of Stuart Brothers, went on to achieve greatness in the building industry. For more than 30 years Stuart was chairman of The NSW Brick Masters' Association until his death in March 1934. He was also a prominent investor in The Austral Brick Company Limited. Born in Scotland in 1859 Stuart emigrated to Australia with his

37. State Records of New South Wales, Register of Companies 1910–1913, Container 11/11785, Clay Industries Pty Limited.

SLIC's Memorandum of Association indicates wide-ranging interests, the object being to:

... purchase, lease and lay out land for building or other purposes and to build on or improve, sell, let or lease, advance money to persons building on and otherwise develop the same in such manner as may seem expedient to advance the Company's interests.

The formation of suburbs under certain rules and conditions as to the class of buildings size or allotments and sanitary arrangements. The erection of houses and buildings on the lands of Company or on lands secured to the Company by Mortgage or otherwise.

To construct, improve, maintain, develop, work, manage, carry out or control any roads, ways, tramways, railways, branches or sidings, bridges, reservoirs, watercourses, wharves, manufactories, warehouses, electric works, shops, stores and other works and conveniences which may seem calculated directly or indirectly to advance the Company's interests ... And last, but not least, to carry on the business of Manufacturers of, and to buy and sell bricks, tiles, brick, earth, etc.[38]

wife, and entered the building industry during the 1880s. In 1899, following the depression years, he left for Britain, Europe and America, but returned to Sydney at the turn of the century, becoming managing director of The NSW Brick Co. Ltd in 1902.

While it did not manufacture in its own name, SLIC seems to have acquired working yards. One such company established in 1913 was Burwood Bricks operating at Cheltenham Road, Croydon, on part of John Dawson's Estate. It was in production in 1914 and for the next six years jostled for space alongside Excelsior Brickworks. Burwood Bricks closed in the 1930s, having supplied the bricks to build many of the houses in the Burwood area during its life.

In August 1914 the company transferred its office to Castlereagh Street, setting up in rooms immediately adjacent to the Builders Exchange, a convenient move as it turned out. For the brickmasters and builders who sat on the board of various SLIC subsidiaries, 12 Castlereagh Street was centrally located and in the same building as most of their other business interests.

A STATE-OWNED BRICKYARD

It was now the turn of the state government to climb on board the brickmaking wagon. In October 1910, Arthur Griffith, New South Wales Minister for Public Works proposed the establishment of a government brickworks that could produce cheap building materials – a move calculated to break the hold of private brick manufacturers. As the

Fossil finds. Lakes in the Sydney Basin during the Triassic Period were inhabited by fish and invertebrates. Many fossil fish have been found in brick pits and shale quarries. This giant salamander-like labyrinthodont amphibian Paracyclotosaurus davidi was found in The Austral Brick Company's pit at St Peters. A mould was taken and the original fossil is now part of the National Science Museum collection, London. National Science Museum

state government was deliberating over whether to proceed with a brick operation, an unusual discovery of dinosaur remains was made in Austral's yard at St Peters. One Mr B. Dunstan of the Geological Survey of New South Wales was alerted to the presence of fragments of enormous ironstone nodules that had been dislodged from the side of the shale beds during blasting. Dunstan reconstructed the nodule which was found to contain the complete fossilised skeleton of a labyrinthodont. Staff of the Australian Museum made perfect replicas and the original nodules were acquired by the British Museum where they still reside.[39] Such finds were not unusual and there were many instances of men uncovering fossils in the course of their work.[40] Most fossils in the brick pits across Sydney were found before 1950

38. Ibid.

39. D.M.S. Watson, *A New Labyrinthodont (Paracyclotosaurus) from the Upper Trias of New South Wales*, Bulletin of the British Museum (Natural History), Geology Vol. 3, 7.

40. One Triassic species of cycad (flora closely related to conifers), *Cycadopteris scolopendrina*, was discovered in a brick pit in the 1880s at St Peters in Sydney. It has not been collected anywhere else.

when pick and shovel were used to break the shale by hand prior to mechanisation.

Meanwhile, back at Parliament House on Macquarie Street, Griffiths' vision of a state brickworks was finding support, especially from the Railway Commissioners, the Harbour Trust, Water and Sewerage Board, and Public Works. Demand from these utilities alone would be sufficient to keep a large brickworks fully occupied. All of this came as no surprise to an executive of the Master Builders' Association, E.H. Buchanan, who dismissed the idea, observing that he did not think that the 'brickmasters would shiver over the Government's statements, although he claimed it was only the first step towards nationalisation.[41]

But first there were financial and political hurdles to overcome. In December 1910 the vote for the brickworks was thrown out of the New South Wales Parliament. Griffith pressed on, instructing his department to investigate sites, as well as plant and production costs. A committee was established consisting of the Under-Secretary for Public Works, the Principal Assistant Government Architect, the engineer-superintendent of the Fitzroy Dock, J. C. Watson, and North Sydney builder S. E. Hutton who was later appointed as the first manager of the brickworks, a position he held until his death in 1926.

To overcome these obstacles, the Labor parliamentary majority voted on a special Act in 1911 which was subsequently passed. Detailed planning commenced in earnest and a 9.5 hectare site was chosen at Homebush Bay. Drilling sample proved the site contained the best shale deposits in the County of Cumberland. The 20 to 40 feet of red clay overlaying the shale was deemed suitable for bricks, tiles, water pipes and other related

41. *Sydney Morning Herald*, 'Brick Masters Won't Shiver' (in relation to Trade Views), 28 October 1910, p. 7.

Summary of startups, 1900-1914

Start dates	Company and location
1900	Federal and Carlton Brickworks, St Peters.
1901	Butcher Brothers, Artarmon.
1902	Warren Brick Company, Alexandria.
1902	Jubilee Steam Brickworks, Unwins Bridge Road, St Peters.
1902–1962	New South Wales Brick Company Limited, Huntley Street, Alexandria.
1902	National Brickworks Company Limited, Normanhurst.
Sep 1905	Auburn Brick Company Ltd.
Aug 1906	Liverpool Steam Brick Works Company Ltd.
Dec 1906	Builders Exchange Limited.
Sep 1906	St Peters Brick Co. Limited (formerly St Peters Patent Dry Press Brick Co, operating from 1908–1950 at Barwon Park Road, Alexandria.
Feb 1907	Adamstown Steam Brick Company Limited.
Mar 1907	Illawarra Fireclay Brick Company Limited.
Sep 1908	Austral Brick Company, operating at King Street, St Peters and on 3 October 1912 was sold to a new company, The Austral Brick Company Limited. Austral Brick Company was wound up at the same time.
Jul 1909	Bondi Brick Company Limited, operating at Bondi.
Nov 1909	Manly Beacon Brick Company Limited.
Jun 1910	Mashman Brothers Limited, founded during 1885 and operating at Kingsford.
Mar 1910	City Brick Company Limited, operating between 1911–1930 at Euston Road, Alexandria.
Oct 1910	Northern Suburbs Brick Company Limited, operating at Gordon Road, Artarmon.
Dec 1910	North Sydney Brick and Tile Company Limited, operating at Artarmon.
Jan 1911	Clyde Brick Company Limited, operating at Clyde and Auburn. The company ceased to operate in 1965.
Mar 1911	South Ashfield Brick and Tile Company Limited.
May 1911	Abermain Brick Company Limited, operating at Main Street, Cessnock.
Nov 1911	Industrial Brick Company Limited operating at Botany Road, Waterloo.
Sep 1911	The Suburban Land and Investment Company Limited.
1912	Speare's Brick and Pipe Works Pty Ltd (formerly Goodsells).
Mar 1912	Gosford Brick Works Limited.
Mar 1912	Great Northern Brick Company Limited operating at Midson Road, Eastwood. Head office listed as 2 Castlereagh Street in June 1935.
April 1912	Parramatta River Brick Tile and Pottery Company Ltd.
Jul 1912	Sutherland Brick Company.
Aug 1912	Hornsby Tile, Pottery and Brick Works Limited.
May 1912	George Foster and Sons Limited operating at King Street, Newtown.
Jun 1912	Enfield Park Brick Company Limited, operating at Enfield and later at Mortdale and Thornton, near Newcastle.
Feb 1913	The Ashfield Brick Company.
May 1913	Bakewell Brothers, operating at Mitchell Road, Alexandria.
Mar 1913	Butcher Brothers and Company Limited, operating at Hotham Parade, Gore Hill.
1914	Ryde Brick and Tile Works Limited, operating at College Street, Gladesville.
1915	Central Brick and Tile Company Limited, commenced operations at Barwon Park Estate from 1919. Moved to St Peters after 1920. Ceased to operate in 1965.

Source: State Records, Container 11/11785, Register of Companies 1910–1913

products. Construction work pressed ahead and on 15 September 1911 Premier McGowen lit the first permanent Hoffman kiln. Before long smoke was issuing from its 150 foot high smoke stack. The kiln, one of three to be built, had 18 chambers each capable of holding 20,000 bricks. Firing on coal and sometimes wood, excellent dry press bricks were produced on machines imported from England.

Due to poor road conditions across Sydney the first production was sent by water barge to a depot at Blackwattle Bay. From here they were transhipped by horse-drawn carts or solid rubber-tyred motor trucks. At a cost of £2 10s per thousand delivered to the site, bricks were very competitive with other building materials. Most of the early production was sent to government projects. The first 100,000 from 1911 may still be seen today in the Marrickville storm water channel, a power station at Homebush Bay, housing in Gloucester Street, Sydney, and the government agricultural farm at Mulgrave.

With the increase in output surplus bricks were sold to the public. By November 1911 the need to expand the plant was recognised with plans to acquire an additional 23 acres of land. Arrangements were made for the acquisition of: five additional Hoffman kilns with a capacity of 140,000 bricks each per week; six open kilns (150,000 bricks per week); nine grinding pans (sufficient to produce 800 bricks per hour); 15 brick presses (10,000 bricks per day); two sets of extra friction and winding gear; and nine sets of elevators, as well as two extra gantries and bridges. It was an impressive complement of working capital.

AN EVENTFUL YEAR

The year 1912 began like any other for St Peters Municipal Council, with the usual run of matters to deal with, both residential and commercial. In a letter dated 20 February Mr Delph Singh wrote to the council complaining about the problem of straying cattle on his land at Tempe, and the need for a pound in the district. The reply, short and to the point, advised Mr Singh that the nuisance was for him alone to sort out.[42] How the situation resolved itself is not known, but the presence of straying cattle tells us a great deal about the semi-rural nature of this part of Sydney at the time.

Accounts submitted for approval to pay Austral Brick Company, April to September 1912

29 April 1912	15s
13 May 1912	10s 6d
26 June 1912	£25 10s
8 July 1912	£26 15s 6d
22 July 1912	£1 8s 6d
19 August 1912	£24 9s
16 September 1912	£10 10s

Source: Minutes of Finance Committee, 1912 Minute Book, Municipality of St Peters A95/141.9

February was turning out to be a busy month for industrial matters. Further down the road Mr Speare, proprietor of the Newtown Brickworks, was being given a hard time by residents complaining of pollution and unbearable amounts of smoke. An abatement notice was issued giving Speare 14 days' notice to fix the problem. But these things take time and it was not until 11 November that he was able to advise plans for a new chimneystack and forced draught apparatus.[43] It may be that Speare's Newtown plant (formerly owned by the Goodsell family) had closed and the works transferred further south to Tempe. According to State Records of New South Wales a company in the name of Speare's Brick and Pipe Works was established and commenced operations at Tempe.

Meanwhile, Austral Bricks was going through another growth spurt with the purchase from the Australian Mutual Provident Society (AMP) of a block of land measuring 821 feet by 200 feet on Canal Road.[44] A new fence bordering its landholdings

42. February 1912 – Minutes of general meetings of the Municipality of St Peters, p. 11.

43. Ibid, p. 110.

44. Valuation Books for the Triennial Period, years ending 31 December 1912 and 31 December 1913.

Cattle on a main road in Camperdown, Sydney area, 1890. City of Sydney Archives

along King Street was under consideration. Entries in the council minute book for 30 May and 6 June 1912 also tell of the company's plans to construct a shed and chimneystack, part of a general enhancement of the brickyard.[45] The minutes of the council's Finance Committee even contain references to the sale of bricks to the council and accounts received and paid.

Another matter that falls within general housekeeping was Austral's request on 19 September for the council to divert the stormwater channel made by them on its land, and for a meeting to be held between the Works Committee and Austral Bricks to discuss the matter.[46] Austral even agreed to pay half the cost of the work provided it was done immediately.[47] Minutes of the council's Works Committee on 13 December reported that:

> *... the Manager (Mr Rogers) stated that he thought his Directors would allow us to run the water on to their ground at some little distance from the Cart Entrance, until some arrangement could be made with Alexandria to carry the drainage down to the creek.*[48]

45. Marrickville Municipal Council Archives.

46. February 1912 – Minutes of general meetings of the Municipality of St Peters, p. 93.46.

47. Ibid, p. 123.

A CELEBRATED LIFE

The year 1912 also marked the passing of Josiah Gentle, an iconic figure in the St Peters-Newtown brick industry. At the time of his death Gentle's estate was valued at over £72,000. The brickworks were left to his three sons, Stanley, Emmanuel, and Josiah Jubilee. His widow, Dinah Gentle, was left £1,000 per year for life, and the family residence, a grand mansion known as The Towers in Dickson Street, was bequeathed to his six daughters.

Josiah Gentle's funeral was the largest seen in Newtown to that date. The procession from Dickson Street to Newtown railway station, where a special train left for Rookwood, included 44 freemasons of the Order of Odd-Fellows in full regalia, employees of the brickyards, and representatives of the building trade. So great was the press of mourners that the procession was still leaving the house as the hearse was arriving at Newtown station.

48. Ibid, p. 123.

Josiah Gentle, the philanthropist and public benefactor, would certainly not have disapproved of the formation in the same year of the Brickmakers, Clayworkers and Brickcarters Employees Union of New South Wales. At a meeting at Trades Hall in Sydney, it was agreed to establish such a union by subscription of 12s per year. Members of the Committee comprised H.R. Bolt (President), F.L. Parker (Secretary), A. McKay, James Bradshaw, F. Gorman, W. Brown, and H.J. Thompson. Fred Parker remained secretary for many decades and was succeeded in turn by Eddie Monson, Alex Sawtschuk and Trevor Melksham, who recently retired from the post of state secretary and national president.

A BIGGER POOL

There are reasons other than council-related administrivia, the passing of Gentle, or the formation of a trade union that mark the year 1912 as significant. By this time set up costs including land acquisitions, equipment, buildings and production facilities had eaten into Austral's initial capital of £20,000. Undoubtedly, robust sales riding on the back of the pre-war building boom would have swelled company coffers. Yet, having built up capital in the business, the original partners may have been looking to supplement this by attracting new investors. Some time during September 1913, the decision was taken to form a new company in order to buy out The Austral Brick Company Limited and its assets and liabilities. This was a move designed to strengthen what was a going concern, not a business on the verge of collapse. Accordingly, when the directors and shareholders met at the company's registered office on 1 October 1912 a Memorandum of Agreement was struck. This time the arrangement was between 'Austral Brick Co. (the Vendor Company) of the First Part, Robert Longfield Sidey of Sydney in the State of New South Wales, Public Accountant of the Second Part, David Robertson Rogers of St Peters near Sydney, aforesaid Company Manager of the Third Part and The Australian Brick Company Limited (hereinafter called the Purchasing Company).'[49]

Memorandum of Agreement

.. whereas since the execution of the within written Agreement, the Purchasing Company has been incorporated in accordance with the intention in that behalf referred to in such agreements and whereas since the execution of the within written Agreement the Shareholders of the Vendor Company have duly passed Special Resolutions for the winding up of the Vendor Company and for the appointment of the said Robert Longfield Sidey as the Liquidator thereof now it is hereby agreed and declared as follows...

Each of the investors signed the Articles of Association.

We, the several persons whose names and addresses are subscribed hereto, are desirous of being formed into a Company in pursuance of this Memorandum of Association, and we respectively agree to take the number of shares in the capital of the company set opposite to our respective names.

Names, addresses and occupation of Subscribers:

One [share] *Charles C. Paterson, builder, Mary Street, Hurstville*

One Walter J. Oliffe, builder, Kogarah Street, Leichhardt

One A.E. Burrows [also spelt Burroughs], *builder, Victoria Road, Bellevue Hill*

One Alfred E. Smith, tailor, King Street, Newtown

One W. C. Turland, builder, Darley Road, Randwick

One David R. Rogers, Company Manager, Silver Street, St Peters

One Edwin Spackman, builder, Silver Street, St Peters

Dated this first day of October in the year one thousand nine hundred and twelve.

Witness to the above signatures – Frank E. Hardman, Silver Street, St Peters.

Source: Australian Securities & Investment Commission, 'Agreement for Sale of Business and Undertaking, dated 9th day of October, 1912', Ref. 004498-19

Thus, the Austral Brick Company Limited was relaunched with a capital base of £100,000 and its predecessor placed in liquidation on 3 October 1912. Any further speculation about exactly who the investors were can be put aside. The new line up included five wealthy builders, all of them experienced in the ways of business and able to contribute more than just capital. The names Turland and Spackman are already familiar to us and it seems reasonable to assume that both men had been there from the outset. Rogers, the company

49. State Records of New South Wales, Register of Companies 1910–1913, Container 11/11785, Clay Industries Pty Limited.

Delegates to the Federated Master Builders' Conference, 1909. A.E. Burroughs is seated at the far right, middle row

manager, had taken an equity position in the former company following his appointment as company manager in 1909.

Paterson, whose name does not appear in correspondence from the period 1908–1912, was probably a new investor, as were Burroughs and Oliffe. Burroughs, well known in building circles, was from Bellevue Hill and for many years served as president of the Master Builders Association. He was also a director of the South Ashfield Brick and Tile Company Limited until October 1926.[50] Oliffe, another builder, was a director of Austral until shortly before his death on 2 June 1927. Spackman, an active freemason, would have tapped into an extensive – and exclusive – network of potential investors; Oliffe and Paterson were both in the brotherhood.[51]

The odd man out was Alfred E. Smith, a Newtown bespoke tailor, who seems to have known a great deal about bricks. How did Spackman persuade Smith to become a shareholder, and had Smith been in it since 1908? A long shot, perhaps, but consider the proposition that the premises of tailors and hairdressers are part business, part public confessional. People like to talk and share gossip. Very likely Smith had a keen nose for a good investment and his customers may have included two local men, Edwin Spackman and David Rogers. With money squirrelled away from his tailoring business, Alf Smith took the advice of Spackman and Rogers and joined six other investors in the re-formed company. Smith hung on to his shares because he turned up to all of the annual general meetings right up until his death.

So it was that at an ordinary general meeting of investors held at Queen's Hall, Pitt Street, Sydney on 21 November 1912, Messrs Turland, Oliffe, Burroughs, and Smith were elected unopposed as directors of the company with Paterson as chairman. An auditor, Mr G.S. Travis, was appointed. Some matters, however, were still outstanding for at the general meeting held on 23 April the following year

50. *Building: a magazine for the building industry*, 12 June, 1912, p. 41.

51. Oliffe joined Lodge Southern Cross 91 meeting at the Masonic Centre, Castlereagh Street Sydney on 14 March 1903, and became a Master Mason on 21 June. He remained a member of that Lodge until his death in 1927.

Annual picnic of the Building Trades Federation of NSW, held at Clontarf in 1914. Note the dress code and sense of dignity. Men such as these were proud of their standing as respectable, skilled artisans. Vince Higgins Collection

a letter was read from Mr John Aske, the former plant manager, regarding an honorarium voted him by the 'old Austral Brick Company'. Aske's request was declined.[52]

ROYAL COMMISSION IN VICTORIA

Meanwhile, concerns about private manufacturers in New South Wales were being echoed in Victoria where the Royal Commission referred to in the previous chapter was announced in May 1913 to investigate the possible existence of a brick combine, brick price fixing, and what should be done to enable users to obtain bricks at a fair price.

The hearings opened on 10 June 1913, and 73 witnesses, generally representative of the building trade and some other interests, were examined in Melbourne, Ballarat, Bendigo, Morwell, Wonthaggi, Sydney and Adelaide. The evidence taken was voluminous but for reasons of economy was never printed. At the outset of the inquiry it was admitted that an association of certain brick manufacturing companies, registered and trading under the designation of 'The Co-operative Brick Company', and popularly known as the Brick Combine, had existed since 1896. In its report the Royal Commission explored the circumstances which led to the formation of the association, its subsequent operations, and its effect generally on the building trade. Rejecting assertions that the brick manufacturers had acted against the public interest, H.F. Young, chairman of directors of the Northcote Brick Company, gave an insight into a practice that had been a feature of the industry for almost 20 years. In short, manufacturers in the association closed certain yards during downtimes, agreeing to compensate inoperative yards by paying a percentage of the earnings of other kilns.

> *We had to compensate them for the taxes paid and for the care taken of the machinery and sundry other expenses, and then, when the trade increased they were opened again. I know that the Hoffman have two kilns still that they have never been asked to light up and the Northcote has been asked to put out a kiln often when there was no demand for the bricks, and that was a great loss to the company.*

> *Neither the Hoffman nor the Northcote have been going full steam since we have been working together. We might have been short for a few days, but the next week we would have the kilns discharging, and*

52. Minutes of directors' meeting, Austral Brick Company Limited.

The Carlton yard of the Federal Brick Company Ltd, at Hurstville, Sydney, circa 1914. The owner, E.S. Phippard, produced large quantities of salt-glazed pipeware at this yard.

have a great deal of bricks, I know now the Northcote Company has about 4,000,000 bricks in stock.[53]

THE COMING OF WAR

Back in Sydney investors in Austral Brick were in a cautious mood when they met at Queen's Hall on 29 October 1913. Noting a downturn in the industry and price-cutting on the part of manufacturers, a motion was voted on to increase cash reserves. 'In the opinion of this meeting it is desirable to make financial provision to meet any possible disintegration of the Brick Trade with a consequent cutting of prices.'[54] The subject came up again some time later at the Ordinary General Meeting on 22 April 1914 when there was further discussion about the need to conserve capital. The following month shareholders agreed to collectively and individually forego any rights to any rebate on bricks bought by them from the company.

By the close of 1913 the building boom had reached its peak with only one new company making its appearance in the immediate pre-war period, the Central Brick & Tile Co Pty Ltd., located on land occupied by the old Barwon Park Estate. The land was extremely valuable, and at £700 per acre was some of the most expensive in the district. In addition to commons and face bricks, the Central made pipes, tiles, marble wash stands and other pottery items. The company's chief engineer occupied Barwon Park House, built in 1828, which at one time was occupied by Thomas Brown, an engineer formerly of Bathurst Street, Sydney, who had introduced the first fire engine to the city.[55] Central Brick & Tile is known to have produced all classes of bricks including the highest quality commons, moulded bricks, and from 1918, roofing tiles.[56]

It would be a mistake, however, to assume that the industry was in the doldrums on the eve of war. Looking back over the period 1908 to 1913 it had been a time of rapid expansion and opportunity. By 1914 Sydney's subdivision and building boom had spread out along the rail and tramlines. On some of these routes electric power was beginning to replace steam. Australia was increasingly prosperous, with new leisure opportunities opening for all but the very poor. The Melbourne Cup and Sydney Royal Agricultural Show drew huge crowds; surfing became popular and voluntary surf life saving clubs spread from Sydney to other states. Events in Europe with the entry of Britain into a general European war changed everything.

53. Royal Commission Into The Manufacture Of Bricks And Brick Prices, Victoria, 1913.

54. Minutes of directors' meeting Austral Brick Company Limited.

55. *The Clay Products Journal of Australia*, 1 October 1940, p. 13.

56. By 1936 the company had a showroom at the Metropolitan Building, on the corner of Hunter and Bligh Streets. Walter West had been a director of the company since 19 2.

The morning after the first battle of Passchendaele, 1917 – Australian Infantry wounded around a blockhouse near the site of Zonnebeke Railway Station, 12 October 1917. Frank Hurley. National Library of Australia

War and Peace 8

Australia at war – Canal Road – William King Dawes – Austral postwar – Peacetime adjustments – Aspirational – Mortgage belt – A matter of taste – Look east, look south, look west – Town planning – Departures – Where paths cross – Excelsior – Riding the boom – Management move – The reckoning

AUSTRALIA AT WAR

When Great Britain declared war on Germany on 4 August 1914 Australia's status as a dominion of the Empire committed the nation to the fight. With few exceptions sentiment was pro-British and cut across party political lines. Following the Labor Party's return to office in the September elections, Prime Minister Andrew Fisher spoke for the whole country when he promised to support Britain 'to our last man and our last shilling'. At no time since 24 May 1905 when Australians first celebrated Empire Day had loyalty to the Crown been stronger.

Throughout the war the Australian economy prospered with guaranteed markets in Britain for wool, metals, wheat, butter and other primary products. Manufacturing was booming due to the demand for munitions and the production of goods that could no longer be imported. Broken Hill Proprietary Limited opened a steelworks in Newcastle. Farmers and woolgrowers cashed in on lucrative wartime contracts offered by British textile manufacturers who were prepared to pay high prices. Trade prospects were rosy, yet the inflationary effects of war were ever-present; prices rose by an average of 20 per cent at a time when real wages had fallen by 10 per cent. Profiteering was condemned, although it was difficult to stamp out, and despite the number of men who enlisted, unemployment stood at a high of six per cent. Many Australians jibbed at the imposition of 6 o'clock hotel closing hours introduced by referendum in New South Wales in 1915, ostensibly to show men at the front that sacrifices were being made at home. Victoria, South Australia and Tasmania soon followed suit when they restricted opening hours in pubs.[1]

Meanwhile the upward drift of rents in Sydney's housing market was stopped in its tracks by a Fair Rents Act introduced in 1916 by the New South Wales Labor Government. Property prices felt the impact immediately and with falling returns investors fled the rental market. Wartime austerity also had a marked effect on construction techniques and housing styles, with the result that by the end of the war rising costs forced builders to reduce the size of a house. Costs also affected building practices. For example the time-consuming technique of applying lath and plaster in ceiling work was gradually replaced by fibrous plaster sheeting that could be screwed or nailed to ceiling joists. Precast concrete lintels rather than brick were used to carry brickworks over doors and windows. In more affluent suburbs the family car had the effect of spacing houses further apart with allowance made for a driveway down the side of the building. Costs were now at a peak, having almost doubled since before the war. According to Robin Boyd an average brick house of 2,000 square feet cost about £1,000 in 1914, whereas by 1921 the price of a similar house had risen to about £1,800.[2]

CANAL ROAD

Down at Austral's yard the directors were growing edgy. Cash flow – or the lack of it – due to a contraction in business was beginning to affect the balance sheet. At an Ordinary General Meeting on 27 October 1915, Harry J. Thompson, a go-ahead builder with a substantial shareholding, failed in his bid to move a motion to reduce the remuneration of directors by one-third of what it had been the previous year. Thompson let the matter drop,

1. It was intended only as a wartime measure, and only in Queensland, Western Australia and Tasmania after 1937 were the bars permitted to open after supper.

2. R. Boyd, op. cit., p. 60.

deciding instead to enlist in the Australian Imperial Force, and we hear nothing more of him until his return from active service in early 1919. There was discord of a different kind at the Ordinary General Meeting of 18 October 1916 when builder John Rigby asked why his company had been refused bricks for a job in Castlereagh Street.[3] In reply Alf Burroughs briefly gave a summary of certain actions of the Master Builders Association that made it appear to be in the interest of the company not to supply that particular job. Unwilling to rock the boat, Mr Rigby said 'he wouldn't press the matter as he was not desirous of causing any friction in the Company.'[4] This curious state of affairs cannot easily be explained, although money matters, possibly a bad debt, may have had something to do with the refusal.

And what of the extent of the Austral empire in 1916? St Peters Council Valuation Book indicates that nothing had changed in terms of landholdings. The company remained the owner of the King Street property and also another on Canal Road – a substantial undertaking, but Austral was still a minnow compared to competitors such as the Bedford Brick Works on Mitchell Road.

WILLIAM KING DAWES

Enter William 'Bill' King Dawes, although not, as yet, in any official capacity. But why now – the war years – and what was the connection with Austral? In 1914 Dawes was just another 16-year-old boy whose family had a long association with the brick trade. Nothing unusual about that. In due course, Dawes achieved great prominence.

The name William King Dawes is fitting for a man who, over the years, took on the mantle of

William Frederick Dawes, circa 1890. Marilyn Keelty Collection

leadership in the Sydney brick industry. As someone who shunned publicity, Dawes was an intensely complex and intriguing man – an enigma. Mercurial, ceaselessly energetic, and constant in his devotion to running a successful business, he demanded efficiency and was ruthless in obtaining results. Dawes is remembered by many for being independent, a loner, and a man who cut a somewhat piratical figure, unafraid to speak his mind or take resolute action where he felt it necessary to protect the interests of the company. Yet he is also remembered for being extremely generous to his workers and solicitous of their welfare. Strange for someone who detested trade unions, but could pick a man out of the gutter and give him a fair go. A contradiction, perhaps, but one that former employees and acquaintances (such as Ron Hagstrom, John Tindall[5], Ross Bell, Charlie Vella, Bob Rushton and many others) can confirm.

3. Rigby later went into partnership with another builder to form Kell & Rigby.

4. The Austral Brick Company Limited, minutes of Ordinary General Meeting, 16 October 1916, Brickworks archives.

5. Three generations of the Tindall family worked as brick carters for Austral Bricks. John Tindall's grandfather, who was born in 1896, started in the brickyards at the age of 12. He also worked at the store of William Frederick Dawes on Princes Highway where he met Dawes' son, William King, who was of a similar age. According to John: 'my grandfather used to tell me that W.K. used to pay the kids to put the spuds in the bag and he would go and have a sleep'. Interview with John Tindall, 7 August 2005, Sydney.

William Frederick Dawes, circa 1908. Standing in front of the stable at his Cook's River Road produce store. Bill Wheatley Collection

Master Bill, the Pirate. Bill Wheatley Collection

Country squire. Bill Wheatley Collection

William King Dawes, circa 1907. Young Bill, with riding crop and cigarette in confident pose. Bill Wheatley Collection

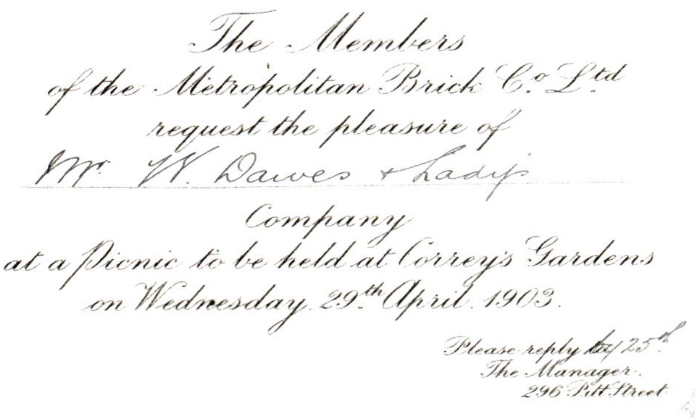

The Members
of the Metropolitan Brick Co. Ltd
request the pleasure of
Mr W. Dawes + Lady
Company
at a Picnic to be held at Correy's Gardens
on Wednesday, 29th April 1903.
Please reply by 25th
The Manager.
296 Pitt Street

Invitation to the annual picnic of the Metropolitan Brick Company Limited, 1903. Bill Wheatley Collection

FOUR GENERATIONS OF BRICKMAKERS

Dawes was no stranger to the brick scene at St Peters. By the time he makes his appearance in this narrative, four generations of the Dawes family had worked the clays along the Waterloo Flats. Born in December 1898, William King Dawes was from a family of three sons and a daughter. The firstborn, Frederick Roy, died of pneumonia in November 1898 at the tender age of three. Then came William King (December 1898), followed by Edna (1907), and finally Jack (1912), who died tragically in April 1943 when his recently acquired motorcar spun out of control and hit a tram.[6]

William Frederick Dawes, the father, was the great grandson of George Dawes, who in 1839 had been the first of the family to settle in Australia.[7] Dawes senior was a big man, huge in terms of girth, although not given to drink. Unstintingly generous he did everything possible for his family. William Frederick was not a brickmaker or yard owner, but for many years was associated with the Federal brickworks at Alexandria, possibly as both an investor and friend to the Phippard family. He was considered important enough by the Metropolitan Brick Company, about which we shall hear in later chapters, to be invited to the company's annual picnic in 1903. He was also active in local politics, serving as Mayor of the Council of Sydenham early in the new century, and as alderman on the St Peters Municipal Council from 1914 to 1920.

There were early signs that young William was developing an independent nature. About the time of his seventh birthday, the practised hand of a professional photographer captured a number of striking photos that reveal something of the boy's character: confident and someone who knew his own mind. What the camera does not reveal, however, is a lazy streak in a boy who from an early age was considered an independent character. And why not? The brickpits of St Peters would have given him every opportunity to indulge his sense of adventure. Across the road, through the gates, or over the fence, a different world beckoned – deep quarries where hard-bitten men worked, blasting and hewing the clays and shales. Down into the pools of water that collected after heavy summer rains before the pumps could be brought to bear; a dangerous sport which resulted in the drowning death of many inquisitive lads over the years. Hardly a rural idyll, but still tracts of land that could be explored – despite parental injunction – to the banks of Shea's Creek and along Cook's River. Thus began young Bill's introduction to the world of the brickmakers.

PRODUCE STORE

So what significance does the family store and its buildings hold for us? The name W.F. Dawes cemented in relief is still clearly visible above the former shop on Princes Highway, one hundred metres down from Sydney Park. Living in and working from this two-storey building William Frederick Dawes supplied feedstuff to the local brick carters. An easy-going fellow, he extended good credit terms to his customers, one of whom was Percy Miller, the owner of several horse studs in the Hunter Valley, and who also kept stables at the Randwick track.

6. During the late 1940s, Jack Dawes' widow, Joyce, married Bill Swann, personal assistant and driver to W.K. Dawes.

7. W.F. Dawes was the second son of Naasson and Harriett Dawes. Naasson Dawes, a potter and also the manager of a pottery works, was the son of Alfred Dawes, the uncle of William Frederick.

Above: W.F. Dawes produce store, circa 1903. Bill Wheatley Collection

Above Right: William King Dawes, circa 1914. Marilyn Keelty Collection

Right: Family picnic – A family outing for the Dawes family.

Back row from left to right: Emma Jane Dawes (mother of W.K. Dawes), William Frederick Dawes Snr, W.K. Dawes and wife Emily. Front row: Jack (brother of W.K. Dawes), Ray Dawes, William Frederick Dawes, Arthur Dawes. Bill Wheatley Collection

Below: Produce store on Cook's River Road, circa 1914. Taken on the eve of the First World War or slightly later. W.F. Dawes borrowed money from the bank in 1909 to rebuild and extend the house and store.

From left: Bob Carveth, a youthful William King Dawes, William Frederick Dawes, Mr Tearnside. Marilyn Keelty Collection

Newtown Demonstration School, circa 1911 – William King Dawes, top row, second from right. There is an air of reticence about the 13-year-old Dawes, posing for the camera with other students from Newtown Demonstration School. Bob Young, a childhood friend and future Chief Engineer for Brickworks Limited is seated cross-legged on the bottom row, fourth from left. Dorothy Armfield Collection

It was from one of the delivery carts that Bill junior made his introduction to the Austral yard while still at school. Like his father he was simply continuing in the family business. Jim Liepins, a retired engineer at Brickworks Limited's City Road workshop adds weight to this: 'Dawes told me that he was helping his father deliver straw to the brick carters even before he finished school.'[8]

From St Peters Primary, William went on to Newtown Demonstration School and at some point was offered a scholarship to Newington College although much to his mother's disappointment he never took it up. Leaving school at the age of 14 young Bill joined the firm of Foley Brothers, whose produce business on Sussex Street was one of Sydney's biggest. Expecting to be paid more than he was worth, Dawes quit and went to work for his father. In these early years a dislike of books and learning closed off the path of formal study and perhaps a future in one of the professions. Never an educated man in the formal sense, unintelligent he most certainly was not. Working in the business, the boy discovered an aptitude for mental arithmetic which was to stay with him for the rest of his life. Employees and friends recall how he would pore over columns of figures jotted down in notebooks, or make calculations on the edge of a newspaper. In the stable yard during the working day, or in the parlour during the gas-lit gloom of early evenings, this was Bill Dawes doing what he liked best – adding up the numbers whether it be profit and loss, production figures, or wages, paid or owing. By this time he was sufficiently mature to be given the responsibility of buying chaff and oats at the Sussex Street Corn Exchange. Recalls Ray Dawes, son of W.K. Dawes:

> *Given a handful of oats he could tell you where it came from – New Zealand, the Riverina, Cootamundra. When he went into racehorses during the Second World War he would talk to the trainer, Maurice McCarten. They would go into the feed room and pick up a handful of straw and grain. Dad could say where it had come from or been grown.*[9]

8. Interview with Jim Liepins, 1 July 2005, Sydney.

9. Interview with Ray Dawes, 10 December 2004, Sydney.

Horse and sulky. Bill Wheatley Collection

George Dawes (1802–1886)

George Dawes = Louise Wratten (1809-1886)
(Marriage 2 in 1838)

George Dawes = Eliza Goodsell (1806-1836)
(Marriage 1 in 1825)

Mary Ann (1826-1883) | Ellen (1831-1900) | George (1832-1867) | Catherine (1834-1916)

Orientia (1838-1903) | Winifred (1840-1928) | Naasson (1843-1917) | Eliza (1844-1937) | Emma (1846-1914) | William Aaron (1849-1906) | Naomi Louise (1852-1935)

Naasson Dawes = Harriet Tye in 1846

Ida (1866-1955) | Alfred George Naasson (1868-1949) | Amy (1870-1949) | William Frederick (1870-1931) | Edward Stutchbury (1874-1941) | Mary Ann (1877-1944) | Reuben (1880-1953) | Winifred (1882) | Isabella (1884-1963) | Lilly (1886-1942)

William Frederick Dawes = Emma Jane King in

William King (1898-1981) | Edna King (1907-1986) | Jack (1912-1943)

William King Dawes = Louise Emily Loveridge in 1916

William Frederick Naasson (1917-2001) | Arthur Manly (1907-1980) | Raymond Trevor (1922-)

Right: Brickworkers at war, circa 1916. From left: Alfred Binksin and his brother, George Binskin. Three generations of the Binskin family worked in the brickyards. Alfred and George were stretcher bearers who served alongside each other during the First World War. They were also assigned to other duties, including manning the listening posts in No-Man's Land.

This unusual gift stood him in good stead, for in later years when Dawes made frequent sorties into Sydney's countryside in search of shale deposits he was able to identify the presence of clays by the ground cover in a paddock. It is now an established fact that the particular characteristics of ground cover such as trees and other plants, including grains, derive their appearance from the underlying soil type. Dawes was too young to join up when war broke out in August 1914 and he continued working the drays which must have been something of a let-down. For a while he indulged his hobby of breeding racing pigeons, although his father disapproved of this apparent lack of direction. What may have galvanised Dawes into action was the discovery that his lady friend, Louise Emily Loveridge[10], a young country girl from Echuca in Victoria, had fallen pregnant. Unable, or unwilling to defy the conventions of the time, and no doubt encouraged by their parents, the young couple married in 1916. Bill and his new bride settled in Silver Street, St Peters, quite possibly in a rented house arranged through a family friend, Ted Spackman. Perhaps married life at so young an age made him restless. At some point he decided to enlist, although much to his disappointment he was still at the North Head training camp when the Armistice was announced in November 1918. Lounging on the cliff top on his evenings off during late autumn of 1918, Dawes must have wondered what life had in store for him.

AUSTRAL IN THE POSTWAR YEARS

For Austral, however, peacetime brought the welcome return in May 1919 of those men who had been on active service with the Australian Infantry Force in northern France. Among them was Alf Smith, Newtown tailor and member of the board, and Harry J. Thompson, builder and fellow director. Thompson established the building firm by the name of H.J. Thompson and Company, and was later joined by his two sons, Harold William, and Charles. Thompson senior served one term as president of the Master Builders Association from 1909.

Four years of war is a long time in business – long enough for a company to lose momentum. At a meeting of shareholders on 20 October 1919, Turland, the chairman, urged those present of the need to recruit as much 'new blood' into the company as possible by way of young builders. His call to action netted a fine catch with the likes of Alexander Stuart, Archibald Howie, John Harrison, and P.E. Jennings, all of whom bought shares, with some subsequently elected to the board. The quest for new blood may help to explain the reincorporation of the Austral Brick Company Ltd on 31 December 1920. It may be that the company's articles of association were changed to accommodate these new developments, although a more complete explanation would require further investigation.[11]

Meanwhile, at Dawes' Silver Street home there was discussion about future prospects. With a wife and a baby to feed, and another one on the way, 22-year-old Bill Dawes was anxious to get ahead, especially in circles where it counted.[12] At the invitation of Spackman he was admitted to membership of Lodge Manoah 41 at Castlereagh Street in March 1920, and later that year became a Master Mason.[13] But change was in the wind and William King Dawes knew where it was heading. With reins in hand, sitting atop the straw and feed delivery cart, Dawes could see further than his father's Suffolk Punch, with its nose thrust deep into a feedbag. He saw the motor car and motor lorry. He saw speed and excitement. Most of all he could see money, for he knew sure enough that a brick carter behind the wheel of a motor lorry could make three times as many deliveries in a day as he could behind slow-moving beasts of burden. Incidentally, the Suffolk Punch

10. Louise died in August 1980 just over 12 months before her husband, W.K. Dawes.

11. Australian Securities and Investment Commission (ASIC), Original Deed Index.

12. The firstborn, William Frederick, was named after his paternal grandfather and started working at the Austral yard in St Peters in 1935.

13. In June 1929 Dawes moved to Lodge Sydenham 610 which met more conveniently at the St Peters Town Hall. Dawes joined Lodge St Stephen Newtown 492 in July 1933 and remained a member of that Lodge until 24 February 1943. He never rejoined Freemasonry.

breed was favoured by brick carters. Smaller than Clydesdales, and with thickset frames and short legs, they were capable of drawing heavy loads which were either on, or close to the ground.

In short, Dawes could see the writing on the wall for horse and cart. He read the signs, and so did his mother, Emma Jane, who knew that her husband had been recently offered several Leyland trucks by business associates. Using money from a savings bank account built up over the years, she gave Bill's father the means to buy the trucks as well as the garage opposite Gentle's Bedford yard.[14] Now the proud driver of one of the Leyland trucks, young Dawes became a member of the Master Brick Carriers Association, and remained so for many years. Jim Liepins confirms the mother's hand in the purchase of the vehicles and Dawes' rise through the ranks, starting as a brick carter for Austral Brick:

> *I heard this story from Fred Mercer*[15] *who was Austral's accountant in the workshop of City Bricks at Euston Road. He looked after purchases and payments, knew the Dawes family well and told me about how Dawes got started. In the first year he did so well that he bought another truck. This was after the war in the early 1920s. In fact he became so well off that later he bought shares in The Austral Brick Company Ltd.*[16]

With motor vehicle traffic building up along the busy thoroughfare of King Street, Dawes and his father installed a couple of petrol bowsers on the footpath. Trade picked up dramatically in 1928 with the opening of Tom Ugly's Bridge across the Georges River.

Yet cartage costs remained high and a load of 1,000 bricks was the carrying limit for most trucks in the 1920s and 1930s. Dawes was careful to price his brick loads at the usual rates. Like other carters switching to motor transport he must have found

Dray horses waiting for a load at the Bedford Brick Works, circa 1915. John Farrington Collection

Horse team, Goodlet & Smith, circa 1915. John Farrington Collection

Leyland truck in the yard of the Austral Brick Company Ltd, circa 1928. The spire of St Peters Church, Cook's River can be seen in the background. Ray Dawes Collection

14. Interview with Naasson 'Rick' Wheatley and Bill Wheatley, 12 November 2005, who heard this story from Emma Jane Dawes, the wife of W.F. Dawes.
15. Fred Mercer started as delivery clerk at the Austral yard on Princes Highway, and went on to gain accounting qualifications. Like many of his colleagues he was also a devout Baptist. Interview with Ray Dawes, 12 December 2004, Sydney.
16. Interview with Jim Liepins, 1 July 2005, Sydney.

Leyland truck showing insignia of The Austral Brick Company Ltd. Government Printing Office Collection, State Library of New South Wales

the new regime exhausting. There was nothing work shy about these men; every consignment had to be loaded by hand at the yard, and then offloaded at their destination. There was no such thing as a 'back' (return) load. When business was slack the carter would stack the truck and wait for an order, whiling away the time over a smoke or a chat with the other delivery drivers. If no buyers were to be had, then he re-stacked the bricks in the yard. For obvious reasons, people bought their bricks from a local yard, which may help to explain why there were so many brickyards around Sydney. Motor lorries with their solid rubber tyres made forays across the city on roads that were poorly maintained and often deeply rutted. Parramatta Road for example turned into a muddy quagmire during heavy rain and was dubbed 'The Dardanelles' by returned servicemen who drove Sydney's delivery trucks in the early 1920s. Wooden blocks covered with layers of tar were used to pave the main thoroughfares. Maintenance was a constant problem although blocks remained in use for several decades. At a later date they were torn up and purchased by brickyards, including Shannon's Brick Works where they were burnt as fuel in the kilns. Making good money running deliveries for the family business and his own one-man operation, Dawes became a local identity in St Peters. He was also a frequent visitor to the home of Edwin Spackman. At some point Dawes made the move from delivery truck to the yard office, working with Spackman who took him under his wing. Dawes' rise to prominence also owed much to the influence of his father in the local pit, but in fairness, he was exceedingly knowledgeable about the brick trade and the economics of running a profitable business.

PEACETIME ADJUSTMENTS

For Australia the war had been a dramatic demonstration of the country's isolation, in terms of both its geography and economic position. The government in Canberra had encouraged economic growth under the umbrella of tariff protection. The early post-war years witnessed a reversal of fortune: a severe slump and wage cuts created hardship, especially in the building industry. Australian workers took to the streets to vent their dissatisfaction. Heightening the fears of many people towards worker unrest was the spread of socialism following the Bolshevik coup in Russia in November 1917. Charges that the Australian Labor Party was tainted by radical ideas were given some legitimacy in view of the fact that its 1921 Federal Conference had set as an objective 'the socialisation of industry, production, distribution and exchange', although a timetable by which it was to be achieved was never stated. The Bricklayers' Union of New South Wales weighed in with a series of pamphlets written by union secretary, Mick Sawtell, who encouraged solidarity among the 'revolutionary working-class movement'. Sawtell, like many others in a similar position, was convinced that:

> *There can be no peace as long as hunger and want are found among millions of the working people ... The army of production must be organised, not only for the every-day struggle with the capitalists, but also to carry on production when Capitalism shall have been overthrown.*[17]

17. M. Sawtell, *Job Control: The Basis of Future Society*, issued by the Bricklayers' Union of New South Wales, The Worker Print, Sydney, 1926. Sawtell's preface to the 1926 edition was dated 11 June 1920 (Adelaide).

Spanish Mission style of architecture

Comments by Sawtell and other trade union leaders, as well as strikes during 1923 by Melbourne police and the teachers of Western Australia were ill omens for those on the conservative side of politics. A militant right-wing organisation known as the White Guard was formed to counter the socialist forces. Under the direction of Sir Brudenell White, retired chief of the general staff of the Australian Infantry Force, the White Guard vowed to resist the spread of Bolshevism. Determined to curb union power – which by 1926 had risen to 60 per cent of eligible workers – Prime Minister Stanley Bruce went ahead with legislation to strengthen the Commonwealth's power to regulate on industrial arbitration matters.

On a lighter note, it was, after all, the age of jazz, a reaction to the horrors of war and the stuffy age of Victoria and its lingering aftermath into the first decade of the twentieth century. Enter the foxtrot, the one-step, and 'flappers'. It was a time for all gay young things to take the stage. For those whose occupation was home decoration, jazz gave inspiration to a new artistry of sorts which found expression in zig-zag lines worked into bathroom tiles, living room cabinets, bookcases, and bedheads. It was even appliquéd on fabrics in burnt orange, green, yellow and black. The Jazz Style, however, was never an architectural expression, and did not last. Instead it was eclipsed by another style which had a far wider impact than many of its predecessors: Spanish Mission.

ASPIRATIONAL

The desire for a place of one's own is not a recent phenomenon, yet what is it that draws people to a

Nature of occupancy of private dwellings in Sydney, 1911

Total number of dwellings - 118,427

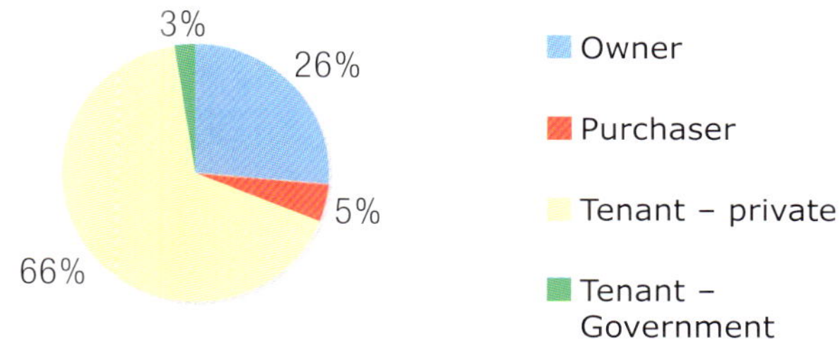

Nature of occupancy of private dwellings in Sydney, 1933

Total number of dwellings - 279,963

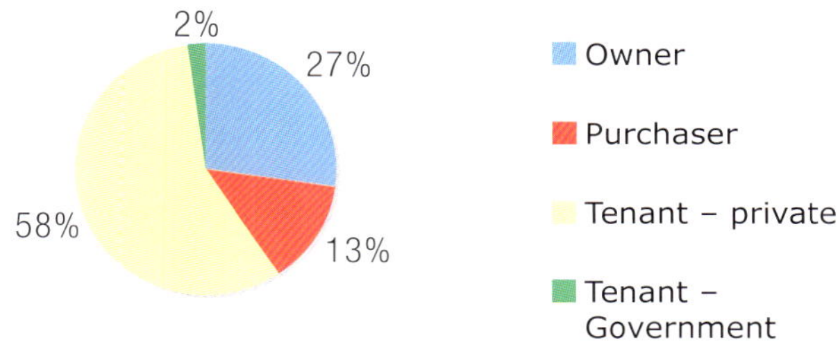

Nature of occupancy of private dwellings in Melbourne, 1911

Total number of dwellings - 114,988

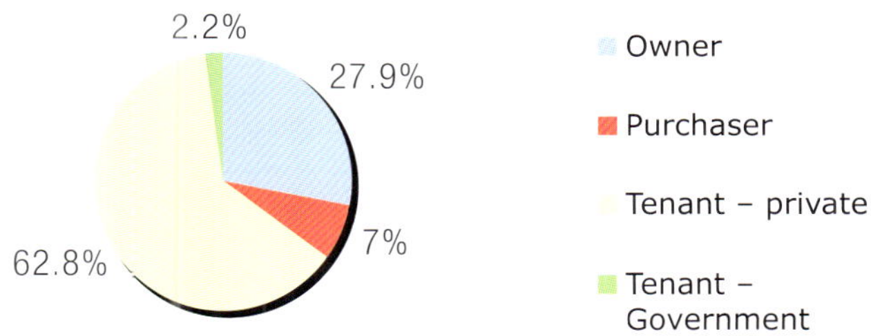

Nature of occupancy of private dwellings in Melbourne, 1933

Total number of dwellings - 229,464

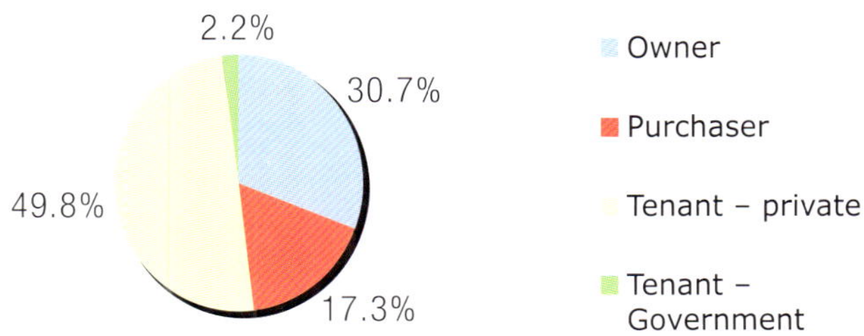

Source: Australian Bureau of Statistics, Census of Commonwealth of Australia, 1911–1971

house and garden? Until the later years of the 1920s home ownership was simply out of reach for the vast majority of people, making it even more desirable. Things were about to change with the war acting as a catalyst.

In Britain politicians declared that the sacrifice of those who fought would not go unrewarded. 'A land fit for heroes' was the catch-cry. These sentiments were echoed in Australia where there was a genuine determination on the part of governments to reward the efforts of returned soldiers. Heightened expectations of an improved standard of living and better housing grew, especially in Sydney, where well over half the population were tenants. The time was ripe for yet another property boom which spread out across the city. More dwellings were built in Sydney in the 1920s than in any previous decade of the city's history, the number in Sydney peaking at 90,131 by 1928. According to historian, Peter Spearritt, this record was not surpassed until the 1950s.

For salary and wage earners the ideal home was a quarter acre lot that came with a single dwelling, preferably beyond the crowded inner suburbs. Municipal councils in Sydney and Melbourne began to stipulate 'Brick Areas' following the example set by local authorities in Perth where brick-built houses retained their monopoly. Most people agreed that dry press and the more expensive, up-market texture brick was infinitely preferable as a building material since it offered superior insulation against heat and cold, low maintenance, stability and longevity. This pleased insurance executives and loss adjustors who rightly saw the humble timber homes as a fire trap. By 1939 many suburbs were given over entirely to brick where weatherboard dared not show its face. Brick reigned supreme, although as the decade progressed the cheaper option of brick veneer began to establish itself as a respectable alternative to full brick. In 1920 concrete roof tiles began to appear, making a dent in terracotta's seemingly unassailable market dominance.

MORTGAGE BELT

The twenties were also a period when new ways of financing spending were being introduced. Through mortgages and long-term loans from savings banks, home ownership came within reach of many more people. Responding to increased pressure on housing stocks and the demand for home ownership, governments introduced two new schemes. The first was in the form of housing loans from 6.25 per cent to 7 per cent offered by the New South Wales State Government Savings Bank. In addition, the Commonwealth War Service Homes Commissions through its War Service Homes Scheme (1918) made available low cost loans to ex-servicemen and their families, either to build new or purchase existing homes.

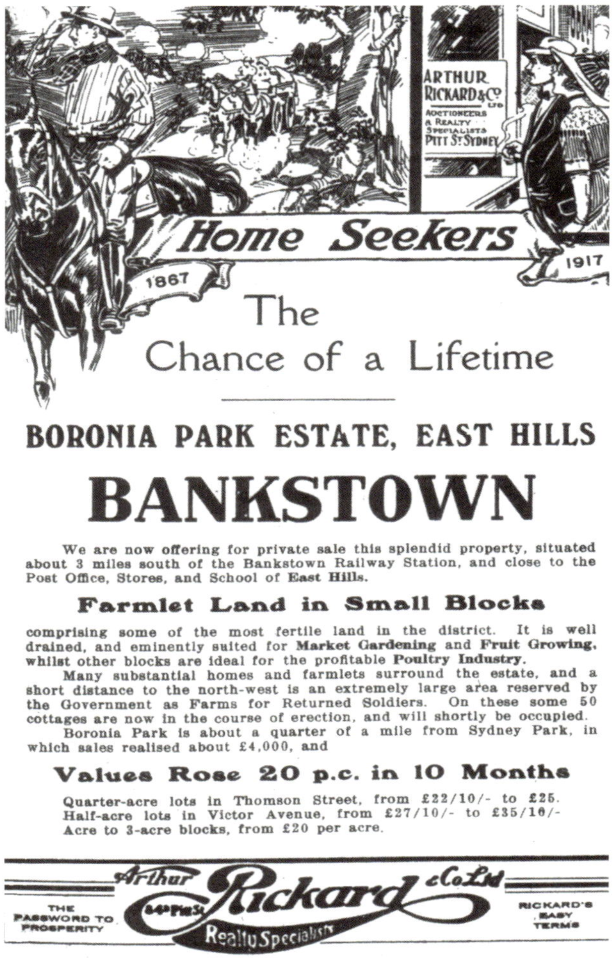

Chance of a lifetime. Realty Review, November 1917

Encouragement such as this enabled the private mortgage market to grow and provide loans up to the value of 70 per cent of the property. Ballot and sale societies were also popular and organised along the lines of a lottery where households contributed on a regular basis. The lucky few who won a deposit could then borrow the remainder of the funds from a bank or building society. At one time Spackman had been particularly active in the market for housing loans as a valuer for the Newtown and Enmore Starr Bowkett Co-operative Building Society at Enmore Road, Newtown. Dawes later became a director of the organisation.

At the same time the transport revolution taking hold across Sydney made possible suburban growth beyond the three-mile radius of the city. Needless to say, speculators anxious for quick profits snapped up land for subdivisions. The Shire of Warringah benefited from the promised improvement in transport and services including the construction of the Spit Bridge over Middle Harbour in 1924 and the Harbour Bridge in 1932. Before 1914 trams and horse-drawn vehicles were the dominant forms of passenger and freight transport. By 1920 the motor vehicle was well on its way to replacing the horse-drawn cart as the main form of haulage.

A MATTER OF TASTE

The home building stampede, however, found no favour with architects, especially Leslie Wilkinson, professor of architecture at the University of Sydney, who complained that:

> ... *Fully 70 per cent of the houses erected today are produced without reference to the profession. Until this state of affairs is altered and until the public appreciate the difference between the beautiful, the good and the horrid, admirable work will continue to be a rarity.*[18]

Wilkinson was right. For the most part owner-builders copied plans from magazines such as *Building* (founded 1907), published by George and Florence Taylor, or the *Australian Home Builder* (1920s), or *Home* (1920-1942) and Florence Taylor's short-lived *Commonwealth Home* (1925-30). Red-tiled roofs, widely regarded today as an attractive feature peculiar to Sydney's roofscape, came in for particular criticism from Wilkinson. In October 1924 even the *Australia* magazine was moved to comment: 'one red tiled roof house is much like another red tiled roof', at the same time deploring the onward march of suburban ugliness. Buyers were blamed, especially women who were considered to be more conservative in their tastes. The Californian Bungalow was the most common house design of the 1920s, examples of which can still be seen in many suburbs, including Lane Cove, Auburn, Ashfield, Rockdale and Randwick. Builders of this type of house disregarded the clean, straight lines of precision, machine-made bricks, and instead covered their buildings with pebble-dash or rough-cast cement.

The Californian bungalow

LOOK EAST, LOOK SOUTH, LOOK WEST

For brick manufacturers boom conditions meant only one thing. Private firms and the State Brickworks at Homebush Bay prospered during much of the postwar period. The Austral Brick Company Ltd, along with many others, took advantage of the rapid construction of new houses initiated by the government to provide low cost housing for returning servicemen. Wunderlich, for example, won the contract to supply roofing tiles for many of these homes and its asbestos cement sheeting was in great demand as a cheap and effective wall and ceiling cladding.

18. Cited in P. Spearritt, *Sydney's Century: A History*, UNSW Press, 2000, p. 33.

Approvals and completion of apartment buildings (flats) in Sydney, 1924–1943

Year to June	Completions	Approvals
1924	852	n.a.
1925	1076 (est)	n.a.
1926	1161	n.a.
1927	204	n.a.
1928	634	n.a.
1929	646	n.a.
1930	363	n.a.
1931	111	5
1932	27	15
1933	63	115
1934	248	335
1934	248	335
1935	n.a.	562
1936	n.a.	520
1937	609	505
1938	544	661
1939	703	727
1940	862	871
1941	745	11
1942	30 (est)	n.a.

Source: Completions – Metropolitan Water, Sewerage and Drainage Board. Approvals – New South Wales Statistical Register

By 1920 Sydney's eastern suburbs were generally built up, although apartment building was able to absorb population increases in the municipalities of Woollahra and Waverley. Unfortunately, progress came at a price; estates of historic significance and beauty were subdivided to such an extent that their original boundaries were no longer recognisable. To the south, along the shores of Botany Bay and the Illawarra railway line, Rockdale forged ahead especially with the release of allotments on the Hoon Estate by the real estate firm owned by Sir Arthur Rickard. The St George area followed suit with local councils quick to organise the installation of sewerage, electricity and gas services. On the other hand, the development of Sutherland was held back by the absence of piped water until the construction of the Woronora Dam from 1927 to 1930 and its completion in 1941. In Sydney's outer suburbs backyard toilets and water tanks were the norm for decades to come so that many houses had more in common with those in country districts.

Meanwhile, to the west of Sydney the housing boom straddled the rail line from Strathfield to Granville, and from Canterbury to Bankstown, in two distinct bands. Due to its vastness, it was many decades before the western suburbs achieved high levels of population density. Even today these urban expanses hardly compare with the eastern suburbs which have the city's highest concentrations of inhabitants per square kilometre.

TOWN PLANNING

Assisting the postwar housing and subdivision boom were municipal reforms enacted during 1919. One early pioneer of urban and civic reform was John Daniel Fitzgerald (1862–1922), Member of the Legislative Council and town planner, who in 1918 observed:

> *Everyone admits that Sydney has outgrown her plan, (if her casual growth can be called a plan), as a child outgrows its swaddling clothes. Visitors from abroad wonder that the great city managed to get through its work as one of the world's greatest commercial entrepots and goods-distributing centres in the face of its hilly contours, narrow streets, its huddled plan, the comparative inaccessibility of its harbour front, the want of adequate means of communication and of transportation between its outlying boundaries on the north, south, west and east … Sydney is choked.*[19]

After becoming minister for Local Government in the Holman ministry in November 1916, Fitzgerald argued strongly for the cause of urban and civic renewal. It was apparent that pressure on housing had reached crisis proportions in 1920 by which time the balconies of thousands of larger terraces had been filled in to accommodate the many lodgers and family members crammed inside. Some of them still remain in place today.

In the course of debates in Parliament House Fitzgerald spoke in favour of a new Local Government Act which would give councils greater power to regulate various matters including the establishment of industries in residential areas. The passage of the Act in 1919 gave the government much greater powers to control new roads, subdivisions and buildings – and the number of houses per acre

19. New South Wales Parliamentary Debates, 9 May 1918.

Apartment buildings in Sydney's eastern suburbs. circa 1928. Photgraphy by David Smyth

that could be erected in a residential area. It also gave the nod to brick veneer. An amendment ensured that municipal or shire councils in New South Wales could no longer reject plans for brick veneer construction of buildings provided they were properly constructed with adequate footings. For some the Act lacked teeth, especially when it came to strengthening planning powers and preventing urban sprawl. Yet many local councils took the opportunity to embark on beautification schemes that could provide more recreational facilities such as parkland and golf courses. Rising suburban land values, given additional impetus by the great many subdivisions, were a boon to councils who received a windfall by way of an increase in rate revenues.

On the other hand, the Master Builders Association saw too much red tape:

> *Sydney cries aloud for homes and more homes for its workers of every class. Building, health and other regulations make it impossible to erect a home at a figure that had any possible relation to the average worker's resources … It was [once] possible … to put up a dozen brick terrace homes on the minimum frontage required for half a dozen detached cottages. Right away, the biggest individual item – the land – was cut in half.* [20]

Despite these concerns, the 1919 Local Government Act did not hold back the building boom of the 1920s. In fact, so great was the demand for new homes and apartments that it created an enormous shortage of bricklayers and carpenters. The Master Builders Association offered practical remedies to meet the demand for skilled labour. During 1924 the establishment of a bricklaying school along the lines of those being conducted in the United States was mooted. Before inaugurating the project the possibility of importing bricklayers from Britain was taken into serious consideration, but eventually it was decided to establish an experimental school at Sydney, and to seek the co-operation and assistance of the state educational authorities. The school tapped into a large pool of 'unemployed lads and young men who had no skilled trade at their command, but who had been at various forms of unskilled labour and in many cases had reached an age when they were being supplanted by young

Index of new building projects in Sydney 1919–1932

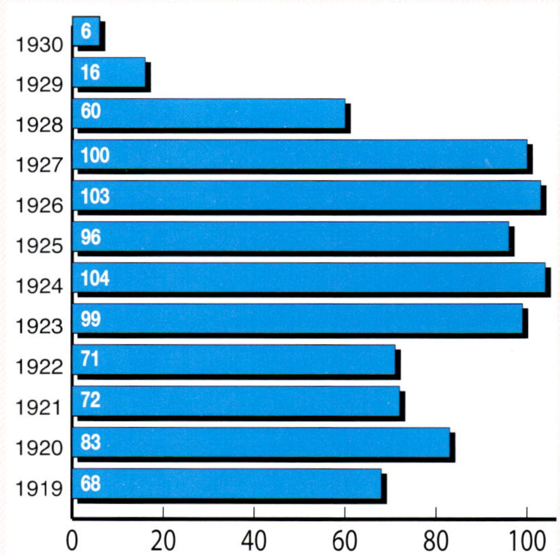

Base year 1927 = 100

Source: C.B. Schedvin, Building and the Trade Cycle in Australian Cities Between the Wars, Paper presented to the 35th Australian & New Zealand Association For The Advancement Of Science (ANZAAS) conference, Brisbane, 1961, p. 4

Principal suburbs for new apartment (flats) construction in Sydney, 1923–1942

	1923–1931			1932–1942		
Suburb	Rank	Number of new flats	% of metropolitan area	Rank	Number of new flats	% of metropolitan area
Sydney	1	972	21.5	7	202	4.4
Waverley	2	710	15.6	3	530	11.5
Randwick	3	627	13.8	1	630	13.7
Woollahra	4	512	11.2	5	456	9.9
North Sydney	5	501	11.0	2	559	12.1
Manly	6	298	6.5	6	389	8.5
Mosman	7	201	4.4	10	171	3.7
Ashfield	8	95	2.1	4	478	10.4
Petersham	9	84	1.8	9	186	4.0
Vaucluse	10	58	1.3	11	78	1.7
Marrickville	11	54	1.2	8	199	4.3

Source: Metropolitan Water, Sewerage and Drainage Board yearly and half-yearly returns of building activity in the board's area. Local government area is synonymous with suburb

20. *Sydney Morning Herald*, 26 August 1925.

boys, and it was concluded that the better course would be to train these young men in the art of bricklaying rather than import bricklayers from elsewhere.'[21]

DEPARTURES

Wartime and the decade of the twenties saw the inevitable disappearance of many older brickmaking establishments. Some operations, such as Daley's brickworks and the Toyer Brothers' pit at Dibble Avenue, Marrickville, were abandoned because they had been worked out. This was the eventual fate of many brickworks in the St Peters locality. By 1923, the nine-acre (3.6 hectare) property of Daley's Standsure Brick Company had been sold to Marrickville Municipal Council, later becoming the site of Henson Park. The Excelsior Brickworks, operating in the Croydon-Ashfield area of Sydney, shared a similar fate. In *Working The Clays: The Brickmakers of the Ashfield District*, local historians Chris Pratten and Nora Peek have compiled a most useful account of companies bearing the name Excelsior. It seems the name was first used in 1890 when The Excelsior Steam Brickmaking Co. Ltd was established in Ashfield, Sydney. The company changed hands during the depression years and was acquired by Henry George Downton, a contractor. Its debts were negligible, although the large sum of £333 4s 9d owed to the Sydney Brick Company suggests that the creditor may have been some sort of central brick selling agency for Excelsior and other brickmakers. By 1918 the Excelsior kilns were fired for the last time and in 1921 the property sold to Ashfield Council which used the site as a quarry and later as a garbage incinerator and disposal area. In 1971 it was filled and playing fields established. The site is now known as the Ashfield Centenary Sports Area.

Considerations other than commercial ones led to the closure of a number of inner city brickyards. Brickmaking was not a popular industry and the opening of the new Burwood works in 1913, for example, was not without controversy. Established works such as the Excelsior yard and the Croydon Steam Brick Works were put under increasing pressure in their later years from local residents who had built houses close to the works as urban development spread, but who objected to blasting, smoke pollution and noise.[22]

Flooded brickpit of the Standsure Brick Company, later the site of Henson Park, Marrickville. Marrickville Local Studies Library

View of the two patent kilns at the Excelsior Brickworks, Croydon, circa 1895. Burwood Local Studies Library

21. Master Builders Association of New South Wales: *Bricklaying and Plastering – The Bricklaying School*, Rare Books Collection, Mitchell Library, p. 3.

22. N. Peek and C. Pratten, *Working the Clays: The Brickmakers of the Ashfield District*, Ashfield & District Historical Society, 1996, p. 74.

WHERE PATHS CROSS

Of growing importance during the decade of the 1920s were the activities of two selling and distribution agencies that worked on behalf of brick manufacturers. The concept of a sales and distribution agency was nothing new and had developed from various attempts by manufacturers in the late 1870s to safeguard their interests through the Brickmasters' Protective Association of New South Wales mentioned in a previous chapter. This body was established in 1879 but wound up in 1893 at the height of the depression. For a short time its role may have been taken over by the Sydney Brick Company. We cannot be certain but it seems probable that the New South Wales Brick Masters' Association was the successor, and thereafter The Metropolitan Brick Company Ltd.

William H. Nicholls, a business partner of Richard Stanton who developed the Haberfield Garden Estate in the early 1900s, was a director of The Metropolitan Brick Company Ltd from about 1902 until at least the early 1920s. Insight into the working of the Metropolitan can be gained from a report entitled 'Brick Companies under the control of the Metropolitan Brick Company' submitted to the Profiteering Prevention Court of New South Wales. The accountant appointed to inquire into the Metropolitan, C.T. Crockett, stated that any person requiring bricks was required to order through this sales and distribution agent, which in turn placed it with the nearest yard. This was a practical measure, taking into account the cost of brick cartage. Crockett also added that:

> *The object of the company is, inter alia, to prevent competition and price cutting [by the manufacturers]. It has certainly proved very successful in this direction, [and] also in the matter of bad debts; defaulting customers cannot obtain bricks from an Association yard [NSW Brick Masters' Association] until they have paid for their former purchases.*[23]

Crockett's report also includes a table of eleven brickmakers affiliated with the Metropolitan Brick Company Ltd showing their profitability, although two yards, City Brick Company Ltd, and National Brickworks, made a sizeable loss.

Certainly, private manufacturers had concerns about profitability at a time (1922) when the State Brickworks was selling commons at 55s per 1,000 at the kiln, considerably less than the 68s per 1,000 being charged by private firms. The special advantages enjoyed by the State Brickworks continued to be a source of discontent, and its competition with private manufacturers was resented. Envy at the superior facilities enjoyed by the State's yard may have been a factor. The photographs on the opposite page were taken in about 1928 and clearly show an enterprise on a massive scale by the standards of the time. Few if any private brickmakers had installed mechanical excavators and trucks in the shale quarry, and certainly on nothing like the scale of the State's yard. It was only after the Second World War that modern technology in the form of diggers, scrapers and excavators was adopted, almost 20 years after their appearance at the Homebush works.

Brick firms distributing via the Metropolitan Brick Company Ltd, 1921

	Max normal output	Capital including goodwill	Profit per 1,000
Auburn Brick Co. Ltd	16,640,00l bricks	£32,978 18s 8d	5s 5d
Bedford Brickworks	8,106,985 bricks	£30,856 6s 6d	2s 2¾d
Butcher Bros Ltd	10,129,165 bricks	£85,097 14s 7d	19s 7d
Canterbury District Brick Co. Ltd	3,640,000 bricks	£39,808 6s 8d	12s 1¼d
City Brick Co. Ltd	2,132.000 bricks	£26,473 7s 7d	Loss of 11s 11½d
Clyde Brick Co. Ltd	6,430,000 bricks	£53,896 14s 6d	19s 8½d
Croydon Steam Brick Co. Ltd	3,120,000 bricks	£19,858 6s 11d	3s 3¼d
Judd's Hurstville Brick Co. Ltd	6.130,833 bricks	£43,732 6s 6d	10s 5¾d
National Brickworks Co. Ltd	6,760,000 bricks	£46,624 0s 10d	Loss of 9s 3d
State Brick Yards	Not available	Not available	9s 10d
Warren Brick Co. Ltd	7,150,000 bricks	£29,637 10s 3d	19s
Wilsons Brickworks Ltd	3,120,000 bricks	£11,905 8s 11d	15s 1½d

Source: C.T. Crockett, Brick Companies under the control of the Metropolitan Brick Co. report to the Profiteering Prevention Court, 1921

23. Letter from Metropolitan Brick Company Ltd to Ashfield Council, dated 24 April 1920. Archives of the Municipal Council of Ashfield.

Aerial view of the State Brickworks at Homebush, New South Wales, circa 1928. This photograph was taken following the installation of electric shovels in the pit and a belt conveyor system to haul the clay and shale to the top of the quarry. Across the Parramatta River with its railway bridge shown can be seen the distant suburbs of Concord West, Rhodes, Rydalmere, Meadowbank, Ermington and Ryde. Vince Higgins Collection

Aerial view showing the Parramatta River and open land that was later developed as the suburbs of Rydalmere and Meadowbank, circa 1928. Overall view of No. 1 and No. 2 works with cottage residence partly obscured behind first chimney on left with general views of abbatoirs and meatworks and Lidcombe and Auburn in the distance. Note the wharf and cranes used to load bricks on to punts. Vince Higgins Collection

Base of Ransome Rapier electric shovel being lowered into the pit at the State yard, Homebush Bay, circa 1928. The depth of the pit is about 30 metres. A fully assembled excavator can be seen at the pit bottom. Vince Higgins Collection

Thorneycroft truck being loaded with excavated shale by an electric shovel, circa 1928. Vince Higgins Collection

EXCELSIOR

Another sales and distribution agency was the Excelsior Brick Company Limited, established in 1920, which bore a name similar to the defunct Excelsior Brickworks at Croydon, although it was a different company altogether. Just like the Metropolitan, Excelsior controlled the sales of participating brickmakers and appears to have been owned by the NSW Brick Masters' Association. Established to 'carry on the business of manufacture and to buy and sell Bricks, Tiles, Brick Earth, Chalk Sand Pipes, Drain Pipes, Pottery, Ornamental Pottery, Earthenware China, Terra Cotta and ceramic ware of all kinds' there is little evidence that it made any bricks at all.[24] The existence of such companies reveals the extent to which brickmakers were bound together by common interest. A directorship in Company A might lead to a seat on the board of Company B. Company C, on the other hand, might own shares in A and B but have no representative on the board of directors. For example, the initial subscribers in the Excelsior were Alexander Stuart, a building contractor of Marrickville, Arthur Scott, a Randwick builder, James Alexander, engineer from Moore Park, Edwin Phippard, a builder and brickyard owner from Randwick, George Hocking, a manager from Canterbury and George Whiteman from Beecroft. Stuart had been president of the Master Builders Association, was president of the New South Wales Brick Masters' Association, and, by 1930 sat on the board of Austral. Edwin Phippard was also a director of Austral from 1912, and Hocking sat on the board of the Canterbury Brick Company Limited.

For a short time during 1928 and 1929, about the time the East Hills railway was being constructed, the Metropolitan yard was established in Lakemba. Set up on 19 acres of land on Penshurst Road, the yard was well-equipped with a large patent kiln, two downdraught kilns and plant and machinery and manufactured about two million bricks in 1930. By the end of the year the plant was closed and never produced another brick.

24. These records are held in the archives of Brickworks Limited and relate to the period commencing with the establishment of the Excelsior in 1920 to its liquidation in 1938 by its then owner, Brickworks Limited.

The fact that a William Nicholls was associated with Metropolitan and Excelsior and shared office premises with the aforementioned at Castlereagh Street points to the two companies having close ties through the Association. Nicholls resigned in January 1922 due to ill health, his successor being Bert Moseley who had started with the Metropolitan as a stenographer and typist in 1903. Moseley was immensely knowledgeable, influential (he, too, sat on quite a number of boards), and was manager of the Metropolitan until June 1936. He was also a devout Baptist. Clearly Moseley had intimate knowledge of the industry for in 1920 he was appointed company secretary of the New South Wales Brick Masters' Association, which was registered as an industrial union of employers, with an interest in all industrial matters such as wages and awards. The Association was also run from 12 Castlereagh Street in Sydney.

Further proof of the complex web of relationships that existed within the brickmaking community can be gleaned from the minutes of a meeting of the Excelsior Brick Company Limited on 11 December 1923. Following a number of resignations from the board E.D. 'Edward' Lanceley, whose father was one of the founders of the North Sydney Brick & Tile Company, was appointed as a director. The meeting also resolved that Excelsior's application for shares in the following companies be granted:

- The North Sydney Brick & Tile Company – 920 £1 shares
- Butcher Brothers Co. Ltd – 920 £1 shares
- Northern Suburbs Co-Operative Brick Company – 920 £1 shares
- National Brickworks Company – 800 £1 shares
- Wilson Brickworks Limited – 240 £1 shares
- Hornsby Tile Pottery & Brickworks Limited – 200 £1 shares

Lanceley, of course, was also managing director of the aforementioned North Sydney brickmaking firm. On 15 January 1924 he was joined on the board by Francis 'Frank' Butcher, managing director of Butcher Brothers, and Alfred Burroughs, a long-time director of Austral and president of the Master Builders Association. The latter stepped up to fill vacancies left by the departure of Stuart and Phippard. It is not unusual for companies, or any other organisation for that matter, to develop interlocking and overlapping relationships. What is surprising is how often the same names and faces appear on the boards of the various brick companies.

RIDING THE BOOM

By the late twenties it was possible to look back on almost a decade of strong growth at Austral. Both the value of production and the work force had steadily increased. The company was in good hands and making big profits from the construction boom sparked by another surge in Australia's population that had reached six million by 1928. Improvements in the form of plant upgrades were made throughout the industry, especially at the State Brickworks at Homebush in Sydney. Bob Hudson, the yard's brilliant chief engineer pioneered containerisation, loading brick directly from the kiln onto pallets placed on a flat-bed narrow gauge rail truck. These were taken the short distance to the wharf at Homebush Bay and sent to Pyrmont on barges where they were stashed in cargo nets and winched aboard coastal ships. In fact, all government buildings up and down the east coast of New South Wales used the same type of dry pressed brick manufactured at the State yard.

At Austral, modernisation of the plant went ahead under the direction of Bob Young, chief engineer at Austral during the 1920s and 1930s and major domo of all the yards owned by Brickworks Limited after its formation in 1934. Young was a personable fellow and would have found time to swap yarns with Bill Dawes, although the two men fell out just before the Second World War. Young left to work at Speare's yard which must have rankled with Dawes who disliked his competitor intensely. The feeling was mutual. Young returned to Austral after the war years. Both of them, however, went back a long way – to St Peters Primary, in fact, where they

Decorative business card holder in press copper, circa 1920. Ron Ringer Collection

had been schoolyard acquaintances. While steam engines remained the main form of motive power, the conversion of plant equipment to electricity and diesel was something that Young had firm control over. He would also have made it his business to keep up with brickmaking technology. For example Young met with local machine manufacturers including the Todd and Williams families in St Peters whose engineering firm, Todd Williams, made brick making and gold ore crushing machinery. At some point during the 1920s the firm was acquired by Robert Shannon, a successful brickmaker and entrepreneur. Shannon's forebears were Irish builders and bricklayers who settled in Sydney during the second half of the nineteenth century. By the 1890s Robert Shannon, a young man still in his thirties, had built up a huge investment portfolio consisting of more than 80 properties along Sydney's Parramatta Road. In 1925 he was operating yards in Merrylands and Wentworthville, and later in Canberra and Brisbane.[25]

25. Interview with David Shannon, grandson of Robert Shannon, on 16 March 2006.

Bob Young was also familiar with the Raynor Side Lifter dry press machine as well as the Marrickville Side Lifter press, both of which were manufactured locally. The Whittaker Dry Press machines in use during this period were imported from Britain and could stamp four bricks in one stroke. Bradley & Craven's grinding pans were also imported from England. The US-made large capacity Incla pans imported at the time are still being made today.

As the Austral business continued to expand, Spackman and the board took the logical and necessary precaution of acquiring more land to ensure sufficient reserves of clay and shale, as well as to provide for any future expansion of the yards. A phone call to Mr Turtle or Bruce Parks at the Newtown real estate office was sufficient to set things in train, and according to council records the early months of 1927 seems to be when it all happened. By April of that year Austral had become the owner of a shop and dwelling on a block of land at 316 King Street measuring 40 feet by 120 feet, purchased from Albert Warnes.[26] The purchase of a substantial holding in a competitor operating close by, The Central Brick and Tile Co. Ltd., was a further mark of Austral's growing confidence. Turtle's guiding hand secured the transfer of another two small parcels of land (51 feet by 120 feet and 57 feet by 118 feet) on King Street into the hands of Austral. It was turning out to be a good year all round. The company manager, David Rogers, finally joined the Freemasons, along with his brother, Len, who would be appointed company secretary in 1929. In February 1927 Archibald Howie, an Austral investor, was elected President of the Master Builders Association. Howie was fourth generation in a family of builders and principal of Howie, Moffat and Co. Ltd. This notable company was responsible for the construction of a number of major buildings, including the Mitchell Library, The Art Gallery of New South Wales, Long Bay Prison, Bond's Cotton Mills and the forts at Middle and South Head.[27] Another Austral investor was Sir John Harrison, an experienced and wealthy builder.[28] Harrison was also a respected philanthropist. Immediately after the end of the First World War he had been involved with the New South Wales Department of Repatriation, building low-cost housing in the sandy coastal region beyond Daceyville in Sydney. The settlement housed wounded soldiers and war widows and became known as Matraville. The scheme captured the imagination of public-spirited individuals such as 'Big John' Harrison. Indeed Harrison paid for much of the skilled labour employed during construction out of his own pocket.

Don Bradman, 1934. Aquaintance of Edwin Spackman. National Library of Australia

Unfortunately this did not prevent him from becoming embroiled in a very public altercation with Archibald Howie who implied that he had been disloyal to the Master Builders' Association during a recent strike by crane and hoist drivers.

26. St Peters Municipal Council, Valuation Book, for the three years ending 31 December 1927.

27. *Sydney Morning Herald*, 18 February 1927, p.14.

28. Sir John Harrison was born in 1866 and died in 1944. He is not to be confused with Sir Eric John Harrison (1892-1974) who was elected member for the Federal seat of Wentworth in December 1931. Dawes was well acquainted with both of these men.

MASTER BRICK CARTERS' ASSOCIATION

Phone : L 1789

108 KING STREET,

ST. PETERS,

Letterhead of Master Brick Carters' Association, circa 1929

Agreement between The Austral Brick Co. Ltd and the Master Brick Carters' Association, circa 1929.

1 Deliveries are to be made at such time and by such mode of conveyance as the carriers may think fit.
2 The rates for carrying shall be the rates set out in the Scale of Rates exhibited from time to time at the Company's Office unless varied by mutual agreement in pursuit of any particular contract.
3 Each delivery shall be deemed to contribute a special contract.
4 The contractor shall be at liberty to undertake any other carrying work and these terms shall not be deemed to constitute the contract or the agent, either sole or otherwise of the company.
5 The company shall be entitled to object to any application without assigning any reason therefore.
6 No officer, servant or employee of the company shall have any authority to vary, modify or alter these conditions but they may be varied from time to time by notice in writing and signed by a director of the company and shall remain in force until withdrawn by such notice in writing.
7 The company shall not be liable to pay for the delivery of bricks upon any other terms than the foregoing.
8 Payment shall be made as agreed upon.
9 Bricks to be delivered under these conditions will be available to the contractor during the ordinary hours of the brickyards.

Source: Handwritten notes made by W.K. Dawes on letterhead of the Master Brick Carters' Association, Archives of Brickworks Limited

Stung by this assertion, Harrison accused Howie of being tactless about work he had undertaken for Smith and Waddington on Parramatta Road, where his company acted to avoid being in breach of contract on a matter for which there was still no explanation.[29] Clearly, there was more to the disagreement than can be ascertained from our sparse records.

Meanwhile, reflecting the growing prosperity of his company, Ted Spackman, a cautious man, felt that years of parsimony and hoarding had paid off. Taking the plunge he moved from the brick grime of inner city Silver Street at St Peters to Frederick Street in the upwardly mobile suburb of Rockdale. At last, fresh sea air! Quite by chance, Spackman found himself the next door neighbour of Frank Cush, secretary of the St George District Cricket Club and later president of the Australian Cricket Board. Cush took in Donald Bradman who had come to live in Sydney as a young man of 20 during the 1928–29 cricket season.[30] The connection was not lost on Dawes, a follower of the game, who persuaded Spackman to accompany him on the long drive down to Melbourne to see Bradman in action. If Dawes' driving habits were anything to go by, then Spackman would have fared badly on the bumpy trackway between the two capitals. Bound together by their association with the brick trade, the two men, nonetheless, were very different people, although they both got on well together.

MANAGEMENT MOVE

Dawes, too, was riding high, and not just on the success of his brick trucks. By the middle of 1926 he had already made the transition into management, evidenced by the court case brought against him for breach of working conditions. His appearance before the Chief Industrial Magistrates Court on Tuesday 19 October 1926 marks the first mention of Dawes in official documents. Out of the blue we discover an obscure reference to him by Fred Parker, secretary of the Federated Brick Tile and

29. *Sydney Morning Herald*, 20 May 1927, p.11.
30. Frank Cush's son, Neville went to school with Arthur Dawes, the brother of William King Dawes.

Pottery Union. Parker noted in his diary that a complaint was 'found against W.K. Dawes of The Austral Brick Company Ltd for breach of Clause 5 (iii) of The Brick Masters' & Assistants Awards, who obliged men to work 11 hours in one shift.'[31] Austral was fined 5s with costs of 4s, which must have given Parker some degree of satisfaction, but left Dawes spoiling for a fight. He maintained a lifelong hostility towards trade unions.

Trusted by Spackman, and recognised as a man of some authority, Dawes was appointed to succeed David Rogers as plant manager. From evidence tendered at a 1940 court case, Attorney General v. Brickworks Limited, Dawes states that his experience of the brick industry commenced in approximately 1927. This was not strictly true if we take into consideration his work as a brick carter over the previous six or seven years, and his presence in the Industrial Court on 19 October 1926. This last date, however, does provide us with a formal starting point and can be verified through Sands' Directory.

Dawes was ambitious, yet it took more than ambition to run a successful company in what was an exceedingly tough industry. He was capable, and had proved it to Spackman and the other directors who presumably would have voted on the decision to appoint him as works manager. Dawes' appointment seems not to have affected his contract with the company as a brick carter. Today this would be regarded as a conflict of interest, although in times past such practices raised few eyebrows. If Dawes harboured further ambitions, then they were in the direction of the board and eventual control of the company. Accordingly when the chairman, Alf Smith, announced at a meeting of shareholders on 13 August 1928 that a number of shares in the Company were for sale, Dawes seized the opportunity. At the next meeting held on 11 February 1929 he appeared on the list of shareholders, a move that paved the way to a seat on the board. Indeed, by his own admission at the 1940 court hearings Dawes stated that he became a director of the company in 1929 and also general manager of a company that owned quarries and offices covering 23 acres of the site off King Street, just down the road from the impressive yard belonging to the Bedford Brickworks which occupied 16 acres at St Peters Bridge.

THE RECKONING

Outwardly, the twenties were full of gaiety and the promise of better things, but other signs were less encouraging. Inflation was on the rise with unemployment hovering around 11 per cent. Cities continued to grow, but Australia's rural population was in decline with both urban and rural poor feeling the effects of the slowing economy. Sadly, this went largely unnoticed by the rest of society. Growing antagonism between capital and labour was creating a bitter mood. To cap it all, the states continued to borrow far beyond their means to repay despite the efforts of the Commonwealth Loan Council. What happened next – the Wall Street Crash of October 1929 – was out of everyone's hands, but how the Austral Brick Company Ltd responded was very much up to one man. William King Dawes was about to hit his straps.

31. Diary notes of Fred L. Parker, Archives of the Federated Brick, Tile & Pottery Union papers, now held by the Construction, Forestry, Manufacturing, Engineering Union (CFMEU) in Sydney.

On the Edge 9

On Wall Street – Sell out? – Down all the days – Open for business – Who's in charge? – The Brick Agreement – Trade War – Praise – Bedford Brick Works – Old wine, new skins – On Homebush Bay – Up for sale – Some important discussions

Cave dwellers of Kurnell. Homeless people took refuge in stone huts and caves in Sydney's bushland during the Depression decade. National Library of Australia

ON WALL STREET

For much of the twenties the United States rode a wave of land and property speculation given impetus by the expansion of easy money. Customers were encouraged to buy on credit or to borrow heavily from banks that stood to earn huge profits from interest repayments. 'Spend now, pay later' appealed to middle America which was determined to enjoy the boom years following the war. Few people, however, realised that in the event of a credit squeeze or a slump in confidence consumer demand would fall, goods pile up and prices collapse leading to deflation. On the other side of the Atlantic Ocean, Britain was struggling with problems of a different nature and magnitude. The country was still mired in debt, the legacy of its wartime borrowing from US financial institutions. Debt levels in Australia were equally burdensome and there were indications from about the mid-twenties that the economy was beginning to contract. Dependent on Great Britain for its loans, the country was dangerously exposed.

Ignoring the growing signs of structural instability in the US economy, share traders were still riding high at the end of August 1929. Despite a fall in prices during September, speculation continued; conditions were ripe for a massive correction to the stock market. A warning shot was fired that month when the Bank of London withdrew its funds from American banks triggering an immediate decline in the value of the US dollar. On 18 October 1929 the market plunged, recovered slightly, with the bubble finally bursting on 24 October, 'Black Tuesday'. By November 1929 the share index had fallen to a new low of 224 points and continued in freefall, reaching a historic low of 58 points in July 1932. In the same period (September 1929 to June 1932)

Southern Cross flying over Sydney Harbour Bridge, 1931. National Library of Australia

Standard and Poor's Composite Index fell by 86.2 per cent and it was not until September 1954 that it passed the level of September 1929, a hiatus of 25 years. The collapse of the New York Stock Exchange precipitated the longest and most severe depression experienced by the industrialised Western world. Once the American economy faltered and the flow of American investment credits to Europe had dried up, the slump turned into a worldwide collapse, commonly referred to as the Depression.

In Sydney the impact was felt immediately, the earliest and most marked indication being a rapid shutdown in building construction. The Master Builders' Association and its board of directors, including president, H.J. Thompson, watched helplessly as the crisis deepened. Even before the New York crash, building work across Australia, especially in the capital cities, had begun to slow. By 1932 it had virtually ceased. Figures supplied by the Commonwealth Statistician illustrate the decline and partial recovery over a five-year period from 1928. The totals shown represent the value of all building operations in Australia including new works and alterations. The one exception was the construction of the Sydney Harbour Bridge, which became the symbol of a city united and a sign that a return to suburban prosperity was possible. Its opening in 1932 was an opportunity for much-needed celebration.

The Depression blighted the lives of millions of Australians leaving a deep mark on the national psyche. Poverty, especially in rural areas, was widespread, due mainly to unemployment which was by far the most crushing aspect of this economic catastrophe. Communities dealt with it as best they could, yet despite the lack of work prospects adversity tended to bring out the best in people. Much later, many recalled the Depression years as a time of great generosity, concern for others, and a growing sense of community. Contrary to popular belief relatively few people were evicted from their homes. There were, however, instances in Australia of shanties which sheltered the homeless, most notably those at Cronulla. Likewise, references to gangs of jobless men moving from town to town ('on the

Value of builcing operations in Australia, 1928–1933

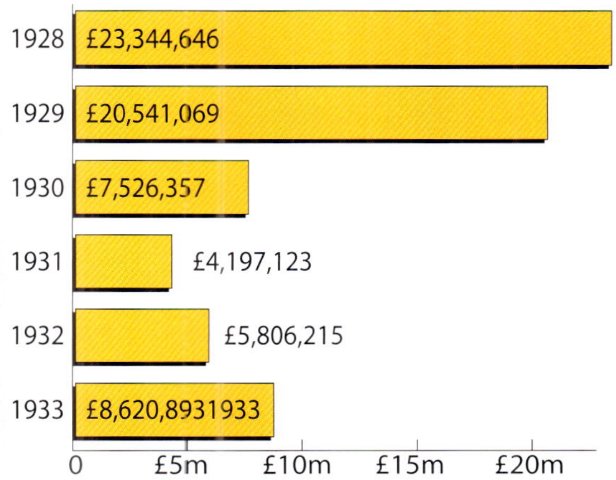

Source: Clay Prcducts Journal of Australia, April 1934, p. 7

wallaby') are exaggerated. On the other hand, in the United States starvation and eviction walked hand in hand and there were few alternatives but to take to the road. Life aboard the American transcontinental railway in empty boxcars became a safety valve for many homeless men whose only hope was to remain on the move in search of seasonal work.

In Australia the vast majority of unemployed men stayed put in their own communities. Relief schemes were limited, although some local councils in Sydney's inner city, including Marrickville Council, extended earlier programmes such as putting men to work on paving roads and footpaths with brick. Some of these paved areas can still be seen in Juliet and Bourne Streets, Enmore, and Victoria Street, St Peters, and in Hillcrest Road, Tempe.[1] Brick roads had particular appeal for the simple reason that installing the bricks was labour intensive and required few skills. Most of these roads were laid in a basket weave pattern, although alternative designs were also used. A local brickyard, the Central Brick and Tile Co. of St Peters, supplied the bulk of bricks used in similar schemes in the early 1920s and continued to do so.[2]

1. Portions of road were sealed with asphalt in the year 2000 for protection and a large section on the east side of the road was re-paved in 1988.
2. Approximately 20 per cent of Marrickville footpaths were brick-paved in the 1920s.

Aerial view of Central Brick & Tile Co Limited. Inset: Front cover of trade catalogue, Central Brick & Tile Co Limited, circa 1929

SELL OUT?

How did all of this affect Sydney's brick industry? Well in advance of the share collapse of October 1929, Sydney brickmakers, like the builders, were beginning to feel the pinch. Price-cutting was driving manufacturers out of business and drastic action with all its radical implications was being considered. Responding to a proposal put forward by Alexander Stuart, president of the New South Wales Brick Masters' Association, the directors of South Ashfield Brick & Tile Co. Ltd voted on 25 September 1929 to endorse a scheme for the company to exchange all of its brick-making assets for 30,000 shares in a central company to be formed at a later date. Everything, including land, plant, machinery, kilns, buildings, brick stocks, maintenance parts, tools, coal and working capital would be pooled along with the assets of other companies who elected to join the scheme. At an extraordinary general meeting held at the School of Arts in Pitt Street on 14 October 1929, shareholders of The Austral Brick Company Ltd were also asked the same question. They were to:

> *consider The Austral Brick Company Limited joining a proposed new Company for consolidation of Brick manufacturing interests, and to pass, or reject the following Resolution to be moved by Mr A. Howie: 'that this meeting of Shareholders approve of such scheme, and authorise the Board of Directors to accept same.*[3]

These were desperate measures indeed and provoked a great deal of soul-searching. Two months later at a meeting of Austral shareholders on 11 December 1929 the resolution was finally put:

> *that this meeting approve of the Company disposing of its brick-making assets to a new company to be formed and registered under the New South Wales Companies Act. The value to be received for such assets when agreed upon to be satisfied by the issue to this Company of fully paid up Shares of the proposed new Company, and authorises the Directors to carry into effect such sale upon such terms and conditions as the Directors may think fit.*[4]

3. Minute Book, The Austral Brick Company Limited, Archives of Brickworks Limited, 14 October 1929.
4. Minute Book, op cit., 11 December 1929.

Howie, a force to be reckoned with in the building and brickmaking industries, failed to gain the necessary three quarters majority for his resolution. With the industry on the verge of collapse it is hardly surprising that members of the Austral board were divided on how best to respond to the scheme originally proposed by Alexander Stuart. Determined to pursue the matter Howie and his faction took legal action. Spackman fought off the challenge unwilling to relinquish control over the company he had founded. Company minute books usually record little or nothing about internal squabbles, but on this occasion we are given a tantalising glimpse into the politics of Sydney's brickmakers. In fact, it should come as no surprise that Stuart's radical proposal was finally accepted almost five years later when the manufacturers voted to establish Brickworks Limited. Anxious not to be drawn into a public brawl at the December meeting of shareholders, Dawes assured those present the company had nothing to fear from price-cutting and the Depression. In a move obviously intended to placate those who disapproved of his brick carting business, Dawes also spoke of his cartage contract with the company and 'produced the same for the members' perusal.'[5] Today this would not be acceptable under the Corporations Act. Even at the time some shareholders considered it to be a matter worth raising, indicating a level of disquiet about the practice.

DOWN ALL THE DAYS

The extent to which the industry's fate hung in the balance was documented by Bert Moseley, secretary of the Metropolitan Brick Company, the distribution agency established on behalf of the New South Wales Brick Masters' Association. According to Moseley in mid 1929 there were 25 companies operating a total of 29 yards across metropolitan Sydney. Not a single yard was idle, and in the course of the year they sold a total of 216,290,422 bricks at an average price of 70s per 1,000 for common bricks and 118s for face bricks. Moseley records a very reasonable profit of 12s 5d per thousand. At the same time the chief men in the industry, including burners, setters and draggers, were earning about the same money, or £5 6s per week. Pitmen labouring in the quarries were on slightly less at £5 4s per week.[6] By early 1930 it was a different picture altogether. Large-scale brick operations were quick to follow idle construction sites into oblivion, and a once healthy and productive industry collapsed like a pack of cards. Yards that remained open could be counted on the fingers of one hand, and included Austral, its near neighbour the Central, and the State Brickworks. Speare continued intermittently for some time.[7] The recently established *Clay Products Journal of Australia* (CPJA) painted a grim

5. Ibid.

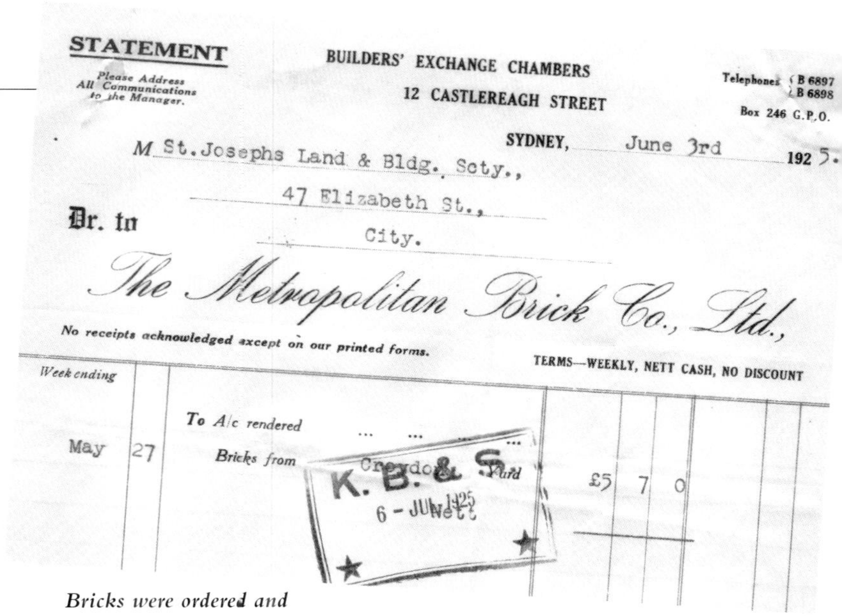
STATEMENT

Please Address All Communications to the Manager.

BUILDERS' EXCHANGE CHAMBERS
12 CASTLEREAGH STREET
SYDNEY, June 3rd 1925.

Telephones B 6897, B 6898
Box 246 G.P.O.

M St. Josephs Land & Bldg. Scty.,
47 Elizabeth St.,
City.

Dr. to

The Metropolitan Brick Co., Ltd.,

No receipts acknowledged except on our printed forms.

TERMS—WEEKLY, NETT CASH, NO DISCOUNT

Week ending			
		To A/c rendered	
May	27	Bricks from Croydon	£5 7 0

K. B. & S. 6 - JUN 1925 Nett

Bricks were ordered and distributed through the Metropolitan Brick Co. Ltd.

Clay Products Association of Australia

Ernest Mashman, (1895–1964), founded the Clay Products Association of Australia in 1931 with Ernest Swain. The association established branches in every State and represented pottery interests to the government as well as promoting the trade. Mashman was president of the association in 1937–38 and continued to be a prominent advocate for most of his life. He was largely responsible for expanding the membership which came to include brick and ceramics manufacturers.

In November 1933 Mashman and Thomas Campbell, the managing director of R. Fowler Ltd, founded the association's magazine, *The Clay Products Journal of Australia*. This was superseded by *Australian National Clay* in 1959.

6. Bertram Moseley, testimony before the court, Attorney-General v. Brickworks Pty Ltd and Others, 1941, Supreme Court of NSW – Inquiry, 1940, p. 465.
7. Speare's Brickworks was owned by its managing director, Ernest Loftus Speare, who worked in his father's brickmaking firm, Newtown Brick Works. 'Speare's' produced dry press face and commons, as well as clay pipes and sold mostly in Sydney's eastern suburbs, the inner west, and south east, but not on the North Shore. Examples of his work can be seen at the nurses' quarters, St Vincent's Hospital and the Royal Prince Alfred Psychiatric Block, an Art Deco corner building. Speare refused to join the New South Wales Brick Masters' Association.

picture. Out of four brickyards in the Canterbury District which was considered to be one of the most productive in Sydney, only one yard had worked for about two months during 1933, having sold a meagre one-third of the bricks produced. The other three yards were idle. Out of twelve yards at Alexandria and St Peters three worked at normal output with nine closed up. For Australia as a whole, the statistics were nothing less than a disaster.

The picture is equally gloomy if we consider the brick, tile, pottery and earthenware industry for the whole of the period, 1908 to 1939.

Bricks, tile, pottery and earthenware, 1900s – 1930s: industry statistics for Australia as a whole

Year	Number of factories	Persons employed	Salaries and wages paid	Value of output £'000
			£'000	
1908	426	5,466	478,000	1,000
1913	467	8,226	938,000	1,888
1920–21	375	9,114	1,701,000	3,599
1928–29	367	8,684	2,036,000	4,593
1929–30	356	7,341	1,669,000	3,686
1930–31	271	3,534	686,000	1,376
1931–32	233	2,815	462,000	952,000
1932–33	274	3,935	652,000	1,464,000
1933–34	316	5,512	880,000	1,983,000
1934–35	334	7,433	1,239,000	2,934,000
1935–36	345	8,512	1,510,000	3,599,000
1936–37	335	9,288	1,749,000	4,112,000
1937–38	360	10,209	2,045,000	4,790,000
1938–39	368	10,484	2,178,000	4,975,000

Source: The Clay Products Journal of Australia, 1 May 1940, p. 15.

At North Sydney only one yard out of six was at work, and then at half its normal output. Another, had stayed open for just six weeks, with three others standing idle. Edward Lanceley, a director of North Sydney Brick & Tile Company, and proprietor of the Ryde Brick & Tile Works, later remarked:

> [The building trade] *is the first to feel the effects of a depression. As soon as a depression comes the people cannot spend their money and do not spend and the brickmaking and the building are the first people to suffer. In 1930 when the depression started price-cutting commenced, and we did not do much of that at all; we lowered our price down to a price and stopped. The price of bricks had reached 73/- per thousand in 1929 just before the onset of the depression. Then it slumped rapidly to half that figure.*[8]

Even Lanceley, a successful businessman and manufacturer, was forced to scale down his operations on the North Shore, an area which had suffered less in the savage price-cutting wars being waged south of the harbour. On 16 June 1932 a special resolution of the North Sydney Brick & Tile Company Limited was passed to reduce the company's capital.[9]

Meanwhile, out of ten yards in the Western District, one had worked during the whole year, one for six months, one for two months, and seven were idle. Some yards had stocks on hand that had been there for three years.[10] In April 1932 only four men were employed at the Burwood Brickworks. The minute book of the South Ashfield Brick & Tile Co. Ltd reports that production stopped altogether on 8 January 1930 due to the large accumulation of stocks of face bricks. In April trade was so poor as to prevent donations to the Women's Hospital. In June the directors took a pay cut, halving their fees, 'owing to the present existing depression of the financial position in the Commonwealth and the paucity of trade.'[11] The following month the company laid off its manager, Mr Cave, solely due to the Depression. Even the State Brickworks on Homebush Bay saw the number of people employed falling from 450 in the period 1928 to 1929 to just 55 in 1932. The chart on page 157, shows the extent of the decline in turnover.

Further west at the Wunderlich plant in Merrylands the situation was also desperate, although earlier diversification into a range of building products enabled the company to weather the storm.

8. Edward Lanceley, Attorney-General v. Brickworks Pty Ltd and Others, op. cit., p. 373.
9. State Records NSW, Register of Companies 1910 –1913, Container 11/11785.
10. *The Clay Products Journal of Australia*, November 1933, p. 31.
11. Minute Book, South Ashfield Brick and Tile Company Limited, Archives of Brickworks Limited, April 1930.

Workers pose for a group photograph at the Carrington Yard, circa 1925. Note the oil cans used to lubricate dry press machinery, also the coal scoop and hook (centre) in the hands of the burner. See Chapter 14 for more details.

The yard owned by the Bakewell Brothers in MacDonaldtown ceased to operate in December 1929 with 700,000 bricks stacked in the grounds leaving no room to spare. The plant remained closed until September 1938, whereupon it opened for a few months, closing for good in November 1938.

Annual brick production as at 30 June, State Brickworks of New South Wales

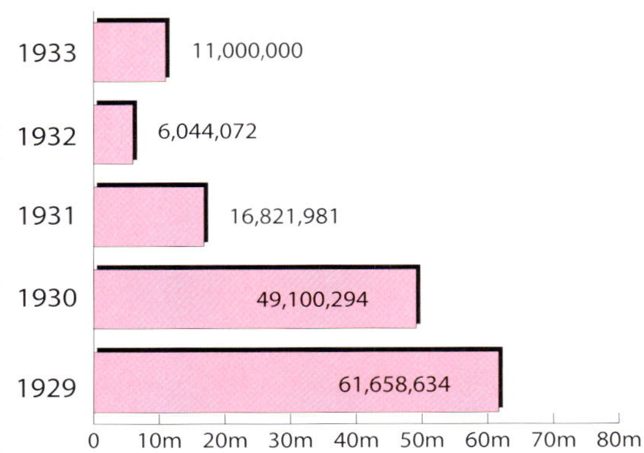

Other companies that ceased active life as brickmaking facilities included the Warren, Carrington, St Peters, Sydney, and Federal. In the case of the latter the only employees remaining were its owner, Fred Phippard, the foreman and a few carriers. In October 1932 Moseley pointed out that two out of forty yards in the metropolitan area were working and then only at half production. Others were selling their accumulated stocks. Some yards were even running out of shale. By 1940 Federal's Alexandria yard had been completely worked out, although the hard nature of the shale hastened its closure due to the extent of wear and tear on the machinery. Marrickville Council acquired the site as a council tip before restoring the land as a public park – Camdenville Park. Ten kilometres to the south-west, the Carlton yard had little more than seven or eight years' reserves of clay and shale, although it continued to operate for a short while following the end of the Second World War.

Closure of a brickmaking yard, however, did not necessarily mean the liquidation of the company and its assets. Most continued to file tax returns and were still on the Register of Companies as going concerns for some years to come. In fact, many received subsidies from the New South Wales Brick Masters' Association and its successor body, the Council of Brick Manufacturers, as compensation for having ceased production.

Times were hard for everyone involved in the brick trade. Fred Parker, secretary of the Federated Union of Brick, Tile and Pottery Workers, made a diary note at the time of the Tile Conference on 22 January 1930:

> "Man to man is so unjust
> That I know not whom to trust.
> I've trusted many to my sorrow
> You pay today, I trust tomorrow."

Workers, brickmakers and builders – all could relate to Parker's gloomy reflections. The *Clay Products Journal of Australia* tried to lighten the mood in its regular joke column, 'From the Pug mill'.[12]

From the Pug mill. – Old lady (to men struggling to move large piece of stone in a quarry): "My good men, why don't you try blasting it?" Workman: "Well, mum, me and my mate's used every blinking word we knows, but it ain't moved it yet." The Clay Products Journal of Australia, April 1935, p. 29. Artist David Dickson

OPEN FOR BUSINESS

Despite a reputation for being a hard man Dawes was loyal to his men, and they to him. Former employees claim that Austral may have been the only brickworks in the country to remain open continuously during the Depression years, although this is difficult to verify. What's more, the men preferred the system of piecework introduced by Dawes after he was appointed manager in 1926. In addition to the award everyone was paid according to the number of bricks produced. For example, Austral's setters (a five or six-man team that stacked the green bricks in a patent kiln ready for firing) were expected to set an individual tally of 6,000 bricks a day. For the three draggers who unloaded the fired bricks from the kiln chambers, it was usual to shift up to 30,000 per day. According to Ray Dawes: 'my father kept piecework going as long as

Hupmobile, circa 1927. Frederick Dawes in the backyard of his produce store, inspecting his son's new 'Hupmobile'. These were built by the Hupp Motor Company of Detroit, Michigan from 1909 until the firm's closure in 1940. Hupp's philosophy was to build tough and durable car at prices the working man could afford. W.K. Dawes traded in the car in 1929 for the more impressive 1929 Buick Silver Anniversary, which carried the Dawes family on Sunday excursions in search of new deposits of clay and shale. Bill Wheatley Collection

12. *The Clay Products Journal of Australia*, April 1935, p. 29 and *Clay Products Journal of Australia*, June 1935, p. 32.

Austral Brick pits at 12.30pm on 23 January 1933. Heavy rain during the summer of 1933 caused serious flooding and halted clay and shale extraction at Austral's quarries at St Peters, Sydney. Brickworks Limited

he could. He fought the union over this. The men would work until they got their tally and go to the local pub, the *White Horse Hotel*. The idea was to get as many bricks out of the kiln in the shortest possible time. Bricks would still be smoking as they were being unloaded.' [13]

Norm Binskin, one of the setters at Austral's Eastwood yard, worked at the St Peters plant during the 1950s and recalls the tally for setters being 5,000 per day. The target for a three-man team of draggers was similar to previous levels set two decades earlier. By the mid-1950s a bonus was paid weekly, but only if the worker had attended every day in that pay week. This more or less compelled an employee to attend work on Friday no matter how drunk he had been the night before, Thursday being pay day.

The application of piece rates was one way to maintain production. Throughout the Depression years, however, Dawes made it his business to keep ahead of the competition. Rather than close the kilns he kept them working, albeit at reduced rates, then stacked the yards with bricks to a height of several metres. With recovery after 1935, and also at the end of the Second World War in 1945, Dawes made a small fortune for Austral, selling bricks at higher prices than would have been obtained when produced. A man on a mission, he was never idle, covering the length and breadth of Sydney in search of clay and shale deposits and also to spy on the opposition. 'It was all bricks, even on family occasions', recalls Ray Dawes of his boyhood years, 'driving out to the opposition every Sunday to see how much land and clay they had left.'[14]

WHO'S IN CHARGE?

In running the yards at close to capacity, Dawes burned a great deal of coal in the kilns from the state-run mine at Bellambi. In fact, if we backtrack for a moment, the galloping rate of fuel consumption may have prompted a nasty tiff at an ordinary general meeting on 17 February 1930 held at the School of Arts in Castlereagh Street. During the proceedings Sir John Harrison and Archibald Howie were elected as directors. It appears they had something important to say. With the support of William Turland they circulated a memo claiming there had been an excess of coal used in the dome kilns, as well as other (unspecified) 'leakages' from the Austral yard. Tension between Alf Smith, who chaired the meeting, and Howie, bubbled to the surface with Smith accusing the latter of trying to blacken his name. Howie shrugged off the challenge, tabling comparative figures for the consumption of coal at the NSW Brick Company over a similar period. But there was more to it than suggestions that coal might be leaving the yard through the back door. Howie also insisted that directors write down the value of the brick stock. Before matters spun wildly out of control, Dawes spoke up: 'I have always strongly urged the plant manager to have a weighbridge installed'. His suggestion appears to have calmed tempers, although not for long. The gloves were off in what was turning out to be a battle for control of the company. Still smarting from the defeat of his motion to effectively wind up Austral several months before in December 1929, Howie was determined to press ahead with the proposal to dispose of Austral's brickmaking assets to a new company.

In the course of the next few weeks Smith took out a court injunction against Turland and Howie, preventing them from calling meetings in the name of The Austral Brick Company Ltd. Behind the scenes there was a great deal of manoeuvring, because on 31 March 1930 at a hastily convened emergency general meeting Howie and Turland threw in the towel and resigned as directors. The resignation of two directors at the same time is a serious matter and in this case illustrates how in-fighting had reached the point where a split was inevitable. Edwin Spackman moved that the number of directors be increased to six and following a vote both Spackman and Maxwell Porter (of the firm Magney, Porter and Weynton) were elected unopposed. Spackman's re-election as a director affirmed those who opposed Howie and his camp. It

13. Interview with Ray Dawes, 25 August 2005, Sydney.
14. Ibid.

Letterhead of The Austral Brick Company Pty Ltd, May 1934, taken from the letterhead of Bedford Brick Works following the acquisition of the former company. Brickworks Limited

is possible that the struggle was only resolved in the Supreme Court of New South Wales.[15]

Dawes himself, however, was not above criticism, and earlier concerns about his personal business dealings were raised once more. At a shareholders' meeting on 11 August 1930 yet another unfavourable reference to Dawes' cartage contract was made. Even Dawes was prepared to concede the cosy nature of the arrangement. Edwin Spackman, chairman of the company, was quick to defend his protégé, stating that one of the factors for the increase in sales by Austral and its associated companies was brought about by the reduction of cartage charges.

THE BRICK AGREEMENT

Where did the industry's professional body, the New South Wales Brick Masters' Association, figure in all of this? Since the turn of the century many of its members had been party to an informal agreement (usually referred to as the 'Brick Agreement'), whereby they promised to maintain price parity with others in the scheme. This gave rise to claims that the industry acted in collusion to fix prices, but the reality was that manufacturers adopted these measures to avoid price-cutting wars. With this in mind, bricks were sold and distributed through an agency, the Metropolitan Brick Company, which took orders from the customer, placing them with the yard nearest the purchase. Travelling salesmen – 'travellers' – made the rounds of building firms and agencies, allowing members to dispense with their own sales force. The Metropolitan also sent out accounts to customers, collected payment for the bricks, and ran credit checks. In return, brick manufacturers paid a levy of 9d per 1,000 brick to cover administrative costs of the organisation. The arrangement was highly successful because it guaranteed against bad debts using a middle man that sold bricks on their behalf. All of this considerably reduced costs, especially cartage rates.

The role played by the Metropolitan Brick Company in organising sales and distribution was integral to the workings of the Brick Agreement. Some years later, Dawes recalled that 'around about 1927 there was some form of brick agreement in existence, and the whole of the trade of the metropolitan area was shared with the exception of yards such as Speare's and a few that were outside.'[16] By 1929, 32 yards were under such an arrangement and as new yards were established they joined the agreement, taking advantage of the Metropolitan's extensive network of sales offices.

15. Fifty years later Jim Millner, chairman of Brickworks Limited, alluded to this bitter row and tussle for control in his remarks at the annual general meeting in 1981. Invoking the memory of William King Dawes, who had died shortly before, Millner paid tribute to the man who had been the guiding force behind The Austral Brick Company Limited and also of his fight through the courts to establish control over the company.

16. William King Dawes, Attorney-General v. Brickworks Pty Ltd and Others, op. cit., p. 441.

The Metropolitan's network was based on organising Sydney into a number of discrete zones or districts including Sydney, North Sydney, Thornleigh (where the National yard and Fowlers' operated), Ryde and Eastwood. The idea was that each yard would keep to its allotted percentage of trade. Where one yard sold or produced more than their allocation then an adjustment was made whereby they were charged an 'excess' which was given to the deficient yard.

All yards in the scheme, including Austral, had a reserve or Guarantee Fund into which they paid 1s for every thousand bricks sold. Over time the fund accumulated and from it the New South Wales Brick Masters' Association was able to purchase the South Ashfield and Burwood and Flemington yards,[17] as well as the assets of Excelsior. In the case of the Excelsior which had been opened as recently as 1928 Dawes might have sniffed a bargain. In March and August 1930 we find reports of Austral having built a sizeable stake in the company through the purchase of shares to the value of £3,409. This was paid for out of monies at the company's credit in funds of the New South Wales Brick Masters' Association.[18] Austral was not the only member to buy up stock. Yet however much it cost to erect the plant at Excelsior and get it ready as a yard, by 1934 it was a dead asset in the hands of members of the Brick Agreement.

TRADE WAR

The Brick Agreement was not legally binding but it did suit everyone because it worked. The arrangement was not to last. As the Depression took hold it became increasingly obvious that the association could not hold things together. Alas, the Brick Agreement was in trouble, with manufacturers choosing to sell bricks at any price. Early in 1930, the industry began to fragment when the first in a series of price wars broke out. Trade had always been cut throat, only now manufacturers strove to keep their yards working despite not being able to pay the men's wages. Not only was closure an admission of failure, it was tantamount to a death sentence; once down they were never likely to get back up again. In quitting the agreement in June 1930 Dawes precipitated a series of damaging trade wars from 1930 to 1932, in 1933, and also at the beginning of 1934. Friendships must have been tested with the previously mentioned Edwin Phippard preferring to be diplomatic about the identity of the chief protagonist in the price war. 'There was very little trade for the brick industry. One or two departed from our Association and the prices gradually got down until they reached the ruinous figure of 38 shillings per thousand.'[19] Clearly he was referring to Dawes.

It was only after the death in March 1934 of Alexander Stuart, president of the Brick Masters' Association, that Clarence Davison, Stuart's successor, persuaded Dawes back into the fold. A softly-spoken man Davidson was nonetheless extremely shrewd and respected for being the diplomat and peacemaker. His calm demeanour was sometimes mistaken for lack of resolve. Nothing could be further from the truth for he was able to talk Dawes round. Austral rejoined the agreement in October 1934 and from then on the industry put aside its competitive practices in order that a reasonable price might be possible for all.

PRAISE

Dawes was busy and not only in the sense of making sure Austral stayed ahead of its competitors. He also published a statement of the company's balance sheet for investors at the ordinary general meeting held on 23 February 1931. This was hardly something to write home about but at the time it may have been an innovation brought about by more stringent reporting requirements introduced by the Parliament of New South Wales. Dawes was

17. Flemington was purchased from the Rabones family from money that was in the guarantee fund. Following the purchase its shares were distributed to members in proportion to the amount they had in the fund.

18. Reported at the Ordinary General Meeting of Austral Brick Company Limited, held on 11 August 1930 at Sydney School of Arts.

19. Frederick Phippard, Attorney-General v. Brickworks Pty Ltd and Others, op. cit., p. 339.

Block boys. No less important to the inner workings of commerce and industry was the makeover taking place along Cook's River Road outside the Austral plant. Major work to widen the highway was underway during October 1930, yet despite the gathering pace of traffic and commercial expansion, it was still possible to cross main roads in the area of Marrickville and St Peters without risking life and limb. Hood Collection, State Library of New South Wales

Going places. W.K. Dawes in his newly-delivered 1929 Buick Silver Anniversary. Dawes had a passion bordering on obsession for fast and powerful cars. Commencing in 1922 his first vehicle was the all-aluminium Vauxhall 30/98, a rare model that would now be worth a huge sum of money. He also owned cars made by the Moon Motor Company. Bill Wheatley Collection

praised by the Chairman, Edwin Spackman, who was keen to remind everyone – again – that the company would not be where it found itself had it not been for Dawes. There were also plaudits from no less a figure than the president of the New South Wales Brick Masters' Association, Alexander Stuart. We can only guess the contents of Dawes' letter to Stuart on 15 May 1931, but the reply was unequivocal in its support for Dawes and the other directors at Austral.

Writing from 'Glenhurst', his substantial home on Yarranabee Road, Darling Point, Stuart gives us an insight into what had really been going on in the brick industry.

> *The contents of your letter of the 15th May are most interesting and I congratulate the Directors of the Austral Co. on their courage and enterprise. The intentions and hopes expressed by you, and in these you no doubt voice the opinions of your colleagues are laudable, and if achieved may materially assist to*

Plan showing Bedford Brick Works and the surrounding yards, circa 1935. – Mitchell Library   Inset: Stan Gentle, circa 1928. Richard Gentle Collection

> *reinstate the Industry into the position it held some years ago when a reasonable profit only was required from the Public.*
>
> *The Industry was run to death, the State Yard was permitted to grow until it became a very serious menace. The conditions and price demanded induced the use of concrete in many forms, enormous profits were taken, and on the other hand I doubt whether one improvement of consequence has been made in brick making during the last 35 years. In fact there has been retrogression. The D.P.* [dry press] *bricks made many years ago were far superior to the bricks now produced.*[20]

What intentions and hopes Dawes and his colleagues had in mind is difficult to gauge, but evidently Stuart blamed the State Brickworks and the negative effect it had on private firms and the industry in general. In fact according to Stuart the Homebush brickworks may even have been a catalyst for the increasing use of concrete in construction during the 1920s, presumably due to its cheapness relative to brick. Yet rather than depress the price of bricks in the period 1912–1929, the State yard may have actually initiated successive price rises and modernisation of productive capacity. As a consequence, Stuart warned, the industry (presumably he was referring to private manufacturers) was undercapitalised and backward, so that bricks pressed in 1931 weren't a patch on those made when dry-press techniques made their appearance just before the turn of the century. Stuart, a builder by profession, an investor in Austral and a big hitter in the brick industry, was in a despairing mood. He saw in Dawes a man who could pull things round, even though the two men did not see eye-to-eye on Austral's go-it-alone policy of price cutting.

BEDFORD BRICK WORKS

Meanwhile Dawes' eye for a bargain was being put to good use through the acquisition of the Bedford Brick Works in the opening weeks of 1931. We know this from archival records held at the Mitchell Library. An entry in the trial balance in Austral's General Journal for April 1931 clearly contains a reference to Bedford Brick Works. Again, at the ordinary general meeting held on 7 September 1931, Spackman gives particulars of the purchase, informing investors that the acquisition was intended to eliminate serious competition, and would stabilise the value of the Austral Company's shares. It would also prevent the yard falling into the hands of another rival.

Industry rationalisation was painful for the yards operating along the Brick Mile at St Peters, especially those that had prospered and enjoyed a long tradition of stable management. Bedford Brick Works fell into this category. Put simply it was now in deep trouble. Since the death in 1912 of its founder, Josiah Gentle, the company had been managed by his three sons Emanuel, Josiah Jubilee and Stanley Josiah 'Stan' Gentle (born 1877). Dawes was on close terms with Stan Gentle, their yards being in close proximity. Estimates of how many bricks were stacked in the Bedford yard vary, but according to Ray Dawes there were as many as five million. The annual shutdown over December 1930 (the yard did not reopen following the Christmas break), the death of Emanuel on 1 February 1931, death duties, as well as stagnant brick sales at the height of the 1929 depression provide some clues as to why the Bedford Brick Works was sold. Perhaps Stan Gentle made a direct approach to Dawes with an offer to sell. Knowing there would be few takers, and loathe to see the family firm disappear, a deal appears to have been struck, possibly over the Australia Day public holiday.

As Ray Dawes put it:

> *It was over a long weekend as I recall. Dad couldn't get in touch with his employers so he took it on himself to buy Bedford Bricks. Come the Tuesday morning it went up the line of command to the Austral chiefs. They said to him, 'you've exceeded your authority and we don't want it'. From there he went to all his friends around the St Peters and Newtown area, including Alf Smith. That's when he dropped the price of bricks and sold the bricks in the yard.*[21]

20. Private papers of Ray Dawes.

21. Interview with Ray Dawes, 12 December 2006, Sydney.

Above: Artistic licence was taken with the draft of this illustration. The four dome kilns to the left of the picture are facing the wrong direction.

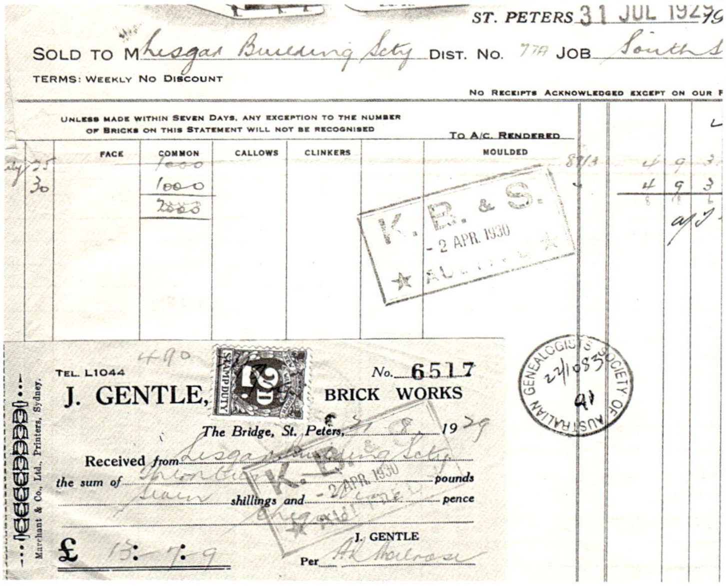

ST. PETERS 31 JUL 1929

SOLD TO M Lisgar Building Socty DIST. NO. 7A JOB

TERMS: WEEKLY NO DISCOUNT

NO RECEIPTS ACKNOWLEDGED EXCEPT ON OUR F

UNLESS MADE WITHIN SEVEN DAYS, ANY EXCEPTION TO THE NUMBER OF BRICKS ON THIS STATEMENT WILL NOT BE RECOGNISED

TO A/C. RENDERED

	FACE	COMMON	CALLOWS	CLINKERS	MOULDED				
30		1000					4	9	3
		2000							

K. B. & S. 2 APR. 1930

TEL. L1044 No. 6517

J. GENTLE, BRICK WORKS

The Bridge, St. Peters, 1929

Received from Lisgar Building Socty

the sum of ... pounds ... shillings and ... pence

£ 13 : 7 : 9

J. GENTLE Per

Left: Receipt made out on letterhead of Bedford Brick Works to Lisgar Building Society, July 1929 for the sale of 1,000 commons. Society of Australian Genealogists

This last comment requires explanation. It may be that the acquisition of the Bedford yard and its millions of bricks triggered the start of another, more damaging price war and the exit of Austral from the Brick Agreement. Many years later a descendant, Richard Gentle, himself a manufacturer of industrial ceramics, met Bill Dawes at the Brookvale plant of Brickworks Limited. Dawes acknowledged that Austral had paid £32,000 for Bedford Bricks which was a great deal of money at the time. Perhaps the Austral board had no choice but to dump bricks on the market in a bid to increase sales and thereby recoup its outlay on the recent purchase. The truth may never be known, but the unusual circumstances by which the Austral Brick Company Ltd acquired Bedford is confirmed by Ron Hagstrom, former manager of the Brookvale brickworks.

At one point Dawes considered buying the yard for himself but this was beyond family means. The sudden death of his father on 12 January 1931 may have opened up one source of funds through inheritance, but the reading of the will caused a rift between Dawes and his sister, Edna.[22] Perhaps he got less than he felt he deserved after having been the mainstay of the family business for so long? In any case the produce store was sold months later. Ill will, however, seems to have been short lived because Edna was employed at the Austral office during the early part of the Depression until 1936. In 1937 Dawes installed Edna's husband, George Frederick Wheatley, as plant manager of the Bedford yard where he worked until 1952.

Proving a direct link between the acquisition of the Bedford yard and the start of a price war may be difficult. Certainly, Dawes' capricious action in

22. The funeral cortege for W.F. Dawes filled more than 70 rail cars which made their way to Rookwood cemetery. It was attended by the Mcyor of St Peters, and a large number of dignitaries, including members of the Sydenham Masonic Lodge.

offering to buy the yard angered the other directors. In the long run, however, it paid for itself many times over; bricks from the Bedford yard were highly sought. In fact, the plant was the last brick factory to close, bringing to an end almost 150 years of brickmaking in the St Peters–Waterloo district.

IMPACT ON AUSTRAL

Back to our story. The purchase of Bedford Brick Works increased Austral's capital base and capacity enormously. Bedford's Hoffman kilns were capable of modernisation and its clays and shales were sufficient for another 60 years' production. What's more, Bedford's bricks were highly rated and the firm is reputed to have made the best chocolate-brown bricks in the state. Upgrades at both the Austral plant and Bedford were made during 1933 with the installation of gas-powered boilers by Australian Gas Light, replacing the old steam-powered dry press machinery and grinding pans.[23] Slowly, things began to turn around helped, among other things, by Austral's directors taking a pay cut which was timely in view of the reduction in the basic wages agreed to by the Industrial Commission. Spackman's report to the ordinary general meeting of 22 February 1932 praised the stronger than expected half-yearly result. The minutes of that meeting also mention Austral's continuing interest as a shareholder in Goodlet and Smith, Excelsior Brick, the Flemington Brick Company, as well as a petrol filling station which the company operated on King Street.

OLD WINE, NEW SKINS

With a trade war still raging throughout 1932, the Brick Masters' Association tried to bring an end to the industry free-for-all. On the evening of 10 January 1933 at the Builders Exchange on Castlereagh Street, an inner circle of industry leaders gathered to discuss the way forward. Present were Alexander Stuart, Reg Allot, Clarence Davison, Fred Phippard, John Brown, Geoffrey Hocking, Bill Dawes, Arthur Charlesworth, Walter West, E.D. Lanceley, and Archibald Howie. On the spot they resolved to call a general meeting for the following week, which would be open to all brick manufacturers in Sydney. The meeting would be asked to vote on a proposal to establish a management committee called the Brick Committee. This would have 'great powers to conduct all matters in a similar manner to that done by the directors of a limited company such as appointing a secretary, or fixing up Bank accounts and other matters.'[24] Yards were encouraged to sell through the Committee and keep with the existing arrangements that grouped them together in geographic areas. Under the scheme yards that had closed would receive an income through their affiliation with the association. The 'rating' received was calculated on the basis of the output of which the yards were capable.

The proposal was accepted with those prepared to lend their support lining up to sign the Brick Committee Agreement. The board of Austral closely followed these developments. At a directors' meeting on 20 February 1933 Chairman Spackman reported that an agreement had been made with all the other yards that would make for harmony and avoid cutting of prices. At subsequent meetings of the Brick Committee manufacturing costs and a forthcoming brick enquiry on the state of the industry being pursued by the NSW Employers' Federation were discussed. In further meetings during February, March and April of 1933 there was also talk of a new industrial association that could more effectively represent the interests of brickmakers. Thus, the idea of a Council of Brick Manufacturers was born. For reasons best known to themselves, many yards preferred to remain outside the Brick Committee, including Central Brick & Tile, Eagle Brickworks, Punchbowl Brick & Pipe Co Pty Ltd, Shannon's Brick, Tile & Pottery Company, Western Suburbs Brick Company, Speare's, Goodlet & Smith, State Brickworks, Manly Brick and Tile Company, Moorview and Bankstown.

In the final outcome Dawes signed up to the Brick Committee, pledging an end to the trade war.

23. Minutes of the Board of Directors, The Austral Brick Co. Ltd, September 1933, Archives of Brickworks Limited.

24. Minute book of The Brick Committee, p. 4. 1933. Archives of Brickworks Limited.

Brickyard, St Peters, possibly the Austral Brick Company Pty Ltd, 1946. According to Ray Dawes, Lindsay's drawing is almost certainly that of the Austral yard at St Peters and was completed after weeks of heavy rain flooded the pit. Artist: Lionel Lindsay. National Library of Australia

Speare's Brickyard, St Peters, circa 1935. Marrickville Local Studies Library

Buoyed by the knowledge that the Austral yard was highly efficient, more so than any other yard in Sydney, Dawes knew that, if necessary, he could outstare the competition yet again. He also produced a lot more bricks than rival firms. Ernie Speare was outside the agreement and intended to stay that way. 'Rather than join the Association I would go on the dole and dig sand,' he later remarked. There was no love lost between Dawes and Speare, the latter being known to undercut by as little as 1s per thousand. In fact, Speare hated the association and all it stood for.[25] Animosity showed itself in petty acts of spite that went on for several years. Fringe builders would go to Speare's and put in an order. If Speare's could not fulfil the order they would go to Dawes who would put them at the bottom of the list, making them wait their turn.[26]

Amidst all the gloom accompanying the Depression years, brickmakers would have been heartened to learn that the proportion of houses constructed of brick had grown steadily since the late nineteenth century. The 1933 census showed that nearly three in every ten houses were now brick – a 20 per cent increase since 1900. Stone houses had passed their peak, the 1921 census showing a total of 79,983, and were now being demolished faster than they were being built.[27]

ON HOMEBUSH BAY

Private manufacturers were not alone in their struggle to stay afloat. Despite being an efficient and modern operation, a far cry from most private operators, the State Brickworks on Homebush Bay was facing a bleak future. Established by the McGowan Government, the yard had commenced in a small way in 1911 and grown in size until by 1935 it had ten patent kilns and about 15 down draught kilns, with a capacity of well over one million bricks per week. The yard possessed water, rail and motor facilities for delivery, and was equipped with the

25. State Records NSW, Testimony of E.L. Speare given on 6 February 1939 to the Royal Commission of Inquiry into charges of fraud and corruption in connection with the sale of industrial undertakings, 1937–38, 1938.
26. Interview with Ray Dawes, 25 March 2006, Sydney.
27. R. Boyd, *Australia's Home: Its Origins, Builders and Occupiers*, Melbourne University Press, Melbourne, 1952, p. 80.

very latest machinery and methods for pit working. Since its inception and up to 1934 it had repaid all the loan capital, achieved by drastically undercutting prices the private manufacturers could charge. Bigger by far than any other manufacturer, the New South Wales State Brickworks competed aggressively in areas south of the harbour, but tended to stay out of markets on Sydney's north shore. The State's yard was the most modern in the southern hemisphere and there had been considerable additions to the plant during 1926 and 1927. In the financial year 1928–29 the Homebush works turned out 62 million bricks of which 50 million were sold to private purchasers. For the last working year of 1934–35 output was down to about 14 million bricks per annum. By this time only about one fifth of the plant was working.[28]

What rankled most for private manufacturers was the unfair competition posed by the government yard. From the date of its commencement it had received government favours and patronage such as free dredging services, cheap railway facilities, government preference as a supplier, and, until later years, freedom from income tax. The owners of private yards believed that circumstances were biased in favour of the state-owned works as it had 'all the money they needed; they have not had to bother about interest charges; they have had an unfailing purchaser for their products in the State Government and other benefits'. Even with these exceptional advantages, the 'business career' at the State Brickworks [sic] had been a chequered one.'[29] This comment echoed the belief of most private manufacturers that the strong union presence in state-run industries ensured thst pay and conditions were over-generous. While others may disagree, government backing enjoyed by the State yard had been detrimental to private brick manufacturers all along, reduced profitability having led to the lack of capital investment in plant and machinery. For the disaffected who jibbed at primitive working conditions the State yard acted as a magnet. Jumping ship was not confined to seafarers who were, by nature, restless men. There was always work in the brick industry simply because turnover was high. Higher wages during more prosperous times stabilised the labour force, but were also an incentive that offset the harsh nature of the work. According to *The Clay Products Journal of Australia*, the extent of this unfair competition may be judged by the following table. The price of bricks manufactured at the State's yard was consistently below that charged by private manufacturers.[30]

Brick production, NSW State Brickworks, 1923–1929

	1923 – 1924	1928 – 1929
Bricks manufactured	44,966,718	62,276,614
Used for public works	9,882,900	10,231,720
Sold to private purchasers	34,826,912	50,800,410
Used at works	52,769	626,484
Stocks on 30th June	478,479	1,652,119
Cost of manufacture per 1,000	£2 6s 8d	£2 14s 2d
Sale price per 1,000		
Seconds	£2 10s 6d	£2 18s 6d
Commons	£2 13s 0d	£2 18s 6d
Face	£4 10s 0d	£5 2s 0d

Source: Official Year Book of New South Wales from 1929–30, cited in The Clay Products Journal of Australia, October 1933, p 7

UP FOR SALE

Up against the wall like any other brickyard, the State Brickworks was living on borrowed time. By quirk of fate the dismissal of Jack Lang, Labor premier of New South Wales, on 13 May 1932 by Governor Sir Phillip Game, a retired Royal Air Force officer, was the harbinger for radical change. Lang's vehement opposition to Prime Minister James Scullin's support for the Premiers' Plan provoked an extremely bitter – and lasting – split between Scullin and other federal Labor figures and New South Wales Labor. The Plan called for debt reduction measures, including spending cuts, all of which were intended to placate British bankers. Lang's refusal to go along with the Plan set off a chain of events that led to his dismissal. In his stead

28. Brickworks Limited, Fifty Year Retrospective, 1934 –1984, Archives of Brickworks Limited.

29. N. Peek, & C. Pratten, *Working the Clays: The Brickmakers of the Ashfield District*, Ashfield Historical Society, 1996, p. 7.

30. *The Clay Products Journal Of Australia*, October 1933, p. 7.

Game appointed the leader of the conservative United Australia Party, Bertram 'Tubby' Stevens, who promptly called a state election for late May 1932. Lang was decisively beaten at the polls and never held office again although he continued for some years in opposition until he was dumped by his own party. Determined to balance the books Stevens identified the remaining state industrial enterprises comprising blue metal quarries, the Monier Pipe Works and the brickworks, as prime candidates for sale. In March 1933 Stevens' cabinet instructed the State Brickworks to restrict sales to government departments. Notwithstanding that trade had pretty much collapsed, this put the yard at a serious disadvantage to private manufacturers. It also created serious congestion at a time when bricks were still being produced, although using only one patent kiln. Stacking space within a reasonable distance of the kiln was at a premium with roughly 2.6 million bricks in stock at the end of 1933. The decision to abandon restrictions in November led to a gradual reduction so that by the end of April 1934 the stockpile was down to 1.78 million bricks. Having taken note of the situation Mr A.E. Barton, Assistant Treasurer, New South Wales State Treasury, recommended that if a buyer could be found then the government should sell at a suitable price. Accordingly, sale by tender was announced.

Casting off loss-making industries was not the only measure the government had in mind. The New South Wales Brick Masters' Association was also under pressure to reduce the basic wage agreed to by the Industrial Commission. Cartage rates were next in line for the chop with the industry (private carters) agreeing to work for less money. On 16 May and again on 22 June Bert Moseley, secretary of the association, wrote to Barton offering to enter into discussions to purchase the State's yard. Moseley argued cogently for the sale of the yard as this letter of 16 May 1933 demonstrates:

> *For years we have had to compete with the State Brick Works, notwithstanding that [it] has used the best part of its profits to extend its works, having been able to do this by not having dividends to pay to Shareholders on capital invested, nor did they pay Income Tax until the last two or three years, all of which would be expected from an ordinary business. This rendered competition very unfair, and while we were able to carry on in good times against this competition, we reached a stage, due to the depression and the falling off of sales, that to sell in competition with the State left us without profit, or being able to provide for depreciation. In addition we were all left with considerable losses, and that is the position at present.*[31]

Apart from anything else,

> *the State Brick Works were of a socialistic character which the Community has never at any time declared itself in favour of, and such Undertakings are against the declared principles of the present Government.*[32]

Thereupon, Moseley offers every necessary assistance if called upon by Mr Barton. Writing again on 22 June 1933, he gets straight to the point. Did the Government want to sell? If so, how much, and on what terms?[33] 'Yes', replied Barton on 3 July, 'the government would sell by public tender, but it would also introduce a system of controls to protect the public from price fixing or the like'.[34] As if no further proof was needed, Barton contacted the manager of the State Brickworks, a Mr Dale, who reported in his letter of 11 July 1933 that 'the average cost of manufacturing and delivering bricks for the financial year ending the 30 June, last, was £4 4s per thousand, and the average selling price £3 1s 9d per thousand.' He then drafted a memo on 11 September to the Assistant Treasurer in which he admits that 'the consumption of bricks by Government Departments and Undertakings has recently dropped almost to vanishing point' and that stacking of bricks in the yard was no longer possible or could justify the expense.[35] At Barton's urging tenders for the sale of the State yards were called for on 27 August 1934, the closing date being 4 October. Accordingly, sale by tender was announced.

31. Letter from B.A. Moseley on behalf of the New South Wales Brick Masters Association to A.E. Barton, The Treasury, cited in Royal Commission of Inquiry into charges of fraud and corruption in connection with the sale of industrial undertakings, 1937–38, Exhibit No. 109.
32. Ibid., Exhibit 109
33. Archives of Brickworks Limited.
34. A.E. Barton, Royal Commission of Inquiry, 1937–38, Exhibit 98a.
35. R.S. Dale to A.E. Barton, Royal Commission of Inquiry, 1937–38, Exhibit 101.

The Government felt that if industry was to be encouraged to revive there must be complete removal of the difficulties and the irritation that was caused by Government encroachment upon private enterprise. The Government trading activities had been commenced by past Governments, ostensibly to cater for Government requirements. Many of these undertakings, however, had spread their activities until they had come, in some cases, openly in direct competition with private enterprise, so that in certain instances they now stood primarily as industrial undertakings and only incidentally as industries for Government convenience.[36]

The Clay Products Journal of Australia was ecstatic, declaring: 'it is good news for the brick-making industry ... that the Government has decided to close down the State Brickworks at Homebush.'[37]

SOME IMPORTANT DISCUSSIONS

Concerned members of the association had much to talk about. On 19 September 1933 Arthur Charlesworth, owner of Sydney Brick Works,[38] who had also taken over from Moseley as secretary of the New South Wales Brick Masters' Association, wrote to all members of the Brick Committee outlining a proposal dealing with the purchase of the State Brickworks. He asked, 'Are you in favour of a tender being submitted for the purchase of the Works by the parties to the Brick Committee Agreement?', In the event of the State yard being purchased it was proposed to form a limited company to purchase the Works on extended terms and time payment of a cash deposit. The company's paid-up capital would be not less than the purchase price of the State, including stocks and the necessary money to carry on the concern, ie £159,306 as per the balance sheet of June 1932.

Sydney Brick Works was owned by the Charlesworth family and acquired by Brickworks Limited. The yard on Unwins Bridge Road was directly opposite Silver Street, St Peters. Marrickville Local Studies Library

Charlesworth went on to explain that those who were party to the Brick Committee Agreement, and who took up shares, would do so in proportion to the percentage held in the Brick Committee. The balance of the purchase money would be found from the Committee's Guarantee Fund and paid over a period of time to the new company until the whole of the purchase money had been paid, at which point shares would be fully paid up. Furthermore, money for a deposit and working capital – £40,000 – would need to be provided by the combined parties based on their percentage share in the Brick Committee. Charlesworth went on to point out that it would also be necessary to alter clause 17 of the Brick Committee Agreement, which required three months' notice should a member retire from the agreement. Thus, parties to the amended agreement could retire only when the whole of the purchase money for the State yard was paid by the new company. The Articles of Association of the new company was to be similar to those of the Flemington Brick Co. Ltd or the Excelsior Brick Company Ltd. Even if the government works were purchased it was not expected, Charlesworth added, to provide a return on any money invested for a very long time. On this last point, history proved him wrong.

Finally, the tender documents stipulated that the selling price of bricks would need to be fixed from time to time, by one person representing the government, one representing the public, and one representing the new company. These conditions

36. Ibid., Exhibit 101.
37. Ibid., Exhibit 101.
38. Arthur Charlesworth was a director of the City Brick Co. Ltd and part owned the Sydney Brick Company with his brother. This firm had been operating since 1900. Charlesworth Snr had purchased a half share in the Sydney Brick Company for £15,000 pounds in 1920, forming a partnership in 1923 or 1924. The yards were operating from 1920–29 at Marrickville, and from 1930–32 at Unwins Bridge Road, St Peters. During the depression years Charlesworth leased the Sydney Brick Company to the Flemington Brick Company owned by the NSW Brick Masters' Association.

didn't appear to deter brickmakers who met on 26 September to vote on the terms of the proposal. Out of 35 member yards 27 were in favour of tendering with a view to purchase, an overwhelming majority of Sydney's brickmaking firms and their representatives.

Whether intentional or not, representatives of the Strathfield and Enfield Brick & Tile Company and Shannons Brick, Tile & Pottery Co. Ltd failed to reply. Neither was a reply received from W.G. West representing Central Brick & Tile Co. Ltd who claimed at the committee meeting that he had not had the time to peruse the Articles of Association of the company mentioned in the proposal. After being handed copies of the articles he left the committee meeting but did not attend the next meeting of the Board of Management. Mr. G. Allibone of Bankstown Brick & Tile Co. also got cold feet. He phoned Charlesworth in advance informing him that one of his directors was in Queensland, and he himself would not be able to attend due to pressures of work. Despite majority support for an 'industry' company the refusal by three major players to join derailed Charlesworth's plans. Reg Allott, owner of the Punchbowl and Sutherland yards, and Spackman and Dawes, representing the Austral and Bedford yards, were not prepared to sign. The sticking point was the length of time their companies would be bound to the Brick Committee Agreement should they decide to quit the arrangement. Apart from this one issue, Spackman and Dawes were in favour and in their reply assured Charlesworth that 'should the present agreement operate as at present, our share of the suggested percentage deposit is available when required.' Unfortunately, without a unanimous vote in favour, the proposal foundered so that Stuart was obliged to write to Barton on 29 September 1933 advising that no tender would be submitted by the combined members of the association.[40] Barton conveyed the news to the government but he remained hopeful the Brick Masters' Association might come up with an offer.[41] For the next 18 months the fate of the Homebush works remained undecided, with only part of the plant still in operation by the end of 1935. Evidently the time was not yet right for Sydney's brickmakers to sink their differences and form a united front. Unable to agree on a proposal that would have neutralised the State Brickworks, each went about their business as best they could until March 1934 when William King Dawes pulled out of the Brick Agreement, precipitating a crisis which, ironically, led to the creation of Brickworks Limited several months later.

Signatories to the agreement to purchase the State Brickworks at Homebush.[39]

S.J. Gentle,	representing Ashfield
W.F. Cropley	representing Auburn
E.C. Leizzio	representing Bakewell Bros
A. Charlesworth	representing Excelsior
A. Stuart	representing Flemington, New South Wales Brick Company, Burwood Brickworks, and South Ashfield Brick Company
F. Butcher	representing Butcher Bros
G.F.W. Hocking	representing Canterbury Brick
G. Crispo	representing Carrington Brick
H.W. Alderson	representing City Brickworks
C. Davison	representing Clyde Brick
J. Hankinson,	representing Rupert Cook's Brickworks
G. Kemp	representing Croydon
Arthur Lyell-Scott	representing Eagles
F.G. Phippard	representing Federal & Carlton
J.H. Gilliland	representing Great Northern & National Brickworks
E. Judd	representing Hurstville
E.D. Lanceley	representing North Sydney and Ryde
J. Brown	representing Northern Suburbs
C. Scott-Waine	representing St Peters
J.H. Charlesworth	representing Sydney Brickworks
J.F. Walker	representing Western Suburbs

39. Source: Matters for discussion, *State Brick Works and Proposal To Purchase*, Council of Brick Manufacturers, 14 October 1934, Archives of Brickworks Limited.

40. Royal Commission of Inquiry, op. cit., Exhibit 102.

41. In 1935 Barton left the employ of the New South Wales Government to become consulting accountant to Bakewell Bros. In September he was appointed a director of the firm and later, was accused of having been in collusion with the brick manufacturers, a charge that was tested in court but rejected outright. It was simply a case of Barton having received a better offer.

Building revival – Trends in brick veneer – Texture brick – Fibro makes its mark – Slums

BUILDING REVIVAL

Recovery and consolidation were slow to arrive. Between 1931 and 1934 the number of new buildings approved by Melbourne's municipal councils never fell below 2,300 annually. In Sydney, a meagre 688 buildings were approved in 1932.[1] More than any other capital city Sydney was affected by the reduction in construction activity. Unemployment in the city's industries had reached 60 per cent by June 1933. Only from the beginning of 1934 did residential house and apartment building in Sydney experience some sort of revival, especially in the suburbs. This was due to the fact that planning requirements were straightforward when compared to those needed for large city projects. In fact, the year 1935 was the best for Sydney's leading real estate companies since the time of the boom during the 1920s. The continuing movement of young families to the suburbs, each in search of their own house and garden, was heartily approved by all sides of politics. Another sign of recovery was the resurgence of building societies which was given impetus by the New South Wales Registrar of Co-operatives. In time, building societies became major providers of housing finance across the country until the 1980s when deregulation of financial markets forced a spate of amalgamations and their transformation into banks. Some even listed on the Australian Stock Exchange.

We should also remind ourselves that recovery was more than just a resurgence in house constructions. In these pre-war years bricks of various types were used in the construction of bridges, culverts, sewer drains, telephone line channels, railway station platforms, footpaths, embankment facings, swimming baths, chimney stacks and boiler settings. By the early 1950s, however, bricks had been largely replaced by concrete in most large-scale engineering work. On the other hand, brick remained supreme

THE CLAY PRODUCTS JOURNAL OF AUSTRALIA

Official Organ of The Australian Clay Products Association Limited

Vol. 7. No. 1. NOVEMBER 1, 1939. Price Sixpence

The bath-rcom is no longer an afterthought!

To-day, modern designs, comfort, convenience and — of course — cost are bath-room earthenware details inseparable from th- discussions between architect and client. Satisfactory agreement is readily achieved when Fowler white earthenware suites, basins, floor and wall tiles, etc., are specified and installed.

When special designs or earthenware fittings in standard colours are desired, R. Fowler Limited will readily submit estimates for your requirements.

A modern bath-room is illustrated above, showing Fowler Ware "Marrick" Suite (Regd. No. 11589, 10927), "Bellevue" Pedestal Basin (Regd. No. 10926), Gargoyle Faucets, Recess Soap and Sponge holders, etc., and logically, Fowler Floor and Wall tiles.

R. FOWLER LIMITED, MARRICKVILLE, N.S.W.; THOMASTOWN, VICTORIA; ALDERLEY, QUEENSLAND.

The oldest active potteries in Australia; established over a century ago.

Growing affluence, The Clay Products Journal of Australia, 1939

Bricks sold by members of the Council of Brick Manufacturers, 1929–1938

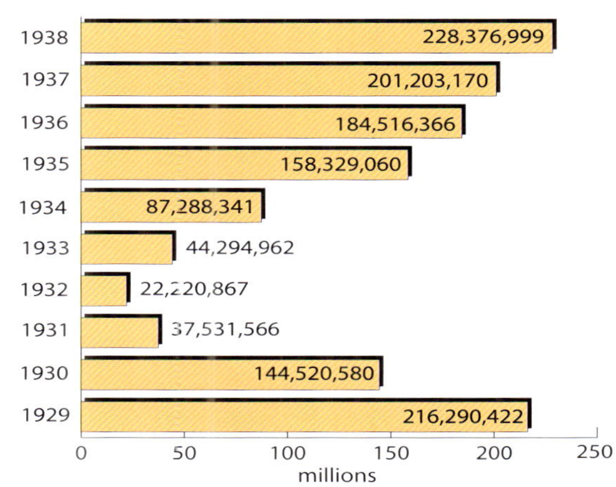

These figures include bricks sold by the New South Wales Brick Masters' Association and its successor body, The Council of Brick Manufacturers

1. *Sydney Morning Herald*, 18 February 1932.

Bricks produced in New South Wales, 1929–1938

Year ended 30 June	Metropolitan	Country	State total
1924	299,444,618	103,635,095	463,080,713
1925	272,544,759	111,681,576	384,226,326
1926	277,725,292	120,401,610	398,126.362
1927	301,211,071	128,853,808	430,064,879
1928	319,111,316	128,558,573	439,669,889
1929	244,255,206	92,902,414	437,157,514
1930	228,091,370	78,963,168	307,054,588
1931	39,280,937	19,420,414	58,701,351
1932	17,563,782	10,956,763	28,520,545
1933	55,703,293	29,910,581	77,613,874
1934	124,595,680	43,478,569	168,075,249
1935	213,346,282	64,351,148	277,697,430
1936	226,407,734	76,852,494	363,280,228
1937	233,047,716	83,670,322	316,718,038
1938	250,762,605	100,876,421	351,639,026

New South Wales Bureau of Statistics and Economics, 1939

in the housing and building construction industries, and continues to dominate even today.

TRENDS IN BRICK VENEER

The growth in building societies and less restrictive lending practices by the banks boosted construction figures and also the start of the long drift towards brick veneer homes. Brick veneer had first made its mark as early as 1850 in Victoria, but in its modern form was used in Melbourne in 1915 on an ordinary suburban timber frame house.[2] The depression decade bore witness to the increasing popularity of brick veneer, a trend that may have owed something to Hollywood. The December 1931 issue of *The Australian Home Beautiful* carried an article, 'Glimpses of Californian Homes', which explored the appeal of brick veneer to Hollywood stars and producers.

> *Brick veneer is common practice in California, indeed it is being increasingly employed through the States. On concrete foundations a timber (Oregon) frame is erected and weatherboards nailed on diagonally to form the bracing. On this building paper is nailed, and outside this a veneer of brick is applied. Exterior treatment varies from tuck pointing to stucco.*

2. R. Irving, (ed) *The History and Design of the Australian House*, Oxford University Press, Melbourne, 1986, p. 199.

At first, brick veneer was regarded as deceptive, yet by the mid 1930s it was fast becoming a standard technique. In fact it had become so general that most people had no qualms about owning such a home. The obvious drawback was that sound travelled from room to room, but this was remedied by the use of fibrous plaster sheeting. One popular product used as a substitute for lath and plaster on interior walls was beaver board, manufactured by the Beaver Board Company in Ottawa, Canada. Imported in large quantities following the end of the First World War, beaver board was supplied in panels, ready for nailing to joists or studding or over old walls. By the end of the 1930s, its market was being squeezed by masonite and canite, products manufactured using wood fibre and sugar cane fibre respectively.

TEXTURE BRICK

The Depression years may have set economic development back for a decade, yet the same could not be said for fashions in brick where there was still an appetite for the new. One such fad that made its appearance from about 1935 was the craze for texture bricks. Before the Depression, the fashion in face bricks had called for shades of blue, liver, reds and browns up to black. With better times, the public demanded a lighter, more optimistic brick

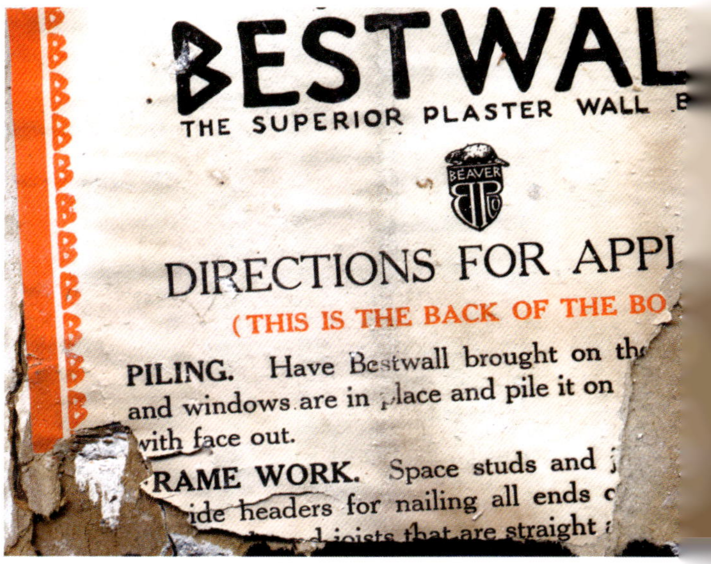

Rare examples of beaver board showing directions for installation, retrieved from a demolition site by Carl Johnson, Hudson, New York State. Carl Johnson cj@mynonurbanlife.com

in white, buff or cream. Texture bricks with their contoured, striated surface were extruded rather than dry pressed and being in relatively short supply were expensive, making them something of a status symbol. Only a handful of manufacturers had installed this new machinery, including Wunderlich, Lion Tiles and Goodlet & Smith. In middle-class residential districts there was a swing towards the use of texture bricks in the construction of flats and cottages. Even in some of the poorer districts, such as the Erskineville Government Housing Scheme, textured face bricks were being used. Some of the success of Goodlet & Smith was due to innovation, and the firm was an early producer of texture bricks. It was also a very successful and profitable operation. Speaking in 1940, Bert Moseley remarked that 'in the last few years the tendency of the public is for a brick which is brighter in hue, and a brick which gives a livelier effect. So much so that they have been looking for changes ever since, and they have now got to a white brick.' Moseley's two sons who were builders found themselves fully occupied building houses out of texture bricks. Jealousy, and the fear of falling market share, may help to explain why dry press manufacturers were so determined to get their hands on independent firms, most of whom made texture bricks.

Dry press bricks passing through a texture machine. See also page 262. Ron Hagstrom Collection

For several years, however, white bricks were anathema to the people living on Sydney's lower North Shore who considered them vulgar in appearance. This made life difficult for Edward and Herbert Lanceley, especially the latter whose high-pitched voice earned him the nickname, 'Squeaker'. Herbert's North Sydney yard manufactured a dry-pressed cream brick using the distinctive local clay. The Lanceleys had few buyers for this product and Moseley later recalled that the Council of Brick Manufacturers agreed to help, making arrangements for Herbert to sell bricks at 2s and 3s less than the nearby yards run by the Butcher family in North Sydney.[3] Much to his delight the situation had reversed itself by 1939. So great was the demand for creams that builders of the Concord Hospital ordered over one million from the Lanceley's North Sydney works, and could never seem to get enough. Reg Allott's Punchbowl brickyard was also struggling to keep up with demand for the cream variety, but unlike the Lanceley brothers (who could mine local clays), Allott was obliged to truck the raw material in from his Sutherland yard.

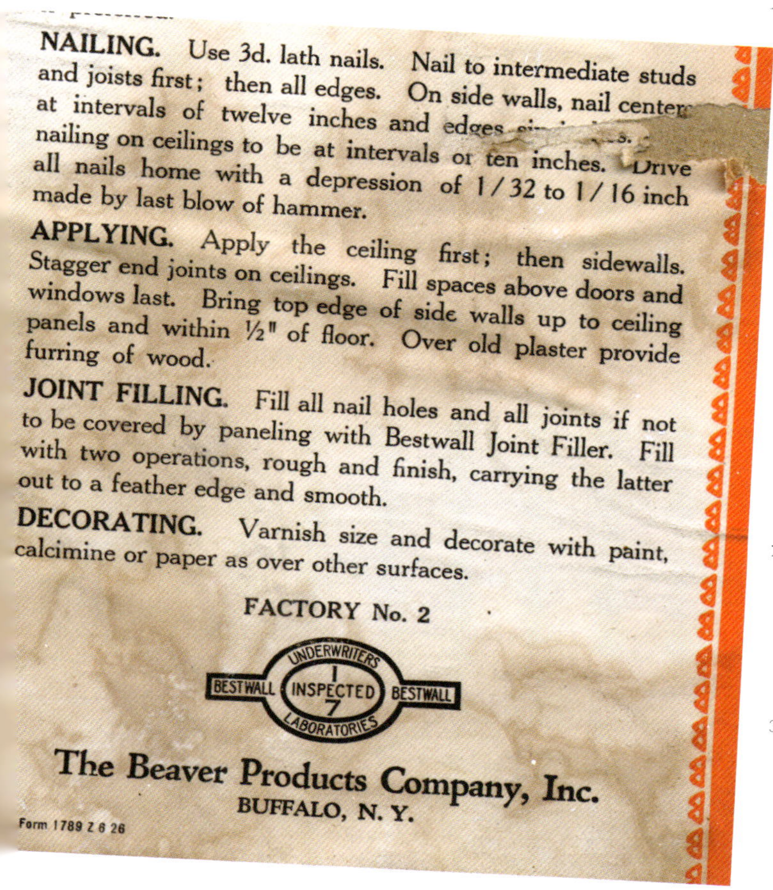

NAILING. Use 3d. lath nails. Nail to intermediate studs and joists first; then all edges. On side walls, nail centers at intervals of twelve inches and edges [illegible] nailing on ceilings to be at intervals of ten inches. Drive all nails home with a depression of 1/32 to 1/16 inch made by last blow of hammer.

APPLYING. Apply the ceiling first; then sidewalls. Stagger end joints on ceilings. Fill spaces above doors and windows last. Bring top edge of side walls up to ceiling panels and within ½" of floor. Over old plaster provide furring of wood.

JOINT FILLING. Fill all nail holes and all joints if not to be covered by paneling with Bestwall Joint Filler. Fill with two operations, rough and finish, carrying the latter out to a feather edge and smooth.

DECORATING. Varnish size and decorate with paint, calcimine or paper as over other surfaces.

FACTORY No. 2

BESTWALL UNDERWRITERS 1 INSPECTED 7 LABORATORIES BESTWALL

The Beaver Products Company, Inc.
BUFFALO, N. Y.

Form 1789 2 6 26

3. Bertram Moseley, *Attorney-General v. Brickworks Pty Ltd and Others*, op. cit., p. 280.

***Above:* Prosperity made a slow return to the brickyards. Men of the Austral yard at St Peters gather for a group photograph in front of Number 1 patent kiln, circa 1935. The brick building to the right was the engineers' room where the main switchboard for the plant was housed. It also doubled as a lunchroom. A small workshop was used to store tools and grinding machines. The boilers that powered the machinery were located at the rear of the building. Pat Binskin**

***Left:* Dry press operators and barrowmen standing in front of the Foster dry press at Austral's main yard, St Peters, circa 1935. Note the steel wheels on the barrow. Brickworks Limited**

Competition was also being felt in another market for clay products. The Depression provoked fierce rivalry between manufacturers of clay pipes and concrete pipes. From early 1935 a stream of articles, correspondence and photographs feature in the pages of *The Clay Products Journal of Australia*. Many of them explore the disadvantages of using concrete pipes over clay pipes and were quick to ridicule concrete. The February 1935 edition is of particular interest with its gleeful jibes: 'This Month's Concrete Pipe Failures' and 'Still They Come', or 'Wants to forget Concrete (Well – Can You Wonder!).'

FIBRO MAKES IT MARK

Brick veneer, concrete pipes and texture bricks were not the only innovations to make their mark. The second half of the 1930s was also notable for the increasing use of asbestos-cement ('fibro') weatherboard used to clad timber frame houses. The Timber Development Association was an active promoter of fibro. One quarter of all house building permits issued in Sydney between 1936 and 1940 were for fibro.[4] Fibro and weatherboard were especially popular in working-class municipalities close to the city, and extending outwards. Many homes in the suburbs of Ryde and Gladesville were built of fibro, continuing as far as Parramatta, Granville, Auburn and the western suburbs around Lidcombe, Regents Park and Campsie and Bankstown. This was cause for concern to brickmakers, especially the Council of Brick Manufacturers, which could do little to counter the growing, albeit small, competition from weatherboard. The trend became sufficiently well established for Dawes to comment that 'fibro is making great ground at the present time. They have improved in some manner. I think they have changed the manufacture of fibro-cement to something somewhat softer, and a thicker material that is more durable. At the present time if you go out into the Yagoona district, which is in the Bankstown area, you will find stretches of homes going up in fibro.'[5]

Leading the fibro charge was the Sydney firm of James Hardie Pty Ltd whose factory occupied a 19-acre site at Camellia. The company claimed its facility was 'an enormously valuable asset to the

4. P. Spearritt, *Sydney's Century: A History*, UNSW Press, Sydney, 2000, p. 65.
5. William King Dawes, *Attorney-General v. Brickworks Pty Ltd and Others*, op. cit., p. 416.

Above: Cottage on Grand Parade, Brighton, typical of the period. Cream-coloured face bricks were produced by Reg Allott's Sutherland yard

Parramatta district' and that its product, Fibrolite, 'was stimulating home building by helping to bring the purchase of homes within the reach of people with moderate means.'[6]

Unsurprisingly, the average cost of a fibro home in suburbs like Fairfield was about £533 compared to more than £1,700 for a brick home in the affluent suburbs of Kuring-gai. This did not sit comfortably with brickmakers who could see that deflation (a period where prices fall due to low demand) during the Depression years had worked to the advantage of one of its rivals. In fact, the dramatic fall in construction costs between 1929 and 1938 gave many families the first real chance to have a place of their own. *The Melbourne Age* newspaper commented on a statistical analysis of residential construction costs in New South Wales.

Based on two typical dwellings of weatherboard and brick, estimates were prepared of the quantities of materials and the hours of labour of different types used in their construction. The statistician, Mr Crawford, who prepared the analysis, noted that it now cost less to build a typical low-priced

6. P. Spearritt, op. cit., p. 66.

The Pipe Salesman

"Put them in your sewers," says concreter to the guy;
You'll find them so much cheaper and better when you buy!
They're everlasting, bore straight, they're always round and true,
Your soil will never murmur on its passage getting through.
Ne'er a man will e'er be wanted, out of the council yard,
With brush and sticks and idle tricks to fill his daily card;
Just look what they are standing—200 pressures up,
No earthenware could stand it, it hasn't got the gut,
I see you still are faltering—why not get them in your line?
You can't discredit them; they last a long, long time."

The Answer.

You're quite a virtuous salesman, intent to sell your goods,
I'll just explain to you again—your pipe won't stand the suds!
Fifty thousand pounds were spent on laying one town main,
They only functioned six months and then came out again;
You talk a lot of bunkum of how your pipes are spun,
You're not aware, I do declare, the mixture comes undone!
Just pick up a round bottle, then in your concrete drop,
Then, with arms outstanding, with palms intent and true,
Lift up your little bottle, just twirl a time or two—
You'll see cement is forward bent, the sand has lagged behind,
The spinning motion you've just done has quite destroyed the bind.

—Edmund Evans, Nth. Qld.

The Pipe Salesman. The Clay Products Journal of Australia, February 1935

People viewing Hardie's Fibrolite display home, Sydney, circa 1930s. This photograph shows part of the Sydney Showground near Centennial Park. National Library of Australia

brick house now than it did in 1929.[7] This was cold comfort to brickmakers whose costs were still far higher than the manufacturers of fibro. Labour accounted for nearly 60 per cent of the total cost of a brick. As mentioned previously, fuel costs were also high, especially when kilns were worked only intermittently. Kilns were just as expensive to run at half speed as they were at full capacity. In fact re-starting a kiln from cold was something to be avoided at all costs because achieving the desired temperature was a decisive factor in brick burning.[8] Detailed figures showing the costs associated with running a brickmaking business were compiled by Austral's chief accountant, Frank Trigg. The company's general ledger shows the main cost elements, including overheads, tax, and the effective capital on which shareholders could reasonably expect a return on their investment. Every brickmaker during this period must surely have been conscious of the disparity in the price of brick and fibro homes and the increasing popularity of the cheaper building material.

Comparing the cost of a fibro home in a working class suburb with a brick dwelling on Sydney's North Shore can be somewhat misleading. A more accurate comparison needs to be drawn between say, a four-roomed worker's cottage using about 20,000 bricks, and a larger home using about 35,000 bricks. Durability, insulating qualities and saleability over the long term also need to be taken into account if we are to compare like with like. Yet for those on modest incomes, using fibro offered huge cost savings since less than 1,000 bricks were used in the foundations (brick piers) and the chimney breast combined. As a result new patterns of construction established themselves across Sydney's municipalities. However, few owners preferred the look of fibro to that of brick and over the years brick has enjoyed wider appeal across the suburbs. For years to come the better class suburbs had their brick areas in which weatherboard was conspicuous by its absence.

7. *Melbourne Age*, 13 July 1938.
8. Fred Phippard in testimony before the court, *Attorney-General v. Brickworks Pty Ltd and Others*, In Equity, Supreme Court of NSW, 1940, p. 339.

A happy occasion. Alexander Stuart, owner of the NSW Brick Company Ltd, was a generous benefactor who lent the company's brick trucks for Sunday School outings, as can be seen here. Carrying children atop a brick truck was not uncommon in those days when there was comparatively little traffic on the road. Today, it would constitute a serious offence. Hood Collection, State Library of New South Wales

St Peters Church Sunday School picnic, 1938. Joan and Pat Binskin are sitting cross-legged to the right of the lady in white standing at the left of the picture. The chimneys of the Austral yard can be seen in the background. St Peters Church, Cook's River

SLUMS

No account of bricks and the building industry during this period would be complete without mention of the squalid living conditions that existed in many parts of the inner city. Attempts to eliminate the eyesore of substandard housing and slums across inner-city suburbs was another factor in the pre-war resurgence of Sydney's building trade and the brick industry. In February 1935 the Master Builders' Association of New South Wales urged a 'continuous and comprehensive policy for the abolition of decadent homes or decadent buildings [as] an essential part of the programme of national and state development.[9] Shortly after, the Bricklayers' and Carpenters' Union climbed on board, calling for 'an immediate campaign to abolish slum areas and to build better homes for the workers.' Speakers at the union meeting declared that besides providing better homes for the workers and abolishing eyesores, such a scheme would assist employment.[10] In a rare instance of capital and labour singing in unison (albeit for different reasons), the government of the day was spurred into action. In December 1937, Premier Bertram Stevens of the United Australia Party outlined a proposed slum demolition and rehousing project at Erskineville, informing the New South Wales Parliament that:

Fibro homes. Photography by David Smyth

> *one half of our people already have their own homes. We aim to see that proportion increasing.... The average man aspires to his own home, particularly the man in poor circumstances. For the home is the cradle of our race... the basis of family life.*[11]

In the final outcome very few people were relocated, but those who made the move found themselves living on estates on the fringes of the city. This scenario became increasingly familiar in many industrialised countries of the western world during the decades of the 1950s and 1960s when programs for urban renewal failed to heed the lessons of the past. Unfortunately most of the new developments lacked pubs, factories, shops, cinemas and churches, as well as the necessary infrastructure that made everyday life agreeable – sewers, sealed roads, stormwater drains, and public transport. The 'dunny man' remained a familiar figure for years to come.

9. *Sydney Morning Herald*, 27 February 1935.
10. *Labor Daily*, 24 June 1935; *Sydney Morning Herald*, 5 July 1935.
11. P. Spearritt, op.cit., p. 64.

Cream-coloured firebrick was used as decorative brickwork in a large number of houses from the late 19th century until about 1940. This mansion on Lang Street, Centennial Park, is a particularly fine example of the use of dry press firebricks in architecture of the late Victorian period.
Photography by Josh Hill

An example of apartments faced with texture brick, circa 1935, eastern suburbs, Sydney. Photography by Josh Hill

Brickworks Limited 11

A house divided – Birth of Brickworks Limited – King pin – Business as usual – The Council of Brick Manufacturers – End of the road – A very satisfactory outcome – More acquisitions – Suburban Land & Investment Company – Price rise – Union fightback – Royal Commission – Loose ends – A matter for the Industrial Commission – Home and away

A HOUSE DIVIDED

For all their resolve to work together the various parties to the Brick Agreement of March 1933 were seemingly incapable of putting individual interests aside for the sake of the common good. Behind closed doors at meetings of the New South Wales Brick Masters' Association there was heated debate about the parlous state of the industry. On 2 January 1934 a new collective agreement was finally struck between a large number of brickmakers, the object being 'to help meet the present abnormal conditions, to save a number of yards from bankruptcy, to stabilise the brick trade, and effect economies in production and distribution'.[1] Not everyone was satisfied especially Dawes and Spackman. On 19 March 1934 Austral's board voted to withdraw the company from the arrangement, sparking another trade war. By mid year price-cutting had reached a new intensity with commons reaching an all-time low of 38s per 1,000. It was nothing short of a disaster, especially in the area south of the Parramatta River. The yards in the North Sydney area and the Sutherland yard of the Punchbowl Brick & Pipe Co. Pty Ltd, situated at a distance from the centre of this renewed war, were more fortunate with prices never dipping below 45s per 1,000.

Austral's departure was about more than just the price of bricks in a perennially depressed market. Dawes left because Alexander Stuart, chairman of the association, had wanted to share the trade among all the yards in the agreement instead of keeping some closed. This cut against the grain for Dawes who judged the time was ripe for a major overhaul of the industry. The decision to quit the collective agreement was a great disappointment, although it was not entirely unexpected. Stuart, meanwhile, was exhausted by his efforts to turn the industry around. An elderly man in his mid-70s, and an industry leader since 1906, he had been in poor health for some time. Before the month was out Stuart was dead, his passing mourned by friends, family and colleagues.[2] Clarence F. Davison, managing director of the Clyde Brick Company for more than 20 years, succeeded him as head of the association. Davison immediately set about the difficult task of getting members to agree on creating a new and binding agreement, although this had to wait until October.

BIRTH OF BRICKWORKS LIMITED

In the meantime an equally pressing matter was yet to be resolved, namely what would be the industry's response to the sale of the State Brickworks? Davison knew that whoever purchased the State's yard would be in the ascendant for years to come. In the interests of its members who comprised the majority of brick manufacturers in the metropolitan area, it was imperative the association be in the running. Yet, without an appropriate business structure it was hardly in a position to make a bid. This, then, was the catalyst for the creation of Brickworks Limited. The new company was registered on 21 June 1934. The issue of seven signatory shares was followed nearly a year later on 14 May 1935 by a further 28,551 issued to various brick manufacturers. These shares had been fully subscribed and paid for in cash by 20 August 1935. Signatories to the formation of Brickworks Limited and its founding directors included Davison who was appointed chairman, as well as Frederick

1. *Attorney-General v. Brickworks Pty Ltd and Others, 1941*, Nicholas C.J., In Equity, Supreme Court of NSW, 1940, Vol 41, p. 77.

2. Stuart owned the New South Wales Brick Company, was a director of Suburban Land and Investment Company Ltd, and managing director of Flemington Brick Company Limited.

England versus Australia women's cricket Test match, Sydney 1935. National Library of Australia

Left: Sydney summer, 1934. – children playing in the sand at the Mick Simmonds' Radio Club picnic. Hood Collection, State Library of New South Wales

Above: Four men modelling hats in Pitt Street, for City Hatters. Hood Collection, State Library of New South Wales

George Phippard,[3] Edwin Spackman, Walter G. West, Edward D. Lanceley and Bertram Moseley. Clifford Minter from the firm of Minter Simpson & Company acted as legal representative.[4] It is to Stuart, however, that credit belongs for it had been his intention all along to rationalise the industry and buy out the State Brickworks, long regarded as a menace to the interests of the brick manufacturers. His death may even possibly have sharpened the resolve of brickmakers to act.

The dependable and widely respected Moseley was appointed company secretary, juggling his responsibilities with other senior positions, including a seat on the board of the Metropolitan Brick Company. Moseley hired like-minded managerial and office staff of the Baptist faith whom he knew to be honest and hardworking. Spackman, himself a deeply religious man, would certainly have approved. No matter how much Dawes favoured employees with a strong Protestant work ethic, he steered well clear of spiritual matters, being content to leave Moseley and Spackman in charge. Besides, with Dawes' combination of ambition and ego he would have made a rather poor servant of the Lord.

KING PIN

Perhaps the most critical appointment at Brickworks Limited was that of managing director, and it was in this role that William King Dawes through force of personality came to dominate the industry. Clarence Davison may have been a vital figure, but it was Dawes who did much of the legwork to bring brickmakers together to form the new business. A great deal of cajoling was needed before old established companies, especially those owned by families on the North Shore, would come on board. As a student attending Sydney Grammar School, Ray Dawes recalls being dropped off by his father

Bill Dawes, 1938. Bill Wheatley Collection

who went on to attend meetings of The Austral Brick Company Ltd at 12 Castlereagh Street.[5] These had become more frequent in the weeks prior to the formation of Brickworks Limited.

The extent to which the hand of Dawes was behind the formation of the new company becomes clear when we consider the stake held by Austral Brick in Brickworks Limited in May 1934, and again in April 1935. Over this period Austral's initial holding of 1,368 shares, the largest block held by any of the companies, increased to 6,304 shares, second only to the Suburban Land & Investment Company (SLIC) with 19,048 shares[6]. As we shall see this company had reasons of its own to be in control of such a large tranche. Over the next few months no fewer than 26 companies, nearly all of which were

3. Phippard was a trained civil engineer who became a builder and brickmaker. According to his nephew, Ned Phippard, his uncle became involved in a number of yards, including the National at Normanhurst, Great Northern Brickworks at Eastwood and Burwood Brick Company.

4. The firm of Minter Simpson was founded in 1890s by Alexander Robert Minter and Telford Simpson. Minter's son, Clifford, was admitted as a solicitor in 1910, enjoyed a strong reputation in commercial law and served as vice-president of the Royal Agricultural Show during the 1930s. Minter sat on the board of numerous companies and was connected with Archibald Howie.

5. Cricketing legend Bill 'Tiger' O'Reilly, the best bowler in Bradman's team, taught business and book keeping to Ray Dawes at Sydney Grammar School. He was also the sports master. O'Reilly left teaching in 1938 to work at Lion Tile Co as company secretary and manager. He advised the Austral office on prospective customers who had been refused credit elsewhere, but who came to Lion to purchase tiles. Dawes remembers O'Reilly as a typical outgoing, loud Australian who liked his beer.

6. Archives of Brickworks Limited.

Advertisement for the Suburban Land & Investment Company. Burwood Local Studies

engaged in the manufacture of bricks and members of the association, took up shares in Brickworks Limited.

Stories abound of Dawes and his high-handed behaviour and explosive temper, although contemporaries and family members agree this was offset by a wry sense of humour. In the early days, especially during the establishment of Brickworks Limited, Dawes trod warily and played the diplomat. Only later, when it was established and he was firmly in control, did he feel confident to flex his muscles. For example during meetings of the brick manufacturers Dawes would brook no opposition. Few were a match for his angry tongue and confrontational business manner. In short, he could be intimidating, which is not difficult to believe in view of the force of personality needed to deal with roughneck employees labouring in the yards. While he may have tempered his colourful language in the presence of female office staff, to anyone else he was prone to let fly with expletive-laden directions as to where that person might go – and remain.

A succession of 'tin can' episodes ensued whereby scandalised brickmakers tried to bring Dawes to heel. Meetings of the Brick Masters' Association provide some insight. Adamant that proceedings should be conducted with some decorum, and offended by Dawes' frequent use of the 'F' word, his colleagues would place a tin can in the middle of the table. For every infraction and foul-mouthed utterance Dawes was fined one shilling. On one occasion this move backfired when the wily Dawes simply dropped a £5 note into the tin and told the startled gathering, themselves no stranger to coarse language, to let him know when he had reached the limit. Fred Mercer, like everyone else, came in for a tongue lashing, and later claimed that Dawes had taken him aside, proffering the advice that 'we live by the eleventh

Above: Chrysler Airflow. University of Michigan, Dearborn and Benson Ford Research Center

Right: Austral Court, Rockdale, Sydney. William King Dawes built a handsome pair of flats in the suburb of Rockdale, one for his mother, and the other for himself. The bricks for Austral Court and Bedford Court were manufactured at the Bedford and Ralford yards. Ron Ringer Collection

commandment here, which is generally understood to mean, *thou shalt not get caught*.'[7] Readers can make up their own mind as to what precisely he meant.

There can be no doubt that Dawes enjoyed wheeling and dealing, especially the wheeling part. A fast car was the perfect vehicle to express his impatience with all that was pedestrian and plodding. William King Dawes was a man on the move. The fact is, by this time Dawes was exceedingly rich and, like many people with new-found wealth, liked expensive toys. Success in business, and a penchant for speed, encouraged him to indulge a long-held passion. In early 1935 Dawes took delivery of another American vehicle, the Chrysler Airflow, noted for its sleek, aerodynamic lines. Five years is a long time in car design, and with money no object, Dawes drove the (by now) stately '29 Buick to York Motors on William Street, Sydney for the last time after collecting it from the smash repair shop. Cash, handshakes all round, a cigarette and a drink sealed the transaction.[8] Like everything else to which Dawes turned his mind, there were no half measures when it came to pumping the accelerator pedal. He frequently wrote off cars, whether driving in Sydney or in country areas where roads were little more than dirt and gravel.

Exasperated and often embarrassed his colleagues were nonetheless inclined to overlook these eccentricities. Regardless of his abrasive personality Dawes was acknowledged to be a leader, and to ditch him merely because he ignored polite niceties would have been an act of extreme folly in such times.

BUSINESS AS USUAL

Down at St Peters Austral continued to go about its business. Tootling along in his company car, the sprightly Bill Brooker, sales representative for Austral Bricks during the 1930s and 40s, made the rounds of the builders' yards and sales agents. Accompanied by Bill Dawes' teenage son, Ray, Brooker was, by all accounts, an effective salesman. Certainly, he was well-liked by Dawes who was keen to learn about his competitors.

7. Interview with Ray Dawes, Sydney, 10 March 2006.
8. The 1934 Airflow and its later variations were short-lived. Production ceased in 1937 due to manufacturing problems and flat sales, nonetheless it was impressively styled. The vehicle's sleek appearance had more in common with cars designed two decades later.

Shareholders in Brickworks Limited, May 1935 to September 1938

Shareholders			Number of shares	Amount paid (£)	Amount paid (£)
Shareholder name	Address	Comment	14 May 1935	28 Jan 1936	20 Sep 1938
The Ashfield Brick Co. Limited	Milton Street, Ashfield.	Owner: R.H. Wilkinson. Stan Gentle was a major shareholder.	609	3,451	6,011
The Auburn Brick Co. Ltd	Auburn.	Owner: Clarence Davison.	648	Nil	Nil
Austral Brick Co. Ltd					25,378
Austral yard	King Street, St Peters.	E. Spackman was a major shareholder.	1,368	7,752	Nil
Bedford yard	King Street, St Peters.		1,068	6,052	Nil
Bakewell Bros. Limited	Coulson Street, Erskineville.	Owner: Arthur B. Bakewell.	516	2,924	3,597
Bankstown Brick & Tile	High Street, Bankstown.		609	3,451	Nil
Butcher Bros and Co. Pty Ltd	Hotham Parade, Artarmon.	Owner: Frank Butcher.	1,068	6,052	7,120
Canterbury District Brick Co. Pty Ltd	Canter Street, Campsie.	Managing director: George Hocking.	1,035	5,865	8,851
Carrington Steam Brick Co. Limited	Barwon Park Road, St Peters.	Owner: George Crispo.	609	3,451	Nil
Central Brick and Tile Pty Ltd	Albert Street, St Peters.	Owner: Walter G. West.	711	4,029	4,740
City Brick Co. Limited	Euston Road, Alexandria.		609	3,451	Nil
Clay Industries Pty Limited	12 Castlereagh Street, Sydney.	Owner: Brickworks Limited.	Nil	3,672	1,954
Clyde Brick Co. Limited	Auburn.	Owner: Clarence Davison.	972	5,508	9,344
Rupert Cook's Brickworks Pty Limited	Mitchell Street, Enfield.		609	3,451	9,015
Croydon Brickworks Pty Limited	Webb Street, Croydon.	Owner: George Kemp.	609	3,451	11,163
Eagle Brickworks Limited	Martin Place, Sydney.	Owner: Arthur Lyell Scott.	609	3,451	4,060
Excelsior Brick Co. Ltd	12 Castlereagh Street, Sydney.	Owner: NSW Brick Masters' Association.	696	3,944	Nil
Federal Brick Co. Ltd	Mitchell Road, Alexandria.	Owner and managing director:			30,722
Alexandria yard	Mitchell Road, Alexandria.	F.G. Phippard.	840	Nil	Nil
Carlton yard	Forest Road, Hurstville.		972	5,508	
Flemington Brick Co. Ltd	12 Castlereagh St, Sydney.	Owner: Alexander Stuart.	921	5,219	Nil
Great Northern Brick Co. Ltd	12 Castlereagh St, Sydney.	Owner: J.H. Gilliland.	609	3,451	Nil
Industrial Brick Co. Limited	Botany Road, Waterloo.			1,512	Nil
Judd's Hurstville Brick Co. Limited	Mortdale.	Owner: William 'Billy' P. Judd.	921	5,219	7,898
Moorview Brick Co. Ltd	Great North Road, Ryde.		609	3,451	Nil
The National Brickworks Co. Ltd	Thornleigh.	Owner: J.H. Gilliland.	921	Nil	Nil
The N.S.W. Brick Co. Limited	Huntley Street, Alexandria.	Owner: A.S. Loveridge and Andrew Kyle.	1,521	8,619	25,127
North Sydney Brick and Tile Co. Pty Ltd	Gore Hill.	Owner: H.G. Lanceley.	1,821	10,319	12,140
Northern Suburbs Brick Co. Ltd	Gordon Road, Artarmon	Owner: Alfred Green who was also Vice Chairman of the Council of Brick Manufacturers.	1,080	6,120	7,200
Punchbowl Brick and Pipe Co. Pty Ltd	Bonds Road, Punchbowl.	Owner: Reg Allott.			8,080
Punchbowl yard	Bonds Road, Punchbowl.		696	3,944	Nil
Sutherland yard	Princes Highway, Sutherland.		516	2,924	4,060
Ryde Brick and Tile Works Pty Ltd	Great North Road, Ryde.	Owner: E.D. Lanceley.	609	3,451	4,060
Shannon's Brick, Tile and Pottery Pty Ltd	Percival Road, Stanmore.	Owner: A. Shannon.	516	2,924	3,440
St Peter's Brick Co. Limited	York Street, Sydney.	Owner: Cecil Scott Waine & George Mitchell.	609	3,451	4,363
Strathfield & Enfield Brick & Tile Co.	C/ Johnson, Waine & Edwards, George Street, Sydney South.		609	3,451	4,060
Suburban Land & Investment Co. Ltd	12 Castlereagh St, Sydney.	Owner: Sir Archibald Howie and			149,632
Burwood yard		Alexander Stuart.	609	3,451	4,060
South Ashfield yard			609	Nil	
National yard			Nil	5,219	
Sydney Brickworks, Charlesworth & Co.	Unwin's Bridge Road, St Peters.	Owner: J.H.A. and J.R. Charlesworth.	609	3,451	4,431
Warren Brick Co. Limited	Mitchell Road, Alexandria.		Nil	142	9,053
Waterloo Fire Brick Co. Limited	Botany Road, Waterloo.		Nil	458	
Western Suburbs Brick Company	Dean Street, Enfield.	Owner: J.F. Walker.	609	3,451	4,060
		TOTALS	28,551	161,789	375,231

Source: Archives of Brickworks Limited.

Meanwhile, the company's day-to-day affairs were being carefully recorded by Len Rogers (brother of David Rogers) in the minute book. Entries in the General Journal for 11 April 1934 refer to the rental of a paddock to T.G. Collett at the Austral site. The sale of bricks to related companies was also noted, as well as the transfer of coal on 12 May to the Bedford Brick Company, referred to since its purchase as the 'Bedford yard'. Mention is also made of a 'Ralford' yard, the latter being a small brickworks consisting of one patent kiln on Canal Road acquired by Austral. The Ralford yard was never an incorporated company, but a subsidiary operation.[9] A succession of entries list creditors such as George Foster & Sons, the Tool & Engineering Company, Union Steel, Australian Alloy Steel, and the Commonwealth Oxygen Company, shedding light on the nature of Austral's dealings with local suppliers. New crushing machinery was purchased and installed at the Austral yard, most probably from George Foster & Sons.[10]

Then on 21 October 1934 we find mention of an investment in a petrol filling station. Was this the same petrol station opened some years earlier by Dawes in the name of the family business, and possibly sold to Austral? Perhaps. Again, shareholdings in various brick companies are listed, including a stake in Goodlet & Smith Ltd (often referred to as Goodlets), the Bedford Brick Works,[11] and the Flemington Brick Company Limited and Excelsior Brick Company Limited.

THE COUNCIL OF BRICK MANUFACTURERS

The first statutory meeting of Brickworks Limited on 2 October 1934 was an opportune moment to discuss the future, including yet another industry 'agreement' and the return of Austral to the fold. Before that could take place, however, the industry would need to settle the fate of the struggling New South Wales Brick Masters' Association once and for all. Two days later on 4 October 25 members of the association met to establish a new organisation covering the metropolitan area. While the association continued to represent brickmakers across the state (about 230 in total), control in the greater Sydney area effectively passed into the hands of a Council of Brick Manufacturers. Each manufacturer was entitled to nominate as a member of the council. At the 4 October meeting a binding, 10-year agreement was made between various companies, firms and individuals (numbering 33) and Brickworks Limited. The document was in two parts and comprised numerous clauses, one being provision for the likely purchase of the State Brickworks by Brickworks Limited. At the same time the council agreed to promote the manufacture and sale of bricks in the state and to foster and regulate trade conditions

> *in the interests of the brick manufacturers as a whole, having due regard to the desirability of providing adequate supplies of bricks and reasonable prices to meet the requirements of the community.*[12]

This was a defining moment for private manufacturers in the Sydney region. Unfortunately, the text of the agreement does not survive, but numerous references to it may be found in the report on brick prices of the Industrial Commission of New South Wales, as well as the findings of Equity Court judge, C. J. Nicholas in *Attorney-General versus Brickworks Pty Ltd and Others (1940)*. Part Two of the agreement established the workings of the council through a management committee subject to control and direction of the council. The committee comprised not less than eight, but not more than 12 members. Dawes, West, Butcher,[13] Allott, E.D. Lanceley,[14] Phippard, Green, Davison,

9. General Ledger of the Bedford-Ralford yards for 7 January 1934 to 29 June 1938, Mitchell Library.
10. George Foster & Sons changed hands in 1978 following its purchase by the engineering firm of William Wallbank & Sons Pty Ltd established in 1934. Wallbank made brick presses and machinery of every kind, and soon became the leading producer of brick and tilemaking equipment.
11. Austral Minute Book, 1936. Minutes of an ordinary general meeting on 28 September 1936 show that loans taken out to pay for the Bedford Brick Works owned by the Gentle family were not finally paid until this date.
12. Cited in 'The Industrial Commission of New South Wales: In respect to the Prices Of Bricks sold by manufacturers operating and selling in the Metropolitan Area and in Suburbs adjacent thereto.' Industrial Commission of New South Wales 1938, Vol A, May 1939, p. 6.
13. Frank Butcher was the director of Butcher Brothers and Company Pty Ltd at Artarmon, a director of Brickworks Limited, a member of the Council of Brick Manufacturers, a shareholder in Ryde Brick Company, and a director of Clay Industries Ltd.
14. Herbert George Lanceley (brother of E.D. Lanceley) served as chairman of the Council of Brick Manufacturers in 1940. He was also chairman of Clay Industries Limited.

Howie[15] and Hocking were all voted on to the committee, which met for the first time on 9 October 1934.

These ten brick manufacturers commenced by ratifying the 4 October agreement and setting a member levy of 6d per 1,000 bricks to cover expenses and running costs. This was imposed for each class of bricks manufactured and on each manufacturing company. Any member who had unpaid shares in Brickworks Limited had to pay an additional levy of 5s per 1,000 bricks to the company. Moreover, each manufacturer agreed not to sell bricks produced below the current sales price. Those who wanted to withdraw would need to pay in full for any shares in Brickworks Limited, as well as the current levy for the balance of the 10 years of the agreement. The high cost of quitting the scheme was deliberately intended to protect existing members. Taken as a whole these conditions were not unduly onerous, but the existence of another clause – an option granting Brickworks Limited the right to acquire any of the brickyards of manufacturers party to the agreement – was to have far-reaching implications for the industry. The meeting of 9 October 1934 also sanctioned the first in a number of price rises.

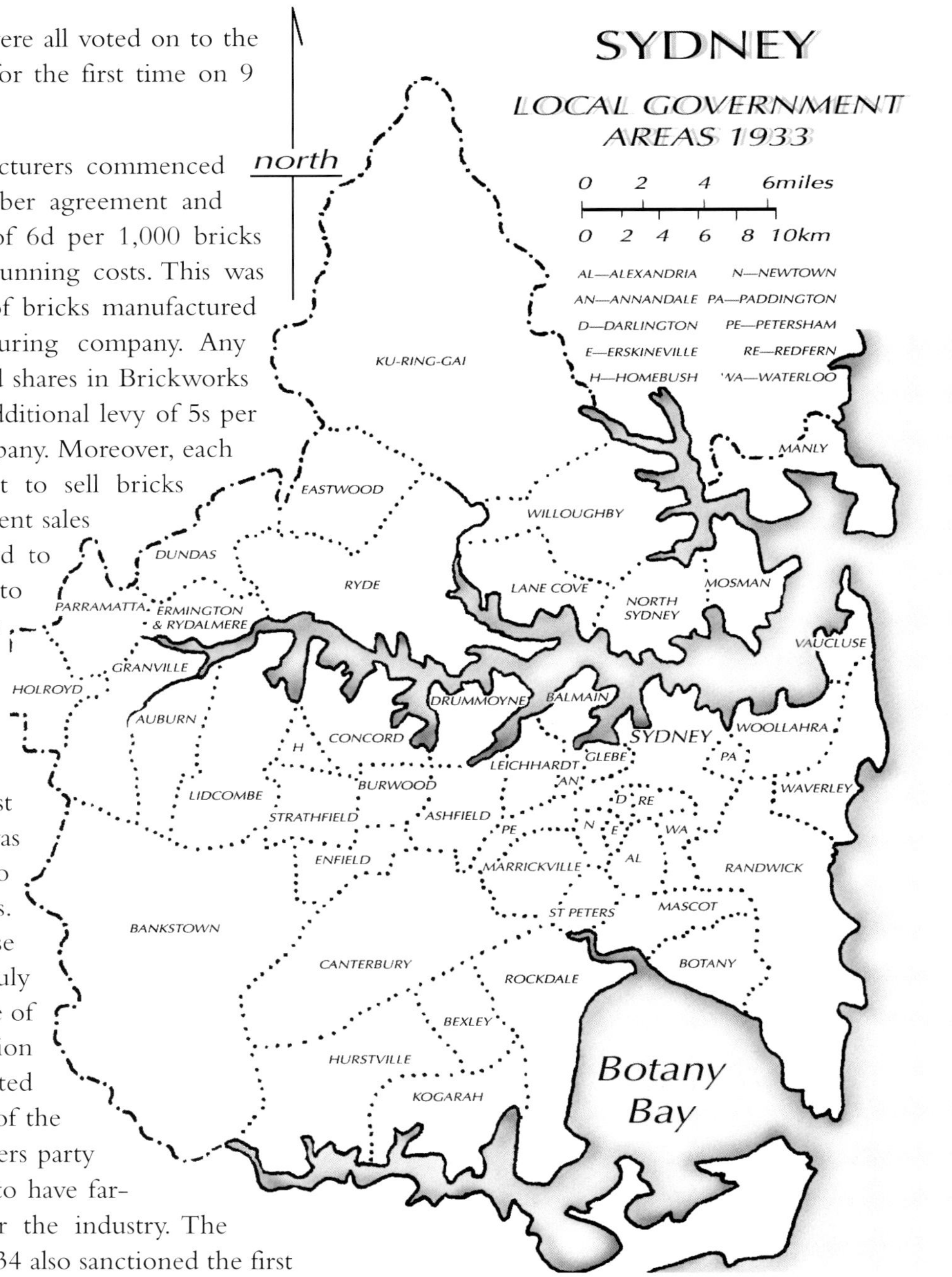

A CARTEL?

One other important measure introduced in these early days was a deal struck between manufacturers competing for business in the same area. At the prompting of Dawes four local agreements were forged which shared trade across the city, eliminating competition almost entirely. These came to light during the inquiry conducted by the New South Wales Industrial Commission in January 1939, although the agreements ceased to operate after June of that year when prices were fixed by the commission.[16] A series of maps tabled as evidence by Moseley showed how the Sydney metropolitan region was organised into zones, or spheres of

15. Howie was director of the St Peters Brick Co, director of Suburban Land and Investment Company, a member of the New South Wales Brick Masters' Association and director of the Metropolitan Brick Company.

16. Fixed at 58s 6d per 1,000 by Mr Justice de Baun in the Industrial Commission on 9 June 1939, and at 64s 11d per 1,000 on 16 October 1939.

influence. The first, and most important was the St Peters agreement dated 20 May 1936. Two others dated 28 May 1937 divided the trade in North Sydney and Manly respectively, and an agreement of 30 August 1937 regulated the trade in Hurstville. Metropolitan Bricks looked after the northern side of Sydney Harbour while Clyde Bricks (Clarence Davison's firm) took over the western part of the city, splitting this with Brickworks Limited. Austral Bricks covered south-eastern Sydney comprising St Peters-Alexandria, the eastern suburbs, inner west and central business district. The agreements also qualified the yards into working yards and non-competitive yards (see table at right).

The so-called 'working yards' manufactured and supplied bricks according to agreed percentages. For example Austral's yards were allocated 53 per cent of the total trade from time to time in the area, but only up to a maximum output of 795,000 bricks per week.[17] On the other hand, the non-competitive yards agreed not to manufacture and sell bricks, except where they were instructed to do so by the council. In many cases they produced no bricks at all from 4 October 1934 onwards, which resulted in plant and equipment being mothballed, sometimes for years. As compensation they received regular payments from a fund into which working yards paid a small levy of 2s per 1,000 on the sale of bricks. Brickworks Limited was able to acquire non-competitive and closed yards and consequently enjoyed a revenue from the levy once the original owners had been paid out either in cash or by shares.

One reading of these facts might lead to speculation that a price-fixing cartel was once again at work. The May 1939 Report of the Industrial Commission came to a different conclusion in stating

> *these agreements have been sponsored by the Council of Brick Manufacturers for the purpose of obtaining greater co-ordination in the industry and controlling competition by the companies in their particular areas and have worked very well. There is nothing contained in these agreements, or any action taken in pursuance of them, which has had the effect of increasing prices of bricks beyond what is reasonable.*[18]

Local Agreements, 1936–1937

Area covered	Signatories	Working yards	Non-competitive yards
St Peters	Austral Brick Co. Ltd.	✔	
	Bakewell Bros Ltd.		✗
	Carrington Steam Brick Co. Ltd.		✗
	Central Brick & Tile Co. Ltd.	✔	
	City Brick Co. Ltd.		✗
	Clyde Brick Co. Ltd.	✔	
	Federal Brick Co. Ltd.	✔	
	New South Wales Brick Co. Ltd.	✔	
	St Peters Brick Co. Ltd.		✗
	Sydney Brickworks		✗
	Warren Brick Co. Ltd		✗
North Sydney	Butcher Bros & Co. Ltd.	✔	
	North Sydney Brick & Tile	✔	
	Northern Suburbs Brick Co. Ltd.	✔	
Manly District	Butcher Bros & Co. Ltd.	✔	
	Clyde Brick Co. Ltd	✔	
	North Sydney Brick & Tile Co. Ltd.	✔	
	Northern Suburbs Brick Co. Ltd.	✔	
Hurstville	Federal Brick Co. Ltd.	✔	
	Judd's Hurstville Brick Co. Pty Ltd.	✔	

Source: Industrial Commission of New South Wales, p.20

The existence of local agreements can best be explained by the need to stabilise the industry and provide a modicum of certainty. If trade increased in any particular locality, Brickworks Limited would have on the spot a fully equipped yard which could be activated for relatively small outlay. In the meantime, the six closed and non-competing yards enjoyed an appreciable revenue. Attempts to regulate trade and fix prices were not confined to the brick industry. Indeed the role of such organisations invites further study. In the tile industry Roof Tiles Ltd was formed in October 1936, replacing the Roofing Tile Makers Employers' Association. The company established price fixing by regulating output for the 14 companies party to an agreement between themselves and Roof Tiles Ltd. One

17. Industrial Commission of New South Wales, op. cit., p. 20.

18. Industrial Commission of New South Wales, op. cit., p. 21.

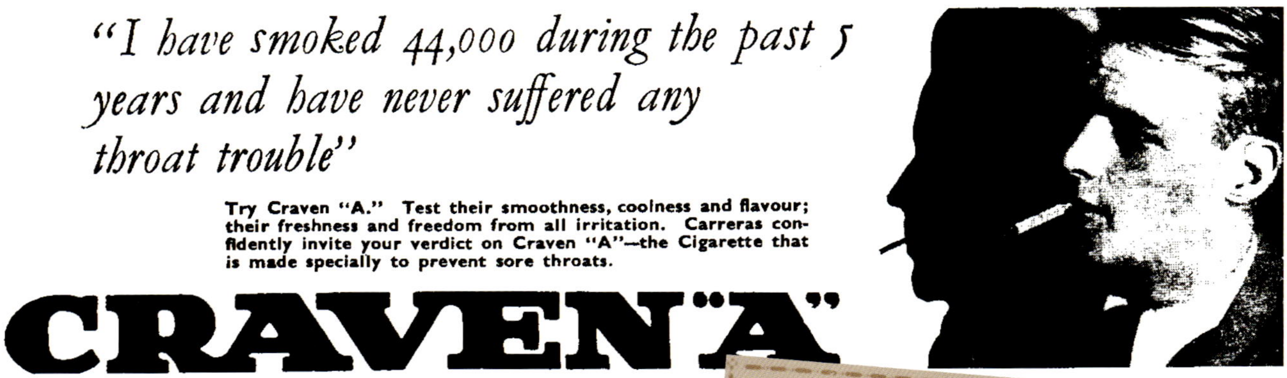

For a generation of men and women who had experienced the war of 1914 –1918, smoking was all the rage

leading manufacturer, Goodlet & Smith, which was not in a production agreement with Roof Tiles Ltd was paid to keep it that way. Mr Justice de Baun who conducted the subsequent enquiry into the price of tiles commented that the company had received £3,031 'although it has not during that time (1936–1938) manufactured or attempted to manufacture a single tile.' He noted that the tile plant was used in the production of other commodities.[19] Was it any wonder in view of low demand and oversupply? The practice of subsidising firms to restrict output made sense for small operators and less efficient manufacturers, but what of the larger firms? To do nothing might well have destabilised the industry through further rounds of vicious price cutting. What is difficult to comprehend is why large firms allowed this to happen rather than drive competition to the wall. Quite possibly, a gentleman's agreement was struck that looks, to all intents and purpose like a monopoly, but was actually no such thing.

BRICK CARRIERS' ASSOCIATION

Brick cartage was another activity that broadly came under local agreements and the new system made it vastly easier to sell and distribute bricks within the local area. It was a fillip to carters who could now more accurately determine rates and offer the cheapest cartage. Most were owner-drivers or small firms with several trucks. By far the majority

Directors of Judd's Hurstville Brick Co. Ltd, Mortdale, on the occasion of its 150th anniversary in 1934. Front cover of Judd's catalogue, circa 1934. Kogarah Historical Society Inc.

19. HLA-Envirosciences Pty Ltd, *A Conservation Plan for the Former Goodlet & Smith Brickworks*, Merrylands, July 1995, p. 22.

Harry Bell's Fargo brick carter's truck showing insignia of The Austral Brick Company Ltd. Ross Bell Collection

were members of the Brick Carriers' Association, the successor body to the Master Brick Carters' Association, established in 1911, but which fell victim to the Depression. The Butlin Archives in Canberra lists them both although, strangely, no records appear to exist. The Brick Carriers' Association, incorporated in 1934, represented the various small companies and owner drivers under contract to Sydney's brick manufacturers. Alexander Robertson is listed in the Industrial Commission's Report of 1938 as the secretary representing the interests of 143 members. Cartage rates between the association and the Council of Brick Manufacturers were fixed by agreement. From October 1934 onwards members of the council strongly discouraged customers from carting their own bricks not only because it interfered with this contract, but also because it disrupted work in the yards. For the most part the system worked to everyone's satisfaction. The carter would arrive with his truck first thing in the morning and having been allocated a delivery would load the bricks and deliver to the various jobs. On a good day he could make three or four runs which might also include carting coal from the Tempe rail sidings to the kilns. Fred Phippard, the widely respected owner of the Federal brickyards, shed light on the business of brick carting.

> *Say a man wants a thousand bricks. We tell the carter to load up 1,000 bricks for the building firm of Howie & Moffatt or Farmer & Company for instance, and give him a ticket made out in triplicate. He signs the butt in our book, which is his acceptance of 1,000 bricks. He takes the other two parts of the ticket and the builder signs one and returns it to him. That is his warrant that the bricks have been delivered and he is entitled to be paid by us, and the third one is kept as a matter of record by the builder. The result is that if the carter takes less than 1,000 bricks, the builder when he counts them would only sign for say, 950 bricks. The carter has signed to us for 1,000 and he is liable to us for that 50.* [20]

Members of the Brick Carriers' Association were trusted men and it was a rare occurrence that loads would need to be counted to ensure honest dealings. This was not always the case with other

20. Frederick Phippard, *Attorney-General v. Brickworks Pty Ltd and Others*, op. cit., p. 336.

This wharf on Blackwattle Bay was used by the Clyde Brick Co. Ltd. Bricks were conveyed by barge from its yard at Clyde in the Homebush area and unloaded in crates by a crane mounted on the north-east corner of the wharf. This was the last of such wharves on this part of the harbour foreshore. A similar brick wharf to the one pictured was located at Vickery Avenue, Rose Bay. Clyde Brick Co. Ltd closed the facility in early 1940. Another such existed at the foot of Wharf Road, Vaucluse and became the 12 foot Amateur Sailing Club facility. Photography by Josh Hill

brick carters who came in for the occasional load. Very often they were found to have left with more than what they had paid for. As a result casual loads were not encouraged. According to Bert Moseley: 'Since it was impossible to count every load – and avoid throwing the yard out of gear for its deliveries – casual loads were few and far between.'[21]

END OF THE ROAD

And what of the State Brickworks, the raison d'étre behind the formation of Brickworks Limited, and whose future was being so hotly debated? On 27 August 1935 advertisements appeared for its sale in Sydney newspapers with a deadline of 4 October 1935 for tenders to be received. One proviso of the conditions of sale was the appointment of a board to fix a fair and reasonable price for products manufactured at the State's yard. This time there was to be no repeat of the fiasco that marred the previous attempt by brick manufacturers to purchase the yard. On 4 October the Council of Brick Manufacturers resolved unanimously that Brickworks Limited would tender for the purchase of the State Brickworks with an offer of £155,000 plus stock in trade and stores. The only other bidder was W.J. Smith who tendered £89,685.

A letter from Arthur Charlesworth reminded members of the council that, finally, the brick trade had an opportunity to purchase the State's yard. Charlesworth went on to urge full support for the government since bids undervaluing the assets would only provoke adverse comment. Hence any tender would need to be on the basis of the stated yard being a going concern, fully equipped to do

21. Bertram Moseley, *Attorney-General v. Brickworks Pty Ltd and Others*, op. cit., p. 334.

one-fifth of the present trade of the metropolitan area. Moreover, 'the price must necessarily be high to keep the competitor from getting it, for should the Works be obtained by any other than ourselves, price-cutting must naturally follow, and from past experience such a disastrous recurrence would eliminate all present advantages.'[22] With £43,000 of levy money held in a trust fund, Charlesworth was confident the council would find an additional amount sufficient to cover the full purchase price. To this end, a debenture mortgage over Brickworks Limited of £100,000 was obtained from the organisation's banker, the Commercial Banking Company of Sydney Limited. Davison actually solicited financial assistance from outside the council and was in touch with independents such as Goodlet & Smith, the city's leading producer of texture bricks. That the latter rejected his approach was due to a recent hostile take-over bid. Just days before the 4 October submission deadline Brickworks Limited had endeavoured to buy a controlling interest in Goodlet. The move was strongly resisted and as a consequence Goodlet refused to join the council, preferring instead to align itself with the firm of Speare's Brick and Pipe Works Pty Ltd at Tempe, Sydney. The unwelcome bid created lasting ill-will which was further aggravated in early 1938 when Dawes and the council tried to bully Goodlet & Smith into raising the price of its texture bricks. Goodlet and Speare must have felt justified in their earlier refusal to join the council as they were apprehensive at the thought of disappearing into the maw of Brickworks Limited.

With little to fear from its rival bidder, Brickworks Limited was informed that its tender had been successful. A final purchase price of £164,073 was agreed, made up of £154,563 15s 6d for the assets and £9,050 3s 9d for stocks and stationery.[23] The assets included three working pits, brickmaking machinery, 25 kilns and 96 acres of land at Flemington, Homebush Bay and the Blackwattle Bay depot where bricks were offloaded by barge.[24] For a short while longer production continued with deliveries being made from Homebush only. The practice of shipping bricks to Blackwattle Bay was discontinued. By 1938 only two out of 10 patent kilns were working at the State Brickworks. In the Sydney region 32 patent kilns were in use out of a total of 64 patent kilns, which was testimony to the lingering effects of the Depression. Until its resumption by the state government in 1947, the Homebush site was sidelined and then closed up.

A VERY SATISFACTORY OUTCOME

The purchase of the Homebush works by Brickworks Limited had been a strategic move that stopped the State's yard dead in its tracks, much to the dismay of its many supporters. Almost three years had passed since the announcement by Premier Bertram Stevens of his government's intention to sell the yard. On 1 February 1936 *The Clay Products Journal of Australia* could finally announce that a very satisfactory outcome had been achieved for private manufacturers. An article penned by a 'Special Commissioner' sounded a triumphal note, describing how, after months of negotiation between the interested parties, the sale to a private company – Brickworks Limited – had become an accomplished fact.

> *For more than twenty years the state works have been turning out bricks in competition with private enterprise. They have had special advantages such as are enjoyed by no private concern. They have had all the money they needed. They have not had to bother about interest charges, and they have had special transport facilities. They have had an unfailing purchaser for their products in the State Government and other benefits. Even with those exceptional advantages their business career was a chequered (sic) one. They made profits in good years, but showed losses in bad ones.*[25]

22. State Records NSW, Exhibit 11, *Royal Commission of Inquiry into charges of fraud and corruption in connection with the sale of industrial undertakings, 1937–38*, 1938.

23. *Brickworks Limited: Fifty Year Retrospective, 1934–1984*, Archives of Brickworks Limited.

24. Ibid.

25. 'State Brickworks sold for £162,025 – Unfair Competition to Cease', *The Clay Products Journal of Australia*, 1 February 1936, pp. 19–20.

Yards acquired by Clay Industries Limited, 1935–1938

Company	Referred to as	Date acquired	Purchase price
Auburn Brick Co. Ltd	Bankstown yard	11 July 1935	£10,250 cash
State Brickworks	Homebush yard	January 1936	£160,000 cash
Manly Brick and Tile Company	Brookvale yard	9 June 1936	£26,250 cash
Excelsior Brick Company	Excelsior yard	March 1937	£47,007 shares and £3,000 cash.
Flemington Brick Company	Flemington yard	March 1937	£49,046 4s made up of £26,752 in cash and 22,294 fully paid £1 shares in Brickworks Limited.
Moorview Brick Company	Moorview yard	June 1937	£12,277
City Brick Company	City yard	September 1937	£23,277 cash
Suburban Land & Investment Company		August 1938	£158,984
Great Northern Brick Company	Northern yard	August 1938	
National Brickworks Co. Ltd	National yard	August 1938	
Burwood Brick Co. Ltd	Burwood yard	August 1938	
South Ashfield Brick & Tile Co. Ltd	South Ashfield yard	August 1938	

Source: Archives of Brickworks Limited

Mr Spooner, State Minister for Local Government, made it clear where his affections lay. 'The wastage of capital, unemployment, the check to the incentive for private industry, all contributed during the last 20 years to the flight of some capital from this State to other places where more settled industrial conditions are conductive to lower prices, profits and employment.'[26] Predictably, the decision to sell the State, and the circumstances surrounding its sale to Brickworks Limited, brought long-standing animosities to the surface. Accusations of corrupt conduct were made in the State Parliament by Jack Lang, the former premier, leading to the establishment of a Royal Commission in 1938. From the viewpoint of either side, each case had its merits. Private manufacturers had been up against the State's yard for more than 20 years, and at a disadvantage since a public enterprise was not expected to make a profit. On the other hand, many private manufacturers were deeply conservative in their outlook and reluctant to invest in new machinery. Who else but themselves could they blame for the backward state of their yards? In contrast, and from a modern perspective the State Brickworks, paradoxically, had proved itself to be innovative and offered healthy competition to private firms. Perhaps a little too much! Then, as now, it is difficult to compete effectively with a commercial opponent who plays to a different set of rules. Regardless of the relative merits of either side, the private sector joined together at the height of the Depression to avert the very real prospect of complete collapse. In the hands of Dawes, Brickworks Limited became the means of salvation, although at the expense of the State Brickworks which was very promptly taken out of the competitive equation.

MORE ACQUISITIONS

Despite signs the economy was on the mend by the end of 1934 the brick industry remained in the doldrums, with too many producers chasing too little business. The long awaited shake-out of those who had survived the initial shock of the Depression was about to take place. Having purchased the State's yard through Brickworks Limited, the Council of Brick Manufacturers proceeded to make a clean sweep of things. With the approval of their owners and shareholders, one company after another accepted Brickworks Limited's offer to buy. These acquisitions were made by Clay Industries Limited (CIL) which was formed on 11 July 1935 to acquire non-operating companies that held shares in Brickworks. Its formation was a legal measure consistent with the rule that a company may not purchase its own shares. Any unpaid calls on shares which Brickworks had purchased were credited to the account of CIL of which Dawes was managing director. The Auburn Brick Co. Ltd was the first to go, followed by many more. Yards and assets acquired by CIL from 1935 to 1938 included a line-up of well-known manufacturers.

26. Ibid.

Above: Men in front of patent kiln, Brookvale yard, circa 1914. Ron Hagstrom Collection
Right: Ron Hagstrom, wartime pilot and brickmaker. Ron Hagstrom Collection

The next brickyard to fall under the control of Brickworks Limited was a minnow by comparison to the State's yard. On 9 June 1936 the Manly Brick & Tile Company at Brookvale was purchased by CIL. On 29 September Brickworks paid an additional £4,494 for a huge stock of bricks (1,676,505) that lay stacked and unsold at the yard. Ron Hagstrom, former plant manager at the Brookvale yard over a 30-year period from the 1950s, joined the firm as a clerk in December 1935. Hagstrom recalls that David Rogers, company secretary of The Austral Brick Company, was sent by Dawes to inspect the Brookvale facility. Hagstrom, incidentally, was the only employee to be employed by the new owners. Brookvale ceased production and remained closed throughout the war years, although in the meantime the clay brick mountain was gradually sold off through the efforts of Ron Hagstrom who manned a sales office at Seaforth, adjacent to the Spit Bridge. Operating as a distribution point for bricks from the Clyde works and other yards owned by Brickworks Limited the depot was kept busy, handling cargoes of bricks carried by barge down the Parramatta River, through the harbour, and thence to Seaforth. Ron Hagstrom worked at the office until 1941 when he joined the Royal Australian Air Force, later transferring to Britain via Canada where he became a pilot with Bomber Command of the Royal Air Force.

The following year on 4 May 1937 the board of Brickworks Limited voted to amend the articles of association to become a private company. It remained so until 29 August 1939 at which time the firm voted to revert to its original status. This interesting manoeuvre may have had something to with a ruling in 1938 by the Commissioner for Taxation which treated Brickworks as a public company for accounting purposes. The adoption of of a proprietary limited status was also accompanied by an increase in capital to £300,000, a move which anticipated the purchase of the entire assets of the Excelsior and Flemington Brick companies.[27] The assets of both these companies were considerable and comprised fully equipped brickworks and shares. Then in September 1937 City Brick Co. Pty Ltd at Euston Road, Alexandria, on what is now the Metromix Plant, was acquired. At the same time CIL on behalf of Brickworks Pty Ltd built up shareholdings in many other companies so that in due course it had acquired almost every major brickyard in Sydney. The driving force behind these acquisitions was none

27. Limited liability ensured that investors were protected in the event of the company becoming insolvent.

Alderman Sir Archibald Howie, MLC (1934 –1941), Lord Mayor of Sydney 1936 –1937. City of Sydney Archives

other than Dawes, whose actions as managing director of Brickworks, and CIL, and as chairman of the Council of Brick Manufacturers (1937–1939), effectively brought a significant part of the industry under the control of the one company, and one man. By the end of the 1930s William King Dawes was the acknowledged 'King of Bricks'. Shares in other companies continued to be snapped up, reaching a total of 108,000 by late 1938. By 4 October 1938 Dawes informed directors of the company that the debit balance incurred by the purchase of the State Brickworks and other yards had been cleared. There was even a slight credit in the account allowing for the first dividend of 10d to be declared.[28] This was a startling achievement in so short a time.

Most of these companies never made a single brick following their purchase. The fact is their owners were only too glad to sell out owing to financial hardship. In some cases they had been already closed for many years. Apart from anything else most firms were heavily undercapitalised and in due course would have disappeared like so many others before. A vast amount of machinery was released by yard closures and used by Dawes to retire obsolete machines or placed in storage, much of it at the premises of City Bricks on Euston Road, Alexandria. 'Never for a moment', Dawes reminded himself as he went out about his business at the Austral yard on Canal Road, 'give the competition an even break'. On the contrary, rather than offer brick machinery for resale to rival manufacturers he instructed the firm's engineers to salvage workable parts, disposing of the rest to scrap metal merchants who immediately hauled it to nearby foundries to be melted down. Within the Brickworks group of companies this 'beggar your neighbour' approach did not apply for obvious reasons; those who had found refuge within the entity were held together by a common understanding – the need to think and act as one. As more and more of the original investors disposed of their interests, Dawes began to flex his industrial muscle. Those who chose to remain outside the fold were up against a ruthless competitor.

SUBURBAN LAND & INVESTMENT COMPANY

At this point the Suburban Land & Investment Company assumes greater importance in our story. The reader may recall that SLIC had been incorporated in 1911 and shared offices with the New South Wales Brick Masters' Association at 12 Castlereagh Street. SLIC was strictly not a brick manufacturer, but rather an investment company that owned a number of brickworks. For several years the firm had pursued its own expansion strategy. For example, on 1 July 1935 SLIC purchased the assets of The National Brickworks Company Limited operating at Pennant Hills Road and Dartford Road, Thornleigh. This included all goodwill, freehold property of more than 25 acres of land in Thornleigh, plant, stock, and 921 shares in Brickworks Limited as well as 14,000 shares owned by The National in the Great Northern Brick Company Limited.[29]

At more or less the same time, SLIC gained control of the Great Northern Brick Company which operated at Midson Road, Eastwood. Incorporated on 1 March 1912 with a capital of £10,000, Great Northern's annual production was about 7 million bricks by June 1937.[30] SLIC's annual report for that

28. Report of the Industrial Commission of New South Wales, 1939, op.cit., p.18.

29. One month before, on 4 June 1935, at an emergency general meeting, a resolution was passed to increase the capital of The National to £150,000 by creating 50,000 new shares of £1 each, and also to increase the number of directors to five.

30. State Records NSW, Clay Industries Pty Ltd, 1935–52, Container 11/11785.

year lists its directors as Howie (chairman), Phippard (managing director), Davison, James Gilliland, and Arthur P. Evans.[31] The company secretary was Raymond Wilkinson, an extremely busy man who also wore the same hat for the Metropolitan Brick Company, South Ashfield Brick Company,[32] City Brick Company (1935), Flemington Brick Company (1936), and Excelsior Brick Company (1936). From December 1936 Wilkinson was also chief accountant for the Great Northern Brick Company. Further research may reveal the full extent of Wilkinson's directorships, but at first pass his curriculum vitae was impressive. Unsurprisingly, the same, familiar names appear on the board of many other companies as can be seen on page 205. The Great Northern was voluntarily wound up in November 1937 and a last dividend of 10 per cent was paid which reflected the good health of the company, the last year of trading being profitable. The Eastwood yard, now wholly owned by Brickworks, continued producing a range of dry press commons and face bricks until its closure in 2003.

With a board of directors who were themselves on the Council of Brick Manufacturers and investors in Brickworks, it was only a matter of time before SLIC fell into line. On 19 July 1938 Dawes foreshadowed its amalgamation with Brickworks by asking shareholders to vote to increase the capital of the company to £500,000 by issuing an additional 200,000 £1 shares. The amalgamation was complete by 9 August when £158,984 was paid for the issue of 149,632 fully paid £1 shares, with the balance of £9,352 paid in cash. SLIC's new owner gained an extra four fully equipped yards, the Burwood, South Ashfield, Great Northern and National bringing to 11 the number of brickyards owned by Brickworks Pty Ltd. At an emergency general meeting of SLIC on 5 October 1938 a resolution was passed that the company be wound up voluntarily. A final meeting was held the following year on 28 June 1939 at the offices of the Council of Brick Manufacturers. Then and there it was decided to place the books, accounts and documents of SLIC in the custody of the Metropolitan Brick Company Limited for a period of three years. It was left to the discretion of the manager of the aforementioned company to destroy them. And indeed they were, with the result that a rich vein of historical enquiry has been lost to posterity.[33] As for Clay Industries Limited, this firm continued until 28 November 1950 when it too, was wound up voluntarily. The liquidators account statement for 25 March 1952 shows £104,981 2s in assets in the name of the company which passed into the hands of its parent, Brickworks Limited.[34]

PRICE RISE

Even before the wave of acquisitions commenced brick prices were on the rise, a clear indication that the economy was beginning to emerge from the Depression. Firmer prices ended a five-year slump, the return to profitability being welcomed especially by those inside the industry. For the customer, however, it was a somewhat different story and cause for complaint, as we shall see.

The new prices introduced in October 1934 were reviewed at a further meeting of the Council of Brick Manufacturer's management committee held on 16 October, and again two weeks later. Members were impatient that the price of commons had

Sales price of bricks at the kiln, October 1934

A. Yards between Georges River and Parramatta River:	
Common bricks	38s per 1,000
Callow and clinker bricks	33s per 1,000
Face bricks	73s per 1,000
Levy on all bricks sold by yards in that area	1s per 1,000
B. Yards south of Georges River and north of Parramatta River:	
Common bricks	45s per 1,000
Callow and clinker bricks	42s per 1,000
Face bricks	90s per 1,000
Levy on all bricks sold by yards in that area	5s per 1,000

Source: Industrial Commission of New South Wales, 1939, op. cit., p. 8

31. Archives of Brickworks Limited.

32. On 15 May 1935 Moseley resigned as secretary of South Ashfield Brick & Tile Co. Ltd. Wilkinson replaced him in accordance with 'the terms of the arrangement now existing with the Council of Brick Manufacturers.' Source: State Records NSW, Clay Industries Pty Ltd, Container 11/11785.

33. Testimony of Lawrence Johnson, an examiner in the Department of Taxation in *Attorney-General versus Brickworks Limited and Others*, casts light on the history of each yard during this period. Johnson compiled detailed records on each yard, including its life as a working plant and its earnings and assets. Unfortunately his figures and tables do not survive.

34. State Records NSW, Clay Industries Pty Ltd, 1935–52, Container 17/8940, Item 16282.

Trade occupations in the brick industry, 1936

KILN WORK

- Burners
- Setters
- Drawers – patent kilns
- Classer
- Yardman
- Coal wheeler
- Loader Out (dragger)
- Wheeler
- Machineman
- Machineman's Assistant
- Assistant under 21 years
- Loftman — Loftman under 21 years
- Panman — working the winding gear; working on crusher (no gear); worker under 21 years

PIT WORK

- Foreman
- Shooter or Jumperman
- Pitman
- Plastic Work

Source: The New South Wales Brick Masters' Association, 28 April 1936, Archives of Brickworks Limited

still not reached the November 1933 level of 50s per 1,000. This can be explained by the fact that a number of yards were still supplying old orders and, having given quotations, wanted to stand by them. On 11 December the prices of brick and contract cartage on the south side of the Parramatta River were lifted to 50s per 1,000 for commons and 45s for callows and clinkers, with the levy changed to 10s for all classes. In the metropolitan area face bricks rose to 100s with a levy of 15s. These prices remained in place until 1 October 1936 whereafter they continued to increase until the end of 1938 when a government inquiry arrested their upward drift.

On 28 July 1936 prices were raised all round by 10s per 1,000, taking effect on 1 October with commons fixed at 60s, face bricks at 110s and clinkers and callows at 55s. The rises were justified on the grounds that the basic wage, manufacturing costs, and awards had all increased. They were not unrealistic. Wage costs in the brick industry were higher than the basic wage set by the Industrial Commission of New South Wales. For example, in September 1937, burners, setters and draggers earned £4 12s 6d. Pitmen – blasters and labourers

Prices of common bricks per 1,000, fixed by brick manufacturers, 1928–1938

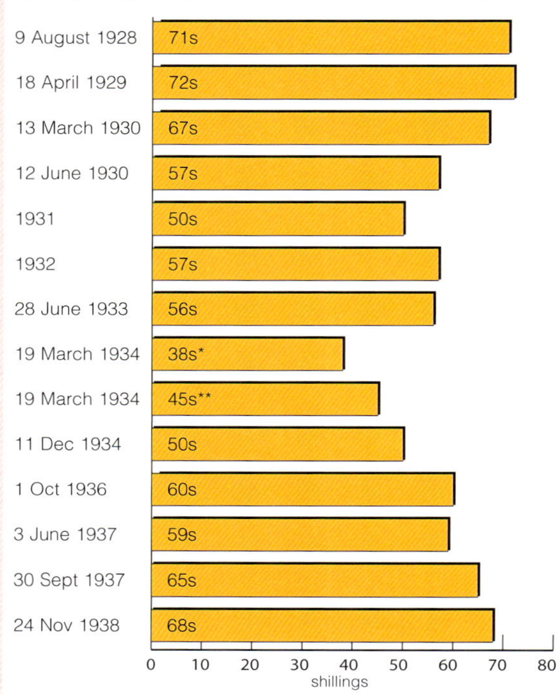

* yards south of the Parramatta River

** yards in North Sydney and Sutherland

Source: Figures supplied by the Council of Brick Manufacturers (Exhibit 50), 1939, op. cit., p. 5

who worked in the quarry – earned £4 10s 6d. On the other hand the Industrial Commission's inquiry into claims of an industry monopoly noted that the basic wage was £3 7s 6d on 26 April 1934, rising to £3 10s 0d by October 1936, a rather stingy increase by any measure. Award rates for the brick and tile industry were negotiated between the Federated Brick, Tile and Pottery Union and the 230 members of the New South Wales Brick Masters' Association. In 1936 the two bodies reached agreement on a new award, and an application was made and accepted, taking effect from 18 December 1936. Curiously the award was rescinded on 3 June 1937 following an appeal by certain manufacturers who were members of another body, the Employers' Federation of New South Wales.

Apart from wages, the price of coal used in the kilns, especially the dome kilns which burnt face bricks, was clearly an issue in calculating the price of bricks.

Bricks produced by all yards in Brickworks Limited, 1934–1936

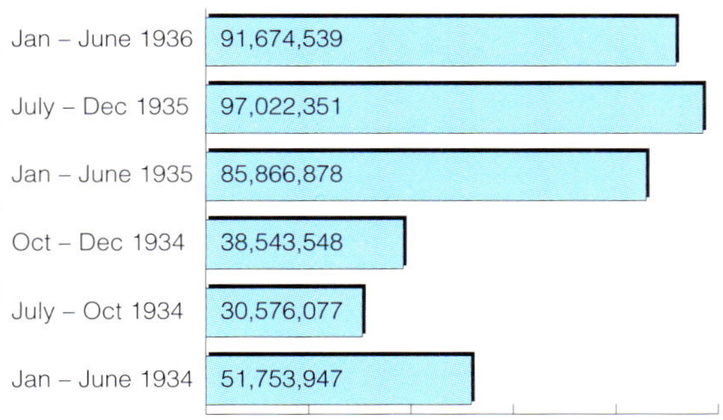

Source: Industrial Commission of New South Wales, 1939, op. cit., p. 5

From 21s a ton in June 1934, the cost of coal used in dome kilns had reached £3 10s 0d by October 1936. Wage costs, fuel costs and the need to return a profit as well as pay government taxes continued to force up the price of common bricks which reached 65s towards the end of 1937.[35] This was all too much. A chorus of complaints from outraged builders and the buying public eventually forced the government's hand. In early 1939 the Industrial Commission of New South Wales commenced an inquiry into prices, which we will come to later in this chapter.

UNION FIGHTBACK

Meanwhile, there was consternation down at the Trades Hall offices of the Federated Brick, Tile and Pottery Workers Union on Sussex Street, Sydney. Price rises, the sale of the State Brickworks and the consolidation of so many brickyards into the hands of just one company, set alarm bells ringing. In July 1936 the union launched its own journal, *The Clayworker*, promising to take the fight all the way to the employers. Fred L. Parker, union secretary, was nominated as the inaugural editor.

> *For the first time in the history of your union we present an official organ. No doubt many members will think that this is long overdue owing to the wonderful organisational journals that are issued by other similar bodies* [an obvious reference to the industry's The Clay Products Journal of Australia] …. *The coming of The Clayworker should be a great boon to all those connected with the industry.*
>
> *The conditions existing in our industry today make unity absolutely necessary … Generally speaking, we are like so many robots living and working for the purpose of giving our all to other members of society who are better organised.*

And in a call to action:

> *Let us be realists and use the knob at the end of the stick that the other fellow is using so effectively. If we do not, while there is yet time, the same opportunity may not be available. The members of this organisation*

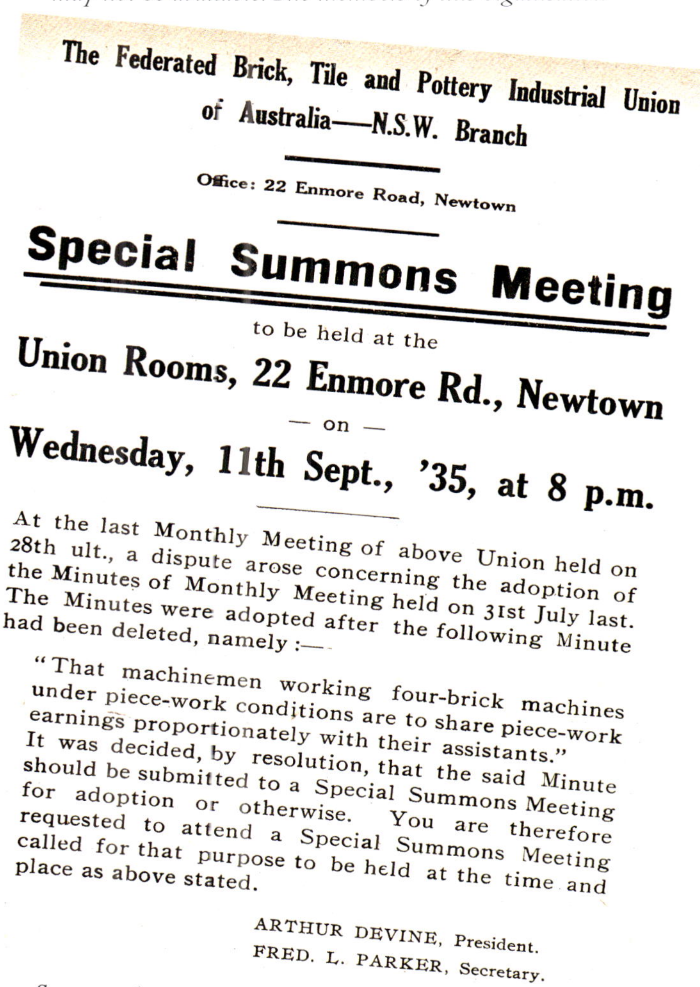

The Federated Brick, Tile and Pottery Industrial Union of Australia—N.S.W. Branch

Office: 22 Enmore Road, Newtown

Special Summons Meeting

to be held at the

Union Rooms, 22 Enmore Rd., Newtown

— on —

Wednesday, 11th Sept., '35, at 8 p.m.

At the last Monthly Meeting of above Union held on 28th ult., a dispute arose concerning the adoption of the Minutes of Monthly Meeting held on 31st July last. The Minutes were adopted after the following Minute had been deleted, namely :—

"That machinemen working four-brick machines under piece-work conditions are to share piece-work earnings proportionately with their assistants."

It was decided, by resolution, that the said Minute should be submitted to a Special Summons Meeting for adoption or otherwise. You are therefore requested to attend a Special Summons Meeting called for that purpose to be held at the time and place as above stated.

ARTHUR DEVINE, President.
FRED. L. PARKER, Secretary.

Summons to a meeting of the Federated Brick, Tile and Pottery Industrial Union of Australia, September 1935. Brickworks Limited

35. By 1938 there were 18 companies and firms operating 24 yards in Sydney. In that year a total of 228,376,999 bricks were sold across the city. Profits were 15s 6d per 1,000, with an average price of 63s 5d for commons and 113s 5d for face bricks.

The Council of Brick Manufacturers, 1938. Archives of Brickworks Limited

Bottom row (l. to r.) A.B. Bakewell (Bakewell Bros. Pty Ltd), J. Hankinson (Rupert Cook's Brickworks Pty. Ltd), G. Crispo (Carrington Brick Co. Pty Ltd), A. Green, Vice-Chairman, (Northern Suburbs Brick Co. Ltd), A. Lyell-Scott (Eagle Brickworks Pty Ltd), J.F. Walker (Western Suburbs Brick Co. Pty Ltd), S.J. Gentle (Ashfield Brick Co. Pty Ltd)

Second row (l. to r.) G. Mitchell (Federal Brick Co. Pty Ltd – Carlton yard), C.F. Davison (Clyde Brick Co. Pty Ltd), G.F. Hocking (Canterbury District Brick Co. Pty Ltd), F. Butcher (Butcher Bros and Co. Pty Ltd), B.A. Moseley - Secretary, C. Scott-Waine (St Peters Brick Co. Pty Ltd)

Third row (l. to r.) A. Charlesworth (Sydney Brickworks, Charlesworth & Co), F.G. Phippard (Federal Brick Co. Pty Ltd – Alexandria yard), W.G. West (Central Brick and Tile Co. Pty Ltd), William King Dawes, Chairman, (Austral Brick Co. Pty Ltd, Brickworks Pty Ltd, Clay Industries Pty Ltd), R. Allott (Punchbowl Brick and Pipe Co. Pty Ltd), E.D. Lanceley (Ryde Brick and Tile Works Pty Ltd), W.P. Judd (Judd's Hurstville Brick Co. Pty Ltd)

Top row (l. to r.) G. Kemp (Croydon Brickworks Pty. Ltd), A.S. Loveridge (The NSW Brick Co. Pty Ltd), E. Spackman (Austral Brick Co. Pty Ltd), Sir Archibald Howie (Burwood Brick Works), A. Shannon (Shannon's Brick, Tile and Pottery Pty Ltd), H.G. Lanceley (North Sydney Brick and Tile Co. Pty Ltd)

contribute a considerable amount of wealth to this country and yet there is no market inquiry through the union as to whom benefits by such combination. You will have no doubt that we happen to be up against a team of the Oliver Twist bodyline bowling variety and their rules. But it is not sporting, and it is certainly not cricket.[36]

36. Unpublished papers of Frederick L. Parker, Secretary, Federated Brick, Tile, and Pottery Industrial Workers Union, Archives of the Construction, Forestry, Metalworkers and Engineers' Union (CFMEU).

Apart from protest, there was nothing the union could do about either the government's decision to rid itself of the State Brickworks, or the resurgence of the private sector.

ROYAL COMMISSION

'Twas the season for inquiries. Still smarting from the sale of the State Brickworks, the former Premier, Jack Lang, accused private brick manufacturers of

Portrait of Jack Lang, oil on canvas by Fred Leist, National Library of Australia

colluding to fix prices. Using parliamentary privilege during early 1937 he made the more serious claim that Brickworks had gone as far as to commit fraud in the purchase of the State Brickworks. A Royal Commission of Inquiry was held which concluded in 1938. After hearing testimony from dozens of witnesses Justice Halse Rogers dismissed the charges. Clearly, Lang had the industry in his sights. As a former real estate agent he must surely have been familiar with private manufacturers and their long-standing relations with builders and developers. Whether Lang was acting on a matter of principle – that government should not divest itself of state enterprises – or whether he harboured ancient grudges, is a matter of conjecture. Certainly, *The Clay Products Journal of Australia*, representing the brickmakers, was incensed over Lang's attack:

> *Nobody with the slightest knowledge of the Lang Plan of broadcasting charges of dishonesty against political opponents and any others capable of being associated with them – in his mind – in possible or impossible nefarious transactions, had any doubt that his accusations against the brick masters who purchased the State brickworks [sic] from the State Government would be found by an independent tribunal after the fullest investigation to be unsupported by any evidence whatever.*
>
> *And so it has turned out. After sittings extended over 38 days, during which Mr Lang was represented by learned counsel engaged at the expense of the Government, that is to say the taxpayers of New South Wales, the Royal Commissioner, Mr Justice Halse Rogers said that, 'There was no evidence of fraud or corruption against Brickworks Proprietary Ltd.'* [37]

Conscious that he could not repeat his claims outside Parliament, Lang called for the reform of defamation laws, although his demands fell on deaf ears.

LOOSE ENDS

Minute books rarely offer the historian more than a hint of the real story, and this applies in good measure to the various ordinary general meeting recorded in those held in the company's archives at Horsley Park.[38] With events dominated by Brickworks Limited, Austral takes a back seat until 27 September 1936 at which time an emergency general meeting was held, and a motion put, and carried, to convert the company from a public entity to a private concern. Other member companies of the Council of Brick Manufacturers followed suit. Captions under a group portrait of the council taken in 1938 indicate the proprietary nature of each member company. In many cases this was a recent development and in marked contrast to their previous status as a public company. Since we know that Brickworks Limited eventually acquired all member yards in the council, a change in articles of association paved the way.

At the ordinary general meeting held on the day after the emergency meeting, the board took pleasure in

37. *The Clay Products Journal of Australia*, 'Purchasers of the State Brickworks cleared of charges of fraud & corruption', 1 March 1938, p. 2.

38. Dawes seems to have favoured the School of Arts in Sydney's Castlereagh Street as the venue for public meetings. Now a fashionable watering hole for city business people, the building has had a chequered past and for many years was allowed to fall into a state of dilapidation.

Brick carters in front of Butcher Brothers, circa 1938. George 'Bunny' Cheney (1935 International), Bernie Keenan (in front of 1929 Chevrolet), Jack Hodge (1935 Fargo), Frank Vandine (1929 Chevrolet), Stow Kirby with hat (1936 International), Reg 'Possum' Butcher, (1935 Ford), George Watts, Harold 'Boy' Butcher (Federal 1935), Jack McFarlane (1935 Dodge), 'Fony' Freeman (1929 International), Alec Morrow (1935 International), Harold Spackman (1927 Chevrolet). Some trucks carried 500 bricks, others 1,000. The 1929 Chevrolet was the first truck with a six cylinder motor. Warwick Gemmell Collection

announcing that the balance owed for the purchase of the Bedford Brick Works had finally been paid in full. Dawes was delighted for he had acquired the Bedford yard on the spur of the moment, and much against the wishes of the board.

There was even talk of a bonus share issue, a sure indication that things were looking up. Other yards had fallen on hard times, although the Bedford was different. Plant and equipment were relatively modern by the standards of the day, and its location allowed Austral to dominate an enormous piece of real estate from St Peters station down to Canal Road at Tempe. According to Bill Wheatley, whose father, George Wheatley, had been appointed by Dawes to run the yard, the Bedford made some of the best liver-coloured bricks in the country.[39] The only dampener was that by this time the softer, more workable shales appear to have been running out at Austral's St Peters yard. There was discussion at board level on how best to deal with the problem of hard shale now being extracted. A hammer mill to process these deeper shales was installed with similar plans to provide the same equipment at the Bedford works. Ultimately, the progressive exhaustion of clay and shale reserves, pollution, obsolescent plant and the move to Horsley Park in 1961 finally put paid to Austral's pits along the Princes Highway. When it closed in 1983 the Bedford yard was one of the last remaining brick plants in the area. Its passing went unmourned by many local people in 1983, yet closure spelled the end of an era dating back to the 1830s.

Throughout 1937 the company's fortunes continued to prosper. Chairman Spackman praised Dawes' continuing success. In 1937 the company made a whopping 52 per cent profit on its brick

39. George Wheatley was yard foreman at the Bedford and was transferred to the Ralford yard in the early 1950s. Interview with Bill Wheatley, 5 May 2006.

Known directorships, 1938

Name of director	Austral Brick Co.	Brickworks Limited	Butcher Brothers	Clay Industries Limited	Clyde Brick Co. Limited	Croydon Brickworks Pty Ltd	Council of Brick Manufacturers	Federal Brick Co. Ltd	Great Northern Brick Company	Flemington Brick Company Ltd	Hurstville Brick Co. Ltd	Metropolitan Brick Co. Ltd	Northern Brick & Tile	Northern Suburbs Brick Co. Ltd.	Ryde Brick & Tile Works	NSW Brick Co. Ltd	Punchbowl Brick & Pipe Co. Ltd	SLIC	St Peters Brick Co.	Sydney Brickworks
Reg Allott																	•			
Henry W. Alderson												•								
Alfred B. Bakewell							•					•								
Frank Butcher		•	•	•			•													
J.H. Arthur and James R. Charlesworth		•										•								•
George Crispo							•					•								
Clive P. Curlewis												•								
Clarence F. Davison		•			•		•		•			•								
William King Dawes	•	•		•			•													
Arthur P. Evans																		•		
Alfred Green		•		•			•							•						
Harold Judd							•		•		•	•								
Stanley J. Gentle							•					•								
James H. Gilliland												•						•		
John Hankinson							•					•								
Sir Archibald Howie		•					•		•			•						•	•	
George Kemp						•	•					•								
E.D. Lanceley		•		•			•						•		•					
H.G. Lanceley				•			•													
A.S. Loveridge							•									•				
George Mitchell							•					•							•	
Bertram Moseley		•		•			•					•		•						
Frederick G. Phippard		•					•	•				•						•		
Edwin Spackman	•	•					•													
Cecil Scott-Waine							•					•							•	
Walter G. West		•					•													
J.F. Walker							•													
Arthur Lyell-Scott							•													

manufacturing assets.[40] It was also the year that 64-year-old Edwin Spackman married Anne Bell at Ryde, embarking on a round the world cruise, which gives us some indication of his wealth. Spackman and his bride enjoyed just 10 years of married life. Ann predeceased her husband who died in 1947.

There can be no doubt that The Austral Brick Company Limited was highly profitable, and figures supplied to the Industrial Commission's inquiry in early 1939 bear this out. For example, E.S. Wolfenden, an independent accountant, examined the accounts of 10 of the leading manufacturers to demonstrate the effective capital employed, their adjusted profits and loss and the number of bricks sold. Wolfenden's figures for the year ended 30 June 1938 revealed a true net profit of £28,022 on an effective capital of £144,382. On the other hand, the official accountant to the Industrial Commission arrived at a net profit of £57,301 on £141,165, a truly remarkable figure.[41] Wolfenden's calculation of effective capital used in the manufacture of bricks included bona fide reserves, bad debts and reserves for depreciation. Net profit was arrived at by allowing for income tax and adjusting the allowance for depreciation. For the year ended 30 June 1929 the annual percentage of profits to capital was about 8 per cent, and for the year ended 30 June 1938, 12.2 per cent.[42] The magnitude of Austral's profitability was not necessarily enjoyed by the rest of the industry. Taking into account all yards (open and closed) net profits were less spectacular.

A damaging coal strike during September 1938 preoccupied Austral, although not as much as the company's inadequate capital base. At the emergency general meeting on 26 September 1938 shareholders voted to increase the capital to £200,000 by the issue of 100,000 new shares at £1 each. In putting the resolution Dawes explained that assets now exceeded the nominal capital which had been used to purchase the Bedford and rebuild another brickmaking unit. In the event of a downturn in trade the company would not be able to meet contingencies, hence the need for adequate reserves.

A MATTER FOR THE INDUSTRIAL COMMISSION

Unfortunately for Brickworks and the Council of Brick Manufacturers public disquiet over brick prices simply would not go away. The Legislative Assembly of the New South Wales Parliament ordered its own inquiry, the latest to be conducted by the Industrial Commission. Instructed to get to truth about the cost of bricks, slates, tiles and timber, proceedings commenced on 29 December 1938, recommencing, after a Christmas recess, on 16 January 1939. The break gave representatives of the various groups with an interest in the inquiry time to prepare. Between 23 January and 16 March 1939 dozens of witnesses took the stand and numerous reports were prepared by those called to give an account of their dealings. A line-up of Sydney brick companies was present, most of them members of the Council of Brick Manufacturers. Fred Parker appeared on behalf of the Federated Brick Tile & Pottery Union of Australia (New South Wales Branch). Goodlet & Smith Ltd, which was not a member of the Council of Brick Manufacturers, was represented separately by a solicitor from the legal firm of Priddle, Gosling, Dalrymple & Sillar.

Counsel for the Industrial Commission did not hold back, and argued that the Council of Brick Manufacturers was using Brickworks Limited as a front to acquire shares and assets in other brickyards or companies. The aggrieved party was Goodlet & Smith and other manufacturers of texture brick who, according to their testimony, claimed they were being intimidated by a threat from the New South Wales Brick Masters' Association and Brickworks Limited to increase prices. Apparently, the threat was made at a meeting on 29 March 1938 attended by Goodlet's managing director, George Fulton Copeland.[43]

40. Spackman in testimony before the court, *Attorney-General v. Brickworks Pty Ltd and Others*, op. cit., p. 517.

41. Industrial Commission of New South Wales, 1939, op. cit., pp. 34-35.

42. Archives of Brickworks Limited.

43. Industrial Commission of New South Wales, 1939, op. cit., pp. 26–27.

Brickworks Limited and the Council of Brick Manufacturers were also accused of operating a brick quota in the interests of controlling the industry. This was strongly contested by the brick manufacturers, especially Moseley who was, by now, a very sick man and under constant medical supervision throughout the hearing. That the council had raised prices to a level above the cost of production could not be denied. It was not, as some have suggested, an attempt to gouge the buying public. Such conspiracy theories simply held no water either at the commission of inquiry or during the case brought against Brickworks Pty Ltd and Others in 1940 by the State Government. In fact the Crown's case against Brickworks (and by association, the Council of Brick Manufacturers) was comprehensively dismissed as we shall see in the next chapter.

In summary Mr Justice De Baun noted that

> *The Council has created a virtual monopoly in the industry; though, in my opinion, there is nothing intrinsically wrong in this so long as the price of bricks is kept at a fair and reasonable price. Indeed, the agreement expressly contemplates that it shall be so kept. The increase from 50s to 60s per 1,000 in October 1934 was more than sufficient to provide for expected increases in wage and other costs ... After the recission of the award by Mr Justice Cantor on 3 June 1937 the price of brick could have reduced by more than 1s per 1,000 ... The fact is no reason at all was given for the increase from 65s to 68s per 1,000 on 24 November 1938.*[44]

De Baun's concerns were two-fold; he reminded the Legislative Assembly in his conclusion to the report that there was a real need for rationalisation of the industry during the early 1930s. For this reason the agreement of 4 October 1934 ensured that the business would survive and return to profitability, although this meant competition was restricted or in some areas eliminated altogether. On the other hand, regular price hikes were cause for concern, and the public deserved a measure of relief. In June

Greville Montgomery (centre) flanked by Bert Moseley (far right of picture), W.K. Dawes (left), possibly Clarence Davison (far left). The man (unknown) standing behind defied social convention by not appearing in a hat

1939 he ordered the price of common and face bricks to be reduced at once by at least 5s per 1,000 with commons fixed at 58s 6d per 1,000. On appeal, the Industrial Commission relaxed this ruling.

HOME AND AWAY

In times of travail nothing acts more like a tonic than a holiday. For the luminaries on the Council of Brick Manufacturers and at Brickworks Limited, rest could not come soon enough, although for most it had to wait. In the second half of 1937 Davison was packed off to England where he met with officials of the London Brick Company Limited. *The Clay Products Journal of Australia* followed Davison's move closely and was happy to report that their man had made an excellent impression abroad, a fact reported in the pages of *The British Clayworker*: 'Mr Davison made friends wherever he went; he was welcomed at every yard he visited, and he is returning home with an increasing affection for the old country and

44. Industrial Commission of New South Wales, 1938, op. cit., p. 593.

a sincere regard for fellow Clayworkers with whom he came in contact.'[45] In fact Davison was astounded by the sheer size of the workings at London Brick's main yard outside the town of Bedford. Mile after mile of countryside was given over to clay mining and brick production using the latest technology. Once forged, the bond between Sydney manufacturers and their counterparts in Britain grew. At the invitation of the Council of Brick Manufacturers, Greville Montgomery, publisher and proprietor of *The British Clayworker*, paid a visit to Sydney, arriving on 22 December 1938 with his wife, just in time to enjoy an antipodean Christmas. Clarence Davison and Bert Moseley were at Circular Quay to greet them as they stepped off the boat. Throughout their stay Moseley acted as guide and 'philosopher' to Montgomery who was whisked off on sightseeing tours, including the Blue Mountains, Sydney Harbour and the Sydney Cricket Ground. This was a little tame for Dawes who preferred the Randwick races where he and Montgomery indulged their penchant for a bet and a whisky. Accommodation and entertainment were provided by Sir Archibald Howie at his Richmond property where the visitors were treated to traditional Christmas fare in the sweltering heat of summer.

Above and opposite: All aboard. Dawes' visit to the Niagara was recorded on a postcard sent to his family in Sydney. **Bill Wheatley Collection**

Dawes was not far behind in the travel stakes, although the Royal Commission of 1938 and the New South Wales Industrial Commission's inquiry delayed his departure for the USA until the following year. Mindful of the backward state of the brick industry, a condition not helped by the long years of the Depression, Dawes was anxious to see how things worked overseas. Far from being an industry-funded junket, his visit was arduous, time-consuming, yet highly instructive. Accompanied by H.W. 'Harry' Thompson, son of H.J. Thompson, a major Sydney building contractor, Dawes set sail aboard the US passenger liner *USS Mariposa* in March 1939.[46] Unfortunately, their departure happened to coincide with a deteriorating international situation, for on 15 March German troops occupied the remaining portion of Czechoslovakia that had not been ceded to Nazi Germany under the Munich Agreement of September 1938. As the clouds of war gathered the pair remained in the USA for four months, carefully examining methods of mass production that had made that country's brick industry a world leader.

One innovation introduced by Dawes, possibly as a result of his overseas visit, was the installation of forced draught fans in kiln chimneys. These enabled the hot exhaust gases to be recycled and expelled more effectively, as well as achieving a faster, more even burn. Forced draught chimneys also made it a

45. *The Clay Products Journal of Australia*, 1 December 1937, p. 9.

46. Ray Dawes served his building apprenticeship under Harry Thompson who was a close friend of Dawes. Thompson carried on the work of his family's building firm.

great deal easier to run the kilns on fuel oil, although Dawes fought shy of making the switch from coal. According to his son, Ray Dawes, he was always apprehensive about its use from a safety point of view, besides which he distrusted the oil companies. Reliance on oil would only make it easier for them to ramp up the prices, whereas coal could be had from virtually any pit on terms that suited the brickmakers.[47]

Dawes sent letters and postcards to his family, as well as to colleagues in the Council of Brick Manufacturers, although very little survives apart from some commercial photographs purchased during a sight-seeing visit to the famed Niagara Falls. Once again, *The Clay Products Journal of Australia* was eager to report Dawes' overseas trips, noting the enthusiastic reception provided by his American hosts. Under the headline 'Sydney Brick Maker Appreciates US reception' the journal reported:

> *Australian brick manufacturers who contemplate a tour of America will read with interest the following extract from the US, 'Brick and Clay Record' for February.*
>
> *'Brick and Clay Record' has received several letters recently expressing appreciation for the consideration shown to W.K. Dawes, manager of the Austral Brick Company Ltd and chairman of the board of directors of Brickworks Limited, in St Peters, Sydney Australia, during his visit to the United States this summer.*

Dawes never forgot the hospitality of his American hosts and gave glowing reports of the frankness and freedom of US manufacturers in imparting information. The firm of J.C. Steele of Statesville, North Carolina, was one of those visited by Dawes, initiating a close business relationship that lasted until his death, and continues to this day.

> *After having had a pleasant run through the States I returned to Sydney in July, and now take this opportunity of writing to express my appreciation and thanks for the reception which you gave me when visiting your State. I feel that everything possible was done to assist me in obtaining information in any shape or form which it was possible to give, and which I highly esteem* [48]

There were more overseas trips ahead but these would be delayed for almost a decade. In the meantime, with another European war just months away, William King Dawes and his fellow brickmakers must have pondered long and hard what new surprises lay in store for the industry. Lucrative government contracts, perhaps, or another slump in trade that would pitchfork the industry into another depression? Time would tell.

47. Interview with Ray Dawes, 21 July 2006, Sydney.

48. *The Clay Products Journal of Australia*, 1 April 1940, p. 13.

The 2/3rd Australian Independent Company at Timbered Knoll, north of Orodubi, New Guinea, 29 July 1943. Photographer: Damien Parer. National Library of Australia

Tobruk burns in the background as Australian troops rest on the heights overlooking the coast town, before advancing to occupy it, January 1941. Photographer: Frank Hurley. National Library of Australia

Frontline 12

Defending the shores – Wartime regulation – William Dawes and Essington Lewis – Magnesite – Day in court – Wartime Austral, wartime Australia – Home truths – Social diversions – The triumph of Brickworks Limited – Time out

DEFENDING THE SHORES

At dawn on 1 September 1939 German troops massed near the border town of Gliewitz were given the order to attack Poland. Sweeping all before them advance units reached the Polish capital, Warsaw, within just seven days. Demands by Britain and France that Germany halt its aggression and immediately withdraw were ignored. With the expiry of the ultimatum on 3 September 1939 Britain's Prime Minister, Neville Chamberlain, declared war. Édouard Daladier, Premier of France, was quick to follow suit. The following Sunday Australia's Prime Minister, Robert Menzies, went on radio to announce that 'where Great Britain stands, there stands the people of the entire British world'.[1] For the second time in 25 years Australia found itself at war. This time no-one had any illusions about what was at stake. Recruitment of a Second Australian Imperial Force to assist Great Britain commenced immediately, although the main problem was where to deploy Australian forces. In the six months it took Germany to consolidate its hold over the former independent state of Poland, the brief respite in fighting had all the appearance of a 'phoney war'.

Canberra took some comfort from comments made by Anthony Eden, the British foreign secretary, who assured the Australian government that Japan would remain neutral for a time, and that Britain would defend Singapore and Australia from attack. Two Australian divisions were despatched to Palestine in early 1940 where they played a valuable part in the defence of Britain's North African and Middle Eastern interests against attacks by combined Italian and German armies. By June 1940, following Germany's invasion of western and northern Europe, Hitler's New Order was beginning to take shape.

1. Wartime broadcast of Robert Gordon Menzies to the nation on ABC radio on 3 September 1939.

The Rt. Hon. R.G. Menzies P.M. of Australia. Broadcasting to the nation the news of the outbreak of war, September 1939. National Library of Australia

WARTIME REGULATION

Before the month of September was through, drastic measures were being taken to put the Australian economy on a war footing. On 25 September 1939, the Price Fixing Commission introduced strict controls, including setting the price of bricks and placing lending restrictions on building cooperative societies. Price controls were only removed in 1955, a decade after the end of the war. Equally stringent controls were imposed over coal allocations to brickmakers who were also fast losing manpower as men answered the call to join up. It was not until January 1942 that brickmaking was officially

Distribution of clay product factories, 1940

	Factories	Employees
New South Wales	176	5,043
Victoria	83	3,255
Queensland	35	738
South Australia	37	764
Western Australia	27	516
Tasmania	10	168

Source: The Clay Products Journal of Australia, 1 May 1940, p. 15

classified as a reserved occupation, although when Austral Brick carter, Harry Bell, applied to join the army in the early months of 1940 he was refused, like others, on account of owning a truck. Instead, Bell was kept busy carting bricks used to build air raid shelters and gun emplacements.[2] Sharing the same trade as Bell was George Crispo, whose father was a brick manufacturer, a shareholder in Brickworks Limited, and the owner of the Carrington Brick Co Pty Ltd which was later acquired outright by Brickworks.

For a short time the building revival of the late 1930s continued. Then in December 1940 National Security Regulations were introduced, requiring consent for the construction of any building costing more than £5,000. The hand of government was progressively extended so that by June 1942 the threshold was just £25 including repairs.[3] The number of buildings completed in Sydney fell from 8,138 in 1940 to 582 in 1943. It was not until August 1945 that wartime regulations were eased and permits no longer required for single unit dwelling houses. By this time a severe housing shortage imposed restrictions of its own making. Yard closures plunged the industry back into depression times, although certain plants were still able to find skilled workers under powers vested in the Commonwealth Department of Labour and National Service and the Director General of Manpower, Wallace C. Wurth.

2. Interview with Ross Bell, 1 September 2005, Sydney.
3. P. Spearritt, *Sydney's Century: A History*, UNSW Press, Sydney, 2000, p. 80.

Austral, it is said, was the only pit at St Peters still making bricks, with the neighbouring yards of City Bricks, Carrington, St Peters and Central closed due to the lack of demand.[4] The City yard, adjacent to the gasworks on Euston Road, Alexandria, was already a storage yard for discarded machinery and spares.

WILLIAM DAWES AND ESSINGTON LEWIS

Too old to enlist even had he wished to join the fight, Dawes became actively involved in the war effort. Indeed, the rise and rise of William King Dawes and his protégé, Brickworks Limited, had not gone unnoticed. In the years before the outbreak of war Dawes had become a familiar figure in the world of industrialists, and not just in Sydney. As managing director of various brick and investment companies Dawes was acutely aware of the importance of the industry to the war effort. His own profile was certainly enhanced by his dominance of the New South Wales Brick Masters' Association and its inner circle, the Council of Brick Manufacturers, both of which had become powerful lobby groups. It was through one man in particular, Essington Lewis, that Dawes gained access to a wider and more influential circle.[5] A brilliant and visionary figure, Lewis was managing director of Broken Hill Proprietary Co. Ltd (BHP) and in 1938 was appointed chairman of the Advisory Panel on Industrial Organisation (1938-40), a government body set up to recommend measures necessary for mobilising secondary industry in the event of war.[6] It is highly likely that Dawes and Lewis knew of each other's existence even before the outbreak of war. Certainly, Dawes' friendship with the likes of Sir Eric John Harrison, the federal Member for Wentworth, and Sir Archibald Howie, would have been useful. Harrison was already well acquainted with Menzies and may even have recommended Dawes as someone who could get things done.

4. St Peters Brick Company was a small yard, operating a few domes and a patent kiln. The Carrington yard was also a small yard.
5. Dawes and Lewis might have known of each other through negotiations for the supply of refractory bricks used in high temperature brick kilns and steel plants. Newbold Silica supplied to a variety of industries where the use of such bricks was essential.
6. Essington Lewis was born in 1881 and died in 1961 following a riding accident.

'I am work.' Essington Lewis 1952 by William Dargie, oil on canvas 102 x 76 cm. National Portrait Gallery, Canberra. Gift of BHP Billiton 2003

In early 1940 Menzies invited Lewis to become Director-General of the new Department of Munitions, giving him a free hand in a far-reaching charter described as unique in Australian administrative practice. In a radio broadcast on 16 June 1940 Menzies urged Lewis to: 'go ahead in your own way and achieve your objectives in the shortest possible time. The Director-General will be authorised to make purchases direct, without tenders or circumlocution ... he will not be limited by Public Service regulations or otherwise in regard to the employment of personnel'.[7] Enjoying direct access to the Prime Minister and Cabinet, a privilege shared only by the Chiefs of Staff of the Armed Services, Lewis overhauled munitions production, increasing output many times over. In terms of plenipotentiary powers, those granted to Lewis were strikingly similar to the mandate granted to Albert Speer in Germany, although the magnitude of Speer's responsibilities were vast by comparison. Lewis's role was expanded in December 1941 when he was appointed Director-General of Aircraft Production; and again in August 1943, when he was asked to take over control of small craft production which by then had been absorbed into the Department of Munitions.

MAGNESITE

At some point Dawes was invited to join a wartime munitions committee, and in due course came to play an important role in resource planning, especially the production of a substance called magnesite. Crude magnesite was the raw material from which a multitude of essential materials could be produced, the most important of these being metallic magnesia. Lighter than aluminium and stronger than steel, magnesia was used extensively in all types of construction where weight was an important factor, particularly aeroplane fabrication. It was also an essential ingredient in the manufacture of tracer bullets. Cut off from its traditional supplies in Nazi-occupied Greece, Britain looked further afield to make up the shortfall. An extensive search was conducted throughout Australia as well as other Allied territories. The Australian government threw its full support behind the project which turned out to be a huge undertaking. By a stroke of good luck and careful prospecting, large deposits of crude magnesite were located about 20 kilometres outside the town of Young in New South Wales. Since processing required firing the raw material in high temperature kilns, enquiries were made and the firm of Punchbowl Brick & Pipe (possibly at the recommendation of Dawes) was chosen by the federal government to undertake the job. A speedy interchange of cables to and from London set things in motion and Reg Allott,[8] Punchbowl's managing director, agreed to excavate and freight the ore to his Punchbowl plant where it was processed to specifications provided by the British Government.

7. G. Blainey, *The Steel Master: a life of Essington Lewis*, Macmillan, Melbourne, 1971.

8. Allott was a former Rugby union forward who played with St George and was also a star performer with Cronulla Surf Club during the 1920s and early 1930s.

In the meantime, a sizeable camp was hurriedly built near Young to house the workers, mechanical excavators and loading equipment, while tipping trucks were despatched to the site. Within three weeks crude ore was being railed to Allott's plants at Punchbowl and Sutherland. Unable to handle such large quantities, Allot subcontracted Dawes to calcine a portion of the crude ore at the St Peters yard of The Austral Brick Company Pty Ltd. When not carting bricks or coal, drivers were paid to load magnesite by hand from the Cook's River rail yards, before taking their trucks the mile or so up the road to either the main Austral plant or the Bedford yard.[9]

On arrival the crude magnesite was stacked into the kilns and fired to 1,000°C, losing half its original weight in the process. For every 18 tonnes of dirt excavated at Young only one tonne of calcined magnesite was extracted. The calcined material, magnesium oxide, was packed in bags lined with a waterproof paper to resist moisture on the voyage to England, and then loaded aboard ships in conditions of maximum secrecy, usually late at night. About 2,500 tonnes of magnesium oxide was shipped each month. One consignment got no further than 120 kilometres from Sydney Heads when a Japanese torpedo sent the craft to the bottom of the ocean. It was later revealed that Allott's company was paid a sum in excess of £250,000 for magnesium oxide supplied throughout the war.[10] By any measure the contract had been extremely lucrative for Allot, and no doubt, for Austral and Brickworks.

DAY IN COURT

As Dawes got to know fellow industrialists in Canberra, and the top-secret magnesite project got underway, there were further trials and tribulations for Brickworks on the legal front. The defeat of Stevens United Australia Party at the state elections saw the newly installed Labor New South Wales Attorney-General bring charges under the Monopolies Act against Brickworks. In October 1940, the Crown alleged that the company 'had for a long time, monopolised the trade in bricks to the detriment of the public both in respect of supplies and prices, and did so with intent to restrain trade'.[11] The period during which the defendants were alleged to have been in breach ran from 4 October 1934 to 9 June 1939. The case was heard in the Equity Division of the Supreme Court of New South Wales where expert witnesses were called ranging from accountants, brick manufacturers, to assessors from the Taxation Department. The books of Brickworks and various other companies were examined in order to prove a link between the payment of levies to the firm and the apportionment of production quotas.

Brickworks and other brick manufacturers denied the allegations, relying for their defence on a clause in the Monopolies Act which held that any combination of producers of any commodity that was reasonably necessary for the maintenance of the industry would not be deemed to be to the detriment of the public. The task before the Attorney-General was to prove that the defendants had acted in collusion and that rationalisation of the industry had been at the expense of the public through the operation of untenable yards. For several months the argument flowed back and forth. Tempers frayed on both sides with Dawes, in particular, refusing to yield an inch under cross-examination. Showing his customary disregard for rules and authority, Dawes clashed frequently with counsel for the prosecution. In his summary, Mr Weston for the defence posed the question:

> *What has happened? A number of traders get together. Three yards are working out of 30. I think in one year the sale was only 11 million bricks, 1/20th of the normal output. It was very low for several years. It was 1933 before there was the faintest sign of*

9. Coal was also carried by barge up the Cook's River where it was unloaded at the wharves.

10. *The Clay Products Journal of Australia*, June 1948, pp. 11, 14

11. *Attorney-General v. Brickworks Pty Ltd and Others, 1941*, Nicholas C.J., In Equity, Supreme Court of NSW, 1940, Vol 41, p. 494.

emergence from the depression, and in 1934 there was still not much progress. The figures of their trade show that. They had to nurse this trade. They had to consider the public as much as they could, and the lack of money. But they gradually brought the price up. The real question is, was it their intention to do the public a detriment?

Most certainly it was, argued the prosecution, alleging that brick manufacturer had deliberately harmed the public interest by raising brick prices as they had done so in 1936, 1937 and 1938. Yet, the assertion simply does not hold water if we consider that brickmakers were merely trying to restore prices to what they had been in October 1929. Moreover, at no time were independent brickmakers coerced into joining the Brick Agreement, or builders compelled to buy only from members of the agreement. Weighing the evidence Justice Nicholas dismissed the action against Brickworks and its co-defendants. The episode left a sour taste in the mouths of the brickmakers who had been obliged to defend themselves against charges of fraudulent behaviour and price fixing on two occasions in as many years.

WARTIME AUSTRAL, WARTIME AUSTRALIA

The case against Brickworks hardly made front page news at a time when Australian troops were active in the Mediterranean theatre of war. On the home front, Austral's contribution to the war effort was substantial. Apart from burning magnesite, the kilns were kept busy making bricks to build blast walls around fuel tanks in the event of a Japanese bombing attack at coastal refineries. Not a likely prospect, but one that was taken seriously, all the same. Having enjoyed three good years, Dawes advised the Austral board against paying bonus shares, instead urging it to conserve resources in view of the deteriorating international situation. Nonetheless, Spackman was still able to report at the ordinary general meeting on 30 September 1940 that trade was good and bookings firm. He also announced the purchase of the Warren Brick Company which had been established by the Curlewis family in 1902. Spackman, Dawes, Smith,

THE CLAY PRODUCTS JOURNAL OF AUSTRALIA

Official Organ of The Australian Clay Products Association Limited

Vol. 10—No. 12. OCTOBER, 1943. Price Sixpence

They who serve must be first served!

Prior customers for Fowler Ware to-day are the men and women of the fighting services. Fowler earthenware and fireclay equipment are important necessities in hostels, hospitals, camps and ships.

When servicemen, and women, have won the Victory and return to civil life they will also enjoy the lustrous, hygienic advantages of Fowler sanitary ware and crockery in the new, modern homes now being planned for post-war living.

● *While serving the national interests first, we are following our usual policy of co-operation with the trade distributors by endeavouring to maintain equitable distribution of essential civilian requirements, as opportunity affords.*

R. FOWLER LIMITED, Manufacturers of Ceramics for over a Century, MARRICKVILLE, N.S.W.; THOMASTOWN, VIC.; ALDERLEY, QLD.

They who serve. The Clay Products Journal of Australia, 1943

and Taylor are recorded as the directors, with Len Rogers as company secretary. On 9 December 1940 Rogers arranged for the transfer of parcels of shares from deceased members of the Curlewis family and others into the Austral Brick Co. Pty Ltd, the idea being to issue new certificates in that name. In all a total of 30,000 shares were transferred although such was the state of the Warren's share register that Austral's directors could not determine with any certainty the real owners. They did, however, resolve that the transfer be approved and the Secretary instructed to issue new certificates under the common seal of the company. The month of December 1940 also witnessed the appointment of

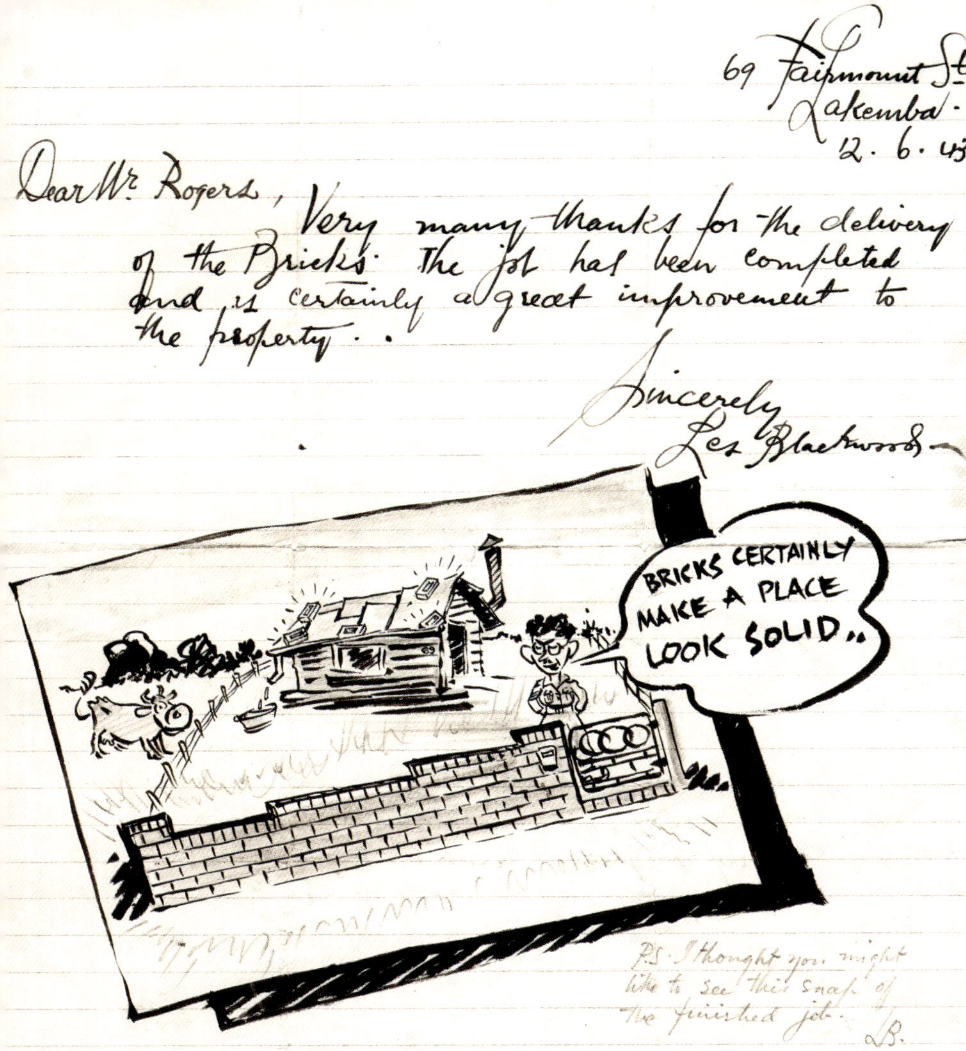

69 Fairmount St
Lakemba.
12. 6. 43

Dear Mr Rogers,
Very many thanks for the delivery of the Bricks. The job has been completed and is certainly a great improvement to the property..

Sincerely
Les Blackwood

P.S. I thought you might like to see this snap of the finished job. L.B.

Something to smile about. Courtesy Ray Dawes

the firm of G.S. Travis as auditors in place of Bogan & Collard, commencing a 50-year association with Austral Brick. Less than twelve months later in September 1941 we find Dawes returning to an earlier concern – the amount of workable clay left in the pits and land held in reserve. At the same time a strike at Austral during late 1941 was an unwelcome interruption, although the men were back at work within several days. Yet, despite wartime restrictions, Austral was still able to make a dividend payment of 4.25 per cent declared for the year to June 1941.[12]

Overseas, the military situation went from bad to worse. By early 1942 the advance of Japanese forces into French Indo-China and the fall of Singapore had caught the imperial powers of Britain and Vichy France by surprise. With Allied forces throughout North Africa in headlong retreat, Singapore lost, and the Dutch East Indies under threat, the prospect of Japanese troops landing on mainland Australia became a distinct possibility. As the crisis deepened, the country's political leaders remained at loggerheads. Prime Minister Robert Menzies appealed for a government of national unity, but this was rejected by John Curtin, leader of the Labor Party. Menzies resigned on 28 August 1941. Within six weeks, the short-lived government of Arthur Fadden, leader of the Country Party, had also collapsed. John Curtin was invited to form a new government on 7 October. Curtin was his own man. In an end of year message published in the *Melbourne Herald*, he announced that circumstances had shifted Australia's dependence from Britain to the United States. 'Australia looks to America, free of any pangs as to our traditional links or kinship with the United Kingdom'.[13] Apparently, US President Franklin Roosevelt was not impressed with this unilateral declaration with all its future implications for the United States.

The following year, in March 1942, the former State Brick Works were requisitioned by the Department of the Navy which converted the kilns into ammunition stores. Brickworks was not adversely affected due to the fact that production had ceased at the plant some years previously, although a claim for compensation was made. A legal battle continued for a considerable time before settlement was finally reached.[14] It was turning out to be a bad year all round. By September 1942 there was little by way of good tidings to report at the ordinary general meeting of Austral shareholders. With restrictions on building starting to bite deep into the company's profits, the directors agreed to close part of the Austral plant due to lack of demand, although limited production continued at Dawes' insistence. An emergency general meeting was held straight afterwards in which a motion was carried to reduce capital from £200,000 in

12. Archives of Brickworks Limited, Board minutes, Austral Brick Company Limited, 29 September 1941.

13. *Melbourne Herald*, 26 December 1941.

14. *Fifty Year Retrospective, 1934-1984*, Brickworks Limited.

£1 shares, to 150,000 shares of 15s 6d each, and return it to the shareholders. At a time when the outcome of the war hung in the balance, the federal government still found the time and inclination to look to the future. In April 1943 a Commonwealth Housing Commission was appointed which was to report to J.B. Chifley, the Minister for Post War Reconstruction. The chief concern was the need to eradicate inner city 'slums' and how to manage the anticipated peacetime housing shortage.

HOME TRUTHS

In contemporary Australia, turning 40 is considered to be a landmark, marking full maturity after the impatient years of early adulthood. Whether reaching such a landmark carried the same weight more than half a century ago is debatable, but in 1938 Dawes had reached the age of 40 and was riding high. The coming of war confirmed his pre-eminence. From brick carter, to plant manager, to Austral director and general manager, Dawes had the world at his feet. He was king – 'King Billy' to those who knew him in the industry. A formidable competitor with a brilliant mind, but a fellow who drove his workers on with threats and curses, nothing escaped his attention. Slackers were fired on the spot. He would think nothing of making surprise visits to the Austral and Bedford yards in the dead of night to make sure that burners stoking the kiln fires had not fallen asleep. Day-to-day running of the yards was the responsibility of the yard manager, tough men who weren't afraid to dish out rough justice to recalcitrants. One such manager was Wal Hadley who telephoned Dawes each morning to report on progress at the Bedford works. Arriving by car, Dawes would stump around the yard on a tour of inspection with Hadley in tow. One particularly hot summer's day Hadley asked Bill Dawes to do something for the draggers who were extracting the still smoking bricks from the patent kilns. Dragging was the worst possible job in a brickyard, with the men having to endure temperatures that could reach 300°C during the removal process. With the men about to walk off Hadley suggested using pedestal fans to get the air moving. Dawes stood at the wicket, acknowledged that it was warm, and then

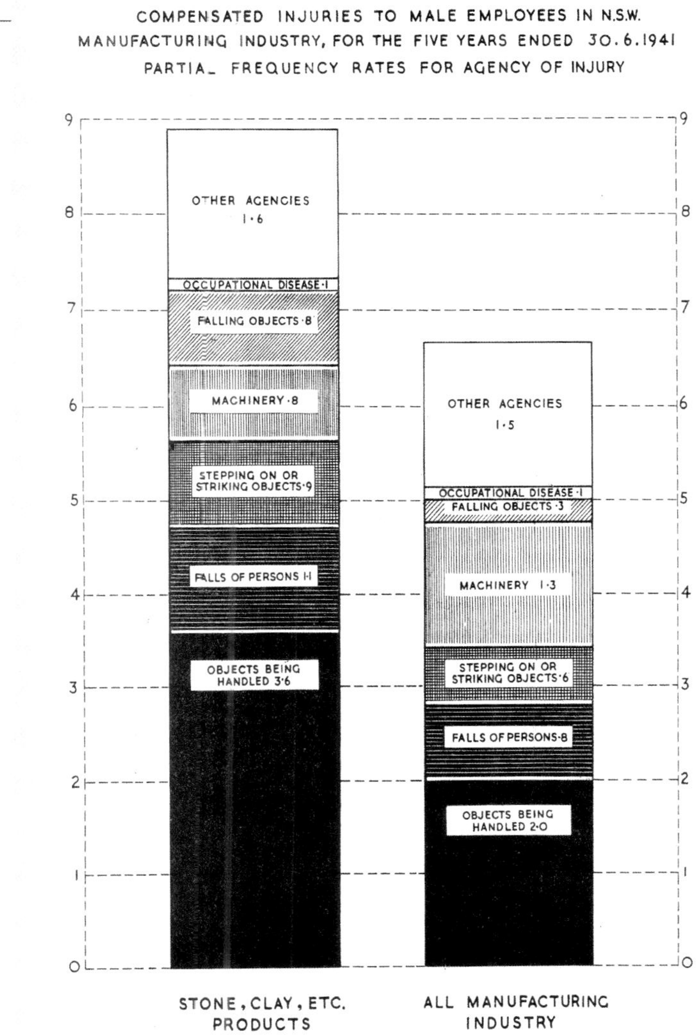

Compensated injuries to male employees in New South Wales manufacturing industry, 1936–1941. Partial frequency rates for agency of injury.

Source: Working Conditions in the Brick. Tile and Pipe Industry, Department of Labour and National Service, Commonwealth of Australia, 1948

made the ludicrous suggestion that the men create their own draught by running in and out with the barrows. The year and date is not recorded but it is probable this incident took place shortly before the end of the war.[15]

George Wheatley who was married to Dawes' sister, Edna, recalled that Dawes senior believed his son was 'too hard on the men', a reference to his employees.[16] This incident, and other grievances may have precipitated another strike at Austral's yards during 1944, which Patricia Binskin recalls

15. Interview with Ray Dawes, 21 July 2006, Sydney.
16. Interview with Bill Wheatley, 1 September 2005, Sydney.

having witnessed as a child. Binskin's father left the pit temporarily to seek other work in order to avoid the stigma of performing scab labour.[17] Dawes was obdurate and refused to make concessions, forcing the men back to work.

His refusal to compromise, and a confrontational manner, deeply affected those around Dawes. Personal relationships began to dissolve. For some years his marriage to Louisa had been under strain, and the couple's increasing estrangement did not go unnoticed, although they continued to live under the same roof at Watkins Street in Rockdale. Domestic tensions were heightened by his cavalier attitude towards drink driving. His son, Ray, recalls an incident that had farcical consequences. Dawes was a regular at the Tattersall Club in Pitt Street, Sydney, during the war years. One night, much the worse for wear, he left the building and climbed behind the wheel of his beloved Chrysler Airflow which was parked on the other side of the road. Gunning the motor into life, Dawes reversed into a stationary Austin 8 Roadster, locking bumpers in the process. After performing a U-turn he made his way home by the usual route along the darkened streets of south Sydney. Well before Brighton-le-Sands he became aware that the Chrysler's engine was straining. Only when Dawes opened the gates to the driveway did he notice the other car, riding high off the ground. Unhitching the Austin he reported it as stolen to the local police.

A standoff also prevailed between Bill and his elderly mother, Emma. Comfortably wealthy, she cared for him deeply but always felt used by her son. Her death in 1951 reveals a very different side to Dawes. His nephew, Bill Wheatley, remembers Dawes weeping unashamedly on hearing the news: 'Our grandma was living with us at Brighton-le-Sands and she died at home. Even as a kid I knew uncle was a hard man and a disciplinarian. There was never any give and take, but I was surprised to see him cry.'[18] Shedding tears over the death of a loved one is nothing unusual, but for Dawes this was significant; at heart he was a deeply emotional man. He also showed great affection towards his nephews, Bill Wheatley and his brother, Naasson, both of whom recall family visits to see their uncle at his comfortable office at the Bedford yard where Dawes surprised the boys and their mother, Edna, with small gifts. Disappearing behind the door of an ante room into a huge safe, he would rummage about, returning a short while later with tins of salmon and other delicacies – things that were supposedly unavailable during the war. Occasionally, the boys were allowed to follow him, and in the dim light of the strongroom they were able to make out large amounts of notes and cash. Some of it was being held as wages intended for the men while other amounts were personal race winnings. More still, as they later deduced, was bookmakers' money entrusted to Dawes for safekeeping; away from the attention of thieves, whether local or those among the many thousands of American servicemen camped at Randwick racecourse.

17. Interview with Pat Binskin, 9 December 2004, Menangle, Sydney.

18. Interview with Bill Wheatley, 1 September 2005, Arncliffe, Sydney.

SOCIAL DIVERSIONS

As building approvals slowed to a trickle, the rapid contraction in brick production left Dawes with more time on his hands. Drink and disappointment make uneasy bedfellows. Perhaps it was the cumulative effect of years spent dragooning industry leaders into a common front through the Brick Committee and its offspring, Brickworks Limited. Or maybe it was disappointment at seeing yet another wholesale collapse in trade, this time caused by war. The truth is, Dawes was inured to long hours and had taken refuge in the isolation that so often accompanies a position of leadership. If the strain was beginning to tell he concealed it well, although rarely from his family. An early sign that things were going awry was a growing fondness for strong drink and expensive diversions in the form of horse racing. Strong drink may have helped to dull the pain of a man trapped in a solitary existence, but it could never be more than a temporary anaesthetic. Dawes struggled with the demon drink until he succumbed to chronic alcoholism in the decade before his death in 1981.

Jack Dawes, standing at the top of Hoffman 2 kiln chimney, Bedford yard, circa 1943. Bill Wheatley Collection

There was no indication during the 1930s that Dawes would fall prey to the heavy drinking which blighted his later years. This habit may have been sparked inadvertently by appreciative clients such as Joseph Scott and Sons, Austral's coal suppliers. Scott's son would visit the family home every Christmas bearing gifts, although the bottle of whiskey he presented, like so many others before, was placed in the drawing room cabinet where, until the war years, it remained untouched. Reg Allott of the Punchbowl Brick & Pipe Company was another unwitting contributor to the metamorphosis that was taking place in Dawes. It was Allott who suggested Dawes buy into racing. Younger than Dawes by several years, Allott was extremely well known in the Punchbowl area, and Sutherland and Cronulla which were just starting to take off as suburbs, due partly to supplies of local brick from his yards. At the suggestion of Allott, Dawes acquired one or two horses and, what started out as a novelty, unleashed a passion that rapidly became an obsession. With a string of racing horses to their names, Dawes and Allott enjoyed success at the Canterbury races employing the services of celebrated trainer, Maurice McCarten who gave the horses regular feeds of kaolin and rock salt. This combination was often referred to as a salt and mineral lick which was beneficial to the health of the animal as well as adding a distinct lustre to the coat of the horse. Kaolin was excavated from a dyke in the earth present in the Brookvale brick pits. The geologist, Edgeworth David, traced this seam from Brookvale to western New South Wales by following the native Australian Christmas bush with its vivid red flowers. The blue grass country in the US state of Kentucky is famous for its racehorses and kaolin deposits. When Brookvale reopened after 1945, Dawes would order truckloads of kaolin to be delivered to McCarten's stables.

There was more to horse racing than merely the thrill of the chase. Huge amounts of money were wagered by massive crowds at Sydney's racetracks despite the Curtin Labor Government having decreed a raceless Saturday every month as an austerity measure. Hundreds of thousands of servicemen, many of them American soldiers who arrived in increasing numbers from early 1942, sustained the wartime racing industry. With the Randwick racecourse now an immense billet, the Australian Jockey Club

Above:* *William Frederick and Arthur Dawes, circa 1943 – W.F. Dawes (left of picture) with his older brother, Arthur, who was a sergeant in the 2/8th Field Company of the Royal Army Engineers. Arthur fought in New Guinea. Marilyn Keelty Collection

Left: Reg Allott

was obliged to transfer the Sydney Cup to Rosehill in April 1942. In smart uniforms, and with money to burn, many of these well-fed, stylish American GIs were out for a good time and had no trouble wooing local girls.

Not one to hold back, Dawes embraced his new-found passion. Using the nom de plume, 'WK Doral', Dawes and Allott raced many gallopers, and co-owned Veiled Threat, winner of the 1942 Sydney Cup. They sold the horse for a tidy sum which went on to win the 1944 cup. Another horse, Sweet Chime, was the finest three-year-old filly of her year in Australia and the mother of Belmont Park, the classic winning mare raced by Dawes. Using a friend, Jack O'Brien, the racing writer for the *Daily Telegraph,* to place bets, Dawes won, lost, and regained a substantial fortune from this point until the early 1970s when his interest waned. He may even have wagered a huge bet on Sweet Chime, but having lost was obliged to settle the debt by selling two flats ('Austral' and 'Bedford', see picture on page 187) in Rockdale that he owned and built during the early 1930s. While Belmont Park was not nearly as good as his mother, Sweet Chime, he was still an outstanding horse, in view of winning two consecutive races at Rosehill on the same day during the early 1950s. This feat has never been achieved before or since, which is hardly surprising since even at the time it was considered dangerous to race a horse in such a way. Sadly, Belmont Park was killed in a stable fire at Kensington several years later. By now a wealthy man, a successful racehorse owner, and a noted industrialist whose companies controlled much of the brickmaking industry in Sydney, Dawes was at the peak of his powers. Mixing with the in-crowd and surrounded by mug punters and adoring females, Dawes was moving into the fast set. Allott seems to have had second thoughts and left Dawes to it, although the two men continued their association. While they could never be described as close friends they did, however, share much in common and, for a short time, were close drinking and gambling partners. Whether it was in the company of friends such as Allott or alone with a whiskey bottle in Austral's office, Dawes cut a solitary figure. The death of his brother, Jack, in April 1943, following a motor accident on Botany and Gardener's Road, heightened this sense of loneliness. Jack's death affected Dawes and his mother profoundly and for a while grief threatened to overwhelm them. Work, and his attachment to Brickworks and Austral, however, filled the void in Dawes' life.

THE TRIUMPH OF BRICKWORKS

From the latter part of 1943 the threat to Australia began to recede as the tide in North Africa, on the Russian front, and in the Pacific began to turn in favour of the Allies. The mid-year mark was also highly significant for Austral Brick. On 15 June the company secretary, David Rogers, wrote to shareholders enclosing a letter from the chairman,

Spackman, proposing the amalgamation of Austral Bricks and Brickworks. Spackman's letter told of concerns about the future held by the brick industry, prompting Austral to appoint a committee consisting of E.D. Lanceley, Dawes, Davison and A. Green. With so little work, and too many pits, further consolidation of the industry was, in their opinion, inevitable. In fact, representatives of several yards had already intimated their desire to amalgamate. What better place to start than a marriage between Brickworks Limited and Austral Brick? Both companies were in sound financial shape at the time of the amalgamation.[19]

The committee recommended that, subject to Commonwealth regulations, a fair and equitable basis of the amalgamation for the purchase of the whole of the share in Austral Brick by Brickworks Limited and Clay Industries Pty Ltd (CIL) would be £50,000 cash, 100,000 fully paid ordinary £1 Brickworks shares, and 150,000 fully paid 5 per cent £1 preference shares. The latter was not to be cumulative until the cessation of hostilities with Germany and Japan. The offer of preference shares would be achieved by a reduction in Brickworks' capital, and the shares would carry no votes in general meetings. All costs and charges would be covered by Brickworks Limited including stamp duty, legal and accountancy costs. Finally, any amalgamation would be conditional on the appointment of Dawes for 10 years, the first five years as managing director, the second five years as general manager, and that he continue as managing director for the next five years.[20]

At the meeting of Austral shareholders on 21 June 1943, Dawes spoke of the desirability of accepting the Brickworks offer for the simple reason that it would capitalise the previous 15 years of effort, eliminate any possibility of a trade war in view of the end to the present trade agreement in October 1944, and that it was the best offer the company was likely to get. With few dissenting voices, those present voted to sell the whole of the shares in the company to Brickworks Limited and CIL on the terms previously outlined by Spackman. Austral investors were certainly not disadvantaged. For each Austral share of a nominal value of 15s 6d they were to receive 10s in cash, leaving 5s 6d per share invested in The Austral Brick Co. Pty Ltd, which issued 98,000 shares. The meeting also voted to renew the trade agreement on a yearly basis.[21] In retrospect, The Austral Brick Company Pty Ltd might well have continued as an independent entity for years to come; it was financially sound, well-managed, and efficient, unlike many of its competitors, and despite earlier attempts to amalgamate, management had stood firm. It was only a matter of time before Brickworks Limited moved to strengthen its position by acquiring a controlling interest in one of its main investors.

Further consolidation took place at a directors' meeting in September 1944 with the sale of Brickworks' share in the Flemington Brick Company. This was agreeable to the Council of Brick Manufacturers which had an existing financial interest in the Flemington yard. Unhappily, this coincided with the death on 25 September 1944 of a prominent and popular figure in the brick industry, Sir John Harrison. Then on 22 December 1944 the directors of Brickworks passed a special resolution to convert 75,000 of its ordinary shares held by its subsidiary, CIL, into 10 per cent preference shares. These were accepted by the vendors of shares in The Austral Brick Company as having a value of £2 each for the purpose of the sale of Austral shares to CIL. Thereafter, the mop-up of Austral's stock progressed in dribs and drabs. As the war in Europe was drawing to a close the directors agreed on 9 February 1945 to exchange 7,500 ordinary shares

19. Brickworks' dividends for the four years 1939 to 1942 had returned an average amount of 7.73 per cent per annum on a paid up capital of £375,231. During the same period Austral Brick had paid out £39,200, an average of 10 per cent per annum. On the other hand, for the year ending June 1943 neither company had been able to pay dividends owing, principally, to the government's restrictions on building.

20. Dawes, had been a director of Brickworks Limited since its inception, and was Chairman of Directors for the period February 1937 to January 1940, with the exception of an interval of four months while in America.

21. At a meeting of the Austral board on 18 August 1944 a new Brick Council Agreement was ratified for a further 10-year term.

held in the capital of Brickworks for 3,750 shares in Austral. During March they purchased shares held in the company by the Council of Brick Manufacturers. Finally, at an emergency general meeting in early April 1945, shareholders voted to amend the articles of association to bring them into line with those of its parent, Brickworks Limited. By early 1945, Brickworks, with the assistance of CIL, had obtained a controlling interest in Austral. A partnership agreement was struck between Brickworks, CIL and Austral, all of whom agreed to pool their assets. Until the wind up of CIL this partnership went under the name of 'Brick Operators'. Dawes was elected chairman of Austral, a position he held in addition to being its managing director. The transfer on 10 December 1945 of shares held by H.W. Thompson in the Warren Brick Company to Austral's directors further consolidated Brickworks' position.

TIME OUT

The war took its toll, especially on industry figures such as Bert Moseley who succumbed to a debilitating illness in September 1944. In the same month R.H. 'Ray' Wilkinson, was appointed Assistant Secretary to the Council of Brick Manufacturers.[22] On 1 May 1945 he became company secretary of Brickworks Limited, an association that was to last until his death in 1962. In May 1945 Moseley sailed for England to recuperate, although in his capacity as Secretary of the Council of Brick Manufacturers, he investigated the methods being adopted in that country for post-war reconstruction. He also swapped notes on brick prices and wage rates with his British hosts. During a five-month stay, Moseley was entertained by the Worshipful Company of Tilers and Bricklayers. Sir Malcolm Stewart invited Moseley to view the works of the London Brick Company Ltd at Stewartby and Peterborough and Mr Morris Whitehouse of Horsham gave him the opportunity of visiting the fine engineering brick works at Southwater.[23]

Dividends paid by Austral Brick Co. Pty Ltd

Year ending June 1939	£62,538 10s = 16.7%
Year ending June 1940	£18,761 11s = 5%
Year ending June 1941	£18,761 11s = 5%
Year ending June 1942	£18,761 11s = 5%
Total	£118,823 3s = 7.5%

Source: Fifty Year Retrospective, 1934 – 1984, Brickworks Limited

Assets of Brickworks Limited June 1943

Shareholders	No. of shares	Amount paid for same
Clay Industries Pty Ltd	79,990	£79,990,00
Ashfield Brick Co. Pty Ltd	12,848	£10,813 10s 6d
Clyde Brick Co. Pty Ltd	16,810	£16,65616s
Goodlet & Smith Ltd	2,467	£4,082 13s
St. Peters Brick Co. Pty Ltd	1,400	£1,228 2s
Carrington Brick Co. Pty Ltd	453	£456 13s
Comm. Banking Co. of Sydney Ltd	1,000	£20,270 16s
Total		£133,500
War Loans		£7,000
Fixed Deposits		£17,000
Stocks of Bricks 6 million–say		£18,000
		Fwd. £175,500

Source: Fifty Year Retrospective, 1934 – 1984, Brickworks Limited

Following his return to Australia, *the Clay Products Journal of Australia* ran an article from the British Clayworker in which Moseley reported his:

> *interview with Sir Malcolm Stewart, Chairman of the London Brick Co. Ltd, who is also largely connected with cement interests. I told him we in Australia were anxious to mechanise our works where possible, because of the shortage of manpower.* [24]

Having spent most of his working life in the industry, and mindful of his recent collapse, Moseley took life easier. A dedicated bowls player, he eased himself into retirement on the greens of Kensington Bowling Club, Sydney, where he played alongside Ernie Swain, the secretary of the Stoneware Pipes Association.[25]

22. Born in Brisbane, Wilkinson's family moved to Sydney. Wilkinson attended Cleveland Street High School, after which he joined the Commercial Bank of Australasia. He later graduated as a chartered accountant with H.W. Gladwin & Co, and worked for several years with MLC Company. In 1935 he joined The Metropolitan Brick Company as company secretary and manager before moving to Brickworks Limited in 1945.
23. *The Clay Products Journal of Australia*, October 1945.
24. *The Clay Products Journal of Australia*, January 1946.
25. *The Clay Products Journal of Australia*, June 1948, p. 7.

All the menu has for him this birthday... April 24, 1945. Artist: J. McClelland, with kind permission of P.I. McClelland

Thus, by war's end Sydney's brick industry had experienced yet another major upheaval, the latest episode confirming Brickworks as the dominant force. With its three yards in the St Peters-Alexandria area – Austral, Ralford and Bedford – The Austral Brick Company Pty Ltd was the 'jewel in the crown'. These yards produced about 32 million bricks a year, while those owned directly by Brickworks Limited manufactured 28 million bricks per annum, taking yearly output to a combined total of 60 million bricks. Extensive reconstruction work being undertaken at the yard of City Bricks, which had lain idle for seven years, later boosted this figure following the plant's return to operation. As Australia celebrated victory over Japan on 15 August 1945 the future looked bright. The transition to peacetime, however, would be more difficult than anyone had imagined.

Home Building 1951 style by Wep. The Australian Women's Weekly

Austerity 13

Picking up the pieces – Gold mine – Homebush resumed – Labour shortage – Working conditions and industrial expansion – Shooters and blasters – In search of Allis Chalmers – The engineers – Success – Cumberland Plan – Winter of discontent – Further consolidation – Modernity – Ahead by a mile

PICKING UP THE PIECES

The end of hostilities in Europe on 8 May 1945 and in the Pacific on 6 August could not have come soon enough. In Sydney homes, pubs, and the pages of newspapers, the talk was all about 'bringing the boys home'. Demobilisation. Then it was jobs, expansion, new buildings, and the confident expectation that good times lay ahead. The only difficulty for thousands of returned servicemen and their brides was the pressing matter of finding a place to live. With banks and building societies approving an avalanche of home loan applications, the demand for bricks rapidly outstripped supply.

Even as the war raged, there had been various attempts to lift building restrictions to ease the housing shortage, but to no avail. In April 1943 Sir Arthur Rickard, the hugely successful real estate agent, property developer and advocate for boosting Australia's population, estimated the shortfall in Sydney to be up to 45,000 houses.[1] So, too, did the New South Wales Labor Premier, William McKell.

> *We 'lost' over 45,000 new houses and flats during the depression. It was not until 1938 that the pre-depression figure was reached again. And then came the war! This meant another six years during which house-building slowed down and then practically stopped. The end of the war found us about 90,000 homes 'short'.*[2]

All but four or five yards in Sydney's metropolitan area were closed up between 1940 and 1945. Add to this the decade-long decline in investment as well as stringent price controls introduced in 1938, and it is not hard to understand why the industry was in poor shape. Who would put money into a

Outside the Melbourne GPO, 15 August 1945. Bruce Howard Collection, National Library of Australia

business that was run-down and vulnerable to so many external forces?[3] Regardless of what private investors, mostly banks, thought about sinking their money into a hole in the ground, strenuous efforts were made to revive the industry, not the least by returned servicemen eager to settle down and start a family. By early 1946 demobilisation was well advanced and the government had given the go-ahead for the resumption of building. This was given impetus by the dramatic increase in marriage rates and the postwar 'baby boom'. The government estimated that brick and tile production would reach 484 million and 46 million respectively by the end of

1. P. Spearritt, *Sydney's Century: A History*, University of New South Wales Press, 2000, p. 83.
2. W. McKell, *Five Critical Years*, Australian Labor Party, Sydney, 1946, p. 100.
3. Building activity is affected by numerous factors, including the purchasing power of the home builder, building costs, demand, population increases, the availability of building materials, labour and production facilities and architectural design.

1951. During 1949–50 alone 70,000 dwellings were commenced and 57,000 completed. In addition to residential housing, buildings such as factories, shops, offices and public works also required bricks and tiles, as did alterations and repairs.[4] Indeed, bricks and brickmaking were making newspaper headlines. In February 1946 the Mayor of Ryde, Alderman C.D. Bondfield, criticised the low brick output, in response to which Brickworks Limited offered Ryde Council a three-year lease on one of its yards. This was a case of Dawes having a dig at the Council, an off-the-cuff gesture telling Ryde Council to mind its own business.

GOLD MINE

Under pressure to kickstart brick production, the state government of New South Wales looked to the industry to spearhead a revival. It also made plans of its own to re-enter brick manufacturing. For the time being, no-one was better placed to lead the way than W.K. Dawes, for when it came to stock on hand and productive capacity he had ensured that Brickworks Limited was in a unique position. Throughout the war Dawes had frequently expressed the view to his two eldest sons, William and Ray, who worked in the business, that he had no intention of being caught wrong-footed. With the revival in trade after 1935 he had made a killing by selling the mountain of bricks stacked in the Austral yards during the Depression years. He was now about to enjoy another windfall. Brickworks was the only company that had sizeable stockpiles of bricks – millions of them lying in the company's yards adjacent to the kilns. Local residents tell of bricks reaching a height of four or five metres, with only a narrow pedestrian gully from the front gates to the office.

That Dawes was sitting on a gold mine was common knowledge to the men working at the various plants, especially Wilfred 'Bill' Swan, the company book keeper and officer manager at the Austral and Bedford plants. Swan rose through the ranks to become Company Secretary of Austral Brick Co. Pty Ltd. He also happened to be Dawes' brother-in-law.[5] Even today Swan is acknowledged as having been a difficult and testy man, an old-time manager who never let an opportunity pass to remind the men who was boss. In a brick-hungry city, yard foremen were in a powerful position to determine when the builder and brick carter received their order, and whether it was supplied in full. With bricks in short supply, there were many who tried to sneak out with a concealed load past the ever-vigilant Swan whose unpopular role as the brickyard policeman earned him the opprobrium of many workers. Brick carters had particular reason to dislike him because most had no written contract and could be sacked on the spot. Altercations were frequent and abuse was hurled at Swan, who provoked some of the men working the kilns to violence. Dawes took a dim view of Swan's offhand treatment of the workers and was known to upbraid his manager publicly. Perhaps Swan kept his job because he was a highly effective minder and factotum. But there was also another side to Swan who enjoyed frequent fishing excursions to Botany Bay with some of the men from the Bedford yard, including young Arthur Dawes. Alan Binskin, among others, recalls that he was a good fisherman who had a knack for knowing where the fish shoaled under Georges River Bridge.[6]

HOMEBUSH RESUMED

As the boats arrived home in the closing months of 1945, packed with war veterans and POWs, tensions between former allies were being played out on the other side of the globe. By 1947 the countries of eastern Europe, occupied by the Soviet Union in its push towards Berlin in early 1945, found themselves with communist régimes imposed by Moscow. In Britain, a Labor government had been swept into office with plans to remodel the economy along socialist lines. Communist parties in France and Italy looked set to take control of government. In Australia fears of Soviet-style communism were on the rise yet again, especially in New South Wales

4. Ministry of National Development, Division of Industrial Development, *Brief Review of The Australian Clay Brick And Roofing Tile Industries*, Commonwealth of Australia, April 1951.

5. Bill Swan married Joyce, the widow of Jack who was a brick carter. Jack was killed in a motor accident in 1943 while driving his Plymouth. The wreck was taken back to Austral's workshop by his grief-stricken brother and covered with a tarpaulin. Swan remained in the service of Dawes for many years.

6. Interview with Norman and Alan Binskin, Menangle, Sydney, December 2004.

where Premier William McKell was nervous about communist influence in the union movement. Public anger over the housing shortage was beginning to spill over into minor acts of civil disobedience. Out of desperation an increasing number of people had resorted to squatting. McKell urged the New South Wales Housing Commission to accelerate the public housing program, and once more the government became a brick manufacturer. On 19 March 1946 it announced the resumption of the original site at Homebush. Legislation was hastily drawn up and passed in the form of a State Brickworks Act 1946, allowing for the establishment, management and control of a State brickworks and the acquisition of lands to erect the necessary plant.[7] Reconstruction at the Homebush site was soon underway, and the plant and kilns re-equipped and refurbished. Over time the second State Brickworks developed into a highly resourced factory with a large staff. For many years it maintained approximately 7 per cent share of the New South Wales brick market.[8] In fact, a large number of men employed by Brickworks Limited and Austral Brick rejoined the state enterprise where pay and conditions were known to be a good deal better.

LABOUR SHORTAGE

Clearing the backlog of brick orders was a lot more complicated than simply increasing production. More or less everything was in short supply, including the coal that powered boilers and fired the kilns, and labour. Manpower shortages in the brick industry were chronic and despite Dawes' work on various wartime committees relief was never forthcoming. In August 1944 efforts had been made through the Department of War Organisation of Industry to release men from the armed services, but with little success. In July 1945 letters had been sent to the Prime Minister, the federal Minister for Works and Housing, the chairman of the New South Wales Housing Commission, the state Minister for Housing, and the Acting Premier, stressing the need for immediate demobilisation of the necessary manpower before the situation became acute. This attempt also failed, with the result that by the end of 1945 yards across Sydney were on a three-day week. The labour shortage also had a direct impact on the manufacture and repair of brickmaking machines. One local firm in the St Peters area, George Foster, was pouring castings only every third working day.

Trainer and owner: Maurice McCarten and W.K. Dawes at Flemington races. Dawes continued his obsession with bloodstock and achieved celebrity status in August 1954 when he raced Belmont Park at Rosehill. Belmont Park had won the Auburn Stakes (a race of 1¼ miles) at 3.17 pm and was entered by Dawes in the following race, the Welter handicap, which started at 3.49 pm.

The Herald, Monday 3 November 1947

Alan Binskin, who worked in the brick industry for over 50 years, was just 18 when he started at Austral's Ralford yard in 1946 under his uncle Ranji Galway, the yard foreman. There was always plenty of work

7. *State Brickworks Act 1946*, The Statutes of New South Wales, Act No. 16, 1946.
8. Public Works Department of New South Wales, Annual Report for 1985–86, p. 20.

The Binskin Boys. Left to right: Jim (Fowlers Pottery), Roy, Alan (Bedford and Ralford), circa 1948.
Alan Binskin moved to the Bedford yard with its Hardy Patent, Hoffman and six downdraught ('dome') kilns in about 1950. He stayed there for 12 months before moving across to the main Austral yard. Alan Binskin Collection

for young men like Alan; attracting and retaining them was difficult on account of the harsh working conditions. War veterans, in particular, were unwilling to put up with overbearing foremen. The work was rough, physically demanding and many of the men who worked in the yards could handle themselves in a fight. They gave as good as they got, and if Dawes was a taskmaster it was because he knew that brickmaking was not a gentleman's occupation. The emotional strain would have been intense and may help to explain why Dawes found an outlet in social drinking, breeding racehorses, and punting.

In the short term the labour crisis was temporarily resolved with the mass migration programmes of the 1950s. By the end of the decade finding suitable labour had become a chronic long-term problem for Australia. State government restrictions on the hire of non-union labour also held back progress. The first step towards freeing up the job market in New South Wales was taken during November 1949 when agreement was reached between the government and the Federated Brick, Tile & Pottery Workers Industrial Union to allow the use of Displaced Persons (DPs) and other European migrants in the clay industry. This, and similar measures at a national level, opened the gates to a flood of people in search of a new life. By 1951 the federal government's migration programme envisaged a yearly intake of about 200,000 immigrants.

Welcomed with open arms by employers, 'New Australians' had no trouble finding work, especially in yards operated by Brickworks Limited. Hard-working and prepared to accept direction, migrants were grateful to be employed after having escaped the dire economic problems of their homelands. Many of them were from southern Italy. Over the years, Ron Hagstrom, plant manager at the company's newly modernised plant at Brookvale employed hundreds of Italians, often through the recommendation of Joseph Vescio who worked at the yard. Today, a glance at the local telephone directory is all that is needed to confirm the ongoing presence of the extended

Vescio family in the Brookvale area, as well as the Pellizons, Turcatos, Bivianos, and the Francos who worked the Eastwood yard. Dawes respected these good natured men and made a point of looking after them. He organised transport from the point of disembarkation at Walsh Bay or Circular Quay to billets close to the brickyards. Using an old bus, he even ferried the men to where they were needed at the various plants. Jim Liepins, himself a migrant from Lithuania, worked at the City yard's Euston Road workshop as an engineer from 1954 until his retirement in 1991. The late Girolamo Gardoni, a market gardener from the northern Brescia region, started at the Eastwood plant soon after stepping off the boat in 1950. From minding the shale crushers, Gardoni graduated to the loft where clay and shale were mixed, was promoted to wheeler, learned the art of burning, and finally attained the coveted, and well-paid position of continuous burner. Gardoni was later joined by his wife and young family. His eldest son, Natoli, was 12 years old when he arrived in Sydney and started work at Eastwood several years later. Following close behind was Caesare, who worked alongside his elder brother, Natoli. This was in the days when the patent kiln was set by hand, and before the introduction of forklifts in 1968. The Gardoni family was a typical migrant family whose story makes familiar reading. A less common scenario was the employment of Aboriginal workers. A small number of men, one known only as 'Black Dick' who hailed from Port Macquarie, were employed at the Austral yard. Black Dick was greatly liked by his workmates at Austral's St Peters yard, and drank with the other men at *The White Horse Hotel*. This was in stark contrast to back home where he would have been banned from hotels on account of his Aboriginality. Black Dick was something of a character. According to Bill Wheatley:

> *One day, Black Dick and a group of draggers refused to work. Bill Dawes came along to do his usual 10am inspections and was told by Wal Hadley (a trusted foreman and troubleshooter) about the stoppage. Bill tried to get the blokes to go back to work, but the retort from Black Dick went along the following lines.*

Harry Miezis (seated) and family, circa 1952. Miezis was a brilliant engineer who worked in the City Road workshop. Jim Liepins Collection

> *'Listen Bill, there was no bloody work in this country until that bastard Captain Cook came along and that was the worst bloody thing that ever happened.*[9]

Dawes took the comment in his stride, but the men still went back to work, receiving a tongue-lashing from Hadley. Apart from the shortage of labour, there were other, more deep-seated problems that had never been properly addressed. Working men with families needed stability, yet by any measure

9. Interview with Bill Wheatley, 25 February 2005, Sydney.

brickmaking was a difficult industry in which to make a living. Time lost due to bad weather or mechanical breakdown created a great deal of anxiety, and when we take into account the discrepancy between the output of experienced and inexperienced labour (the former on piecework, the latter on day rates), take home pay could fluctuate wildly.[10] In early 1949, for example, higher than abnormal rainfall flooded many of the brickpits at St Peters, throwing hundreds out of work. While training schemes existed for those who assembled houses, there were none for the men who produced the materials. Little wonder that without sufficient inducements such as continuity of employment, sick leave, and amenities, the brickmakers found it difficult to attract and retain skilled men, and relied increasingly on migrant labour.

WORKING CONDITIONS AND INDUSTRIAL EXPANSION

By 1947 the labour situation prompted the Commonwealth Department of Labour and National Service to conduct a survey of a sample of brick, tile, and pipe works in New South Wales and Victoria. In its report published the following year, the department remarked that many of the industry's problems arose from the small degree of mechanisation and reliance on hand methods, which made for arduous and dirty work, and a high accident rate. Norman Binskin, brother of Alan, started at Austral's St Peters yard in 1950 at the time George Keep was yard foreman. Recalls Binskin, 'It was very labour intensive – even towards the very end of its life in 1983 there was very little mechanisation in the Austral pits.'[11] Data in the 1948 report published by the Commonwealth government clearly showed that clay industries experienced a high accident rate, a phenomenon reflected in premiums charged by insurers under the New South Wales Workers' Compensation Acts. At this time insurance rates were based on the insurers' experience of each industry as a whole. Although an individual firm may have had a good accident

Average annual employment in New South Wales, 1948

Judging by the total number of people employed, the clay products industry was the most significant employer by industry sector in New South Wales.

Industry	Number employed
Bricks, tiles, firebricks and fireclay goods	3,554
Earthenware, china, porcelain and terracotta	1,076
Glass (other than bottles)	1,959
Glass bottles 1,039	
Modelling, etc. 43	
Cement and cement goods	1,229
Lime, plaster and asphalt	1,495
Stone cutting and polishing	694
Total	11,089

Source: Working Conditions in the Brick, Tile and Pipe Industry, Bulletin No. 12, Department of Labour and National Service, Commonwealth of Australia, 1948, p. 7.

record, it was customary in Australia for it to pay the same premiums as a firm with the average accident rate. In contrast, insurers of workers' compensation in the USA maintained an Experience Rating Plan whereby the premiums were modified according to a firm's accident history.[12] The absence of such plans in Australia offered no encouragement to firms that wanted to provide a safer workplace. It would take time for increasing mechanisation to improve working conditions. Rather than wait the government suggested immediate measures such as improved safety practices, personnel training, and the provision of amenities such as washrooms and canteens.[13] It was years before the owners of many privately-owned brickyards installed such facilities, and the issue was a constant source of friction between employers and the brick and tile workers' trade union.

Developments in Sydney's brick trade, its problems and its triumphs, were part of a much bigger picture – the growth of industry across the municipality of Alexandria. Mid-way through the

10. Piece workers were paid by how much was produced. Day rates offered no opportunity for additional payment regardless of the effort.

11. Interview with Norman Binskin, 9 December 2004, Menangle, Sydney. George Keep relinquished his position as works foreman in 1958 and was succeeded by Arthur Hoppit. Keep's brother, Fred, also worked in the yard.

12. California Safety News (San Francisco: Department of Industrial Relations), March 1946, p. 12, cited in Working Conditions in the Brick, Tile and Pipe Industry, Bulletin No. 12, Department of Labour and National Service, Commonwealth of Australia, 1948, p. 7.

13. Little research was conducted in the early postwar years into the application of scientific advances to brick and tile making industry. Technical advice was provided by Australian representatives of overseas manufacturers. Private consultants with overseas experience, especially the Commonwealth Scientific and Industrial Research Organisation (CSIRO) offered specialised knowledge.

Manual operations in an Australian brick works. Working Conditions in the Brick, Tile and Pipe Industry, Commonwealth of Australia, 1948.

Above: Operating a Foster's dry-press brick machine
Left: Shale deposit prior to blasting

war, Alexandria Council published *The Birmingham of Australia: 75 Years of Progress* which claimed the existence of 550 factories in the municipality. The development of the Alexandria, Waterloo, St Peters industrial conurbation had been a long time in the making, commencing in the early 1840s with the establishment of brickworkings, abbatoirs and tanneries. Industrial activity, much of it on a large scale, provided employment for local people and commuters, serving to underpin community life. It would be a mistake to idealise the lives of working class people who struggled to make ends meet, and where drink was an unwelcome visitor to many households. Housing was mostly substandard; living in the shadow of the factory chimney meant a constant round of cleaning. Fine dust cast a pall over the district and the health of its inhabitants. Despite the obvious drawbacks, narrow laneways and tightly packed terrace housing brought people together, first as neighbours and then as communities. Today, many older residents in these rapidly gentrifying suburbs lament the passing of such an era.

SHOOTERS AND BLASTERS

The labour crisis, as well as technical improvements in the processes of brick and pipe making in the post-war period, obliged private manufacturers to change many aspects of their operations. Throwing its weight behind the Department of Labour's concerns about the lack of mechanisation, *The Clay Products Journal of Australia* called for the standardisation of certain operations. What better place to start than in the quarries where most yards still relied on brute strength to load shale into skips. One process – blasting – had changed little from the early days and remained largely unaffected by the changes that were soon to overtake the collection and hauling of dirt to the top of the pit.

The first step in the brickmaking process commenced in the quarry where blasting removed over-burden, which was also known as spoil. This could be extremely dangerous if not conducted properly. Generally, a series of blast holes 3 centimetres in diameter and approximately 10 metres deep were drilled about two metres back from the pit face. Sticks of gelignite were loaded into the bottom half of each hole and a series of controlled explosions at intervals of one second ensured a better shattering of the shale and its mixing with clay overburden. Even the most experienced shooter could not prevent huge slabs of shale falling more or less intact to the ground. Shooters were licensed and specialists in their field. One such was Jack Crawford who travelled between the yards owned by Brickworks Limited. A 20 tonne wall of shale would be broken down further ('popped') by small explosive charges. With one strike of a sledge hammer an experienced breaker could shatter the smaller lumps into manageable chunks. These would be shovelled into skips running on a series of rail tracks radiating outwards from a central loading platform called a chinaman. Skips were pushed by hand and upon reaching the chinaman, the load was tipped into a larger rail skip and the material power-hauled to the loft and bins.[14]

Bill Wheatley recalls that on one occasion the shooter blew a charge placed in the wall only a matter of yards from the boundary fence and footpath opposite St Peters Anglican Church, on the Princes Highway.[15] When the dust had finally cleared most of the path could be seen at the bottom of the pit. This incident provoked an angry response from the local council and the Department of Main Roads. Not everyone's complaints were taken seriously. Laurel Horton, a long-time resident of St Peters and archivist at the aforementioned church, recalls one spectacular blast that showered the church with rocks. Soon after, Len Rogers received a visit from the rector who was fobbed off with excuses. After being told of the incident Dawes asked Rogers to give the rector £20 for church funds and also to buy ear plugs, a joke in rather poor taste.[16] The proximity of the quarries was a problem for local residents and remained so until their closure. It is also well to remember that by 1950 many of the brick pits along the Princes Highway had been worked for decades. In the case of Austral, activity over 40 years had created a series

Rail truck being hauled from the pit into the machine room at the Eastwood yard, circa 1975. Brian Binskin

of enormous canyons close to heavily populated areas. As early as 1947 Dawes could see the drift of things, remarking to his son, Ray, that once the pits were finished, the holes in the ground would be worth far more than any brickworks.

IN SEARCH OF ALLIS CHALMERS

Whether due to interruptions caused by wet weather, blasting, or the shortage of labour, the use of hand-drawn trucks made it difficult to supply enough material to keep the brick presses and kilns running to capacity. Unfortunately, mechanical equipment capable of replacing manual labour had been requisitioned by government agencies during the war and was still not readily available. Approaches

14. *The Clay Products Journal of Australia*, September, 1948 p. 13.
15. In the mid 1950s the pit crew at Austral's St Peters yard comprised Jack May, an experienced pit boss and shooter, two breakers (Ossie Miller and 'Black Dick'), a dragline operator, a bulldozer driver and a truck driver.
16. Interviews with Laurel Horton on 2 December 2004, and Bill Wheatley on 25 February 2005, Sydney.

Front-end loader filling tiptruck at the Sutherland yard of the Punchbowl Brick and Pipe Company Pty Ltd owned by Reg Allott, circa 1948. Brickworks Ltd

to the Allied Works Council about the release of equipment failed to provide any satisfaction. At this point brick manufacturers took matters into their own hands. In February 1946, a large and representative gathering of the building industry met at Brickworks Limited's Burwood yard to witness a demonstration of an Allis Chalmers tractor fitted with a bulldozer blade. The firm of Messrs Tutt Bryant Pty Ltd, distributors for Le Tourneau (Aust.) Pty Ltd, heavy earthmoving machinery, were also present. It soon became clear that a bulldozer could load as much clay and shale in 70 minutes as would normally be handled by six to eight men in a full working day. In just over one hour enough material for 20,000 bricks had been taken to the top of the pit for the brickmaking machines, a rate of loading and transportation of brick material four times greater than previously obtained by hand labour. In the case of the Burwood works, the tractor and dozer took the place of 10 men and 17 skips in the bottom of the pit and maintained a constant stream of material to the pit top for over 12 months.

Within a short time a number of Sydney brickyards had introduced tractors equipped with Angledozers or Tiltdozers, as well as purpose-made Tournaloaders. In larger brick pits such as the Austral and Bedford yards at St Peters where production requirements were very high, link belt shovels with a capacity of three quarters of a cubic yard were installed to load the material into waiting trucks. The portable loading ramp – a moveable chinaman – was another useful innovation which could be easily moved around the pit to reduce the distance for pushing by a dozer. Other labour saving machinery was purchased including tractor-compressors used to drill blasting holes in the shale.[17]

17. *The Clay Products Journal of Australia*, September, 1948 p. 14.

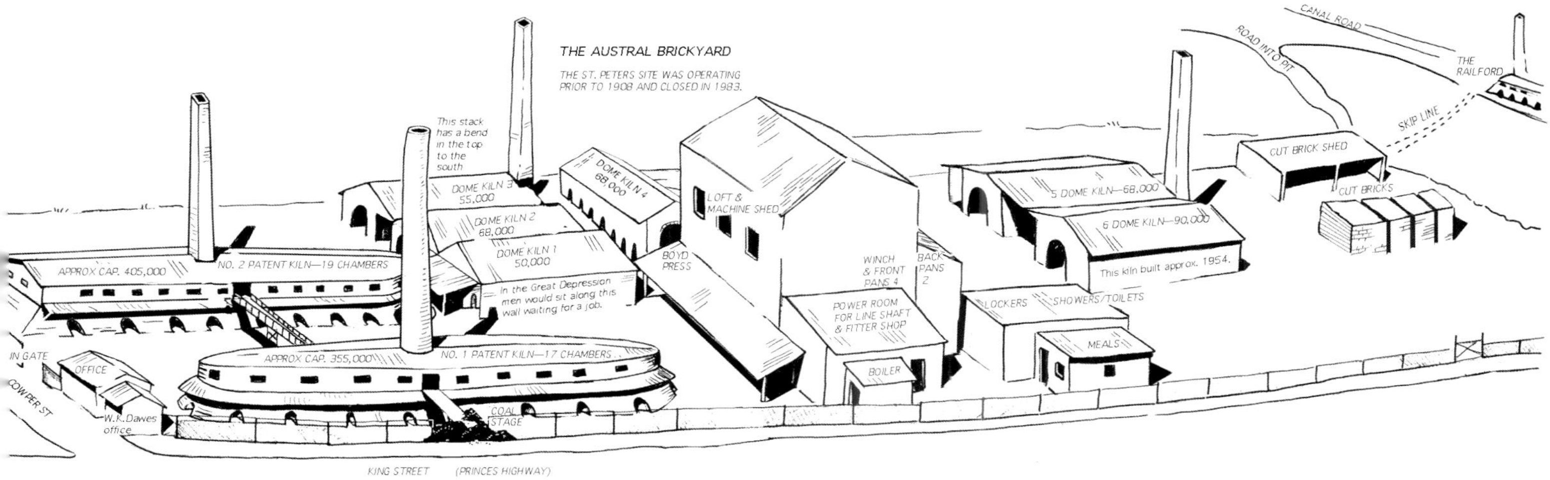

The main yard of Austral Brick, St Peters, circa 1955. David Dickson

The choice of equipment varied with the size and layout of the pit, but was only made possible by the number of machines coming onto the market from the disposal of war service equipment such as International Loaders and Caterpillars.[18] Meanwhile, the inquisitive Dawes, who always enjoyed travelling, scoured the country for bargains, land purchases, or to seek the advice of colleagues in the Melbourne brick industry. On numerous occasions he flew out of Sydney Airport at Mascot and was blasé about where he parked his motor cars. Flying to Brisbane in early 1947 his latest acquisition, an Armstrong Siddeley Typhoon sports coupe, found a temporary home in the carpark reserved for foreign diplomats. Why dump his car in such a place? Answer: because he could.

Apart from acquiring earthmoving equipment Dawes and his fellow directors paid the contracting firm of J. Hennessy to load the clays at Brickworks' Eastwood plant. Hennessy's Model 60 Tournaloader was an over-the-head type that dumped the shot material into trucks that drove up a dirt road to the surface. Such a machine could shift some 160 tons of shale and clay each day. Although this quantity was insufficient for the firm's capacity of about 40,000 bricks of varying grades per day, it could still handle much larger quantities than other machines. In the nearby pit of Butcher Brothers whose managing director was Frank Butcher (he was also a director of Brickworks Limited), a second Tournaloader owned by Hennessy performed the same task under almost identical conditions. The only difference was that the excavator dumped its load into a motor truck which then deposited the dirt into a skip, some 150 metres away, and from there it was hauled to the top. At the Clyde Brick Works in Sydney, one tractor and an angle-dozer were sufficient on the bottom of the pits to keep firebrick-making machines working.

The greater use of machinery in the post-war years resulted in a number of brickyards drawing on shale deposits some distance from their own yard. Burwood was no exception, and shale was carted from the company's South Ashfield pit for use at the Burwood yard. The South Ashfield shale was of a better quality, and softer, than deposits at Burwood. In the words of Ray Dawes the shale gave the Burwood bricks 'a little bit of a lift in colour'.

THE ENGINEERS

Without highly skilled engineers who determined how a kiln operated and thus its output, brickmaking plants would have come to a standstill. Bob Young was chief engineer at Austral's St Peters yard. Bob Hudson was his counterpart at the Bedford yard. Over the years the working relationship between Young and Dawes developed into a lasting friendship, despite a major falling out in the immediate post-war years. Young stormed out of the yard and for a short time went to work for Ernest 'Ernie' Speare, although he returned to Austral Brick a short while later. From the time of his retirement in the early 1950s, Young remained in close touch with Dawes until his death several years later. Bob's younger brother, Bill, took over as engineer although he never seemed to make the same impression on his

18. *Ibid*. p. 15.

employer and his workmates. Two other engineers at the Austral yard noted for the technical know-how were 'Bluey' Harris and Frank Rawlings who worked alongside Young after the war right up to the 1950s.[19] As for Hudson, he was an extremely valuable asset and a brilliant engineer who had instigated modernisation of the State Brickworks at Homebush during the late 1920s. Despite finding a place at Austral's Bedford plant following the buyout, moving to the relatively backward yards at St Peters must have been difficult to come to terms with.[20]

For more than 40 years engineering support for Austral's St Peters operations had been provided by the firm of George Foster, located within shouting distance from the Bedford yard. Foster was the main supplier to the brickyards in Australia, building and servicing everything from dry-press machines to crushers and pit trucks, skips and grinding pans.[21] It may have been the case that Bill Dawes owned shares in George Foster & Sons, which was a private company. Over a period of 18 months or so, commencing in 1951, the relationship between Dawes and Foster appears to have cooled. The catalyst may have been the development of the Foster's Boyd Machine, a press adapted from the American Boyd model. Brickworks Ltd had ordered five such machines from the USA, but the design was improved by Boyd and adapted to suit Australian conditions. When the first of the machines was delivered, installed and tested at the Austral Brick yard there was real enthusiasm, at least until Bill Dawes received the invoice. The bill was far more than he anticipated. As one of Foster's major customers, Dawes may have taken umbrage about this and other perceived slights, such as the late delivery of repaired machine parts. Without doubt, this was aggravated by union demands for higher wages. Worsening relations between Foster's management and its workforce led to a series of damaging strikes which eventually brought the company to its knees in the early 1970s. Whatever the reason, Dawes put his builder son, Ray, to work erecting an enormous engineering workshop at the City yard's Euston Road premises during 1953. Each of Brickworks' plants retained its own engineering workshop, although the City yard became the main focus of operations, serving all the yards in the Brickworks group. The first foreman employed by Dawes at the Euston Road workshop was Indrikis 'Harry' Miezis, a highly trained engineer who had fled Soviet-occupied Latvia, leaving behind a large engineering shop. Somehow or other Dawes managed to gain access to passenger manifests from the shipping lines. Scouring them for the names of engineers, he would then approach migrants with offers of work.[22]

Foster continued to manufacture machine presses and grinding pans for a short while longer. Towards the end of the 1950s the company granted Brickworks Limited the rights to manufacture spare parts for presses and pans. This may have been the beginning of the end, perhaps the final nail in the coffin being the loss of business to manufacturers of extrusion equipment being installed in Brickworks' new tunnel kiln operations at Horsley Park. The gradual descent into oblivion must have been painful to the Foster family who progressively sold off machine parts and castings. Of 250 people employed at the St Peters works during the 1960s, only 16 were on the payroll by the early 1980s when the firm was purchased by William Wallbank. Wallbank developed its own dry-press machines and continues to manufacture them to specification for local and overseas clients.

SUCCESS

One brickmaking firm that stood out as a progressive was that of Goodlet and Smith located in the western Sydney suburb of Merrylands. The company's modern facilities were mechanised so that four men and a boy in the pit, without working flat out, were able to feed the works with enough

19. Interview with St Peters' resident and former brick carter, Bob Brown, 30 December 2004, Sydney.

20. Interview with Vince Higgins, 9 September 2006, Sydney. Higgins' father was store manager at the State yard for much of the 1920s until its buyout and closure by Brickworks Limited.

21. The firm of George Foster manufactured noteworthy brickmaking machinery at its St Peters and Balmain foundries. This included the Foster Press, Foster Top Lift machine, Foster Side Lift (short and long) machine, Foster Crushers, Foster Pans (9' and 10').

22. Interview with Jim Liepins, former engineer at the Euston Road workshop of the City yard, 1 July 2005, Sydney.

BL
RHW.PC

BRICKWORKS LIMITED

HEAD OFFICE:
TELEPHONES { BW 8375 / BW 8376

12 CASTLEREAGH STREET
SYDNEY

7th December, 1953.

Top: Notepaper used by the City yard and Brickworks Limited, circa 1955

Left: Robert 'Bob' Young, chief engineer of Austral Brick in his office at the City yard, Alexandria. Dorothy Armfield, circa 1955

Below: Bob Young at his retirement party. From right to left: W.K. Dawes, wife of Bob Young, Bob Young. Dorothy Armfield Collection

Bottom: This 165 H.P. auxiliary diesel engine at Punchbowl Brick and Pipe Company Pty Ltd was similar to those maintained by Young and his workmen at the Austral yards raw material for more than five million common bricks and 750,000 face bricks a year. Another modernisation success story was that of Reg Allott whose Punchbowl plant in 1947 was turning out 10 million bricks while its stoneware pipe operations was running 14 kilns. Allott's Sutherland yard was producing 81 million bricks per year compared with just six million in 1927. Both plants were in the process of being upgraded with a total of 26 new kilns planned for the Sutherland yard alone. Allott was one of the more progressive owners whose capital investment in machinery boosted output and take-home pay. Exclusive of office staff, 135 men worked at Punchbowl and a further 100 men at Sutherland. The introduction of the 40-hour week from 1 July 1947 following months of industrial unrest took the shine off things for Allott as productivity slumped. From 11,000 bricks per day per machine the new award reduced output to 10,000 each, although this seemed to have been boosted by further mechanisation and an increase in Punchbowl's workforce.[23]

CUMBERLAND PLAN

Against the backdrop of change and continuity in the various yards of Brickworks Limited, the state Labor government announced a slum clearance programme. The main game, however, lay not in fixing broken housing in the inner city, but in developing the middle and outer suburbs of Sydney where demand for new homes was at its most intense. Here, land was still relatively cheap and mass housing schemes, both public and privately funded, were feasible. Slum clearance, and the broader question of how to facilitate Sydney's growth, was also the subject of a report commissioned by the government in 1947. The Cumberland County Plan Report (1948) set out guidelines for the orderly development of post-war suburbs, and reflected concerns over suburban sprawl and the virtually unregulated subdivision boom of the 1920s. It also singled out the example of five large estates in Bankstown, subdivided in 1923 and totalling over 4,000 lots. Real estate agents had sold land on the pretext that major projects, such as railways and tramways were to be undertaken, but the reality was that a quarter of a century later, they remained 'vast stretches of waste land, useless to the holder and equally useless for farming purposes.'[24] The report warned of an undesirable concentration of employment and administration in the city of Sydney, traffic congestion in both the city and suburbs, conflicting land uses due to inadequate zoning and a serious shortage of open space.'[25]

The plan was bold, and needed to be in view of the patchwork of planning measures that had been applied by city and state administrations for half a century. Had its recommendations been adopted they would have zoned business, entertainment, education, and housing, with urban districts encircled by a green belt, a swathe of open countryside. A coordinated rail and road network was intended to link the city with outer urban and rural districts. Instead, it was all too much for government departments and authorities, including the Department of Housing, the Department of Main Roads, and the Metropolitan Water, Sewerage and Drainage Board. Bureaucratic inertia, and inaccurate population projections for the Sydney region delayed the plan's start until 1951, although in the final outcome very little was achieved. The failure to act decisively was a lost opportunity for Sydney, which continues to pay a high price.

WINTER OF DISCONTENT

Town planners and municipal authorities were not alone in their desire to make Sydney a better place in which to live and work. The impetus for change was also being felt in the field of industrial relations. We have already seen that the introduction of the shorter working week did not make everyone happy, especially employers, for it came at a time when the manpower shortage had become acute. As a result of shorter working hours the output from the brickyards across New South Wales was reduced overall by about 10 per cent, although production had recovered by December 1948. The year 1948

23. *The Clay Products Journal of Australia*, June 1948, pp. 11, 14.

24. Cumberland County Council, *County of Cumberland Planning Scheme Report*, CCC, Sydney, 1948, pp. 68–72.

25. Ibid, p. 77.

marked another milestone for Dawes who turned 50. By now middle-aged, he remained a driven man who was determined to keep Brickworks Limited in front. For one director and close friend of Dawes, Harry Thompson, it was, sadly, the end of the road. On 14 December 1948 Thompson suffered a massive heart attack and died suddenly while on business activities. Ray Wilkinson was appointed to fill the vacancy on the board.

LIGHTS OUT

Labour unrest was never far below the surface in these early post-war years. Discontent over pay and conditions, especially on the coalfields of New South Wales, finally came to a head on 27 June 1949 when a national coal strike was declared.[26] Unable to secure regular supplies of coal, gas and electricity, producers were hard hit. Bunnerong Power Station, not far from Allott's Punchbowl works and Austral's St Peters yards, failed on several occasions. The state government and press laid the blame squarely at the door of communist activists in the trade union movement. The problems, however, were more deep-seated than radical agitation would imply. Blackouts were frequent due to excessive demands being placed on outdated power generation equipment, as well as the use of poor quality coal. From householder to industry, no one was spared by the coal strike, including the state of Victoria, a major consumer of Hunter Valley coal. The communist weekly, *Tribune*, accused nervous state and federal Labor governments (1949 was an election year) of 'completely exposing themselves before the Australian working class' by throwing 'their full weight behind the coal-owners against the miners.'[27] Others disagreed, with the *Sydney Morning Herald* and the *Sunday Herald* weighing in heavily blaming the communists and the 'few wreckers who can clog the works for millions.' Builders and brickmakers were unanimous in their condemnation. D. Stewart Fraser, executive director of the Building Industry Congress, claimed that every day of the strike meant the 'loss of 80

Alfred Dawes, died, 1950

In September 1950, Alfred Dawes, grandson of one of the pioneers of the clay industry in Australia, and an uncle of William King Dawes, died. Alfred worked first as a boy in a St Peters brickyard in 1880. Six generations of the family had been engaged in the brickmaking industry in New South Wales.

Alfred Dawes was predeceased by Edwin Spackman, the founding managing director of The Austral Brick Company Limited, who died on 6 August 1947, age 84 years. Spackman's death removed yet another link with the past.

homes, one and a half million bricks, 135,000 tiles ... 2,000 tons of cement and 39,000 square yards of asbestos cement sheets.'[28]

The *Clay Products Journal of Australia* accused the communists of 'having succeeded in closing down every pottery, and brickyard and other enterprise concerned with clay in Australia.'[29] Determined to avoid further disruptions, Allott and many other brickmakers ordered diesel-powered generators to be installed at their works.[30] By late July the strike was broken by the use of armed troops who had been taken by train to work the open cut mines in the Hunter Valley and Lithgow regions. The miners of New South Wales were the last to go back on 15 August, after a seven-week stoppage. Public anger at mass layoffs, electricity cuts (at one stage, power

26. The Austral Brick Company Pty Ltd used coal from the Burragarang Valley near Lithgow in New South Wales. Interview with Arthur Sherman, 16 November 2004, Sydney.

27. *Tribune*, 29 June 1949.

28. *Building & Engineering*, 25 July 1949, p. 63.

29. *The Clay Products Journal of Australia*, July 1949, p. 2–3.

30. Ibid, p. 2–3

was rationed to one light per household), and Cold War fears of a communist threat from within, led to the defeat of Prime Minister Ben Chifley and his Labor Party in the December elections. Riding on a promise to deliver prosperity – cars, houses and jobs – a Liberal-Country Party coalition swept into office where it remained until December 1972. Between Brickworks Limited and the Federated Brick, Tile & Pottery Workers Industrial Union, however, there was comparatively little friction. Union membership was a condition of working in the yard, although very few had any direct contact with union delegates. They simply paid their dues and went along with whatever was decided. Despite all of his bluster, Dawes had a grudging respect for the union secretary, Jim Parker, and his successors Eddie Monson and Alex Sawtschuk. Complaints about working conditions were usually settled by a combination of face to face meetings, mutual compromise, or over a drink at the local pub.

FURTHER CONSOLIDATION

Back to business. As managing director Brickworks Limited and Austral Brick William King Dawes had both hands full, although he managed to lighten the load by rationalising yards that were either unprofitable or were running out of shale. In May 1946 the land and assets of Croydon Brick Works Limited were sold and the company wound up on 10 September 1947. As compensation, shareholders agreed to divide 5,000 fully paid ordinary shares of £1 each in the capital of Brickworks Limited.[31] In the same month Dawes was given a proxy to represent Austral Brick at any meeting of Brickworks, no doubt in preparation for its eventual buyout by the latter in 1950. The Warren Brick Company Ltd on Mitchell Street, Alexandria, was another company that appears in company correspondence during this period. Purchased in 1940 by Austral Brick, the 'Warren' was skirted to the left and right by Austral's Bedford Brick Works, and the NSW Brick Company Ltd on Huntley Street. Having ceased to operate during the Depression, the site was badly neglected, although Dawes retained the pit for its clay reserves and land value. In January 1946 Alexandria Council drew the company's attention to the flooded pit and its dilapidated fencing which posed a danger to local residents. Locals remember it for the huge blue eels that lived in the murky waters of the abandoned pit.

Dawes briefly considered, but rejected, an offer from the NSW Brick Company Ltd to buy the yard. Then on 28 June 1949 a portion of land was resumed by the New South Wales Housing Commission. The Warren still exists today in name only and from about 1950 seems to have played a minor administrative role in handling shares for Brickworks Limited. Another long running issue was settled in 1950 when the courts finalised compensation for the resumption of the Homebush Bay brickworks. The amount was fixed by the Land Valuation Court at £182,120, giving Brickworks a capital profit of £108,513 due to the assets having been consistently depreciated. All found, the company made a very handsome profit on its Homebush purchase since compensation had also been received for a very substantial portion of the plant compulsorily acquired by the Commonwealth during the war years.[32] In July 1950 Brickworks used £75,000 from its Homebush Bay Resumption Account to repay preference shares that had been offered to finance the original purchase.[33] With large cash reserves to its name the company was in good financial shape, the company's Capital Profits Reserve Account being flush with profits from the recent sale of the Excelsior and Flemington brickyards.

In 1949 there were changes to local government with the absorption of St Peters Council by neighbouring Marrickville Municipal Council. A similar spate of amalgamations was taking place across the city. The disappearance of St Peters Council was symbolic in that it was the end of an era that had witnessed the establishment and growth of Austral Brick, and

31. May 1946, Minute book of directors meetings, Brickworks Limited.

32. Fifty Year Retrospective, 1934–1984, Brickworks Limited.

33. The cancellation of preference share capital enabled Brickworks to issue 124,769 ordinary shares at par, 2s payable on application and the balance of 18s on 17 January 1951.

its final acquisition by Brickworks Limited. The year 1950 marked a new chapter when Brickworks acquired 100 per cent of the company in September, making it by far the biggest and most powerful single brickmaking entity. The company now owned a total of nine brickworks comprising the City yard, the Brookvale yard, the Burwood yard, the Eastwood yard, and the Thornleigh yard.[34] In addition, its recently acquired subsidiary operated the Austral yard, the Bedford Brick Works and the Ralford yard at St Peters. To this tally can be added the neglected site of the Warren Brick Co. Ltd in Alexandria. Meanwhile, Clay Industries Pty Ltd, which had been set up in 1935 to acquire brickworks on behalf of Brickworks Limited had served its purpose and was wound up. A first distribution of 25s 6d in cash per £1 share was made to shareholders on 19 April 1951. The share of Brickworks in this distribution amounted to £101,993 12s 6d.

Criminals Darcy Dugan and William Meares at the Supreme Court, Sydney, 14 February 1950. Fairfax Photos

For Austral the decade was about to end, literally, with a bang. In September 1949 two career criminals, Darcy Ezekiel Dugan and an accomplice, William Meares, escaped from Long Bay Gaol where they had been serving time for armed robbery. Recaptured a short time later, they again managed to escape, this time from a cell below Central Local Court, Sydney, where they were being held prior to sentencing. As the city endured heat wave conditions the two men embarked on a 62-day rampage during which they held up various banks and commercial premises, snatching the payroll from the City yard over at Euston Road. A week later it was the turn of the Austral yard. The men working in the pit had not the slightest inkling that anything was amiss except, that is, for Ernie Buckley who was on his way to the office to collect a box of order slips. Hearing a commotion within Buckley peered through the window and spied Dugan and Meares pointing .38mm revolvers at a number of people who were spreadeagled on the floor, including the company secretary, Len Rogers, the accountant, Fred Mercer, Bill Swan, Bill Reagan, Dennis Downie, and Bill Brooker who was Austral's sales chief. Buckley could make out Dugan arguing with Dawes who was still smarting from the previous week's hit. 'I'm not getting on the floor, because if I do I won't bloody well get up again', yelled Dawes to Dugan who stared back, telling him to sit in the chair and shut

34. All of these yards had been acquired during 1935 and 1936. See Chapter 11 for details.

W.K. Dawes relaxing at the home of his son, Arthur Dawes, 32 Archbald Avenue, Brighton-le-Sands in the early 1950s. Bill Wheatley Collection

his mouth. As Dawes took his seat, Dugan stuffed wads of cash into a bag. Whatever he stole, according to Bill and Naasson Wheatley (nephews of Dawes), there was a huge amount more in the strongroom.[35] Norm and Alan Binskin believe that Dugan may have been tipped off by one of the brick salesmen who disappeared shortly after. Len Rogers was told to buy weapons to defend the office staff, although the pair of .32mm revolvers he purchased lay in the drawer at Austral Brick for 20 years without anyone bothering to license them. They were never used.

MODERNITY

The exploits of Dugan and Meares were a somewhat unorthodox beginning to the 1950s and came at a time when Sydney's brick industry, and especially Brickworks Limited, was about to experience a technological revolution that went far beyond the use of bulldozers in the shale pits. The long march of dry press would continue for several decades to come, but it was about to be overtaken by the widespread adoption of tunnel kilns. Meanwhile, one of the lingering effects of the post-war brick shortage was the switch to alternative building materials. Fibro cement sheeting became even more popular due to its cheapness and ease of installation, especially among the growing army of do-it-yourself owner-builders. Indeed, Sydney became the fibro capital of Australia. Of the 58,000 housing permits granted between 1948 and 1953, half were for fibro dwellings and nearly half for brick dwellings, and just 3 per cent for weatherboard. Spiralling building costs meant that even middle class households chose fibro rather than brick, especially if it meant being able to afford an extra bedroom. Fibro's attempt to usurp its main rival was short lived with the growing affluence of the late 50s and 60s and the preference for double brick dwellings. However, a longer term trend was beginning to emerge. At the 1947 census 80 per cent of Sydney's private homes had brick walls, whereas by 1971 the figure was 66 per cent.[36] Apart from fibro, the Australian brick veneer home was another response to the shortage of bricks and rising building costs. Its outer wall was of single brick construction tied to a timber studded frame, usually hardwood, pine or imported Oregon. The lining on the inside of the house consisted of plaster and other types of wallboards. Roofs were constructed of either terra cotta or cement tiles, and in later years, colour-bonded steel and aluminium roofing planks. Despite the interest shown in brick veneer comparatively few houses were built using this technique prior to 1950. An early pioneer in the field of mass housing and brick veneer homes in particular was the A. V. Jennings Group of Companies, located at Springvale, Melbourne.

The decision to build in fibro or brick was influenced by price and the availability of building materials, yet the decision to own a home of one's own might be regarded as an act of patriotism. Levels of home ownership leaped dramatically, although by 1954, the percentage of brick homes built in Australia had

35. Interview with Bill and Naasson Wheatley, March 2005, Sydney.

36. P. Spearritt, op. cit., p. 98.

Private dwelling – material of outer walls, Sydney and Melbourne, 1911–1947

Material of outer wall	1911		1921		1933		1947	
	Sydney	Melbourne	Sydney	Melbourne	Sydney	Melbourne	Sydney	Melbourne
Stone	4,553	1,730	5,229	2,078	4,651	1,498	4,655	1,313
Brick	87,768	55,683	130,342	71,767	215,505	95,719	250,902	122,591
Wood	24,648	56,219	34,467	80,563	51,890	127,025	48,166	133,817
Other	833	859	2,315	3,081	511	2,277	370	2,102
Unspecified	625	497	969	512	506	611	951,571	497
Concrete	–	–	–	–	1,580	1,496	2,519	2784
Iron	–	–	–	–	1,018	236	687	649
Fibro cement	–	–	–	–	4,302	602	16,985	2,462
Total	118,427	114,988	173,322	158,001	279,963	229,464	325,235	266,289

Source: Australian Bureau of Statistics, Census of the Commonwealth of Australia, 1911–1971

Nature of occupancy of private dwellings in Sydney, 1947

Total number of dwellings – 325,235

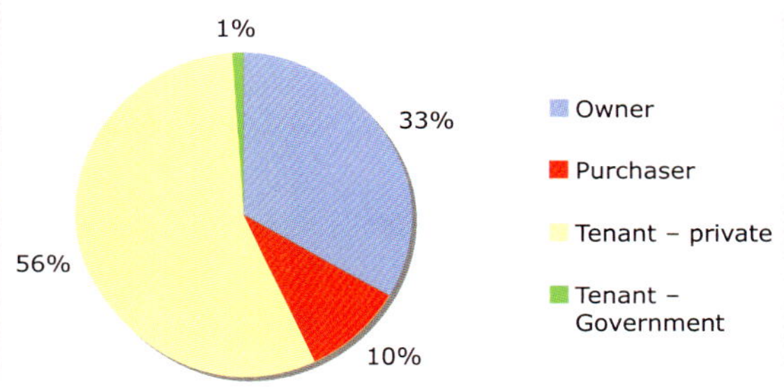

Nature of occupancy of private dwellings in Melbourne, 1947

Total number of dwellings – 266,289

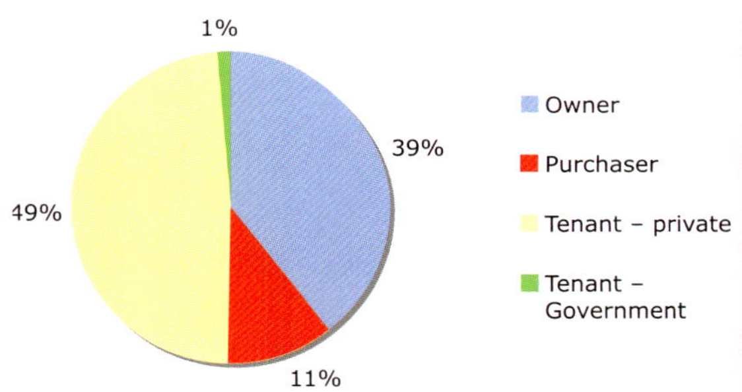

Source: Census of Commonwealth of Australia 1911–1971.

dropped to an all-time low of 25 per cent of homes of all types of construction. In June 1954 there were 2,343,421 occupied private dwellings and of these 1,121,814 were owner-occupied. The increase in such dwellings was 238,789, or more than 33 per cent since 1947.[37] These figures were achieved despite a minor recession and downturn in the building trade during 1952. The Burwood yard of Brickworks Limited was closed temporarily in June of that year but was reopened 18 months later when the economy rebounded.

Home ownership was also touted as an antidote to creeping socialism. *The Clay Products Journal of Australia* was delighted to report:

> *If it be true that one of the best safeguards against development of Communism is the possession of a stake in one's country, Australia should be far removed from the Red Menace so far as its internal growth is concerned.*[38]

Throughout 1954 the Sydney sales boom continued with heavy demand for building materials and a steep rise in the level of commercial and industrial building activity.[39] The heavy demand for brick and tile shows up in the consolidated statements of assets and liabilities for Brickworks, which reveal healthy dividends to shareholders and a substantial amount remaining in the form of surplus profits. The return to profitability was a welcome turnaround following

37. *The Clay Products Journal of Australia*, February 1956, p. 29.
38. Ibid, p. 29.
39. *The Clay Products Journal of Australia*, January 1954, p. 35.

the Depression and the war years, however, the long term problem of raw material supply was beginning to make itself felt. Sydney's inner city brickyards at St Peters were running out of accessible clay and shale deposits. To make up the shortfall, dirt was trucked from the pits at Ashfield where bricks were no longer produced, although this could only be a stopgap measure.

Australian production of clay bricks, 1953–1958

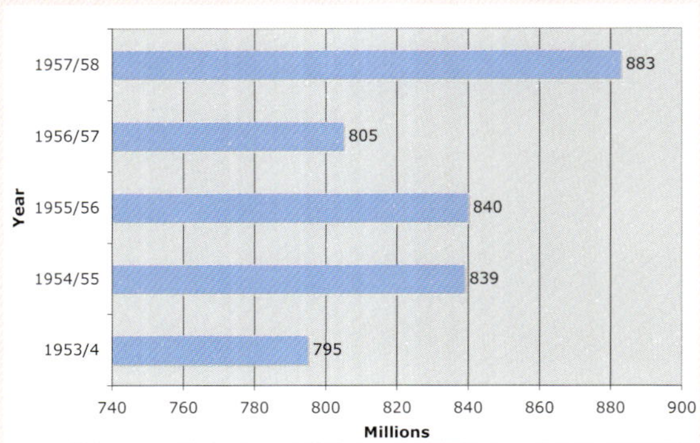

Source: Survey of Manufacturing Activity in Australia, Department of Trade, Commonwealth of Australia, cited in The Clay Products Journal of Australia, August 1959, p. 21

AHEAD BY A MILE

It was steady-as-she goes as the nation approached the mid-decade mark. Prime Minister Robert Menzies, anxious to retain Britain's affection, hosted a visit in 1954 by that country's young monarch, Queen Elizabeth II and her husband, Prince Phillip. Despite never having set foot in Great Britain, many Australians regarded that faraway place as 'home', turning out in huge numbers to demonstrate their enthusiasm and loyalty. At one point the royal train passed close to Brickworks' National yard at Thornleigh, which was served by a spur running off the main Sydney to Newcastle train line. The yard's burners and draggers, who had stripped down to loin cloths in the sweltering heat of summer, walked the short distance along the track to witness the passage of the royal party. Wal Hadley, the plant manager who had been earlier sent by Bill Dawes to pull the Thornleigh plant into shape, was enraged at the larrikin display but there was little he could do.

Corner, King Street and Market Streets, Vic Solomon Collection

Ahead by a mile, Brickworks Limited in 1955 was a cashed up and smooth-running enterprise. To its few private investors the company was the fairy-tale magic pudding, offering seemingly boundless riches. Brickworks and Dawes fashioned their own version of the pudding from the Great Australian dream of home ownership, rising prosperity, and the tidal wave of migrants. You only had to turn the dish, whistle twice, and it changed to whatever you wanted. Things could only get better.

Above: State visit to Australia of Her Majesty Queen Elizabeth II, February 1954. National Library of Australia

Below: Trams loading passengers after a race meeting at Randwick Racecourse circa 1950. Even by this time many thousands of people were being transported to and from the races by tram. Vic Solomons Collection

Aerial survey of St Peters, circa 1943, showing local brickyards. Department of Main Roads

The art and craft of dry press brickmaking 14

From pit to loft – Pressing the bricks – Work and play – Wheelers – Setters – Water smoking – Burners – Draggers – Firing bricks in a downdraught kiln – The right colour

FROM PIT TO LOFT

Brickmaking in the decade between 1945 and 1955 stood on the eve of radical and far-reaching changes, which eventually brought the long reign of machine-stamped bricks to an end. By the early 1960s new extrusion techniques and tunnel kilns were transforming the industry. Dry press bricks burned in patent and downdraught kilns, however, were still the dominant form and their manufacture deserves further consideration before we leave this period behind.

In a handful of Sydney brick pits, the excavation and carting of clay and shale was still heavily reliant on manual labour and horse power as late as 1959. Most pits, however, had already introduced bulldozers and bucket grabs in the decade following the end of the Second World War. On the other hand, mechanisation in the machine shops – cavernous sheds with skillion roofs made from corrugated iron – had been in place for over 70 years. It was here that dirt was ground and bricks pressed. After being winched to the top of the quarry the rail skip containing clay and shale entered the ground floor level where the load was tipped into a hopper set above a grinding pan. One such grinder, the Foster underdrive pan, was favoured by the industry, although other types were in use. The jumbled lumps and debris were then gravity fed into the top section of the pan where a small quantity of water was introduced to ensure proper adhesion of the dirt during pressing. Two giant slave wheels crushed the material which passed through mesh grates into a box chute on the underside. The useable material entered the loft by a vertically placed bucket elevator and was dumped on the floor next to a steel chute, directly above the press machine. The loftman's task was to ensure a consistent mix of clay and shale, although more often than not he would simply let the dirt continue its journey. Before the introduction of crushers in the pit and agitating auto feeders on the hopper chute, shale and ironstone would often become snagged in doors and grinding wheels. Ironstone in particular was a great nuisance. If a blockage occurred the operator stopped the machine by pulling a control lever which shifted the drive belt onto a static pulley. He would then climb in and clear the obstruction using a sledge hammer and crowbar. This was extremely dangerous. In the case of Eddie O'Sullivan at the Eastwood yard it had lethal consequences. O'Sullivan was crushed to death when the control lever vibrated to the 'on' position.[1]

PRESSING THE BRICKS

At the bottom of the drop the shale mix entered a brick-shaped die box lined with plates of case hardened steel. Dry press machines manufactured by George Foster were generally fitted with two boxes, whereas those made by the English firm of Platt Brothers before the First World War had four. The firm of William Wallbank produced a single box dry press machine at its factory in Auburn, Sydney.[2] Liner plates were manufactured by the firm of BHP Pty Ltd and engineered to withstand extreme pressures. They were good for about 250,000 bricks but eventually buckled. Some manufacturers tried to economise by fabricating cast iron liners, but these were noticeably inferior.

Weighing several tonnes, Platt and Foster machines were an impressive sight and stood more than three metres high. Each could stamp 2,800 bricks an hour up to a daily maximum of about 20,000

1. Interview with Briar Binskin, 20 March 2007, Sydney. Many pits began to encounter ironstone by the end of their working lives. Ironstone was found at deeper levels in the shale beds, which spelt the end for many yards since it was simply too hard a substance to crush.
2. William Wallbank established his engineering firm of William Wallbank & Sons Pty Ltd in 1934. In 1978 he purchased the plant and equipment used by G. Foster & Sons in the manufacture of brick-making equipment. Wallbank became the leading producer of equipment for mechanising the production of bricks and tiles.

Machine operators at the Excelsior yard, Croydon, New South Wales, circa 1911. Note the hollow depressions (frogs) in the face of each brick stacked on the barrow. With mass production, the distinctive brick markings favoured by manufacturers gradually became a thing of the past. The yard foreman with fob watch can be seen standing at the right of the picture

units per day. The first downward strike of the top head expelled the air out of the clay (known as de-airing), causing the head to bounce. It would strike once more in order to complete the cycle. As the green bricks were ejected from the mould the machine operators scattered a handful of sand which prevented the bricks from fusing together during the firing process as well as improving their shape. An excess of silica within the clay mix itself produced soft bricks referred to as 'rubbers'.[3] Edging bricks and the unusual shapes required by builders and architects took longer to produce and greater care was taken during the press. For these smaller runs different shaped liners were used, most yards having an old machine that could press squints, bullnose, ovolos and many other interesting shapes. Properly serviced, a dry press machine had a lifetime of between 70–80 years. Most were withdrawn from production well before they had reached this age and their parts used to repair those still in working order.

3. D. Yeomans, *Construction Since 1900: Materials*, Batsford, London, 1991, pp. 37–38.

Brick sizes

Differences in clays and shales and the effect of the burning process, tended to produce variations in the size of bricks.

Yard	Brick type	Size manufactured
Austral	Common	$9\frac{1}{8} \times 4\frac{1}{2}$
Bedford	Old size (circa 1960) face	$9\frac{1}{4} \times 4\frac{9}{16}$
	Common	$9\frac{3}{16} \times 4\frac{9}{16}$
	Foster machine face brick	$9\frac{5}{16} \times 4\frac{9}{16}$
Ralford	Common	$9\frac{3}{16} \times 4\frac{1}{2}$
Brookvale	Clayton machine until 22 February 1965	
	Front	$9\frac{19}{32} \times 4\frac{9}{16}$
	Back	$9\frac{17}{32} \times 4\frac{1}{2}$
	Clayton machine after 22 February 1965	
	Front	$9\frac{5}{8} \times 4\frac{9}{16}$
	Back	$9\frac{1}{2} \times 4\frac{7}{16}$
Burwood	Common	$9\frac{1}{8} \times 4\frac{1}{2}$
City	Common	$9\frac{1}{2} \times 4\frac{9}{16}$
Eastwood	Common	$9\frac{1}{4} \times 4\frac{1}{2}$
Punchbowl	Common	$9\frac{1}{4} \times 4\frac{1}{2}$
	Chromatex face	$9\frac{9}{16} \times 4\frac{5}{8}$
Sutherland	Common	$9\frac{1}{4} \times 4\frac{15}{32}$
	Face	$9\frac{9}{16} \times 4\frac{1}{2}$

Source: Jim Liepins, retired foreman and leading hand at the City yard engineering workshop of Brickworks Limited, February 2007

Main drive at Punchbowl Brick and Pipe Co. Pty Ltd, (Punchbowl yard), showing auxiliary diesel power unit and leather drive belts manufactured by the Mangrovite Belting Company Limited of Lane Cove, Sydney

DAILY GRIND

The line shaft driving the press machines was started at seven o' clock in the morning. At 9.30am the men stopped for 'smoko', a break of about 15 minutes during which they downed a mug of tea, ate a snack and checked the machinery. Oil bottles on the top, side and rear of the machines were refilled, the bottom die bolts checked and excess dirt from the charger board removed. Lunch was at midday. Serving each press was a machine driver, his offsider and as many men (wheelers or barrowmen) needed to deliver bricks to the kilns. Usually, the offsider would be expected to wheel a barrow during busy times.

The deafening thud of half a dozen brick presses, and the rumble of grinding pans added to the general brickyard cacophony – a monotonous clank of belt-driven machinery, shouting, laughter and the factory whistle. At the Eastwood yard the brick presses could be heard at the railway station, a kilometre away as the crow flies.

WORK AND PLAY

Regardless of technological advances, brickmaking was hard work and sorted the workers from the shirkers. If a man, or boy, didn't pull his weight then they were unlikely to survive in a highly integrated work environment where production and bonus piece rates relied on effective teamwork. Layabouts soon got the message and changed their ways, or were told not to come back. Arthur Sherman was just 15 years of age when he started mixing clay in the loft at the Ashbury brickyard in the suburb of Canterbury. Sherman travelled the distance from his St Peters home either by pushbike or pony. Recalls Sherman 'you could go to the brickyards and get a man's job with wages to match provided you could do the work. If it rained the plant stopped which meant you didn't get paid. It was hard on young blokes with families.'[4] He even did a brief stint at The Central Brick and Tile Co. Pty Ltd on Albert Street, although disaster befell the plant in 1951 when it burnt to the ground.

4. Interview with Arthur Sherman, 20 January 2005, Sydney.

Brickyard views at St Peters, 1946, State Library of New South Wales.

Top: Patent kilns 1 & 2 at the main Austral yard showing overhead walkway connecting the two kilns. This was used by the burners and their offsiders to wheel slack coal into the roof space above the kilns. Coal was then fed vertically onto the bricks below, feeding the advancing fire

Above: Main yard of The Austral Brick Company Pty Ltd showing unloading of the two downdraught kilns and the main Patent kiln. The horse and dray are backed up to the downdraught kiln which is being emptied of ash. The wicket has been part removed on the other kiln to allow the bricks to cool prior to dragging

Top: View of the Bedford Brick Works from Mitchell Road, circa 1938. The Steelo factory, the manufacturer of a steel wool abrasive can be seen in the right foreground. Bill Wheatley

Above: Looking down into the Austral pit which was shared by The Central Brick & Tile Co. Pty Ltd whose four chimneystacks and kilns can be seen at left. Bottom right: barely discernible is the image of a man pushing a loaded shale truck towards the chinaman at the end of the rail line

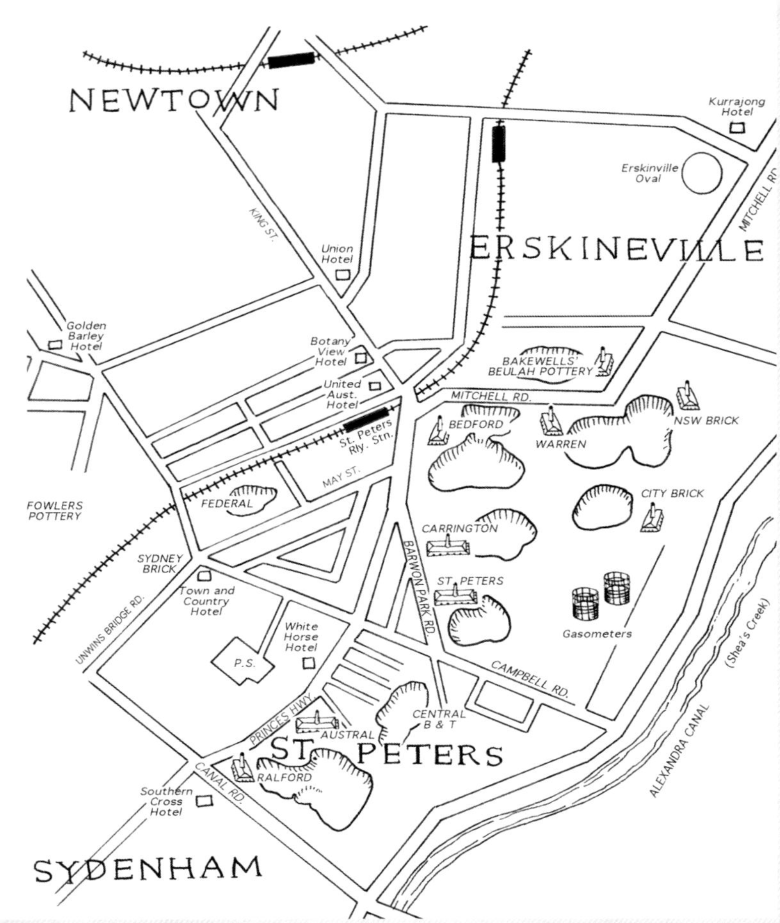

Brickyards and hotels in the district of St Peters, circa 1945

> *During the early hours of Saturday, 28 April 1951, a disastrous fire destroyed the works of the Central Brick and Tile Co. Pty. Ltd, St Peters, Sydney. Six fire brigades fought the blaze for many hours, but the whole of the sheds, tile machinery, brick machinery and drying racks of the patent kiln, were destroyed. About 12 ½ per cent of the Sydney metropolitan output of terra cotta tiles was lost due to the fire.'* [5]

Alas, the fire signalled the end of a career that stretched back to 1913 when it was established at Albert Street, St Peters. In its heyday the 'Central' produced the highest quality common face and moulded bricks as well as roofing tiles. In 1936 it even boasted a showroom at the Metropolitan Building on the corner of Hunter and Bligh Streets, Sydney. A founding investor in Brickworks, the 'Central' eventually became a private company.[6] Production appears to have ceased by about 1965. In 1954 Sherman moved to Austral's St Peters yard where piecework paid better money, and at the age of 29 was promoted to yard manager. Like many other figures in the brick industry, Sherman knew his trade and gave no quarter to men who weren't cut out for the job. A brickyard meant hard 'yakka'; only the fittest survived his scrutiny.

There were, of course, lighter moments to be enjoyed in that great social leveller, the pub. Workers from the Austral yard frequented *The White Horse,* whereas gentlemen from the Bedford quenched their thirst at either *The United Australia* on the corner of Lord and King Street, *The Botany View* on Darley Street, St Peters, or *The Union*. The beer was cold and valuable drinking time proscribed in the days of the six o' clock swill. Those who rented one of the many houses owned by Austral Brick in Albert Street and Victoria Street, adjacent to the Central Brick & Tile, could walk or stumble the short distance home after a blinder. Some returned bruised by the bar-room drubbing sustained during a fistfight with workmates.

At a time when football was still an amateur sport, many of the local players found employment at the Austral's yards which offered plenty of opportunity for exercise and exertion. Reginald 'Snowy' Dowton, one of Australia's top wrestlers during the 1950s, was employed at various yards owned by Brickworks Limited. Dowton would thrill crowds at the Leichhardt stadium, although his day job was less glamorous. This included work at the Austral yard, the National yard at Thornleigh (as a foreman during the early 1970s), the Eastwood yard (as a burner) and a spell as plant manager at the City yard with Tom Smith.[7] Between 1957 and 1960 Dawes employed Dowton, ('Bill' to his workmates), as personal chauffeur for his brand new Silver Hawk Studebaker, the latest in a long line of cars. Off to Tattersalls Club in the city for some liquid refreshment, Dawes was a generous employer, and slung huge tips to ever-helpful bar staff and the doorman. Or down the road to meet with like-minded punters at the Kent Street Pigeon Club. It was only a short walk between rest stops, but why walk when you have a car and driver?

5. The *Clay Products Journal Of Australia*, May 1951, p. 3.
6. The yard was part of Brickworks Limited but was no longer operating, probably due to the fire. Source: J.S. Hoskings, Brickworks in the Sydney region, including the County of Cumberland, Australian Structural Clay Products Works, Division of Building Research, CSIRO, Melbourne, April 1955.
7. Reginald Dowton was trainer to the wrestling team that contested the XII British Empire and Commonwealth Games in Perth in late November 1962.

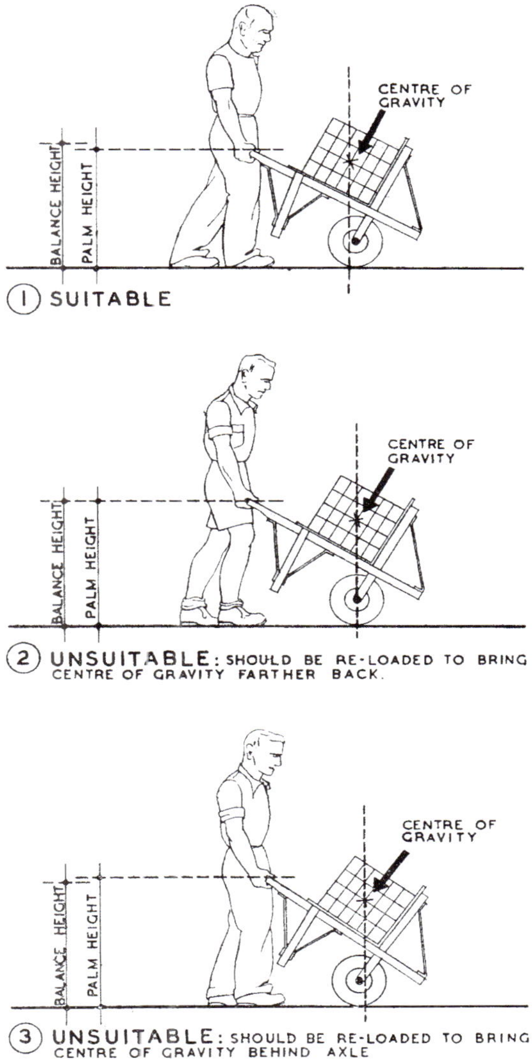

Above: Platt and Company dry press brick machine, circa 1953. The machine operator can be seen stacking green bricks on a barrow, ready for the wheeler to run the load to one of the kilns

Right: Wheelers claimed it was simply a matter of balance as this illustration shows. Department of Labour and National Service, Commonwealth of Australia, 1948

WHEELERS

About 70 metres from the machine shop stood the kilns, which were no great effort to reach unless one happened to be pushing a barrow laden with 52 green bricks weighing about 300 kilograms. Stacking dry press bricks onto a barrow as they came off the press might appear to be a simple task, but if they were not correctly positioned and the boys properly taught, a considerable number could be damaged. Bricks with 'one good side and one good end' were considered to be saleable. Until the introduction of forklift trucks at the end of the 1950s, wheelers were assured of a job.[8] Most barrows were equipped with a steel wheel that would run on a metal plate about the width of a railway sleeper, to prevent the load from becoming bogged. By the early 1960s pneumatic tyres had replaced the steel wheel, although for those men who emptied the kilns of fired bricks (draggers) steel wheels continued to be used due to the presence of red hot embers on the floor of each chamber.

Working in pairs to the speed of the press, the barrowman would load up and set off at a running pace, leaving the full barrow outside the wicket. He then collected an empty barrow and returned to the press, repeating the process until the day's tally had been reached. Men employed on piece rates, whatever their job description, worked on a 'do and go' basis. Once the tally had been met they were free to go home.

8. In 1949 forklifts were still to make their mark, besides which they could not be driven directly into a Hoffman-type kiln with its series of low, narrow openings (wickets) around the perimeter.

Brickmaking operations at the Punchbowl and Sutherland yards of the Punchbowl Brick and Pipe Co. Pty Ltd, circa 1955

SETTERS

Setters were men who worked in the kilns, arranging the bricks in stacks (benches) ready for baking. Setting bricks was a skill usually taught by an experienced 'leading' hand and, like wheeling, was performed manually until the introduction of forklift trucks in the late 1950s. An efficient team was a well-paid team which aimed to complete the tally early. Like any other job in the yard, if a newcomer was not up to the task they were told not to come back.

A patent was usually divided into 22 or 24 chambers, each holding about 22,000 common bricks.[9] Before a chamber could be set the sand floor was raked level and hot embers removed by the man responsible for feeding the kiln fire, the coal wheeler. Four stacks of bricks occupied each chamber and comprised a leader with two wings at either side. Occasionally, a six-bench setup was used in the yards owned by Brickworks Limited. This innovation is said to have been introduced by William King Dawes who wished to speed up production during the postwar building boom. A full complement of six setters would only be used on two days each week; five setters remained the norm on the other days. The lead bench was aligned by the kiln boss who would drop a plumb bob from the overhead fire hole, which acted as a point of reference for the outside benches. A measuring rod was used to mark off a narrow gap between each brick and bench, as well as between the outside bench and the sloping walls of the kiln. The gap allowed a draught of hot air and flame to circulate during burning. As each bench took shape the men would stand on their wheelbarrows in order to place bricks in the last few rows directly beneath the roof of the chamber.

Manpower required to run the Eastwood yard of Brickworks Limited, 1950–1960

Management	1 foreman 3 office workers	4
No. 1 Patent Kiln	3 draggers 5 setters 4 burners 1 coal wheeler	13
No. 2 Patent Kiln	3 draggers 6 setters 4 burners 1 coal wheeler	14
Face kilns	5 classers 5 setters 4 burners	14
Machine shed	1 winchman/panman 2 panmen 2 loftmen	5
Machine shed	5 machine drivers 11 offsider/wheelers (depending on distance)	16
In the yard	1 sandman/ash cleaner 1 general yard hand (putting up dome wickets, cleaning kiln flues, etc)	2
In the pit	1 leading hand shooter 1 truck driver 2 breakers 1 navvy driver	5
Maintenance	3 fitters 1 greaser	4
	TOTAL	**77**

9. The maximum number that could be set by hand each day was about 30,000 bricks.

Right: Hand setting face bricks in a downdraught kiln at the Punchbowl Brick and Pipe Co. Pty Ltd, Punchbowl, circa 1955. Brickworks Limited

Above: Team of setters at the Eastwood yard of The Austral Brick Co. Pty Ltd, circa 1983

Finally, strong brown paper was glued across the face of the bricks, extending to the walls. The seal was temporary and until the fire progressed it ensured that hot air within the kiln was 'pulled forward' towards the green bricks. Sometimes the burners, who fed the fire by dropping small lumps of coal down fireholes made an arrangement with the setters. The latter would put up the papers, while in return the burners would set their last barrow at the day's end. With the chambers sealed, the external wickets were blocked by two widths of burnt bricks plastered with mud, which by all accounts was an extremely labour-intensive practice.

WATER SMOKING

Before the full heat of fire could be raised, moisture was removed through a gradual process of evaporation known as steaming or water smoking. This stage lasted about two or three days and was assisted by the circulation of hot air throughout the kiln. Steaming caused the entire mass of bricks to shrink slightly, the amount being gauged by a metal measuring rod poked down one of the firing holes. An experienced burner would be able to determine the amount of shrinkage (about six centimetres), simply by taking a quick look down the fire hole. Generally, it was a case of the greater the shrinkage, the darker the finished product. If moisture was not fully evaporated then the atmosphere in the kiln became saturated, which robbed the bricks of their load-bearing qualities. During the steaming process temperatures in the kiln were allowed to reach between 650°C to 850°C, at which point carbon present in the clay and shale was burnt off.

THE AUSTRAL YARD AT EASTWOOD, SYDNEY

The diagrams on the following pages shows the workings of the main patent kiln at the Eastwood shortly before the yard was closed and production transferred in 2003 to Bowral in the Southern Highlands. A team of setters using forklifts could place 70,000 green bricks per day, or 350,000 per week. On Saturday morning an additional 20,000 were set, making a weekly total of 370,000 bricks. Each chamber within the patent kiln had four rows of fire holes, a standard feature of this type of kiln. From Monday to Friday burners 'moved' the fire

Expanded view of a patent kiln at the Eastwood yard (circa 1955), showing setting and dragging configuration prior to the introduction of forklift trucks

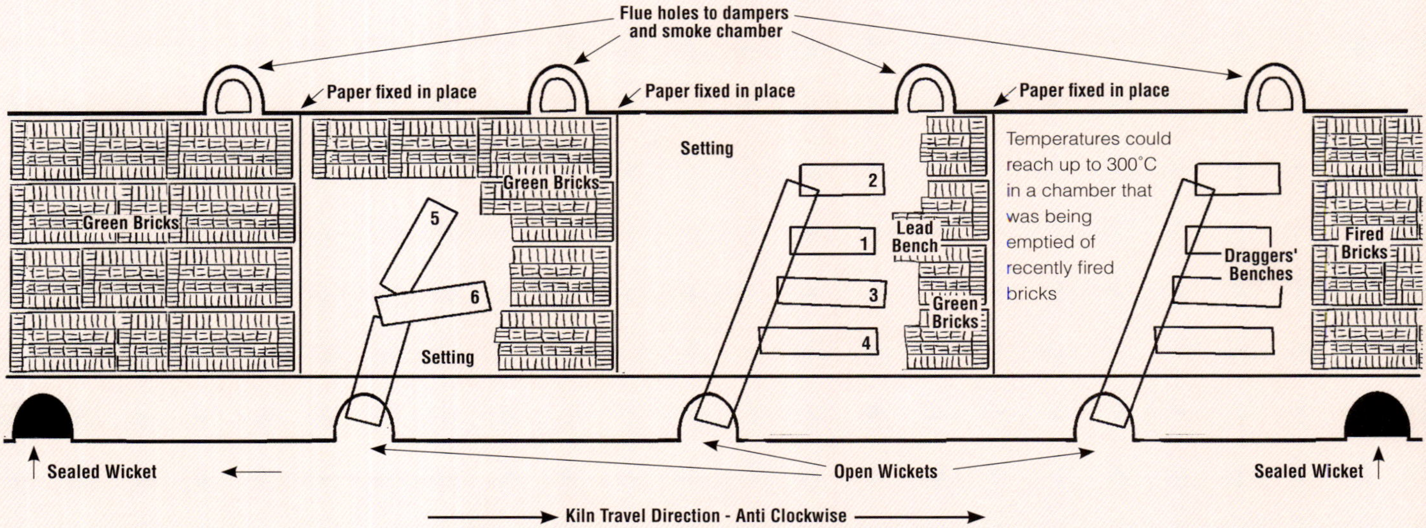

Setters and draggers worked in close proximity to one another. Steel plates were laid on the kiln floor, permitting easier access for the setter's barrow (rubber tyre) and the dragger's barrow (steel wheel). Note the forward position of setter's bench No. 1 – the leading bench.

forward at the rate of nine rows per day, making a total 16 chambers. When possible they took an extra row forward. Over the weekend the forward moves were adjusted to keep the fire in the best position, burning off the papers in a set pattern in order to ensure there would be one holding paper left on Monday morning. Patent kilns at all yards owned by Austral Brick and Brickworks Limited were fired in an anti-clockwise direction, whereas at the State Brick Works at Homebush, kilns were fired in a clockwise direction.

Controlling the rate of burn ensured the kiln was ready for the re-commencement of production on Monday. Burners tried to maintain the best outcome for the fire and the burning of the ware, which meant that temperatures inside the kiln could still be as high as 300°C during setting and dragging. Working ahead of the setters, forklift drivers cleared the kiln at the rate of just over three chambers each day. On Saturday just over one chamber was cleared.

Water-smoking

This term was originally used to describe the early stages of firing on account of a combination of white smoke and steam which issued from the chimney. As much as 100 kilograms of moisture per tonne of bricks was driven off during smoking. Care was taken to ensure that the surface of the goods was not spoiled by allowing steam to form under pressure which caused splitting.

BURNERS

Firing bricks in a patent kiln was a specialised procedure conducted by highly experienced kiln operators called 'burners'. Among other tasks, they carefully controlled the speed of the burn to avoid distorting the shape the bricks. Working above the kiln in the space created by a gable roof made of corrugated iron, the burner went about his business in suffocating heat. When the desired shrinkage had been achieved he would block the fire holes feeding the bottom of the chamber and continue to feed the top fire. Assisting him was a coal wheeler who dumped barrowloads of pulverised coal ('slack') around the small cast iron pots over each zone. To reduce the number of round trips, the wheeler might fit 'hungry boards' to his barrow. With a hook in one hand to raise the pot covering the top of each feed hole, and a scoop in the other, the burner poured the slack down between the bricks. Regular deliveries were made to the coal stage or bridge which stood about five metres above the ground on the outside of the kiln.

Settling on top of each bench, the coal particles rapidly increased the temperature causing the bricks to glow red hot when heat was attained. About one metre above floor level temperatures reached 1,025°C, whereas the maximum on the floor itself

Typical scenario each Wednesday morning.
Patent kiln at the Eastwood yard of Brickworks Limited, 2003

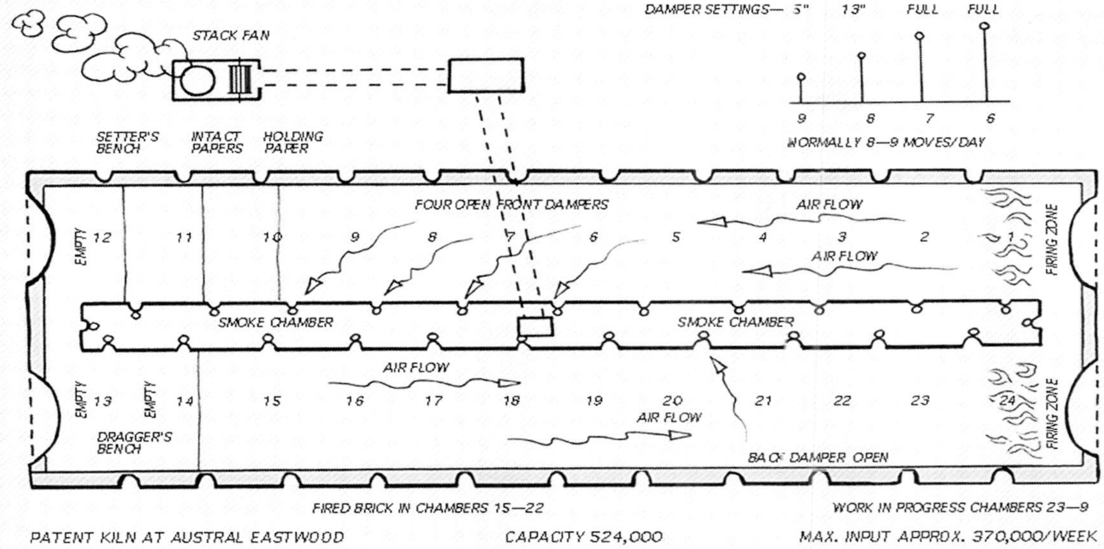

Until the late 1950s when oil burners were installed, the patent kiln at the Eastwood yard was fired using coal. Fork lifts were introduced in 1968 at the time the kiln was enlarged, and in 1978 the oil burners were converted to gas. These three innovations combined to increase the rate of throughput.

was usually no more than 1,080°C. Controlling the rate of fire were five dampers (metal trapdoors attached to long steel rods), which were raised in advance of the fire to create a positive draught. Using the spent hot air at the start of the firing cycle made the patent kiln highly economical to run. Setting, firing and dragging were a 24 hours-a-day operation and continued year-round, apart from a break over the Christmas period when repairs were made to the chambers.

Towards the end of the firing process the dampers on the tunnels leading to the stacks were closed depriving the kiln of draft. A successful 'reduction' created immense quantities of thick black smoke that belched from the chimney and also found its way through cracks in the kiln. Jets of flame would also spout from weak spots in the walls during this stage. Exhaust gases and smoke from each kiln were carried away by underground flue to a nearby chimney, which could reach as high as 40 metres. They were familiar landmarks across Sydney. The last surviving stacks can still be seen at the former Bedford Brick Works in Sydney Park, St Peters. They may now be considered heritage items, but in bygone days locals had good cause to complain about discharge from chimneys and the haze of pollution created by brickmaking activities. On windless days it must have been unbearable, especially for thousands of housewives rushing to bring in the washing before it was smothered in grit and soot.

DRAGGERS

Dragging – removing the burnt brick – was the final stage in the brickmaking process and the least sought after job. Special clothing was not provided, but most draggers entered the kiln wearing an old woollen overcoat which could shield them from the intense heat. A flap of leather worn on each hand, and referred to as a 'cot', offered some protection against burns from the hot bricks.[10] As the bricks nearest the wickets were cleared, a space was cleared in which temperatures hovered around 300°C. Salt tablets and water were always close to hand.

Each load was then wheeled into the yard and stacked ready for collection by the brick carters.[11] Dawes expected the three draggers and six setters for each patent kiln to remove and/or set a weekly tally of 150,000 bricks. Sherman recalls that:

> *working in the heat of a patent kiln in continuous operation was an ordeal, even for the old hands who*

10. Draggers made cots using cheap, poor quality leather. The term is derived from the English word, 'cot', meaning finger stall or sheath.
11. Carters were differentiated by the type of bricks they carried, eg face carters and commons carters. Carters of face bricks were paid more to carry their more valuable load

What makes a good brick?

Good brick is unlike concrete and stucco in needing no maintenance or surface treatment. It is improved by weathering and even looks the better for the passage of five hundred years. It is often more durable than natural stone. Having been hardened by intense heat, bricks are very good at resisting fire. Thus the metal in steel-framed buildings may be buried for safety in brickwork.

A thoroughly good brick should be regular in shape, texture, and colour, equally and perfectly fired throughout, and should be free from all cracks and flaws – including hair-cracks – with sharp arrises, and should give out a clear ringing sound when struck either with a stone, another brick, or a piece of metal. For many purposes, however, it is unnecessary to insist upon all these qualities. Any hard and well fired brick will suffice for foundations and internal work which is to be subsequently covered, and for such purposes rougher and cheaper bricks are frequently the more useful, affording a better key for plastering than those with a smooth surface, and often being better weight carriers than soft, well-finished facing bricks.

Sandy and absorbent bricks should not be used in foundations, nor in external walls likely to be exposed to water or driving rain. Such bricks are generally soft and do not weather well, being frequently underfired; and by retaining moisture they encourage the growth of lichen and climbing plants, which all gather and retain damp.

Soft, underfired bricks (chuffs) are valueless. No brickmaker with a reputation to lose will sell them, preferring to pass them through the kiln a second time, or to crush them for sand or grog. On the other hand, a remarkably non-absorbent brick, heavily pressed and highly fired, may have too smooth a face to adhere readily to mortar, especially in summer time, in spite of a good wetting.

Over-fired bricks will melt and run together forming 'burrs' which are useless except to be broken up for road metal or concrete. Faulty bricks are more often met with amongst those which are hand made, hack dried, and clamp-fired, than amongst those which are machine made, chamber dried, and kiln-fired.

Source: Modern Brickmaking, Alfred Searle, Scott, Greenwood and Sons, London 1920

wore long flannel trousers which offered protection from the heat. Setters and draggers consumed large quantities of water and took salt tablets to ward off muscle cramps. Draggers commenced at 6 am and worked through until 2 pm on the first three days of the week, ie Monday to Wednesday, when the kiln was still relatively cool. You would take more than the tally out for the day, because by Friday it would be too hot. That's why on Thursday and Friday we started at 6 am and finished three or four hours later. [12]

All up, it was filthy, hot and dangerous work and until showers were installed in the mid-1950s the men went home – or to the pub – covered in sweat and grime.

FIRING BRICKS IN A DOWNDRAUGHT KILN

All yards had a number of dome or downdraught kilns where smaller quantities of high quality face bricks – approximately 75,000 per kiln – could be fired. After being fully stacked, this open-ended type of kiln was sealed by burnt bricks and mud. Downdraughts had set brick floors with cross flues

12. Arthur Sherman, *op.cit.*

Roof space directly above the chambers of the Eastwood yard patent kiln showing the steel damper. This device would be vertically raised or lowered to close or open the flues, depending on the required rate of burn. Circa 1995. Brian Binskin

placed every 20 centimetres. These were connected to larger side flues under the bag wall which vented to a main flue and thence to the chimney stack. Built into each side of the kiln were fire holes, which allowed coal to be fed into a combustion chamber that directed the heat upwards from side fire boxes. Heat rose over the top and down through the set bricks into the flue, hence the expression 'downdraught'.

When loading such a kiln the setters would leave a two-centimetre gap between each bench of 3,000 bricks. Green bricks were also stacked to avoid touching others at the front and back. Every row had access to the flu, allowing hot gases to circulate freely. Every few months draggers removed the floor of the kiln, allowing the draggers to re-line it with a fresh batch of green bricks.Wheeling down the ramp into the bottom of the kiln required a degree of skill; at the end of the ramp was a narrow steel plate laid over the sand floor that ran up to the actual bench. As the chamber filled the ramp was removed and bricks were thrown down to the setters.[13]

Lumped coal was delivered by truck and dropped along the sides of the kiln. Burners shovelled fuel into the fire holes using a long-handled, square-mouthed shovel. Each fire hole took up to 12 full shovels of coal as the kiln reached a high temperature. Before temperature recording equipment was introduced a steel measuring rod was poked into the kiln at three different positions along the crown. Full fire was maintained until the required shrinkage was achieved, at which point the coal feed was terminated. Generally, the temperature could be raised in increments of 100°C by adding one heaped shovel load of slack. A dome kiln would burn about 60 tonnes of coal in the seven days it took to steam, fire and smoke reduce the bricks. Wear and tear, and the constant expansion and contraction of the brick walls took their toll. Regular maintenance was essential, and in the case of Brickworks Limited this was carried out by Ray Dawes, a qualified builder whose company employed a team of specialist bricklayers and carpenters.

13. When forklifts were introduced, these bottoms were laid in cement by bricklayers to bear the weight of the fork carrying its two tonne load of brick into the kiln.

John Farrington unloading the patent kiln at Eastwood, circa 1985. Brian Binskin

Regional differences

Melbourne brick manufacturers did not make dry press bricks due to the absence of local shales. Instead, wet-press bricks stamped on a Platt produced stiff plastic bricks with a much higher water content. While the drying process took longer, firing was much the same. Variations on the Foster dry press and Platt press included the Anderson, Lotus and Mitchell, all of them English made and over 100 years old. Anderson presses, originally from England, were made in Melbourne by W. Anderson & Sons for many years. The firm of J. C. Smale acquired Andersons in the early 1980s.

THE RIGHT COLOUR

In broad terms burnt clay bricks can be classified into four main groups, taking into account the presence of iron oxide, alumina and lime which are known to exert powerful, modifying effects.

Downdraught kiln at the former Bedford Brick Works on Princes Highway, St Peters, circa 2005. Ron Ringer

	Composition of brick clays	Shades of fired brick
GROUP 1	Relatively high iron oxide content of between 5 to 9 per cent, low alumina content of 10 to 20 per cent, small quantity of lime.	All shades of red up to vitrification – the higher the temperature, the darker the colour.[14]
GROUP 2	Lower iron oxide content of 4 or 5 per cent, small quantities of lime (CaO – about 6 per cent) in the form of chalk, low alumina content.	Red bricks are produced at full firing, however by reducing temperatures in the later stages of firing, bricks of straw yellow colour are produced. Others have a pink or red, or even a grey tinge.
GROUP 3	Moderate quantities of iron oxide (1.5 per cent to 3 per cent), but a high percentage of alumina (25 per cent).	Brown – buff bricks.
GROUP 4	Low iron oxide and high alumina content found in pure clays (kaolin).	Cream bricks are produced when clay is burned at about 1,050°C. White bricks are produced at higher temperatures. If kaolin contains a minute proportion of iron compounds, the burned clay has a pinkish or reddish tinge.

14. F.H. Clews, *Heavy Clay Technology*, British Ceramic Research Association, London, 1969, p.201.

Brick colour depends partly on the composition of the clay and shale mix and partly upon the conditions of the firing. The firing of common bricks in patent kilns was not usually carried beyond 1,100°C, and it was in the final stages that the process of oxidation occurred. Clays with a high iron content produced bricks containing a variety of red tints. Other factors influencing colour included the fineness of clay particles, the degree to which they were mixed with the clay, the composition of the clay and associated minerals and the temperature during firing.

In his authoritative study, Clews states that:

> *In almost every instance the colour may be attributed to the presence of iron compounds in the clay. Iron produces different colourations, depending first of all on its state of oxidation. In the course of firing, the iron minerals present in a natural clay – whether oxides, carbonates, sulphides or sulphates, or whether ferrous or ferric – are normally converted to ferric oxide Fe_2O_3 by the beginning of the vitrification period. This oxide generally gives rise to various shades of yellow and red.*[15]

15. F.H. Clews, *op. cit.*, p. 200.

Brick patterns and colours are often very striking, as this series of photographs reveals. *www.mayang.com/texture*

W.K. Dawes visited brick factories during his visit to the USA in 1938 when he met with the firm of J.C. Steele. These photographs show dry press bricks being surface-treated to achieve a textured effect. The machine is the US-made Boyd Toggle Press. The Austral Brick Company Pty Ltd and Brickworks Limited did not manufacture texture bricks until a modern plant was opened at Horsley Park, Sydney in 1961. Ron Hagstrom

On the other hand cream burning shale was quarantined from ordinary clay to avoid contamination. A mix of clays produced a brick with red spots. Contamination with ironstone caused black spots. The City yard of Brickworks Limited on Euston Road made the white or cream brick.

A WIDER RANGE OF BRICK COLOURS

Downdraught kilns, unlike patent kilns, could produce a great variety of brick colours, depending on how the kiln was fired. The June 1934 edition of *Building* carried an article regarding scientific colour control. The article stated that clays which normally burned to a red colour develop the clearest red when burned under oxidising conditions. If, however, a reducing atmosphere was introduced near the end of the firing operations, then a wide range of colours could be produced, including red, brown, chocolate, black and the lighter-coloured chromatex bricks. Mottled bricks and red bricks came from the centre of the kiln. Blue-black bricks were from the outside of the set, near the fire holes. Callows, underfired bricks from the bottom centre of the downdraught kiln were called 'DK callows' and far superior to the poor quality callows dragged from a patent kiln.

Additives such as lime, iron oxide or manganese dioxide were widely used during the grinding and mixing process to achieve coloured effects. Sodium chloride (common salt), zinc and manganese were also introduced into the fireboxes at or near the end of the firing operation. In the case of extruded bricks metallic colouring agents, glazes, enamels and other slips were applied by spray or roller. Even flecks of ground coloured glass were embedded in the surface of the green brick. Buff-burning clays, especially those with high lime contents, could produce yellows, browns and greens. In the case of dark blue-purple brick, traditionally used in engineering works, a reducing atmosphere towards the end of firing ensured the increased melt formation characteristic of ferrous oxide. The blue appearance was the result of manganese in the presence of iron, or iron oxide to the extent of more than 10 per cent, fired at a high temperature. Light blue bricks required a pure clay to which has been added phosphates of lime and alum.

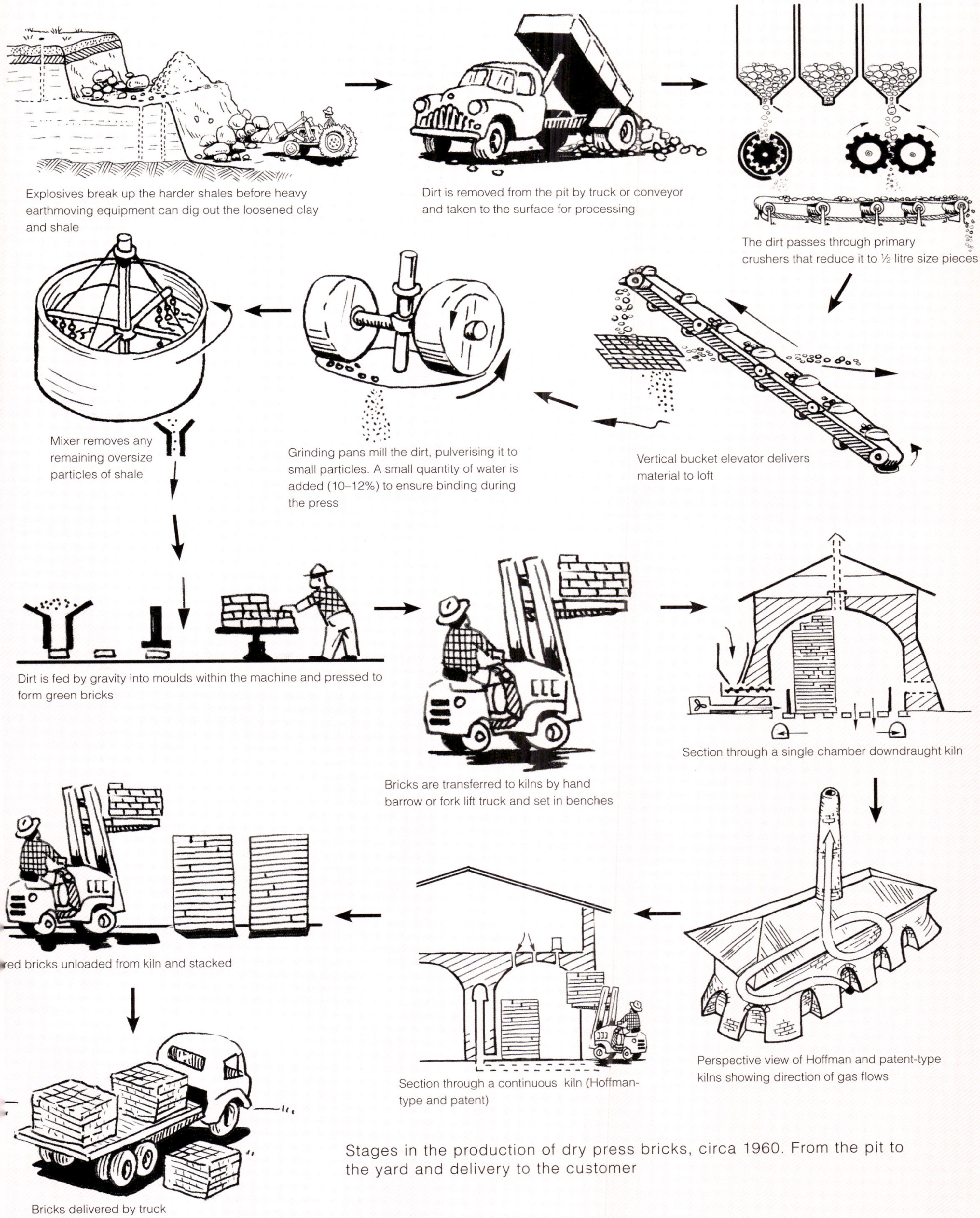

Stages in the production of dry press bricks, circa 1960. From the pit to the yard and delivery to the customer

The Brookvale yard of Brickworks Limited was closed in 1998. Oil on canvas by Jane Bennett

Great leap forward 15

Where to now? – A drawn out affair – Sydney City Council – Cooking with oil – Never had it so good – Money to spend – On the move – Overseas contacts resumed – Wallgrove Road – Diversification – More fact finding

WHERE TO NOW?

"Reports of my death are greatly exaggerated", exclaimed the 19th century American novelist, Mark Twain, upon hearing that his obituary had been published in the *New York Times*. Wishful thinking on the part of his critics, or another example of never letting the facts get in the way of a good story? Probably both.

In 1955 similar observations were being made about the life and times of dry press brick. The reader will recall that by the early 1860s new technology was beginning to put the squeeze, both figuratively and literally, on traditional brickmaking. In the first wave of industrial innovation, the more progressive, and cashed up, brickmakers were installing machines that ejected a column of clay and then sliced it into brick-sized slabs.[1] Mass production was here to stay. Extrusion methods were quick to establish themselves, particularly in Victoria due to the abundance of alluvial clay deposits which required very little processing. Two decades later, the introduction of dry press machinery was also making its mark, especially in New South Wales where shale was the dominant form of raw material. By 1900, hand moulded bricks were almost a thing of the past, although in country districts a few practitioners still clung to the tradition. Half a century later the dominance of dry press was being challenged by the latest trend – extruded texture brick, favoured for its contemporary look and feel.

Extrusion methods in the 1950s were remarkably similar to those that produced the plastic and semi-plastic wire cut bricks of the previous century. The main difference was in burning which took place in a rectangular-shaped tunnel kiln. From beginning to end, the process took less than a week. First, the cut, green bricks were delivered by conveyor belt to a static bench where they were hand stacked on a steel-framed chassis, a low platform which ran on rails. The cars were lined with fire-resistant refractory brick. When fully stacked the kiln car was parked in a drying chamber for about 48 hours during which time moisture was evaporated from the brick using hot exhaust gases vented from the tunnel kiln. Following this, the cars were transferred to the adjacent kiln where they passed through a series of zones which became progressively hotter. The hottest of these was the firing zone. After being thoroughly cooked the bricks entered another section of the kiln where the temperature was reduced. After three or four days the burnt bricks emerged from the opposite end of the tunnel and were allowed to cool before being stacked. Modern tunnel kilns are either side-fired or top-fired kilns, the latter producing a greater degree of temperature uniformity throughout the kiln.

Patents had been taken out on rudimentary tunnel kilns as early as the 1880s, although the technology took years to perfect and was better suited to the production of ceramic ware. The depression years of the 1930s and the Second World War delayed its introduction to Australia, although some European brickmakers, especially those in Germany and Italy, were known to be using tunnel kilns in the 1920s. This was due to the size of the market for bricks and the widespread availability of soft clays.

1. The action of the wires mounted on a strong frame was almost identical to that of a wire cheese cutter.

Dressler muffle tunnel kiln, 1915. The English firm of Dressler Tunnel Kilns (established by Conrad Dressler) was an early manufacturer of tunnel kilns. The one pictured here was built for J.H. Barratt & Co. for use at their Boothen Tile Works in Stoke on Trent. The muffles can be seen on either side of kiln car. Fuel oil was injected into the muffles and burner. Exhaust gases were vented from the end of the kiln to protect the ceramic ware. Muffle kilns to a modern design were built in Melbourne during the 1960s. Bill Dennis Collection

A DRAWN OUT AFFAIR

As part of a push to modernise Australia's clay industry in the post war years, both the CSIRO and the Commonwealth Government urged brickmakers to instal tunnel kilns. The response was not entirely positive. In fact, a row broke out between traditionalists and modernists who became embroiled in the pros and cons of dry press versus extruded.[2] The ceramics industry had no such qualms, and being conscious of the need to manufacture a uniform product installed kilns that were both fuel efficient and better suited to smaller production runs.

In mid 1949, B. Winston Cookson representing Gibbons Bros Ltd of Dudley, Worcestershire paid a visit to Australia. The firm was one of Britain's leading kilns and furnace builders and the maker of carbonising and producer gas plants.[3] A meeting with Cookson confirmed what Dawes had known for some years, and what others close to him had been saying – that tunnel kilns and the new extrusion methods were the way of the future. His son, Ray Dawes, a kiln builder and contractor, urged Dawes to build modern plants, pointing out the obvious limitations of the firm's Hoffman and patent kilns. Even mechanising those kilns by using forklift trucks to stack the bricks was out of the question, due to the extensive work (laying of concrete paths) needed to provide access.

For brickmakers up against the shortage of skilled labour and the need to remain competitive, the extrusion process and tunnel kiln firing held great appeal. What if bricks could be manufactured using a process that minimised input costs and maximised profits? The argument in favour of extruded bricks was difficult to fault. For a given level of production they required the lowest capital expenditure. Better still it was possible to obtain the higher compressive strengths necessary for high rise buildings and other structural brick constructions which began to make their appearance. Extrusion also made for bricks of a lighter weight due to perforations created during manufacture. Moreover, it was easier to produce different sizes of bricks and apply surface treatments prior to burning to obtain the required colours. The soft extrusion process, used almost exclusively in Europe, could be applied to a wider range of materials than the stiff extrusion process commonly used in the United States.

In 1955, neither Brickworks Limited nor Austral Brick made extruded bricks, whereas two other Sydney firms did, namely Clark Kilns and Lion Tile and Bricks at Enfield. These firms used patent kilns to fire their bricks, and being highly competitive were a great concern to Dawes, although eventually Lion Tiles ran out of clay and shales and was obliged to sell up.

2. *A Conservation Plan for the Former Goodlet and Smith Brickworks, Merrylands*, HLA-Envirosciences Pty Ltd, July 1995, p. 21. See also 'Brickmakers and Official in Animated Clash: tunnel versus Hoffman kiln', *The Clay Products Journal of Australia*, Vol 18 (1), p. 13. 1950.

3. *The Clay Products Journal of Australia*, July 1949, p. 27.

It would be another decade before Brickworks opened its first tunnel kiln operation, far to the west of Sydney among the cow pastures of Wallgrove Road in Horsley Park. The slow migration to modern technology had much to do with the considerable amount of shareholder capital tied up in Hoffman-type kilns. The problem, as always, was the initial outlay of capital needed to establish a tunnel kiln operation. Machinery needed to be imported from either European or US manufacturers at great expense, which may explain the reluctance of small, family firms to commit, hence, their willingness to sell out to a competitor. The high cost of innovation helps to explain why smaller companies, and not just those in Sydney and New South Wales accepted offers to sell during this period.[4] In a further twist, the removal of price controls on bricks in 1955 did little to encourage innovation. Consumer prices based on a mark-up of just 15 per cent left little or nothing for plant development. In the meantime the increasingly expensive and scarce commodity of manual labour prevailed.

The City yard of Brickworks Limited, Euston Road, Alexandria had its own group of carters. The Austral, Ralford and Bedford yards of The Austral Brick Company Pty Ltd at St Peters used a ticket system to ensure that brick inventories were evenly maintained across the three yards. Drivers would tear a coloured ticket out of a book before loading their trucks. Austral yard: white ticket. Bedford yard: pink ticket. Ralford yard: yellow ticket.
Bob Brown Collection

SYDNEY CITY COUNCIL

The boom in trade was a double-edged sword. Increased sales raised output, yet rising production costs for traditional dry press bricks drove up prices. Dawes and other directors on the board of Brickworks must have pondered long and hard about ways to improve efficiency. Despite a run of impressive profits the cost of commissioning a modern plant would have pushed the firm deep into debt. Being a private company, the only realistic way to raise large sums of capital would have been to borrow from the bank, something that Dawes wished to avoid. Ultimately, the company was obliged to raise funds from its shareholders, in the process becoming a publicly-listed company on the Australian stock exchange as will be explored later. At this critical moment Sydney City Council approached the company with an offer to purchase the 10 acres of land owned by one of its associates, the Carrington Brick Co. Ltd, on Barwon Park Road. The town clerk's letter of 1 September 1954, asking if it would be willing to sell the land for use as a refuse disposal site, was rejected outright. Undeterred, the office of the Lord Mayor came up with alternative proposals in February 1955, suggesting it might be possible for the Council to take over the brick pit gradually whilst the company continued to take brickmaking material from the unused portion. The reply was still 'no'. Another offer to negotiate made during December also fell on deaf ears. Clearly, the vast swathe of land bounded by the present Sydney Park was of great interest to the council, which already owned the adjacent lands of the former Vulcan Brick Works and the St. Peters Brick Company.[5] Convenient access to the land of the Carrington yard through the partly filled Vulcan pit made ownership an even more attractive prospect. Discussion continued into 1957 but to no avail, prompting the council to change tack and initiate resumption of the entire property.

4. Interview with Peter Mahony, Sydney, 12 October 2005.

5. The City council had acquired St. Peters and Vulcan pits as a result of the amalgamation of the former municipality of Alexandria with the City Of Sydney on 1 January 1949

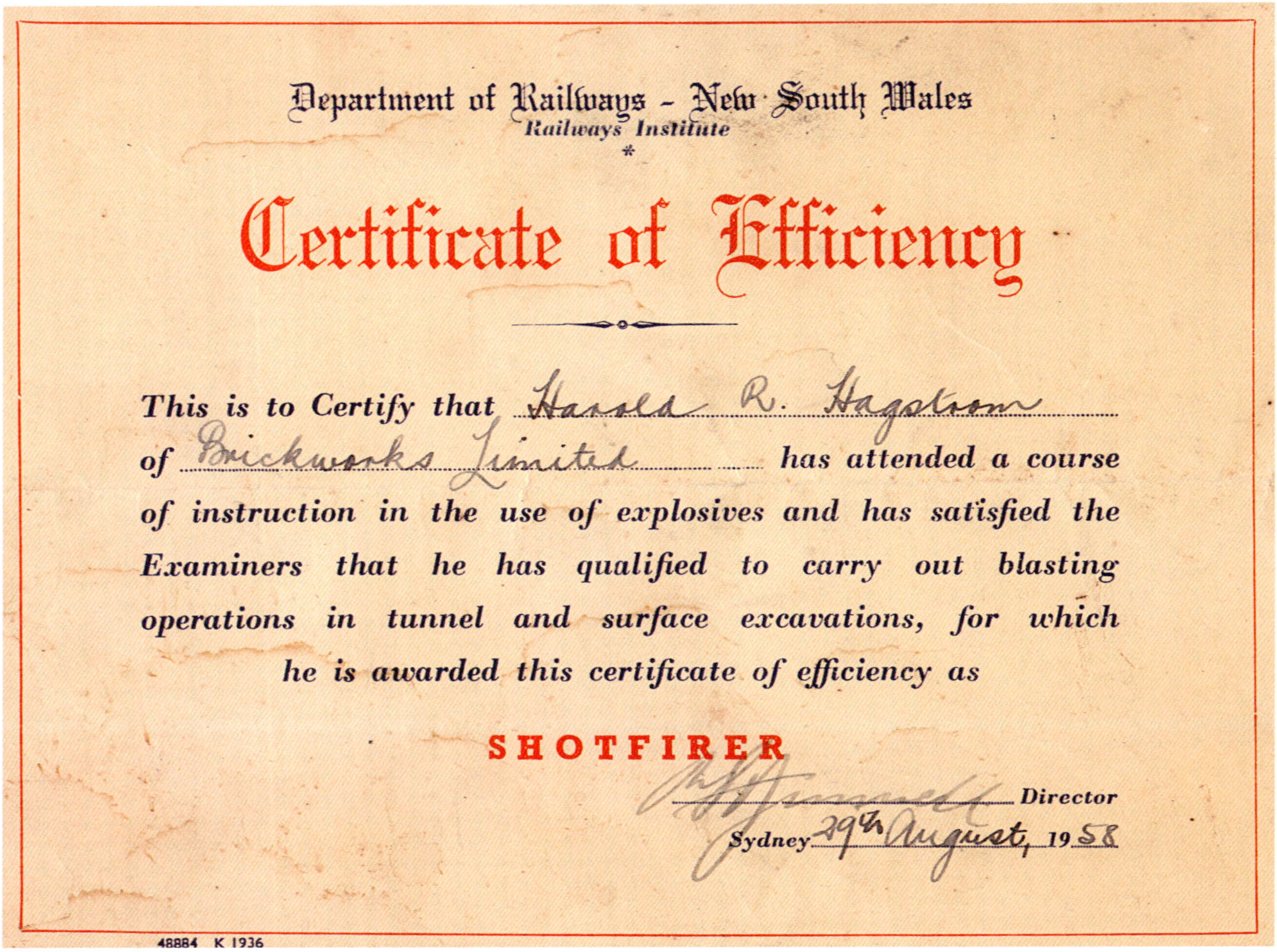

Department of Railways – New South Wales
Railways Institute
*

Certificate of Efficiency

This is to Certify that Harold R. Hagstrom
of Brickworks Limited *has attended a course of instruction in the use of explosives and has satisfied the Examiners that he has qualified to carry out blasting operations in tunnel and surface excavations, for which he is awarded this certificate of efficiency as*

SHOTFIRER

Director

Sydney 29th August, 19 58

48884 K 1936

Certificate of Efficiency in shot firing awarded to Ron Hagstrom in 1958 by the New South Wales Department of Railways.
Ron Hagstrom Collection

COOKING WITH OIL

Regardless of the threat of resumption and its implications there was still a business to run. In the course of 1956 a new head office for Brickworks was opened in Cheltenham Road, Burwood. Over at the Brookvale yard experiments were well underway on oil-fired burning, the first to be conducted in New South Wales according to Ron Hagstrom, the plant's former manager. Dawes had always been apprehensive about moving away from coal, fearful that the oil companies would use their monopoly position to push up fuel prices. Taking the plunge he invited a team of engineers from Caltex, an oil refiner and manufacturer of petroleum products, to rig oil burners in one of the downdraught kilns.[6] These early attempts were disappointing. Even after three separate burns the bricks emerging from the kiln had a greyish hue. An inquisitive man by nature, Hagstrom decided to apply himself to the problem that had so far escaped the attention of industry specialists. He sought advice from experts at the London Brick Company, from technical specialists in the USA, and ceramic engineers at the University of New South Wales. In his quest to identify airflow patterns in patent and downdraught kilns, Hagstrom studied the Venturi Effect which describes air flow dynamics in tunnels and chimneys.[7] Dawes was impressed and gave him free rein to experiment. Hagstrom dropped shredded pieces of car tyres down fireholes during the burn, while simultaneously opening and closing selected dampers. Time measurements were taken from the moment of combustion to the emergence of black smoke from the stack. By a process of trial and

6. No evidence of oil-fired burning remains at the former brickworks at Brookvale near Manly. On the other hand, small circular apertures can still be seen in the ruins of a downdraught kiln at the former Bedford yard at Sydney Park.

7. Named after the Italian physicist Giovanni Battista Venturi (1746–1822).

error, and with the assistance of others, Hagstrom discovered that reducing the heat in the kiln was a critical factor in cooking with oil and determining the colour of the burned brick.

On one occasion during these tests Ray Wilkinson, the company secretary, was on his way to Brookvale. The elderly gent noticed a dense pall of black smoke billowing over the yard several miles away. Convinced that Hagstrom had set fire to the district he covered the remaining distance along Mona Vale Road at breakneck speed with the accelerator pedal pressed flat to the floor. When the bricks were dragged from the kiln everyone agreed they had been cooked to perfection, the colour being magnificent. In subsequent burns the oil injectors were calibrated to avoid excessive smoke, although it was not until the introduction of pneumatic automatic stokers in both patent and downdraught kilns that dark smoke was completely eliminated.[8]

Oil was more efficient than coal, which was bulky, burned inefficiently and tended to block the external fire holes long after it ceased to be fed into the firing chambers. Using oil the kiln cooled down rapidly once the burn was stopped and the feed valve closed. The results were impressive; drying and steaming times were substantially reduced. Dawes needed no further convincing and promptly announced that all downdraught kilns in the company's yards across Sydney would be converted to oil firing. A contract was arranged with Caltex which secured cheap, bulk supplies of heavy residual oil from the newly completed cracking plant at Kurnell.[9] On the other hand patent kilns continued to use coal to produce common bricks, the exception being the Eastwood plant which was converted to oil in the late 1960s then gas in 1977.

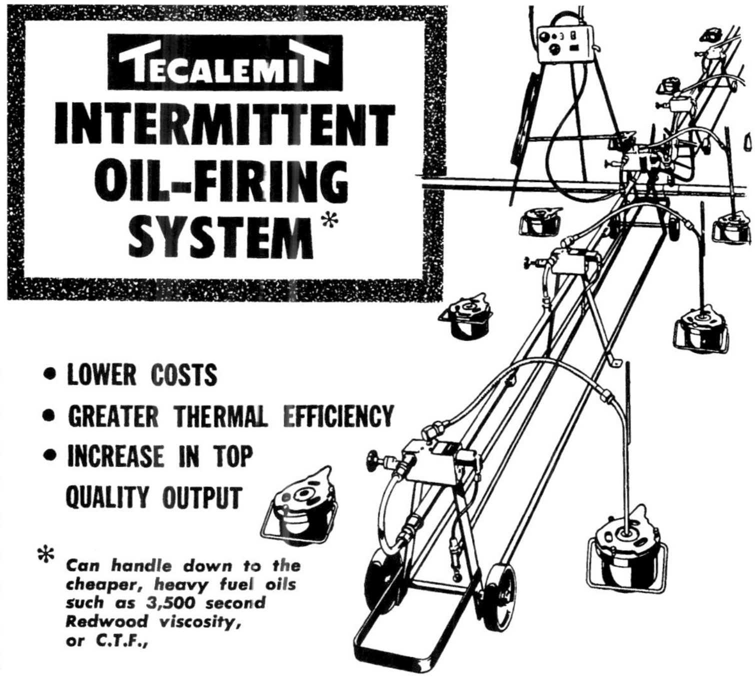

The Tecalemit oil injection system (1962) used atomised sprays and nozzle injectors. The copper fuel pipe fed a measured amount of fuel oil into the top of the patent kiln through an injector attached to the firing hole. W.K. Dawes hoarded oil in huge purpose-built tanks at the Wallgrove Road site

NEVER HAD IT SO GOOD

The switch to oil burning might not have been in the same league as the launch in the following year of the Russian satellite, 'Sputnik', but it was a technological milestone nonetheless. The year 1957 was a time for optimism with Britain's Prime Minister Harold Macmillan painting a rosy picture of the postwar economy. 'Most of our people have never had it so good', he remarked to hearty applause at a Conservative Party rally. Australia too was enjoying economic prosperity. Brickworks was busily refurbishing its plants to meet demand. Work commenced at Burwood Brickworks on four new downdraught kilns to burn face bricks.[10]

8. Pneumatic stokers were first installed in patent kilns by the State Brick Works in the late 1940s and attracted widespread interest among local and inter-state manufacturers who sent employees to learn the new techniques. The first Brickworks Limited plant to instal this equipment was the Brookvale yard in the early 1950s.

9. Caltex (Standard Oil of California and Texas Co.) established a new subsidiary company known as Australian Oil Refining Pty Ltd (AOR). Following an Act of Parliament approving the refinery and the site, construction began in December 1953.

10. Burwood Brickworks was acquired by Brickworks Limited from the Suburban Land & Investment Company in 1935. It closed temporarily in June 1952 due to a temporary downturn in the economy, reopening in the second half of 1953. Production was allowed to decline following the opening of new plants at Horsley Park, Sydney. By 1975 the 5.5-hectare site was closed and acquired by the New South Wales Department of Planning. It was purchased by Burwood Council in 1995 for use as a tip.

Happy days are holidays. The Propert caravan, 1956. Growing prosperity during the 1950s and 60s encouraged many Australian families to take beach holidays. Caravan parks and camping grounds were an affordable alternative to the more expensive hotel and motel accommodation. National Museum of Australia

In the early months of 1958 refurbishment of the City yard at Alexandria commenced in preparation for full mechanisation that took place over the next two years.[11] The makeover, comprising a rebuild of the kilns and installation of oil firing, was completed in February 1959 at a cost of £350,000, which at the time seemed alarmingly high. Notwithstanding, mechanisation of the plant reduced Brickworks' largest cost factor – labour – by at least 40 per cent. A similar plant constructed from scratch, and producing 16 to 18 million semi-dry press bricks annually, would have cost about £1 million at 1959 prices. As it was for many years to come the City yard on Euston Road was able to produce about 10 million bricks per year, with a noticeable improvement in the quality of the finished product.[12]

11. Consolidated Statement of Assets and Liabilities as at 30 June 1958, Brickworks Limited.
12. Chairman's report to Annual General Meeting of Brickworks Limited, October 1959.

MONEY TO SPEND

Always on the lookout for opportunities, Dawes cast his gaze further afield, northward to the rapidly growing city of Brisbane and the Rochedale brick factory owned by the Southern Electric Authority (SEA). Established in March 1953 to produce bricks for the authority's engineering works, the yard operated three small downdraught kilns with an annual output of about 2.5 million bricks. Why not pay a visit and find out if it was prepared to sell? Approaching Brisbane from the south, Dawes and his son, Ray, decided on the spur of the moment to call in at the 300 acre site. Fishing for information they soon discovered from the yard manager and foreman that the authority intended to sell the plant at a price far less than Dawes was expecting to pay. Later, a deal was reached at the SEA's Brisbane office and the brickworks changed hands for the bargain price of £29,500. Rochedale Brickworks joined the stable of Austral Brick companies and was renamed The Austral Brick Company. Four new kilns were erected on the site with a start made on a fifth, boosting production by early 1961 to 17 million bricks.

Production doubled under the plant's newly-appointed manager, an accountant by the name of Gordon Given. An educated, polite and good-natured man, he was the very opposite of Bill Dawes.[13] For over a decade Given put up with his employer's increasingly erratic behaviour, especially his insistence on being met personally at the city's airport at any time of the day or night. His patience exhausted Given left in 1970 to join Boral where he became mentor to up-and-coming talent such as William Emmett.[14] Austral's loss was Boral's gain.

Estimated output of bricks by major works in the Sydney area for 1957

Works	Production (millions)	Principal source
State (Homebush)	50.0	
Austral	26.0	
North Sydney	22.0	
City	20.0	
Bedford	19.0	
Clark Kilns	15.0	
Speare	15.0	
Thornleigh	14.0	
Punchbowl	12.0	
Clyde	11.0	Ashfield
Eastwood	10.0	Shales
Canterbury	10.0	
Northern Suburbs	10.0	
Ashfield	8.0	
Sutherland	8.0	
Burwood	7.0	
Cook's	6.0	
Goodlet and Smith	6.0	
Hurstville	6.0	
Kellyville	2.0	
Enfield	10.0	Bringelly
Lion Tiles – Tucknotts	4.0	Shales
Bankstown	4.5	
Brookvale	10.0	Hawkesbury
Heathcote	5.0	Shales
Warringah	2.0	

Source: F. C. Loughnan, The Origin, Mineralogy and Some Physical Properties of the Commercial Clays of New South Wales, School of Mining, Engineering and Applied Geology, Geological Series No. 2, UNSW, 1960, p. 213, Table 6. The table is based on figures supplied to the School of Mining by the Secretary of the New South Wales Brick Manufacturers Association

ON THE MOVE

Excited by the acquisition of Rochedale, Dawes returned to Sydney determined to push ahead with a hugely ambitious project – setting up a tunnel kiln operation. This required an immense amount of planning, money and land far removed from residential areas. Redeveloping the company's St Peters and Alexandria yards was out of the question, and besides, their future was about to be decided. Earlier discussions between Sydney City Council and Brickworks over the Carrington yard had come to nothing. Talks continued in earnest and over the course of 1958 the two parties hammered out something far more radical. Not only the Carrington, but also everything contained within the boundaries set by Mitchell Road, City Road, Campbell Street, Barwon Park Road and Princes Highway, would be resumed. The ensuing agreement also included quarries being worked further down the road at the Austral, Ralford and Central yards, representing more than 80 acres, a significant amount of land by any measure. At the Austral board meeting on 24 June 1958 the directors agreed to dismantle the Carrington and transfer workable assets to other yards, a clear indication that a decision on land use was imminent. A year later, on 26 May 1959, they voted on a contract of sale for the disposal of land at £10,000 per acre, payable over 20 years. The contract itself was not executed until the end of the following year. Council minutes for 6 October 1959 record that:

> *approval be given to the acquisition by purchase from Brickworks Limited and its subsidiaries of the Carrington Pit, Warren Pit, Bakewell's Pit, Bedford*

13. Given had been a prisoner of the Germans during the Second World War at Stalag Luft III, a special camp for captured pilots, and notable for a mass breakout of POWs known as the 'Great Escape'. He was actually in the tunnel when the Germans discovered the escape plan.

14. William 'Bill' Emmett worked at Rochedale and at Wallgrove Road, and was appointed CEO of Monier and General Manager of the New South Wales State Brick Works at Homebush. Later, he returned to his home state of Western Australia and managed he State Brickyard at Armadale.

Pit, Austral Pit, Central Pit and Ralford Pit, for the purpose of garbage disposal, for the sum of £800,000 on terms providing for the payment by the Council to the Company of annual payments of £40,000 for a period of twenty (20) years, free of interest. [15]

Garbage disposal would be allowed in sections of the quarries no longer worked for clay and shale. The usual constraints applied: Brickworks Limited would take care of the buildings, maintain the fences and ensure that the pits were kept free of water. An annual rental of £1,000 would be charged for each of the Austral and Bedford pits. Furthermore, no industry other than brickmaking was to be carried on. Shortly after the agreement was signed the council approved the first annual payment of £40,000 to Brickworks.[16]

In view of the fact that the land was subject to compulsory purchase, it is difficult to see what else the boards of Austral Brick and Brickworks could have done. What the government giveth, it can take. For the council, the depleted shale pits at St Peters and Alexandria were the ideal place to dispose of garbage, a by-product of Sydney's rapidly growing population.[17] For Brickworks and Austral it freed capital tied up in portions of land that were no longer being used. In the long term, a combination of economics and ecology put an end to brickmaking in the brickyards of St Peters. Despite having adequate reserves of clay and shale, the plants used old technology that was highly polluting and beyond modernisation. The exception was the City yard which was not included in the council agreement. It continued to operate as a going concern until it closed in June 1981. A lesser known fact which also had a strong bearing on the fate of the St Peters yards was the decision by the County of Cumberland Planning scheme to earmark the land for the construction of a major highway. Despite years of dithering by a succession of government ministers and the Department of Main Roads, the so-called Southern Expressway never went through. The idea was finally abandoned in the late 1970s, by which time brick production in the St Peters district had more or less ceased.

OVERSEAS CONTACTS RESUMED

With contract negotiations safely in the hands of Ray Wilkinson and the legal firm of Henderson & Taylor, Dawes embarked on a search for the latest in extrusion machinery and tunnel kilns. In June 1959, following months of preparation, a party of four comprising W.K. Dawes, Ray Dawes, Bill Young (the chief engineer), and Peter Murphy, set off from Sydney's Kingsford Smith airport on the first leg of a world tour. Dawes had arranged for Murphy, a fuel engineer to be seconded from the Joint Coal Board. Murphy's brief was to 'ensure that problems relating to fuel combustion, with particular reference to possible reduction of the emission of smoke and soot, should be investigated by an expert who is conversant with the local coal to our Industry'.[18] Murphy was a valuable asset and continued to provide advice on smoke reduction measures for a number of years.

Arriving in England, the party enjoyed a brief stay in London before heading off to renew old friendships at The London Brick Company, whose operations were spread across an enormous part of the countryside outside the town of Peterborough. The relationship between the two companies had been established during Clarence Davison's visit in 1937 (see Chapter 11). Bertram Moseley had also met Sir Malcolm Stewart, the owner and founder, during his tour of the works immediately after the end of the Second World War. Unfortunately, Stewart died in 1951, robbing Dawes of the chance to meet him in person. On numerous occasions the two men corresponded and spoke by telephone. It is interesting to speculate what they would have made of each other had they ever met.

15. Town Clerk's Files, 'Various brick pits in the St Peters and Alexandria areas of the City – Acquisition for disposal of refuse by controlled tipping,' Minutes of City of Sydney Council, Tuesday 6 October 1959, TC2390/58, p. 525.

16. Ibid.

17. Putrescible garbage was disposed of in the bottom of the pits and compacted by its own weight and graders. Years later, during the final stages, clean fill was used to provide a 'cap', followed by several metres of earth and dirt.

18. Annual Report, Brickworks Limited, 1959.

Bucket grab from Marston Vale, The London Brick Company. Sir Malcolm Stewart established the London Brick Company in 1900 following the purchase of the Fletton Brick Company. Operating in Marston Vale, Bedfordshire, it soon became the largest brickmaker in the country. About one third of all brick houses in England used bricks from The London Brick Company.

In the early 1920s the company commenced manufacture of the distinctive London stock bricks made from yellow clay found in Kent. Reclaimed London bricks are highly sought after for decorative and conservational use. Demand peaked in the 1950s and 1960s with the rebuilding after the Second World War. The London Brick Company was taken over by Hanson plc in 1984, which continues the tradition of brickmaking at the original site in Stewartby, built at the expense of Sir Malcolm Stewart who was also a noted philanthropist. Courtesy: Forest of Marston Vale Oral History Project, funded by the UK Heritage Lottery Fund

Rugged up against the bitingly cold winds that sweep in from across the North Sea at this time of year, the group set off on a tour of Britain's industrial countryside. They were fascinated by the new Swindell Dressler tunnel kilns, which were reputed to be the best on the market.[19] From the very moment he saw them in operation Dawes made up his mind to transform manufacturing at Austral. In fact, he could hardly wait to visit the firm's headquarters in Pittsburgh, USA. [20]

With a full itinerary there was no time to waste. Brick plants and engineering firms were high on the list.[21] Along the way, Ray Dawes was quick to show local builders how to lay brick walls starting from the inside and working out. This Australian technique reduced handling which impressed his British hosts who took up the idea. From England the four men travelled to Europe, visiting France, Germany, Poland, Spain and Italy where they investigated the latest trends in the industry. Unfortunately, this was not a great success. Ray Dawes recalls the European leg was prompted by his father's concern that the type of equipment being supplied by British manufacturers was unsuitable for Australian conditions.[22] Lacking tact, W.K. Dawes did nothing to endear himself to industrialists, many of whom were highly educated men and members of their country's manufacturing élite. Anecdotal stories suggest they refused to take him seriously when he placed large quantities of banknotes on the table during discussions.

19. William Swindell & Brothers was a furnace company established in Pittsburgh in 1850. Dressler Tunnel Kilns was a kiln company started in England in 1908 by Conrad Dressler. American Dressler Tunnel Kilns, Inc. was established by Englishman Phillip Dressler who migrated to the USA in 1915. Both American companies joined forces in 1930 and became known as Swindell Dressler Corporation based in Pittsburgh, USA.

20. Dawes made exhaustive enquiries regarding tunnel kiln construction, and on several occasions visited Melbourne where he inspected a tunnel kiln (the biggest in Australia) built by Mario Ronco, probably for Brick & Pipe Industries.

21. Appointment book of Bill Young. Papers of John Armitage, barrister-at-law, Sydney.

22. Interview with Ray Dawes, Sydney, 12 October 2005.

J.C. Steele, industrialist and Mayor of Hazard County, Kentucky, 1930–1933

Leaving Europe empty-handed, the party flew to the United States where Dawes made straight for Swindell Dressler's Pittsburgh facility. After witnessing further demonstrations he was ecstatic, convinced that face brick produced in US-designed tunnel kilns was far superior than any produced either in England or at the yards of Brickworks Limited. Assuming an output of 40 million bricks per year, it became apparent to his American hosts that two tunnel kilns would be needed, and not one as originally envisaged. It was mainly due to the influence of W.K. Dawes that Swindell Dressler went on to build a number of modern plants for Brickworks, including clay pipe kilns for Punchbowl Brick and Pipe Co., and also for Clifton Brick, Selkirk Brick and Nubrik of Melbourne. Indeed, Swindell Dressler is still active in the Australian market. Tunnel kilns were not the only item on Dawes' agenda. Mindful of Ron Hagstrom's Brookvale experiments, he was anxious to purchase multi speed induction fans that could be installed in chimney stacks and thus ensure a more efficient burn.[23]

From Pittsburgh, the group flew down to Statesville in North Carolina to visit the firm of J.C. Steele, the manufacturer of extrusion machinery.[24] Dawes had already met its owner, Preston Steele, during his first visit to Statesville in March 1939. Preston Steele had been stationed in New Guinea on war service and is known to have spent time in Sydney where he and Dawes forged a close relationship. In the late summer of 1959 the latter took up residence in Statesville where he learned about extrusion methods and picked the brains of his mentor and host. As with everyone he met W.K. Dawes made a lasting impression. Ensconced in one of the town's well-appointed hotels, he became a firm favourite of the African–American employees. From bellhop to barman, everyone received a tip, which were huge even by the standards of the day. Known for acts of generosity in Sydney where he occasionally returned home with down and outs for a feed, Dawes may well have sympathised with the plight of Southern blacks who had been very much in the news following recent challenges to the segregation laws. Since few people in the Deep South had the slightest notion where or what Australia was, Dawes was mistaken for an Englishman. This made things difficult for visitors from that country. Long after his departure, hotel staff lavished attention on surprised English guests who, without exception, were believed to be exceedingly generous.[25] The pickings were invariably slim.

WALLGROVE ROAD

Returning to Sydney in time for the company's October annual general meeting, Dawes was eager to report on his findings, and also make a start on detailed planning for the expansion into red texture brick production. With demand beginning to outstrip supply, and a barrage of complaints about noise and pollution spilling from the company's inner city brickyards, finding suitable acreage became a

23. Following a heated argument with W.K. Dawes (punches were thrown), Ron Hagstrom resigned from Brickworks Limited in late 1959. He was employed at the Parklea plant of North Sydney Brick & Tile Company, owned by Douglas Lanceley. He returned to Brookvale in 1966 and retired in 1988.

24. James Columbus Steele founded J.C. Steele in 1889, in Statesville where he operated a small brick factory. Steele set up a small engineering shop and foundry to make machine parts. Today, J.C. Steele & Sons is the largest producer of heavy clay products machinery in the United States. The company's Australian representative was Barry Seward whose Melbourne business, Seward Nominees, imported and distributed Steele's machinery.

25. Interview with Ray Dawes, Sydney, 12 December 2006.

Troopers of the 21st Light Horse (Riverina) Regiment riding through the dust during squadron drill, Wallgrove, circa 1939-1940. Military camps and large numbers of troops were a common sight in the district in the war years. The one pictured here is on lands owned by the Commonwealth Department of Defence at Wallgrove, near to Horsley Park. Argus Newspaper Collection of Photographs, State Library of Victoria

priority. This precluded anywhere within a radius of 45 kilometres of the Sydney GPO due to the rapid development of new suburbs.

Dawes left the details to Bruce Parks, the estate agent and valuer from the Newtown firm of E.A. Turtle, which handled property and land purchases on behalf of Austral Brick and Brickworks Limited. An expanse of land west of Blacktown at Wallgrove was identified by Dawes, who drilled core samples proving the existence of huge reserves of clay and shale. The proximity of Wallgrove to Sydney's growing suburbs in the west would also mean a reduction in haulage costs, making it an opportunity too good to miss. It is interesting to note that by 1960 the building boom in outer Sydney was consuming approximately 60 per cent of the entire production of Austral Brick's city yards.[26] In fact, the close proximity of so many large-scale inner city building projects had been a powerful argument in favour of the recent mechanisation of the City yard at Alexandria, which continued to operate for another 20 years.[27]

Throughout the closing months of 1959 and into 1960, Dawes, his son, Ray, and Bruce Parks went on a buying spree. Anxious to avoid publicity which would have encouraged vendors to inflate prices, private sales were transacted and parcels of land purchased from anyone willing to sell, including dairy farmers and the owners of an airstrip. In this way 1,375 acres were acquired in the Blacktown district, containing reserves sufficient for at least 200 years of brickmaking, the clays providing 90 per cent of colour requirements. Further acquisitions brought the total to over 2,000 acres by 1967.

Construction work on Austral Brick's new texture brick plant at Wallgrove commenced in early 1960 with the laying of the concrete foundations and the erection of the main shed. This was the realisation of Dawes' long-held ambition to create a thoroughly modern brick plant, the first of its kind in Australia, and certainly a world leader. Installation of the two kilns followed soon after, and was supervised by Fred Petrie who was flown out from the USA for the occasion. The estimated cost of the plant

26. Chairman's report to Annual General Meeting, Brickworks Limited, October 1959.
27. In 1959 it cost 3s per mile to cart 1,000 bricks. The average load weighed 4 tons.

Arthur Dawes (third from right) on a visit to The Pine Hall Brick Company Inc at Winston-Salem, North Carolina, November 1960. Marilyn Keelty Collection

was £800,000, just short of the issued capital of £864,506 of Brickworks Limited at that time. Add to this the outlay on land, and the cost of the new project topped £1 million.[28] This was a prodigious sum of money, although approximately half of the new expenditure was financed from inner reserves accumulated through a decade-long run of profits.[29] During October 1960 the company announced the issue of 432,253 Unsecured Notes of £1 to raise additional finance.

At the same time the Wallgrove site was being acquired, about 250 acres was purchased at Duffy's Forest. This secured a ready supply of raw materials needed to produce Austral's 'cream tone' bricks, for which there was a large and increasing public demand. In addition, an area of 315 acres at Quaker's Hill was purchased during 1961 and the land held as reserve for brickmaking purposes. This was resumed by the Housing Commission of New South Wales in 1973. The shopping excursion was not confined to Sydney. In the Mount Gravatt-Rochedale district, an additional 125 acres were purchased, bringing the total area of land attached to the Rochedale plant to 445 acres by 1961. Dawes made sure that the plant could not be overshadowed by residential developments, and in subsequent purchases increased Austral's holdings to about 1,200 acres. Much of this was later developed and sold off at a considerable profit. [30]

Barely nine months after returning to Australia from the USA, Dawes left the country once more, this time with Ray Wilkinson and his youngest son, Arthur, who had been in charge of the Rochedale plant from its inception. Departing Kingsford Smith Airport at the end of July 1960 with a brief case stuffed full of technical papers and an exhaustive list

28. Interview with Bill Emmett and Chairman's Address to the Annual General Meeting, Brickworks Limited, June 1961.

29. For the financial year ending June 1960 Brickworks Limited recorded a net profit of £127,223, the highest to date (Annual General Meeting, October 1960).

30. Interview with Lyndsay Wilkinson, former employee at the Rochedale plant, 13 September 2006.

Dawes family gathering at Sydney airport prior to the return of Fred Petrie and his wife to the USA, April 1962.
Left to right: W.K. Dawes with Carole Dawes (daughter of Ray Dawes), Arthur Dawes and his two daughters (Corinne and Marilyn), Ray Dawes and his wife, Sadie, mother of W.K. Dawes (off picture), unknown (possibly a kiln contractor), Lourna Dawes, Fred Petrie and his wife, unknown. Marilyn Keelty Collection

of questions, Dawes made straight for the Swindell Dressler office in Pittsburgh before heading down to North Carolina in late September to see his contacts at J.C. Steele. Attending church on Christmas Day, a rare occurrence for Dawes, but unavoidable in the circumstances, new friendships were struck, especially between Dawes and Carl Zeppenfeld, of Swindell Dressler who was invited to oversee the installation of the new plant on Wallgrove Road, Horsley Park.[31]

Shortly before leaving Sydney, Wilkinson had written to Ray Dawes informing the young building contractor that he had been elected to fill the board vacancy left due to the retirement of C.F. Davison, a director of Brickworks since 1934.[32] Dawes senior, and his three sons, William Frederick, Ray, and Arthur, were now all members of the board, giving it the appearance of being a family owned company. This was most certainly not the case although the Dawes family amassed a substantial shareholding over the years. At more or less the same time, Brickworks Limited sold the Clyde Brick Company Limited to Newbold Refractories Limited, including shares held in the company by The Warren Brick Company.[33] Dawes had a prodigious appetite for knowledge and this latest trip allowed him to resolve many of the issues to do with the new texture brick plant that was rapidly taking shape. Sadly, just weeks after returning to Australia, he was dealt a serious blow when Ray Wilkinson and his wife were seriously injured in a car accident. Wilkinson spent seven months in hospital but never recovered and died on 22 June 1962 from multiple injuries sustained in the crash.[34] Lloyd Willis was appointed as acting secretary.

31. Carl Zeppenfeld was a bricklayer by profession and worked his way up in Swindell as a construction supervisor. He was based in Sydney for a number of years and oversaw most of the plant construction at Wallgrove.

32. Letter to Raymond Dawes, from the company secretary, Brickworks Limited, dated 29 July 1960. Archives of Brickworks Limited.

33. Board minutes, The Warren Brick Company Limited, 26 July 1960, Archives of Brickworks Limited.

34. Wilkinson was appointed manager of Metropolitan Brick Company Limited in 1936 and joined Brickworks in 1945 as company secretary. He was offered a directorship in 1947 and in 1952 became chairman of the Brick Manufacturers Association of New South Wales. Wilkinson had a nervous temperament. In addition, he was a devout Baptist and teetotaller, which irked Dawes, especially on social occasions. On one particular visit to the Palm Beach home of Mervyn Richardson, the inventor and owner of Victa Mowers, Dawes was accompanied by Wilkinson and Hagstrom. Like Dawes, the elderly Richardson enjoyed a drink. Whilst the two men attacked a newly-opened bottle of whiskey Wilkinson stayed with orange cordial.

Punchbowl brick and pipe carter, circa 1938. Horse-drawn brick carts were a rare sight by this date

DIVERSIFICATION

With construction work underway at Wallgrove Road, Brickworks Limited branched out in yet another direction. In the course of January 1961 Punchbowl Brick and Pipe Company Pty Ltd was acquired, and paid for by the issue of 324,000 20s shares, bringing the company's paid up capital to £1,188,506.

The Punchbowl group of companies, owned by Dawes' long-time racing confederate, Reg Allott, comprised Heathcote Brickworks, Kirrawee Brickworks and Punchbowl Brick & Pipe Works. Allott's firm was a leading manufacturer and supplier of vitrified clay pipes as well as bricks. Anxious to safeguard the future, he came to an agreement with Dawes and let the company go at a very reasonable price.[35] Reg Allot was fortunate in having a dynamic and youthful company secretary, Albert Burgis, who had joined the firm as company secretary in 1959, and later appointed general manager of the pipe division in 1963. Competition from the firm Porter Galbraith Hanson in the market for clay pipes was keen and Burgis, Dawes and Allott were aware that the Punchbowl plant in particular was in need of modernisation. Following the sale the company was renamed Austral Pipes Australia Pty Limited. Meanwhile, back at St Peters in the winter months of 1961 Austral Brick was tying up some loose ends. In July the mortgage on land formerly owned by Bakewells on Mitchell Road was discharged and the remainder of the property sold to Sydney City council. This was followed one month later by the sale to the Council of another parcel of land for $110,000 at the Carrington yard.[36]

Reginald Allott was a wealthy businessman whose various brick, pipe and tile plants supplied high quality products to the growing Sydney market. During the late 1950s he acquired Speare's Bricks at St Peters and sold it to Brickworks Limited along with his other factories in 1961. Allott and his wife had one daughter and no apparent heir capable of taking over the running of the firm. Reg Allott and his chief engineer, S.J. Eddy, became directors of Brickworks. Both men resigned in 1965, Allott due to ill health. He died soon afterwards.

35. Interview with Albert Burgis, Sydney, 15 November 2004.

36. Board minutes, The Austral Brick Company Pty Ltd, July–August 1961.

Above: View of Punchbowl yard from Belmore Road

Below: Aerial view of the Punchbowl yard

Transferring the fired pipes from the beehive kilns to the stacking area, Sutherland yard, circa 1959

MORE FACT FINDING

In March 1962 Dawes made yet another overseas visit, leaving behind the torrential rain and bad weather that for many months had gripped the city and hampered work on the Wallgrove plant. Included were Bill McGregor, Bill Young's successor as chief engineer of Brickworks who was interested in the manufacture of fire brick and refractories, and Samuel 'Jim' Eddy, the chief engineer from the Punchbowl works who was keen to learn more about US methods of manufacturing salt-glazed clay sewer pipes.[37] Arthur Dawes, who would be the inaugural manager of the Wallgrove plant, was also in the party.

The four men flew to Zurich where they visited brick plants noting 'some good machinery', and thence to Hanover in Germany before motoring to Dusseldorf. London was the next stop where they met with executives of London Brick[38] and finally to Manchester where they toured the operations of Platt Brothers, the engineering firm. Judging by postcards received by relatives the strain of travel was beginning to tell. Leaving the north of England they flew to Pittsburgh, and then Statesville. Knowing little about clay pipes, Dawes was determined to come to grips with every aspect of the Punchbowl operations and how it could be improved. Through his hosts, Preston Steele, and Owen Cross, the managing director of the Swindell Dressler Corporation, Dawes arranged visits to two plants in the South that manufactured salt-glazed pipes.

> *I do not want to be just an ordinary visitor to each plant. If possible I would like to stay till I actually learnt (sic) it in principle. The salt glazed pipes are for our Punchbowl section. The fire bricks will be my personal baby (if any). That is why I could not take that much time of any of your staff. All I want is a large car and a good driver.*[39]

Then a confident prediction which was later proved wrong following the introduction of heavy duty plastic pipes.

37. Bill McGregor was a learned man who had taught in technical colleges and had previously been employed at Bulli Brick and Tile.

38. Like Brickworks, Punchbowl Brick & Pipe Co Pty Ltd developed a close working relationship with London Brick which supplied it with 18 and 24-inch terracotta pipes when the company phased out production of these pipes.

39. Letter from W.K. Dawes to Preston Steele, 30 January 1962, Archives of Brickworks Limited.

***Above:** View of the Sutherland yard from the Princes Highway*
***Below:** Aerial view of Sutherland*

Pipemakers at work at the Punchbowl yard, circa 1939

> *I am certain that as far as NSW is concerned pipes will always be salt-glazed and if I am proved wrong it would give me great pleasure to destroy say £25,000 worth of brine tanks.*[40]

Dawes plied Steele with a myriad of questions on the manufacture of such pipes using tunnel and shuttle kilns. 'Can you load a kiln with large diameter pipes – put a small diameter pipe inside and successfully glaze both pipes? We cannot do so down in this country.' These, and many others, including queries on fire bricks and the availability of local materials for semi-alumina brick (also known as firebrick or refractories), can be found in his letter. Dawes wanted to produce high grade fire refractory bricks rather than rely on expensive imports or locally-produced bricks. The company already made the bulk of low grade refractories for its tunnel kilns which were satisfactory for temperatures from 1,150°C to 1,500°C. Air setting mortars used to secure the joins between firebricks lining tunnel kilns were also on Dawes' mind, as well as other types of refractory mortar such as Cement Fondu. Similar to Portland cement, this was imported from France and used to seal floors and wickets, as well as crowns in patent kilns. Fondu could withstand direct heat up to 1,300°C and Dawes was interested to know its constituents.

40. Ibid.

His letter to Owen Cross took up an earlier offer to spend a couple of days at the Davis Fire Brick Plant at Oak Hill, Ohio to explore how to make these vital components and thereby save money. 'I think you understand what is necessary', Dawes remarked at the end of his five-page letter to Steele.

> *I want the answers, and I am happy to pay for them. You are the only People I know that can do this job for me, that I have complete confidence in. Do not spare the expense. The correct answers can make or break the whole project.*[41]

And on a final note that lends itself to speculation:

> *I have now had my 63rd birthday. I believe you may still remember my 62nd birthday at the John Pick Motel, Montgomery Alabama.*[42]

41. Ibid.
42. Ibid.

CALL ON GREG. NOLAN AT THE SOUTHERN CROSS HOTEL KING & MARY STREETS

The NUTSHELL

RIVERVIEW HOTEL KING ST., TEMPE
M. O'BRIEN, Proprietor
Telephone: LB008
Only the Best Kept

DELIVERED FREE EVERY WEEK IN EVERY HOME

FRIDAY, January 20th, 1961. Phone L1950

Circulating in St. Peters, Sydenham, Tempe and surrounding Districts.

Pipe machines in operation, Punchbowl yard
Inset: Masthead from the Nutshell, a newspaper that covered local news for the St Peters – Tempe district, 1961

Returning home a month or so later, Dawes inspected progress at Wallgrove, then met with the board to discuss plans to float Brickworks Limited on the Australian Stock Exchange (ASX). In a move that attracted little fanfare, the firm was listed on 30 June 1962, when, for the first time, its shares could be traded publicly.

In September 1962, the new plant at Wallgrove was officially opened and production commenced. Arthur Dawes was placed in charge of tunnel kiln operations. William 'Bill' Riley, an accountant who had worked in the firm of Travis & Travis[43] was appointed administration manager and accountant, and remained in these roles until he resigned in 1976. Eric Illingsworth was the plant superintendent and Jim Reid became the chief engineer. One year later in September 1963 a second kiln was brought into production. Such was the demand that contracts for a third kiln were let in June 1964. Feeling the strain of the previous three years, Dawes was hospitalised for most of December 1963, while at the same time the Ralford yard was finally closed and the men transferred to the Bedford plant.[44] It was all go with texture brick creating a pot of gold for Brickworks Limited.

43. Travis & Travis was established by George Farthing Travis and his son, Geoffrey George Travis in about 1951. Prior to this date two firms of similar name were operating, that of George Seville Travis & Co, (great uncle of Geoffrey Travis), and the firm of George F. Travis (father of Geoffrey). George Seville Travis M.B.E. was a significant shareholder in The Austral Brick Company Pty Ltd. George Farthing Travis resigned in the mid-1950s due to ill health, leaving the business in the hands of his son, Geoffrey. The great grandfather of Geoffrey Travis owned the Liverpool Brick and Tile Company and was prominent among Sydney brick manufacturers. Source: *Australian National Clay*, Vol. 2, No. 12, October 1961.

44. W.K. Dawes was a major benefactor of St Vincent's Hospital, Darlinghurst, Sydney and when hospitalised was accommodated in a room of his own (Interview with Jim Liepins, Sydney 12 June 2006).

The use of clinker brick in the NSW Department of Lands at 1 Prince Albert Road, Queens Square, Sydney
The growing popularity of clinker bricks among architects in the late 1960s and 1970s was such that Austral Brick could not keep up with demand. Photographs by Josh Hill

Texture takes off 16

Open all hours – The accidental brick – Do it yourself design – Fast lane – Taking care of business – Seeing red – New plans, new plants – A family affair – The long goodbye

OPEN ALL HOURS

The opening of Austral Brick's first extruded texture brick plant had a profound effect on the company's fortune. The factory proved conclusively that tunnel kilns were able to fire near perfect bricks. Mottled colours, callows and clinkers were unknown, and smaller batches could be made by adjusting the brick mix. Just as important as colour and texture, the load bearing qualities of extruded brick proved to be superior to dry press. Demand was insatiable for the new-look red texture bricks, which made them an expensive commodity. Chocolate-brown bricks, made by the addition of manganese, were also immensely popular, a throwback to earlier in the century when liver-coloured bricks made their appearance. Yet, the new-fangled extruded texture bricks did not please everyone. Some bricklayers and builders were reluctant to use them claiming they required a greater quantity of mortar on account of the holes left by the die, and took a longer time to lay. In the early months sales figures languished which worried Dawes since the company had been obliged to borrow in order to finance the new plant. According to Jim Liepins, texture bricks manufactured at No. 1 Plant used a 10-hole die but the switch to a three-hole die allayed concerns.

The impact of new technology on marketing strategies was noticeable. Austral Brick could now offer a wider range of bricks to suit almost every taste. As a result promotion campaigns became a great deal more sophisticated. Samples were mounted on display panels at sales offices, which encouraged customers to demand – and expect – a wider range of colours and textures. 'Rock face' brick developed as a taste and has its origins in the attempt to add value to reject commons and face bricks by passing them through a rock facing machine with carbide-tipped blades that chipped off the edges. Lighter coloured bricks became fashionable once more with the revival in popularity of cream texture bricks used extensively on the interiors of large public buildings and schools. Cream bricks could even be produced using the dry press method and fired in downdraught kilns.[1] Cream burning shale was supplied from quarries owned by Brickworks Limited at Illawong and Menai. and Duffys Forest.

Huge quantities of the company's textured bricks went to new subdivisions including Sylvania Waters and Bankstown.[2] Despite competition from texture bricks the dry press variety still represented about 75 per cent of the market by 1965. Many home builders preferred the solidity of a double-brick home and the relative cheapness of common bricks, especially when combined with the stylish elegance of traditional face bricks. In fact, so great was the demand that even the State Brickworks found it difficult to keep up.[3] Unfortunately, dry pressed bricks lost out. Textured brick triumphed due to its 'modern' look, load bearing capability, and demand fed by Sydney's rapidly increasing population. The texture craze began to lose steam in the early 1980s as tastes changed yet again, the latest fad being sandstock bricks.

THE ACCIDENTAL BRICK

One development embraced by architects during this period was the use of clinker brick in building work. Clinker became fashionable, especially among homeowners in the middle class suburbs of Sydney's leafy North Shore. The unusual, rustic appearance of clinker struck a chord with sentimentalists who

1. The two dry press yards at Brookvale and Heathcote turned out millions of cream bricks. During the 1970s and early 1980s the Eastwood yard used dirt from the pit at Terrey Hills to make small quantities.
2. Mention is made of Sylvania Holdings Limited in the board minutes of The Austral Brick Company Pty Ltd held on 25 November 1966. This company, owned by Brickworks Limited, was used to acquire land for the development of Sylvania.
3. With its clay reserves beginning to fail, the State Brickworks became increasingly reliant on dirt trucked from Menai, prompting it to investigate clay and shale deposits owned by the State Tile Works at Blacktown.

hankered after a past that was long gone. Regardless of the inconvenience to the bricklayer (building a feature wall was an ordeal), clinkers made a statement about our continuing attachment to the past.

Jim Liepins, a retired engineer who worked at the City yard of Brickworks, believes that clinker bricks were first produced commercially at this yard. He relates an incident where a burner had fallen asleep under the influence of alcohol. Clumsy attempts to rekindle the fire in the downdraught kiln created an uncontrolled fire which burned the bricks. Dumped in the pit as waste the rejects caught the eye of a visiting architect who considered them to be unusual. Soon after, the market for clinker took off. The majority of clinkers were produced at the St Peters yard of Austral Bricks in the patent kilns which were run at extremely high temperatures to achieve the desired effect. Eventually this forced the refractory bricks lining the interior walls of the kiln to 'pop', compromising the structural integrity of the building. [4]

TOO LONG IN THE OVEN

In plain terms, clinkers were bricks that had been overcooked and distorted due to the combustion of carbon within the highly compressed mass of clay and shale. The surface of a clinker was often vitrified. Quite rightly brickmakers saw them as a nuisance – wasteful of materials and energy and with marginal resale value. On the other hand, they understood little about the scientific conditions that created clinker brick.

Ron Hagstrom, who was manager of the Brookvale yard, cajoled his former wartime colleagues who were now pilots for the Australian airline, Qantas, into scouring US and British universities for technical notes and reports on brick burning. Through a series of experiments he came to understand that clinkers could result if temperatures were inadequately regulated during a stage of burning called 'preheat'. Like every other brickmaker, Hagstrom knew that the desired colour, especially for face bricks burnt in a downdraught kiln, was determined by a number of factors. These included controlling the intensity of heat, the duration of the burn, and the presence of certain minerals in the clay and shale. In this regard, the process of oxidation and reduction was critical. By alternating the reducing–oxidising cycle it was possible to fire a range of different coloured bricks – from light reds down to dark and chocolate browns.

Reducing and oxidising

Regardless of whether a brick was burned in a continuous or downdraught kiln, carbon is eliminated during the pre-heat phase (500°C) of the firing cycle. Special care needed to be taken before the temperature was allowed to reach its maximum. If the process was not properly monitored then intense heat built up in the centre of the brick and resulted in the combustion of carbon and bursting of the brick. The removal of carbon was achieved by opening the dampers which created conditions for an oxidising fire. This allowed the brick to breathe in oxygen and exhale carbon. Such a fire burned with a long flame.
Once carbon had been expelled the dampers were lowered and the heat increased by a process called reduction. During this phase traces of iron or aluminium present in the clay and shale were burned to give bricks their distinctive colour (red for iron or cream for aluminium present in kaolin clays). In a reducing atmosphere the fire burned with a short flame that was orange to bluish in colour. See also p. 260.

Hagstrom's experiments at the Brookvale yard demonstrated that CO (carbon monoxide) recorders were indispensable to the oxidising–reducing process. His investigations also confirmed that downdraught kilns burned more efficiently if the fire was kept hot through regular feeds of coal shovelled into the external fire boxes. In short, Hagstrom found a way to control the burning properties of Sydney clays and shales in brickmaking. Unfortunately, for continuous and downdraught kilns alike, a long-standing industrial practice allowed for only three feeds per shift instead of the five preferred by brickmakers. Whilst it would have been preferable to burn Newcastle coal (the size of a tennis ball) with its high calorific heat value, this would have required five feeds. Clearly, this was out of the question unless another man was employed, which would have been prohibitively expensive. Instead, lesser quality lumps of Lithgow or Western coal were used to burn face bricks, but this affected the intensity of the fire. Common bricks were fired using the cheapest, and slowest burning coal supplied by the coalmines of Nattai and Bulli, south of Sydney.

4. Interview with Jim Liepins, Sydney, September 2006

Flats, flats, flats, Sydney home units. These apartment blocks were constructed during the 1980s, yet they are still representative of the type of building design favoured by property developers. Photography by David Smyth

If Hagstrom could do nothing about the industrial award, he could certainly focus on improving the efficiency of the kilns and the rate of burn. With this in mind he installed instruments to record the amount of CO_2 emitted during the firing process. Exhaust gases containing CO_2 were reticulated through a water filtration system and mapped on a circular disc of paper by an ink printer. The progress of the burn translated into a series of loops similar to the petals of a flower. The tip of the loop indicated the presence of CO_2 which reached its peak during a coal feed. Hagstrom deduced that an erratic shape was a clear indication that the fires were not being fed at the prescribed intervals and that the process of oxidation and reduction was being interrupted. The obvious conclusion to draw was that some of the burners were falling down on the job. This was manna from heaven for Dawes who ordered the installation of heat recorders in every yard. With the switch to oil and gas which required fuel injectors, the inattention of coal burners became a thing of the past – at least in theory.

DO IT YOURSELF DESIGN?

Almost overnight textured bricks became the darling of developers, especially those who erected walk-up three and four-storey apartments on quarter acre blocks. Sydney's urban landscape began to change yet again, although not necessarily for the better. Sadly, a combination of cost cutting and the penchant for stark, functional design cast a shadow over the city's inner and outer suburbs. In 1961 a Strata Title Act made it easier for owners or investors to purchase individual flats or 'home units', thus opening the way for what some people considered a period of unquestionable ugliness in taste and design. By about 1965 the new boom in flats surpassed the previous peak during the mid to late 1930s.

In his 1960 book, *The Australian Ugliness*, Melbourne architect Robin Boyd takes a swipe at indiscriminate development across the Sydney metropolis:

> *the recent growth of Sydney is mainly confined to three zones. Out west, the wooden villa, of Villawood Zone from Liverpool north through Parramatta, the*

Housing development on the artificial James Cook Island, Sylvania Waters, New South Wales, 1996. The development of Sylvania Waters was promoted by The Austral Brick Co Pty Ltd on reclaimed land adjacent to the Georges River during the 1960s. Richard Woldendorf Collection

> *Tom Ugly Zone… where the Georges River opens to Botany Bay. The really depressing parts of Sydney, however, are in the North Shore Executive Zone… beautiful country. Out through Frenchs Forest and along the spine of Pittwater (is a) wild scramble of outrageous featurism.*[5]

His disdain for popular taste was all well and good, but for the migrants wanting to start a new life, a brick and tile home on inexpensive land on the western plains was the best they could do. They were also homesick, and like most migrants, they tended (for a while, at least) to cling to the past, which was reflected in the choice of building styles and construction materials. For example, the Federation house reveals just as much about our Anglo-Celtic heritage as it does about the availability of dry press bricks at the turn of the century. Boyd offered no alternative for those with limited means, neither does he seem to have championed the cause of attractive, utilitarian house design for working class people. Leaving aside value judgements on personal tastes it is an axiom that creeping suburbia continues to be driven by economic prosperity, migration, housing affordability and the private car. Unfortunately, as mentioned in previous chapters, the lack of basic amenities reflected poorly on local and state governments. Today, the failure to plan suburban development around an integrated, city-wide bus, rail and road network is having a serious impact on the quality of life of Sydney's inhabitants. Such concerns were being heard during 1959 and 1960 when the future of trams on the city's streets was being questioned. Unable to counter the obvious advantages of the private car, the government of the day felt that it must respond to the problem of increasing congestion on Sydney's narrow streets, so that (reluctantly) trams were withdrawn in February 1961. An article in the *Sydney Morning Herald* mourned their demise, pointing out: 'the best

5. R. Boyd, *The Australian Ugliness*, Penguin Books, Melbourne, 1968, p.106.

A 'corridor' tram on the La Perouse line in Anzac Parade near Kingsford shortly before trams ceased operating on 25 February 1961. Vic Solomons Collection

argument for the trams was their ability to move a Randwick racecourse crowd of 92,300 in half an hour. Over 2,300 fully loaded buses would be required to shift the same crowd.[5] The rail network continued to expand, yet no comparable form of mass transit with all the benefits of the tram has been introduced in nearly 50 years.

More painful than the loss of cheap, efficient and fast transport was the demolition of Sydney's inner city 19th century streetscape. The wrecker's ball had an insatiable appetite for much that should have been revered. By the early 1960s the city's central business district was in the grip of a demolition derby, a race to the finish which has left Sydney all the poorer. Redevelopment became the prevailing ethos, and with successive state governments either unable or unwilling to intervene, it is little wonder that public sentiment eventually turned against the developers. The Cumberland Plan (see Chapter 13), like earlier attempts at infrastructure and strategic planning, failed to make any real difference and so have subsequent schemes. The more things change ...

FAST LANE

The middle years of the 60s decade was a mixed bag all round. As the appeal of post-war beat and band music began to wane, a generation of teddy boys, bodgies and boogie woogie jivers was pairing off, getting married and having children. A new culture of boppers was about to tear the place apart. Enter rock music, the Beatles haircut, short skirts, Cuban heels and nylon shirts. And if you lived in beachside Australia or west coast USA there was plenty of surfing fun to be had as part of a sub-culture whose way of life scandalised sober-minded parents. Panel vans, ornately upholstered or plainly fitted out with

Brickyards out of production by 1963

* Original investor (1934) in Brickworks Limited and member of the New South Wales Brick Manufacturers Association	Original investor*
Ashfield Brick Co. Pty. Ltd	✔
South Ashfield Brick Co. Pty Ltd	N/A
Bakewell Bros Pty Ltd	✔
Bankstown Texture Brick Co.	✔
Butcher Bros. & Co. Pty Ltd	✔
Carrington Brick Co. Pty Ltd	✔
Central Brick& Tile Co. Pty Ltd	✔
Croydon Brickworks Pty Ltd	✔
Excelsior Brickworks	✔
Federal Brick Co. Pty Ltd (Alexandria)	✔
Federal Brick Co. Pty Ltd (Carlton)	✔
Flemington Brickworks	✔
Moorview Brickworks	✔
North Sydney Brick Tile Co. Pty Ltd (No. 1 Plant)	✔
NSW Brick Co. Pty Ltd	✔
Rupert Cook's Brickworks Pty Ltd	✔
Ryde Brick & Tile Works Pty Ltd	✔
St Peters Brick Co. Pty Ltd	✔
Strathfield & Enfield Brick Co.	✔
Sydney Brickworks, Charlesworth & Co.	✔
Walker Benson Pty Ltd	N/A
Warren Brick Co. Ltd	✔
Western Suburbs Brick Co. Pty Ltd	✔

5. 'Today a Part of Old Sydney Will Die,' *Sydney Morning Herald*, 25 February 1961.

Chopper and GI, 1967. Richard Calmes, combat photographer in the US Army, took this photograph during a lull in fighting in South Vietnam

just a foam mattress, remained a winner for the next quarter of a century. The introduction of the bikini, welcomed by men and women alike, was too much for some. One gentleman reader of the *Australian Women's Weekly* remarked that:

> *Bikinis offend a man's sense of feminine beauty, making women look like animated sausages tied in the middle. With even perfect figures (which are rare), they're not half as attractive or smart as neat one-piece swimsuits. Let's hope that this summer our girls try to attract men on the beach with less meat and something more petite.*
>
> *'Male Factor' (name supplied), Glen Iris, Vic.*[7]

Beach outings were few and far between for Dawes who preferred the company of a surprisingly wide range of people, including State and Federal politicians and judges, as well as fellow businessmen. Many of his friendships went back to before the start of the Second World War, having grown from a love of horse racing. Many prominent judges in Sydney kept racehorses. Dawes was well acquainted with Wilfred Robert Dovey, a judge in the Supreme Court of New South Wales, and the father-in-law of a future Prime Minister, Gough Whitlam. Like Dawes, Dovey was a keen racehorse owner and a foundation member of the Sydney Turf Club. He also sat on the committee of the Australian Jockey Club. It was through Dovey that Dawes made the acquaintance of a young man, Gough Whitlam, during the late 1930s. 'He was a nice boy, that's all I'll say.'[8] And that's all he did say, or was prepared to say.

Another mover and shaker in the business world was Henry 'Harry' Tancred, the chairman and managing director of Tancred Bros, one of the country's largest meat wholesalers.[9] Like Dawes, Tancred bred racehorses. Off the track, Dawes made a point of attending Tancred's invitation-only Tuesday luncheon club at *The Hotel Australia* on Martin Place, organised for Sydney's élite, including the Liberal Premier of New South Wales, Robert 'Bob' W. Askin. Another friend, Adolf Basser, was a wealthy Sydney jeweller and generous benefactor of the University of New South Wales. Tancred, Basser, Dawes and many others used the services of the trainer, Maurice McCarten, and celebrated jockey, Neville Selwood. Less flashy was the company of Ron Baker, a brickmaker and the owner of the firm of Baker Brothers which operated plants at Bowral and Maitland, New South Wales. Baker's dry press bricks were noted for their quality and were highly sought after by architects. By the late 1960s about one third of Bowral's production was going to the Canberra market, a similar percentage to the growing suburb of Campbelltown, and a third to local building works.

7. Letterbox, *The Australian Women's Weekly*, 29 November 1961.

8. 'Australia's biggest brickworks is also a major long term property investor', *The Australian Financial Review*, 5 September 1973, p.3.

9. Henry Eugene Tancred (1897–1961) was vice-chairman (1945¬53) and chairman of the Sydney Turf Club (STC) from 1953 until 1959. Under Tancred the STC introduced the photo-finish camera in 1946, replaced the wire starting system with barrier stalls in 1947, and pioneered the use of electrical timing devices and patrol films. The first Golden Slipper Stakes, for two-year-olds, was run at Rosehill in 1957.

TAKING CARE OF BUSINESS

Socialising, horseracing and fast cars may still have played a large part in Dawes' life, but his obsession was the Austral-Brickworks empire and new commercial opportunities, especially property development. Exhausted pit lands owned by Brickworks at Beacon Hill were levelled and roads constructed to prepare the land for subdivision. This marked the beginning of a lucrative new era for a company that was sitting on extremely valuable swathes of land within the Sydney metropolitan area.

During 1965 production at Wallgrove continued at full capacity, although it struggled to keep up with demand. A third kiln was completed in June, which brought the plant's output to 1.125 million texture bricks per week, making it the largest single unit brick producer in the southern hemisphere. By this time over £2.5 million had been invested at Wallgrove plant, which continued the trend noted previously of investing more in this plant than the paid up capital of the parent company.[10]

Meanwhile, the New South Wales Clean Air Act of 1961 was having an impact on the industries clustered in the districts of Alexandria, St Peters and Marrickville. In fact, it was one reason among many that brought production to an end for these inner city yards. Dawes despatched two of the company's officers on another overseas visit in search of less polluting types of kiln for the manufacture of common bricks. The year 1965 also marked the passing of J.C. Charlesworth, an industry stalwart and the last of the Charlesworth family that had been so prominent in the St Peters brick business. On 9 February 1965 Austral Brick issued new pay rates for burnt brick hackers and strapping operators at its Wallgrove plant, ensuring that this intensely physical work was well paid.

Old grievances were also beginning to resurface. In July, Tom Edwards, secretary of the Brick Manufacturers Association, wrote to Askin. Edwards spoke of radical changes in the industry brought about by world-wide advances in brickmaking technology (extrusion methods and tunnel kilns), the near exhaustion of raw materials in the inner metropolitan area, and the high cost of cartage. He went on to claim that, once more in its long history, the State Brickworks was undercutting prices in the dry press market. Following the removal of price restrictions in 1955 increases had been applied simultaneously by the State yard and private manufacturers. The State withdrew from this informal arrangement in 1962 which, according to Edwards, left private firms at a serious disadvantage. His complaint was mostly justified, although in fairness, the State yard was equally vulnerable due to its total reliance on dry press brickmaking.[11]

With its order book full, one might have thought that Brickworks had least cause to bemoan such pricing tactics. The reality, however, was that it was becoming increasingly difficult for any private manufacturer to sell face and common bricks in competition with the State Brickworks. In the company's annual report for 1966, Dawes angrily remarked, 'they do not pay municipal rates and enjoy concession rates for workers' compensation and other insurance which are not available to private concerns. They enjoy priority for the supply of bricks to most State Government projects and then compete with private enterprise for the remaining business at unprofitable prices.'[12] Edwards suggested that Premier Askin follow the commercial lead set by the Commonwealth Government in its brickmaking enterprise at Yarralumla, Canberra.[13] Askin was openly pro-development and lent a sympathetic ear. The imbalance between the State yard and private enterprise was redressed the following year when the Minister for Public Works placed the State yard on a direct competitive basis with private enterprise. From now on, it would be obliged to pay tax on

10. Annual Report, Brickworks Limited, 1965.

11. Tom Edwards died during his term of office in 1983. Edwards had worked in the industry since the 1930s and had been employed at the head office of Brickworks Limited in Burwood. Edwards was succeeded by Robert Rushton.

12. Annual Report, Brickworks Limited, 1965.

13. The operation was run along the lines of a private business under the prevailing accounting régime. As a matter of course, it paid taxes and was expected to turn a profit, albeit modest.

Delegates at the union's Federal conference pay a visit to the Blacktown plant of Calsil Ltd, circa 1975. Left to right: Fred Smith (union organiser), Eddie Monson (union secretary), Bill Haigh (brick worker), unknown, Jack Wood (brick worker), Milton Gowing (brick worker), Pat Conchie (Secretary of WA branch)

profits in lieu of income tax, and land tax would also be taken into account. These measures obliged the yard to increase brick prices which levelled the playing field, although not as much as private manufacturers would have liked.

SEEING RED

On the evening of 6 July 1965, as Tom Edwards was penning his letter to Askin, discussions of a more ideological nature were taking place in the rooms of The Federated Brick, Tile and Pottery Industrial Union of Australia at 22 Enmore Road, Newtown. Contentious items – political hot potatoes – had found their way onto the agenda, but were deftly side-stepped by the management committee. A suggestion by the Australian Soviet Friendship Society (a front organisation funded from Moscow) that the union send a personal greeting to the USSR–Australia Society on the occasion of their fifth anniversary was knocked back for 'ethical' reasons. Likewise, a request by the left-wing Builders' Labourers' Federation for one of its members, John McNammy, to relate his impressions following a recent visit to China and Vietnam was refused, 'owing to the pressure of business'.[14] Considering the emotionally and politically charged atmosphere surrounding Australia's involvement in the Vietnam War, it was a wise decision. Much to their credit, Fred Parker, the union secretary, and his deputy, Edward 'Eddie' Monson, deliberately steered the Federated Brick, Tile and Pottery Industrial Union of Australia away from political entanglements that would have been to the detriment of employees, employers and customers alike. This is not to imply that industrial relations at Brickworks were always tranquil. Labour disputes over the supply of wet weather gear, boots and hard hats stopped work in the pits at Wallgrove in early 1963. Fred Parker and Bill Swan, Dawes' factotum, managed to smooth things out, and averted other stoppages by virtue of having a cordial relationship. This was small fry compared to the spate of wildcat strikes and bitter disputes that characterised the Australian workplace during the 1960s and 70s.

14. Minute books of The Federated Brick, Tile and Pottery Industrial Union of Australia, Thursday 6 July 1966. See archives of the Construction, Forestry, Mining and Engineering Union (CFMEU).

As for Dawes, he disliked union shop stewards and council inspectors equally, and when it came to chasing them out of the yards he did not hold back. Plant managers were under strict instructions to escort them off the property, and if Dawes was present he made his point with typically Australian eloquence. The six-week-long strike of brick carters at Wallgrove in 1966 reveals that the ageing Dawes did not always get his own way. Ross Bell, a carter, recalls how it started.

> *One day Bill Swan put on another truck [from a haulage company based on the South Coast near Wollongong] without consulting the other carters. Up till that point there was job security on the yard. All of us were sub contractors with the exception of a few trucks owned by Austral. We all went on strike and then received a letter of dismissal from Swan who then went on holiday. By the time he returned everyone was on strike.*[15]

Dawes was furious. On one visit shortly before the strike was called off, he leapt out of his car and ripped a fibro placard from one of the striking carters. Stamping on it with his feet he yelled contemptuously at Harry Bell (the father of Ross Bell), 'I'm eating caviar, while you're eating grass.' Came the reply, 'And I'll be eating grass when you're pushing up the grass.' In the heat of the moment, tempers flared, although the exchange was an aberration; the two men had known each for nearly 40 years and had always enjoyed good relations. John Tindall, another brick carter on the picket line, acknowledged the men were afraid of Dawes because he was such a hard man.[16] Backchat was a rare occurrence.

The strike was a nuisance, but not a commercial disaster. Like previous crises, Dawes continued to run the kilns at full production and simply stacked the yard. In fact, a visiting delegation of Polish brickmakers was astounded at the pace of work set by men loading the rail cars. The lure of bonus rates for piece work, as well as their desire to impress foreign visitors, may help to explain the flurry of activity. Presumably, the official archives of Poland's post-communist state retains the record of this unusual visit. Yet, despite the return to the work the strike rankled and encouraged Dawes to establish a company-owned brick carting business, Rochedale Transport Pty Ltd, in September 1968. Based at Rochedale, the aim was to reduce the reliance on subcontractors. Most of the trucks operated by this subsidiary were transferred to Austral Brick in 1998. Things came full circle in 2007 when the operation was wound up and the driver-employees given the option of buying their trucks.

Union picnic day, Bronte, circa 1966. Fred Parker, union secretary (centre), playing darts with Eddie Monson in white shirt to the right of the picture.

The Brickmakers and Brick Carters Employees Industrial Union of New South Wales was formed in 1902. In 1920 the union split into two and The Federated Brick, Tile and Pottery Industrial Union of Australia was created. At this time the New South Wales branch was administered jointly with the Victorian branch. From 1920–1927 a Mr Carroll and Fred Parker rotated the positions of secretary and assistant secretary. In 1927 Fred Parker was appointed manager of Brookvale Brickworks and became secretary of the union following Carroll's death in 1928. Parker remained in the position until he retired in 1969.
Eddie Monson officiated as secretary from 1969 to 1984. Alex Sawtschuk continued until 1995 and was succeeded by Trevor Melksham who retired in 2006.

15. Interview with Ross Bell, Sydney, 10 September 2005.
16. Interview with John Tyndall, Kurnell, Sydney, 7 August 2005.

NEW PLANS, NEW PLANTS

The installation of a third tunnel kiln at No. 1 Plant at Wallgrove in 1965 went a long way towards meeting the insatiable demand for brick, but it was still not enough. In September 1967 Arthur Dawes made another trip to the United States to meet yet again with his colleagues at JC Steele and Swindell Dressler. The visit had a dual purpose: to report first hand on the successful commissioning of the first tunnel kiln for vitrified clay pipes at Punchbowl Brick and Pipe in Sydney, which came on line in July 1967, and to purchase new equipment. A huge order was placed, sufficient to allow for the conversion of the Rochedale operations in Brisbane in 1968 from dry press to tunnel kilns. Eventually, three Swindell kilns were built at Rochedale, the first in 1968, the second in 1970 and the third in 1972. The kilns and associated equipment had a capacity of 55–60 million texture bricks a year. It took several years, however, to establish strong sales in the Brisbane market which, like Sydney, showed some initial reluctance to embrace the new style of brick. Perhaps the single most important factor working in favour of the industry was the rapid development along the Gold Coast. It was here that high-rise apartments and brick veneer cottages began to sprout. Brisbane's growth as a major business centre also attracted a flood of job-seeking interstate migrants, all of whom needed somewhere to live.

Arthur Dawes also arranged for the delivery and installation of equipment for a new plant under construction at Wallgrove. No. 2 Plant was opened in April 1968 and a second kiln completed in November 1969, bringing the number of kilns in the two plants to five, representing a capital outlay of \$12 million.[17] Just how robust the building industry was can be gauged by the number of bricks being produced at Wallgrove – in excess of 2 million per week. Bricks produced by other yards in the group swelled this figure.

A FAMILY AFFAIR

For those who remember him Arthur Dawes was liked and respected for the person that he was – kindly, thoughtful and willing to listen. In the immediate post-war years working for his father was a hit and miss affair, with a brief interlude as the owner-driver of a taxi cab. Over the years he proved himself capable and was appointed manager of the Rochedale plant and manager of the Wallgrove operations from September 1962. Like his father he had a penchant for shiny, fast cars and took great delight in covering the distance between Sydney and Rochedale at breakneck speed. Seated in his brand-new Holden Monaro, and sporting a tweed jacket with leather elbow patches, the moustachioed Arthur looked the epitome of the suave city slicker. The daily journey to Wallgrove Road gave him a regular adrenalin surge. Leaving the family home at the crack of dawn in order to dodge the traffic he would arrive at the plant in time to catch the changeover in work shifts. Swapping the Monaro for a golf buggy, he became a familiar sight, trundling around the enormous sheds on tours of inspection.

Brickworks Limited consolidated net profit, 1956 –1968

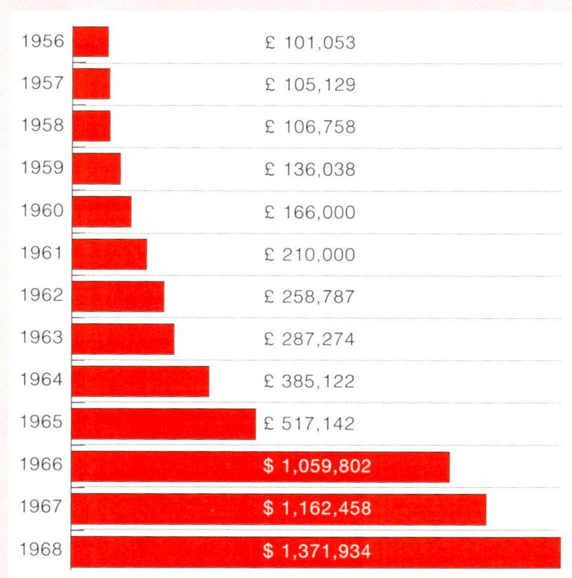

Archives of Brickworks Limited

17. Annual Report, Brickworks Limited, 1969. p.3.

New installations at the Punchbowl yard, 1967

An even-tempered man who was slow to anger, Arthur managed the various plants reasonably well, but according to a number of acquaintances and former employees he disappointed his father. Like others, Arthur ran scared of his father. The fact that he was a grown man and happily married with three daughters was irrelevant. Dawes senior was a bully and someone not to be trifled with, yet he could be as kind as he could be cruel. Arthur's death in 1979 from cancer was a blow for W.K. Dawes who had always wanted his youngest son to take over the company following his own retirement. Meanwhile, his eldest brother, William Frederick Dawes, who had trained as a fitter and turner at one of the St Peters yards before the war, gradually worked his way up and was put in charge of all dry press operations for Brickworks and Austral Brick in 1963 or thereabouts. Like his two brothers, William visited Spain, Germany, France and England and had a particular interest in prefabricated brick panels made by European manufacturers. The technology was never transferred to Australia, although today Austral Brick manufactures architectural terracotta cladding, called Terracade™. [18]

18. Interview with Ray Dawes, Sydney, 10 April 2007.

Unfortunately for him, William left school at an early age during the Depression years and was ill-equipped to deal with the responsibility of running so many plants. Young Bill was a kind and generous family man who doted on his wife and children, but he lacked confidence. Considered by many to be arrogant, the reality is that he felt ill at ease with the men, earning the nickname 'Grunter', for his monosyllabic response to questions. Certainly, he was not a dynamic businessman, and felt out of his depth with tunnel kilns and automation. Tensions between William and his two brothers surfaced frequently. It is quite possible that jealousy may have sullied relations, especially towards Ray who was the better educated of the two.

THE LONG GOODBYE

By 1968 Sydney's brick industry had come a long way. Such is the nature of new technology that once applied it rapidly upsets the status quo. In less than ten years tunnel kilns and extruded bricks had knocked dry press brick off its pedestal. Just as machine pressed brick in the late nineteenth century had created a new market, so too did modern texture brick. Every manufacturer, with the exception of those who had found a niche market in specials and face bricks, bowed to the increasing demands of home builders for wider choice.

The eyes of the world were on the NASA Apollo 11 expedition to the Moon in 1969. The Australian Women's Weekly

New installations at the Punchbowl yard, 1967

Down at St Peters the view across the Waterloo Flats was crowded with industry, canal-side warehouses and uncertainty. The company's dry press yards continued to produce millions of bricks each year, but the writing was on the wall. From his office at the Bedford yard one could sense Dawes' growing frustration at the unwillingness – or inability – of local government and government departments to recognise the essential nature of the industry. Old, cranky and having turned 70, he was dismayed by the government indecision and indifference.

> *Land owned in the St Peters area which has been traditional brickmaking land since the turn of the century was originally zoned 'industrial', then later 'open space'. After appeal it reverted back to industrial zoning.*
>
> *Now a country road is to proceed through both areas where we operate, which will seriously affect four yards. It appears the needs of the motorist to have a straight road will prevail over the needs of the community to have cheap building material for housing available within a reasonable distance of markets.*[19]

There was also pressure from the New South Wales Department of Public Health Air Pollution Division, which was obliged to act against the district's polluting industries.

> *If these government authorities are so confident that they know all the answers they could demonstrate them to us by making the necessary alterations to the State Brickworks, and prove that smoke emission can be substantially reduced and good bricks manufactured. We would welcome any such demonstration. We have spent considerable sums of money converting plants to oil firing, mainly at the instigation of the Clean Air Division, increased our operating costs because of higher fuel costs, and appear to have satisfied no one. The State Brickworks have done practically nothing to reduce smoke emissions and continue to belch smoke in the Sydney atmosphere with apparent immunity.*[20]

What more could be done? The yards at St Peters had served their purpose and were reaching the end of their working life, although it would be another ten years or more before they were finally laid to rest. In the meantime, something was going on behind Dawes' back and that something was not very nice. When you make it your business to know everything and everyone, as Dawes had during his 50 years in the trade, then men in bowler hats nosing around was enough to set alarm bells ringing. There was trouble at t' mill.

19. Annual Report, Brickworks Limited, 30 June 1968.

20. Ibid., p.4.

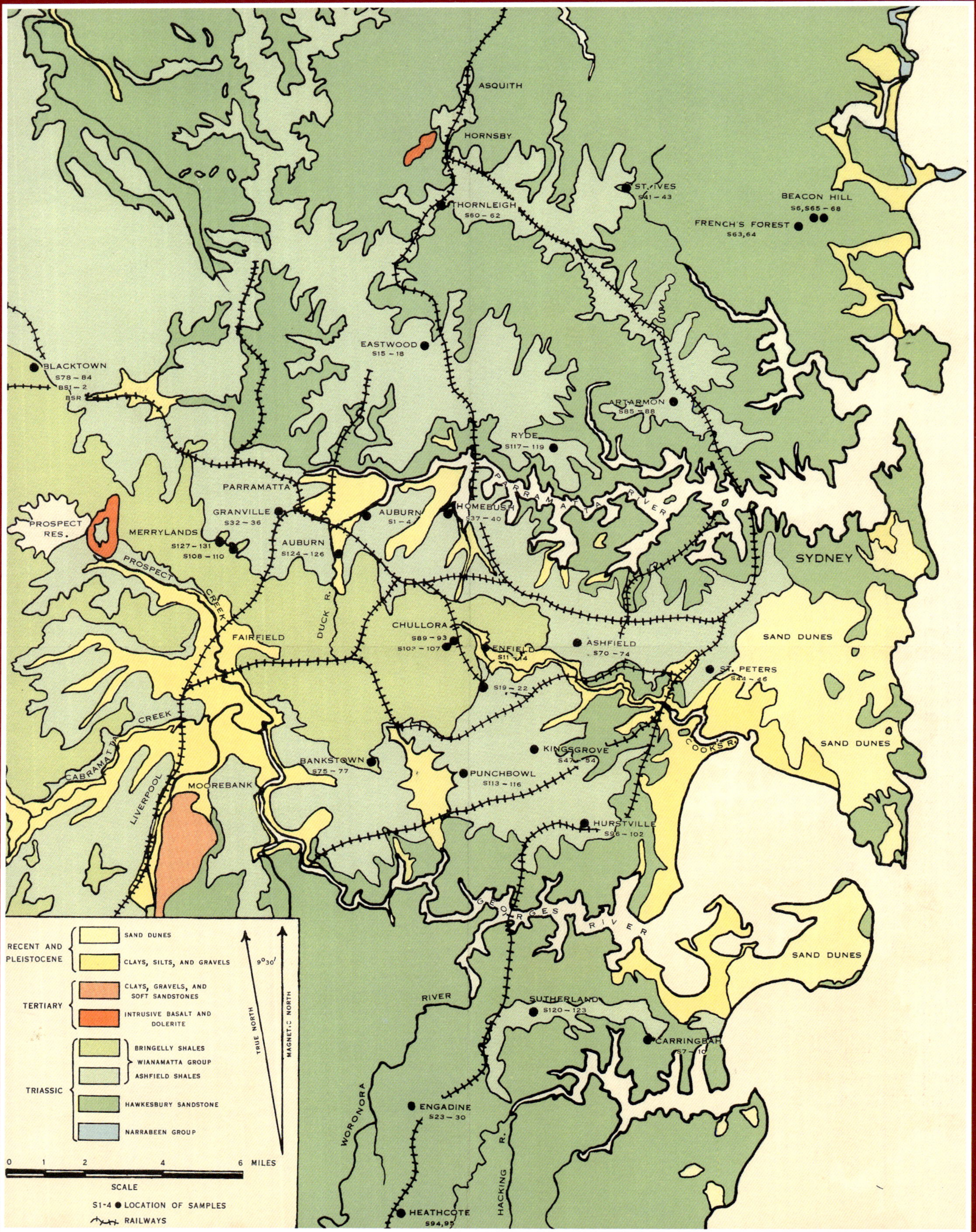

Geological map of Sydney metropolitan and adjacent areas based on Willan (1923, 1925), showing the important clay bearing formations and locations of the samples

It's time 17

Watch your back – Washington H. Soul Pattinson – Raised eyebrows – Getting to know you – New plant and equipment – Land – Storm clouds – Mystery share buyer – Please explain – War in the Middle East – Better news – The Dismissal Rationalisation – From quarter acre block to home unit – End of an era – Death of W. K. Dawes

WATCH YOUR BACK

People get married for all sorts of reasons: love, family tradition, fear of being left on the shelf, or as a matter of convenience. When Brickworks Limited met Washington H. Soul Pattinson for the first time in early 1969 it wasted no time popping the question. And surprisingly, the answer was 'yes'. But why now, and for what reason, did these two eligible parties unite in what turned out to be a most successful and mutually beneficial union? In early 1969, with brick sales from its various plants continuing to flourish, all was going well, yet Dawes was a worried man. Unwelcome news was beginning to filter down the grapevine from his friends at The London Brick Company. Those pin-striped City gents in bowler hats mentioned previously were none other than corporate raiders from the UK firm of Slater Walker. For the first – and last – time in his life Dawes and his creation, Brickworks Limited, were being stalked.

Several years later he confided in Peter Mahony, a brickmaker with nearly 50 years' experience, that he would have done anything to repel the unwanted advances of Slater Walker.[1] What Dawes had in mind was a most unusual plan which he announced at a crisis meeting of directors. A selection of companies listed on the Australian Stock Exchange was being drawn up, he informed his worried colleagues. The idea was to find a business with a similar market capitalisation that might consider a share swap of about 20 per cent for each party. This would make it difficult, if not impossible, for anyone to take over either company. In the weeks that followed Dawes and another director, George Travis, the former auditor for Brickworks, sifted through the pages of

Newbold General Refractories Limited was a leading manufacturer of refractory bricks and a major supplier to BHP. Originally established in 1908 by the Newbold Brothers, the company fell prey to a hostile takeover in March 1974. Newbold's large asset base included substantial land holdings. In the face of a deteriorating earnings it was acquired by Manufacturing Resources of Australia (MRA), which was 50 per cent owned by Hooker Corporation.

The Australian Financial Review and held confidential discussions with the company's share broker. The hunt was no easy task for it required background research on likely partners, their company structures, shareholders, and leadership. At the top of the list was the firm of Howard Smith Limited with its interests in shipping, transport and hardware, but when approached it dismissed the idea out of hand.[2] This was just as well, since the name Howard Smith is no longer with us. In October 2001 the company was delisted following compulsory acquisition by Wesfarmers Retail Pty Ltd.

WASHINGTON H. SOUL PATTINSON

We cannot be sure of the precise date, but in the early months of 1969 George Travis put forward the name of Washington H. Soul Pattinson, or

1. Established in 1964 by James Slater, an accountant for Leyland Motors, and Peter Walker who was a member of Parliament and a highly influential figure in the Conservative Party, the firm gained notoriety during the mid 1960s for purchasing undervalued companies and stripping their assets.

2. In 1854 Captain William Howard Smith formed William Howard Smith and Sons Pty Ltd and began transporting people and supplies to the goldfields. In 1901 the company became Howard Smith Company Ltd, thence Howard Smith Limited in 1914.

James Millner, circa 1995

'Soul Patts', a pharmacy chain controlled by the late James 'Jim' Millner.[3] The company is now an investment house with substantial shareholdings in coal, pharmacies, telecommunications, investment banking, fund management and media (a TV station). Millner was from a Beecroft family and had earned a formidable reputation as an astute investor and businessman. He later recalled: 'initially, we were approached by a broker who said that Brickworks Limited was looking for someone like ourselves – a like-minded partner that was conservative and afraid of takeover.'[4] The broker he was referring to may have been Patrick Brothers.

George Travis then met with Millner at the Eastwood Rugby Club. During the tête á tête Travis laid Dawes' carefully prepared cards on the table, explaining how a share swap would offer protection for both firms. Millner had a nose for a good investment and was struck by the hugely profitable business of Brickworks. He agreed in principle and the two men exchanged handshakes before going their separate ways. Several days later, Dawes went to meet Millner at his office at the 1886 Soul Pattinson's shop in Pitt Street. With his sons Ray and William, Travis, and the company solicitor in tow he liked what he saw. Millner's Spartan office arrangements – desk, chair, filing cabinet, lino – was a good omen. 'No extravagance there,' mused Dawes, whose hidey-hole at St Peters was equally monastic. Like his grandfather before him, Jim Millner always reckoned the thicker the carpet, the thinner the dividend. In fact, he took a dim view of showy displays of wealth. Once during the 1970s, Alan Bond invited him to dinner at the Bond mansion. Millner was chairman of Queensland Mines at the time, and Bond was eagerly courting him. There were other guests. 'I've never seen such ostentation,' Millner recalled. 'I made a mental note never to do business with that man. I made my excuses and quietly left. I've always been very suspicious of flash people.'[5] On the other hand, had he been around eighty years earlier then he might have thought differently. The room upstairs from Millner's office happened to be where a young, and exceedingly attractive Helena Rubinstein developed some beauty products in the early 1890s.

Having weighed each other up the two men got down to business, with Millner voicing similar concerns that his company might one day find itself having to fend off unwelcome advances.[6] Dawes' impression of Millner was borne out some years later by Bill Emmett, former manager of the Wallgrove operations. Emmett went to see Jim Millner in the early 1970s on company business. 'The Pitt Street office was very old fashioned, with a wooden desk, a few chairs, and memorabilia. His secretary was an attentive elderly lady who busied around. You would never have thought that here was a man who made millions for his companies and shareholders.'[7]

3. James 'Jim' Millner (1919–2007) was the eldest of three boys and grew up in Cheltenham, Sydney. His mother was the daughter of the founder of Soul Pattinson. His father was a decorated war hero who returned to Australia from active service in 1919 with the Military Cross. He went on to establish Commonwealth Imperial Gases (CIG) in Australia.

4. Interview with Jim Millner, Sydney, 14 October 2004.

5. Lenore Nicklin, The Heart of Soul, *Bulletin with Newsweek*, 12 January 1998, Vol. 117.

6. Mergers and acquisitions were frequent occurrences throughout the 1950s and 1960s, although most of them were friendly affairs such as the amalgamation of the pastoral companies, Elder Smith and Goldsborough Mort in 1961. Hostile mergers were still few and far between before the 1970s. The acquisition of Drug Houses of Australia by the local subsidiary of the British 'raider', Slater Walker Securities, in 1969 and the TNT raid on Ansett Transport Industries in 1972 signalled what was to come.

7. Telephone interview with Bill Emmett, 12 May 2007.

RAISED EYEBROWS

The exchange of shares took place on 23 August 1969 and was widely reported in the financial columns of the daily papers. Both companies were roundly criticised and the arrangement attacked by stockbrokers who claimed that 'bricks and pills didn't mix'. In reality, the swap was a remarkable coup and provided a two-way flow of investment income, as well as security. In November 1969 the annual report of Brickworks Limited endeavoured to put an end to press speculation.

> *The brick industry is one closely geared to the building industry which is notorious for violent fluctuations of trade. Your directors considered it would be a wise precaution to have assets outside the brick industry to cushion it against building trade fluctuations. Some years ago Brickworks Limited did have a share portfolio but disposed of if to raise additional finance for construction of the Wallgrove plant.*
>
> *Washington H. Soul Pattinson & Co Limited shares provide a substantial source of future income from diversified activities not associated with the Building Industry, as apart from operating a chain of retail chemist shops, a pharmaceutical manufacturing plant and a wholly owned finance company, that company has an investment portfolio of many leading industrial and mining shares including large holdings in Western mining Corporation Limited, Broken Hill Proprietary Co. Limited, Broken Hill South Limited, and Peko Wallsend Limited.*
>
> *Because of the extremely liquid position of Brickworks Limited the Directors also felt that it was very vulnerable to takeover bids which might not be in the interests of shareholders generally.*
>
> *We might also mention that despite newspaper hints of rumours of takeover offers, they are completely without foundation. The investment in Washington H. Soul Pattinson & Co. Limited was not made with a view to Brickworks Limited acquiring any short term benefit therefrom. It has been considered by the Directors as essentially a long term form of insurance for the future.*[8]

This was certainly true, but it was most definitely the case that a hostile take-over was on the cards, according to those who witnessed these events. Clearly, it made no sense for the board to alarm shareholders, which may account for the lengthy statement in the annual report for 1969.

GETTING TO KNOW YOU

Following the swap Jim Millner took a seat on the board of Brickworks, although his first appearance as a director did not take place until 7 October 1970. The occasion was a meeting of directors of The Carrington Brick Company, the investment subsidiary of Brickworks. Millner was present at the extraordinary general meeting of 8 April 1970 which voted to issue seven million new shares of $2 each, increasing the company's capital to $20 million.

Strangely, Dawes spurned a reciprocal offer to sit on the board of Soul Patts, claiming to know little about pharmaceuticals, and that in any case he was flat out running brickyards. The reality was that William King Dawes was a controlling personality, and what lay beyond his control held little interest. And was it really a partnership of equals, in view of the fact that Brickworks Limited had initiated the share swap? Perhaps he had convinced himself that things would continue in much the same way as before. As we shall see his refusal to engage set in motion a chain of events which stymied any succession plan Dawes may have been contemplating for a family member to take the helm. Overall, the relationship with Millner was one of mutual respect, but it never progressed beyond business. The two men were 20 years apart; Dawes had just turned 70 whereas at 50 Millner still had his best years in front of him. In terms of education and family background there could not have been more of a contrast, but as businessmen, they were peas in a pod – straight shooters, innovators, driven and strategic. They could see over the horizon, and neither of them were sentimentalists.

8. Annual Report, Brickworks Limited, 1969, pp. 4–5.

Millner was anxious to learn all that he could about brickmaking and applied his immense capacity for mastering detail. There were meet and greet tours of the company's brickyards. Ron Hagstrom at the Brookvale yard recalls a visit from Millner. 'Old Bill brought him over to meet us. We found him an excellent bloke and all got on like a house on fire.'[9] In the first few years following the share swap Millner also made a point of visiting brick manufacturers during his overseas trips to England,

Advertising material, Austral Bricks, circa 1995. No. 1 Plant produced bricks in the mid-price range, including Manor-house and Federation bricks, Heritage bricks and pavers. It also specialised in rumbling and kiln flashing, a process whereby the atmosphere in the tunnel kiln is changed to produce richer coloured bricks. Brickworks Limited

Europe and North America on behalf of Soul Patts. In time he became highly knowledgeable about bricks and brickmaking. This was apparent at the monthly board meetings which, he insisted, should be held on site, especially at the Wallgrove plant.

One thing that did change was the recording of production statistics. Before the Millner era Dawes logged production, sales figures and diary jottings in notebooks which, regrettably, have not survived. The board worked from prepared figures that, even for the time, lacked detail, although this did not seem to have been detrimental to the business. Regardless of how meticulous Dawes' note keeping may have been, Millner began to move the company away from its reliance on this somewhat idiosyncratic and inadequate practice. Yet there was a limit to what Millner could do, or insist upon – Brickworks was an equal partner and as long as Dawes was in control, he could only go so far.

NEW PLANT AND EQUIPMENT

With the events of 1969 behind him Dawes had every right to feel satisfied. In early January 1970 he set off for England and the USA accompanied by his son, Arthur, and the chief engineer, Bill McGregor. In England the three men inspected a number of factories, one of which may have been the firm of Bason & Sons Limited, a manufacturer of automatic setting machines which were still in development phase.[10] In due course six of these enormously expensive and technically complex machines were installed in the various plants at Wallgrove, one of them being No. 3 Plant which was opened in early 1972.

Dawes returned to Sydney, satisfied that 'our modern brick and pipe plants, kilns and material handling facilities are well up to world standards.'[11] Work was also nearing completion on additions to No. 2 Plant, bringing the number of tunnel kilns to five at the two existing Wallgrove plants. When fully operational these kilns would be capable of producing

9. Interview with Ron Hagstrom, Sydney, 12 September 2006.

10. Bason & Sons Ltd manufactured industrial machinery for use in brick plants. The issue of a US patent (U.S. Pat. No. 3,480,161) to Bason for a setting machine, as recently as 25 November 1969, seems to indicate that the firm's invention was still untested in the market. See www.freepatentsonline.com.

11. Annual Report, Brickworks Limited., 1970, p. 3.

Aerial view of No. 3 Plant of Austral Bricks at Horsley Park. Brickworks Limited

a combined total of 2.5 million bricks per week. Planning approval was also sought to construct an additional mill building and patent kiln at Eastwood with a weekly capacity of 300,000 common bricks. Unfortunately for Brickworks Limited, objections by the local council over pollution concerns (by now an increasingly common complaint) led to an action in the Supreme Court which further delayed work. In Queensland a second high capacity kiln at Rochedale was completed with full production of 450,000 bricks per week commencing in November 1970. Indeed, such was the pace of development along Queensland's brick-hungry Gold Coast that in February 1972 directors approved a third tunnel kiln at Rochedale.[12] According to historian, Peter Spearritt, the year 1971 coincided with one of the high points in Sydney's post-war building boom, with more dwellings completed than in any other year of the city's history. It also marked another landmark, for in that year Sydney became the first city to have more flats (17,431) than houses (15,514) completed. The pendulum between flats and houses swung back and forth during the 1970s. In 1976–77 the pattern reverted to the 1960s with more houses (11,752) than flats (6,813) completed.[13]

PLANT 3

Unfortunately, under its manager, Anthony 'Tony' Padden, Wallgrove's No. 3 extruded brick plant struggled to reach full production, although through no fault of his own. The saga relating to this factory commenced in late 1970 when blueprints

12. Board minutes, Brickworks Limited, February 1972.

13. P. Spearritt, *Sydney's Century*, University of New South Wales Press, Sydney, 2000, p. 105

were finalised and construction was scheduled to begin early in the following year. Objections from the local council and the State Planning Authority delayed the start of work until July 1971. By late 1973 the plant and its two tunnel kilns were still incomplete due to innumerable industrial stoppages and equipment failure. Mid-way through 1974 it was up and running although it soon became clear that the automatic setting machines from Bason & Sons were never going to work as intended by the manufacturer. The highly temperamental gizmos refused to be coaxed into performing their task which enraged Dawes.

Above: Bill Dennis, Manager, Australian Operations, Swindell Dressler International Company, 2007

Right: Bill Emmett (right), Chief General Manager of Bristile Limited, and John Bubenicek, posing for the camera (circa 1990). Bubenicek was the inventor of the Ceric Hydro kiln and the tile cassette system used today for the manufacture of roof tiles

William 'Bill' Dennis, a tunnel kiln engineer, caught the backwash of this expensive failure soon after joining Swindell Dressler's Australian office in 1974. In that year Dennis accompanied Sam Tricase, the US managing director of Swindell Dressler, on a visit that took in the St Peters yards. Dennis recalls the single storey administration block. 'There was Bill and an old lady who must have been his secretary. I can remember thinking what a character he was. One eyebrow was sewn up with very heavy black fishing line, giving him the appearance of a pirate.'[14] Dennis sat quietly until asked about his previous employment, which happened to be the firm of Gibbons Bros Ltd. The latter enjoyed a business relationship with Bason & Sons Limited.[15] Dawes reacted like a bull to a red rag and left Dennis in no doubt what he thought of Bason and anyone connected with the company. Acutely embarrassed, Dennis did his best to blend in with the woodwork until it was time to leave.

Sadly, in the decade prior to his death in 1981, this sort of outburst from Dawes became increasingly common. Shortly after, John Shergold, brickmaker and former managing director of Melbourne's Brick and Pipe Industries Ltd, met Dawes on a fact-finding visit to Sydney with his father.[16] The two men found him much the worse for drink and aggressive to boot. Since the brickmakers of Sydney and Melbourne stayed off each other's patch and had no particular axe to grind, the Shergolds may have had the misfortune to catch him on a bad day. In contrast, the brick king showed every courtesy to visitors from J.C. Steele and Sons in April 1972. Dawes insisted that all hospitality should be extended to the visitors 'because they [J.C. Steele and Sons] had been so helpful during his visits to the USA.'[17]

14. Interview with Bill Dennis, Melbourne, 28 June 2006.
15. Gibbons Bros Ltd of Dudley in Worcestershire, England, was a firm of kiln makers and had close ties with Swindell Dressler.
16. After graduation, John Shergold joined a firm of chartered accountants and for the next 11 years worked on the account of Brick and Pipe Industries. The firm invited him to become the managing director and he remained in the position for 17 years. At the same time his father, Albert Shergold, was appointed chairman.
17. Board minutes, Brickworks Limited, February 1972.

LAND

Meanwhile, as Sydney's suburban sprawl gathered pace the importance of defining and locating deposits of the underlying shales belonging to the Wianamatta Group became critical. (See Appendix 11 for more information on the geology of the Sydney Basin.) Brickmakers were not the only parties concerned about future supply. The New South Wales Department of Mineral Resources and Development[18] recommended that:

> *Definite areas free of conflicting land use constraints .. be set aside to ensure adequate future supplies of certain raw materials.*[19]

Between 1971 and 1972, 10, fully cored diamond drill holes were sunk by the Department of Mines. Additional information was derived from subsequent diamond drill holes for coal exploration, the Metropolitan Sewerage and Drainage Board, and private companies. In all, 58 fully cored drill holes were sunk and the shale tested for its ceramic properties. At the same time, Dawes continued to prospect for clays and shale, having obtained a mineral exploration license for the Bilpin area in Sydney's north west.

It should come as no surprise that the acquisition and disposal of land had long been a part of the brickmaking business. On several occasions Dawes commented to his son, Ray, that in future years the value of the land occupied by the company's yards would be as significant as the brick business itself. His prediction was borne out by the extensive landholdings which are now owned by Brickworks, underscoring their importance in the company strategy of diversification. In the meantime, the trade in land was brisk and conducted on behalf of Brickworks by Bruce Parks, the real estate agent and valuer. Under discussion at the directors' meeting in October 1969 was the approach made by Sutherland Shire Council to acquire 18 acres of the company's land at Menai for road building purposes. The Menai site held deposits of cream-burning clay but with significant reserves at nearby Heathcote, it was agreed to sell the land, provided the company was able to retain land on either side of the proposed road.[20] In February 1970 a further 10 acres of land was put up for sale at Kirrawee, near Sutherland. Two years later at the October 1971 board meeting Dawes informed his colleagues that good, reasonably priced land was available in the Brisbane area and on the Central Coast. Austral Bricks also owned at least 318 acres of land at Quaker's Hill in Sydney's north-west. This land, valued at $18,000 per acre, was resumed by the New South Wales Housing Commission in 1974.

Brickworks was not the only manufacturer with plans. On 19 August 1969 the State Brickworks purchased approximately 200 acres of land at Doonside near Blacktown, which had been owned by the State Timber Works. Buoyed by continuing demands for its dry press bricks, the new yard was close to Liverpool, Parramatta and the west, and better placed to service the rapidly expanding market.[21]

Location and book value of land owned by Brickworks Limited, June 1972

Location	Book Value	Valuation	Increase
Brookvale	$24,812	$60,000	$35,188
Burwood	$118,000	$683,000	$565,000
Eastwood	$185,600	$448,750	$263,150
Thornleigh	$88,000	$150,000	$62,000
Thornleigh	$230	$113,500	$13,270
Alexandria	$665,000	$1,700,000	$1,035,000
Ashbury	$104,000	$906,650	$802,650
Duffy's Forest	$140,000	$300,000	$160,000
Bayview	$55,000	$112,500	$57,500
Rochedale	$129,939	$90,060	-$39,879 9
Total	$1,510,581	$4,464,460	$2,953,879

Source: Minutes of board meeting held at 32 Cheltenham Road, Croydon, on 29 June 1972. Archives of Brickworks Limited

18. This went through various changes in name to become the Department of Mineral Resources and Primary Industries.
19. Chris Herbert, Geological Survey of New South Wales, Bulletin 25, Department of Mineral Resources and Development, 1979.
20. Board minutes, Brickworks Limited, October 1969.
21. Statements of Account and Balance sheet of the State Brickworks for 1973–74, Annual Report. vol.1 1975–76, p.336. In September 1972 the new factory at Blacktown commenced production, and it was officially opened on 6 December 1972. The layout and construction of the plant were based on modern automatic equipment and incorporated the latest design in tunnel kilns.

By late 1972, however, there was growing concern among stockbrokers and financial writers who appeared not to understand the strategic and tactical importance of land to brickmakers. As Dawes pointed out:

> *In the opinion of the Directors land should only be sold when the brickmaking material has been exhausted, or, when for other reasons, it becomes impossible to use it for brickmaking purposes.*[22]

A deep seam of cream-burning kaolin runs beneath Duffys Forest in New South Wales. The Australian Waratah plant (Telopea speciosissima) thrives on kaolin soils which extend far to the west. W.K. Dawes used aerial surveys to identify valuable clay reserves in this district

DUFFYS FOREST

As mentioned previously the need to procure land for future developments was not only a problem for the clay products industry. One other consequence of Sydney's growth was the urgent need to find a suitable location for a second airport. The search lasted for over three decades until it was decided to shelve plans altogether during the late 1990s. For a short while (long enough to spook Brickworks) the area surrounding Duffys Forest was considered. The reader will already be familiar with land in the district having been purchased earlier in the decade as a source of clay and shale, especially the cream-burning variety.

In January 1969 the Commonwealth Government established a committee to advise on the airport needs of Sydney. Duffys Forest was one of 11 sites under investigation.[23] Over the next 18 months there was a steady stream of correspondence between the Department of Civil Aviation and Brickworks about the possible sale of land to the department. By September 1971 the committee had narrowed the search to four possible locations, including sites in Richmond, Somersby, Duffys Forest and Wattamolla. On 22 December 1971 Duffys Forest site was ruled out because of its potential impact on Kuring-gai Chase National Park and drainage problems on land adjacent to the proposed runway. This came as a relief to the board of Brickworks Limited which received a letter from the Department of Interior advising that the Minister of Civil Aviation would not be proceeding with the establishment of the aerodrome at Duffys Forest.[24]

STORM CLOUDS

Like weather patterns, the economic fortunes of a country can fluctuate, sometimes wildly. From about June 1970 a combination of credit restrictions and wildcat strikes in the building industry began to have a serious impact on trade. The following year was no better with a stoppage called at the Doonside (Sydney) yards of PGH Limited in November 1971 by the Federated Brick, Tile and Pottery Workers Union. The former president of the union, Alex Sawtschuk, recalled that 'it was sparked by low wages. We asked for $10 a week extra, but this was refused. I went out to Punchbowl [the plant owned by Brickworks] to ask for support, but they didn't want to join.'[25] The matter was resolved by an award increase of $8 flat and productivity bonuses which were spread across the industry. At the Austral yards renewed trouble was brewing among the brick carters, independent haulage contractors who were responsible for deliveries. The issue was over a request from Mr Hewitt, secretary of the master Brick Carters' Association, for a brick truck and driver to be assigned permanently to Wallgrove. Bill Swan had always made these appointments but was on annual leave. The company secretary, Lloyd Willis, asked that the matter be settled upon Swan's

22. Annual Report, Brickworks Limited, 1972, p. 4.

23. House of Representatives Hansard, 15 September 1971, p. 1,344.

24. Board minutes, Brickworks Limited, February 1972.

25. Interview with Alex Sawtschuk, Sydney, 28 October 2004.

Porter Galbraith Hanson – PGH

PGH had its origins in the mid 1920s when roof tile manufacturer, Maxwell Porter, joined forces with David Galbraith to establish a terra cotta roof tile plant in Doonside, Sydney. In 1949 the public company, Porter and Galbraith Ltd, was formed to acquire the roofing tile manufacturing company which then operated as Porter and Galbraith Pty Ltd. In 1958 the two men purchased a brickworks and merged with Hanson Consolidated. After several changes in ownership PGH was acquired by CSR Limited in 1995.

return. On 23 November 1971 he wrote to the men urging them to think twice before embarking on a strike. The strike went ahead regardless and when the company sought a compromise to break the impasse Hewitt refused to back down. Dawes was incensed by the actions of the brick carters. Of the brickmakers' union, however, he had nothing but praise. 'I have been in the industry for 40 years and 98 per cent of the industrial trouble has been caused by other influences such as coal strikes, oil strikes, metal trade strikes and all that sort of thing.'[26]

Unfortunately, strikes became a depressingly familiar scenario during the 1970s, along with upheaval of a different kind. The pace of social change, a product of postwar affluence and the generational shift, was taking place faster than anyone could imagine – or accommodate in so short a space of time. In Australia and across the western world the anti-Vietnam War protest movement gathered momentum. In Queensland, Premier Joh Bjelke-Petersen declared a state of emergency in response to violent protests

Brick trucks owned by the Bell family. from top:

At the Euston Road yard of City Bricks. Brick trucks were still being hand loaded when this photograph was taken in 1965. 'Texture Brick Plant' signified that the carter was assigned to the Wallgrove site

The yellow stripe on this 'Inter' truck (circa 1975) was introduced by Bill Swan to enhance the appearance of brick trucks that carried bricks from Wallgrove plants

The so-called 'Bell' grab on the back of this brick truck illustrates the transition away from manual unloading. Bricks were stacked by hand on the truck, then unloaded by the mechanical arm which was designed by Ross Bell. Harry Hardfoot, an engineer from Chipping Norton, fabricated the device in 1970

Ford brick truck, 1980

Ross Bell Collection

26. 'Australia's biggest brickworks is also a major long term property investor' *The Australian Financial Review*, 5 September 1973, p. 3.

against the 1971 Springbok tour. Bearded men, long hair, appalling fashions and fondue were 'in'. Parents born in the early decades of the century and their conservative values were 'out'.

The election of Gough Whitlam as Prime Minister in December 1972 convinced Dawes that the country was on the downward slope. Almost overnight the new government enacted a series of sweeping reforms including the introduction of universal health, legal aid, wage indexation and free tertiary education. It also withdrew Australian troops from Vietnam, opened diplomatic relations with Communist China, ended National Service and banned all future sporting tours of Australia by South Africa. Unfortunately for Whitlam, a series of scandals and ministerial dismissals, including the Juni Morosi episode and the Arab Loans Affair, called into the question the integrity of the Deputy Prime Minister, Dr Jim Cairns – and ultimately Whitlam himself.

MYSTERY SHARE BUYER

Brickworks continued to thrive following the share swap with Soul Patts, increasing its sales and maintaining profits. Like everyone else, however, it soon found itself caught up in a prolonged recession. By early 1973 the poor state of the economy was being felt in the industry through the collapse of brick sales and reduced profits. Brickworks was more fortunate than other manufacturers. Helping to smooth results in the lean years was a flow of income from a growing investment portfolio of shares and commercial bills.[27] This war chest received an unexpected boost as the share market continued its slide. In mid-September Dawes instructed a Sydney firm of stockbrokers, Messrs Bernard Curran, McHugh & Co, to purchase stock on behalf of the Austral Brick Co Pty Ltd. Acting on impulse, he went on a buying rampage, snapping up shares in MIM Holdings Ltd, CSR Ltd, Aquila Steel Ltd, Western Mining Corporation Ltd and BHP Ltd. The magnitude of these purchases had the unusual effect of temporarily reversing the downward slide of stocks. The city was rife with rumours of a 'mystery buyer', although try as they might financial writers and market pundits were unable to discover his identity.

The instincts of a gambler may have spurred Dawes into his share-buying exploits, but this sort of swash-buckling behaviour was plainly risky. The word cavalier springs to mind. The foray into the market lasted only a few days and was halted by the board of directors as soon as they discovered what Dawes was up to. To a hastily convened meeting of directors on Tuesday 26 September 1973 he sheepishly explained that he had acquired shares in the aforementioned companies to the value of $2,058,466 because the market was 'so low'. Dawes tried to allay concerns by offering to desist from any further purchases and to refund cheques drawn by the Austral Brick Co Pty Ltd to cover both the company's and his own purchases.[28] Before the meeting closed Millner, Travis and Taylor were instructed to make enquiries with the brokers to determine the facts and report back. Two days later the board reconvened, this time on Millner's home turf – the Pitt Street office of Washington H. Soul Pattinson. Sensing that he had gone too far, Dawes resigned as managing director, but this was shortlived. A little over a week later the annual general meeting re-elected the 74 year-old 'King Billy', with one shareholder describing him as 'the Essington Lewis of the brick industry'. In fact, 'no-one mentioned the share spree – even after Mr Dawes exhorted: "Anything else, gentlemen, you'd like to discuss – remember, you only get one chance a year."' When Brickworks Limited sold these investments in 1986 no one complained: the company made a profit of $10.26 million on the shares.[29]

PLEASE EXPLAIN

In challenging Dawes, the board sent a clear signal that they would not tolerate this sort of behaviour. Corporate regulation has come a long way in Australia over the last 35 years and such activity

27. These included Newbold General Refractories Ltd (400,150 shares), Australian Gaslight Limited (13,200), North Broken Hill Ltd (85,000), Washington H Soul Pattinson Co. Ltd (47,797), Hamersley Iron and Brisbane City Council (commercial bills). Source: Archives of Brickworks Limited.

28. Board minutes, Brickworks Limited, 26 September 1973.

29. *Daily Mirror*, 7 October 1986.

today would have serious consequences. Dawes could feel the wind and from where he stood on the deck it was no longer filling his sails. Remarked Albert Burgis:

> *I recall that Bill Dawes progressively stepped out of the picture and his son took over. Young Bill ran the Wallgrove plant. Millner was on the board, but he left the job of running Brickworks to Bill Dawes Jr.*[30]

Dawes was yesterday's man but wise enough to know it. He was old, irascible, and surrounded by men who had grown old with him, a rollcall of names that harked back to the 1930s – Laurence Taylor, Bill Swan, George Travis, Harry Bell, and many others. 'Once age creeps up on you', he told Robert 'Bob' Rushton who joined the New South Wales Brick Masters' Association in 1972 as industrial officer, 'you are put in a separate room and given a secretary. You aren't involved in the business as much as when you were younger.'[31] The fact is, Dawes was deeply unhappy with just about everything, and was more conscious than ever that he was losing his grip.[32] He worried about industry price cutting, oil price rises and costly pollution control measures which were estimated to add about $7 per thousand bricks, strikes, the state of the nation, failed equipment, his crumbling marriage, health problems and labour shortages. He was also worried about market advances being made by his main competitor, Clark Bricks Limited.[33] Local councils who were unsympathetic to the needs of the brick industry and politicians, especially those in Canberra, raised his ire. Dawes mentioned these concerns to a journalist whose article appeared in *The Australian Financial Review* two weeks before he [Dawes] embarked on his share buying plan.

> *Politicians should remember that, like other members of the community, they are paid to do a job, and to produce results. They have a grave responsibility because their legislation today is the basis on which we build the Australia of tomorrow. Surely it is time members of parliament stopped pandering to pressure groups merely to ensure their own re-election, and instead honestly face and endeavour to find answers to some of the problems confronting us today.*
>
> *Our Company has shown its confidence in Australia over the last three decades. After the end of the Second World War, we were requested by the Government of the day to expand brick production to meet the growing needs of the Building Industry. This we did, and have continued to do. In the last ten years we have expended over $20 million in the construction of new works, increased production, improved our product, and endeavoured to meet the demand for bricks at all times.*[34]

He was particularly scathing about the newly-established Prices Justification Tribunal.

> *I would not go in front of a Labor Government that is playing around like a lot of hens and chickens for their pleasure unless the law is enforced.*[35]

Nothing appeared to be working, especially for builders in search of bricklayers. An article in the June 1973 edition *Australian National Clay* highlighted the acute shortage. The Housing Industry Association blamed the situation on the prolonged failure to train sufficient bricklaying apprentices. New South Wales had trained 115 in the financial year to 30 June 1973, with Victoria training just 67, the other states being in a similar predicament.[36] And what was the Prime Minister going to do about labour shortages in the industry, which were forcing brick plants to operate at less than 40 per cent capacity?

> *You see, politicians 12 to 18 months ago were worried because 70,000 people were unemployed. None of them has ever woken up to the fact that they have 70,000 unemployables. During the last ten years there have been large variations in the rate of building. Most of the downturns in the building industry have been deliberately engineered by the Commonwealth Government to handle and solve current problems.*[37]

30. Interview with Albert Burgis, Sydney, 15 November 2004.
31. Interview with Bob Rushton, Sydney, 18 October 2004.
32. William Dawes separated from his wife of nearly 60 years in about 1975.
33. At the time Clark Bricks was owned by Blue Metal Industries Limited, a subsidiary of Boral Limited.
34. 'Australia's biggest brickworks is also a major long term property investor,' *The Australian Financial Review*, 5 September 1973, p. 3.
35. Ibid.
36. *Australian National Clay*, Vol. 4, No. 3, June 1973.
37. Annual Report, Brickworks Limited, 1971, pp. 6–7.

Brickworks Limited, consolidated profits 1967–1981

Source: Figures compiled from the annual reports of Brickworks Limited

Prices of bricks at kiln, 25 June, 1973
(to be charged on deliveries from 1 October 1973)

Extruded bricks	Kiln prices per 1,000
Red texture, wire cut, smooth	$72.55
Salmon texture, wire cut	$75.55
Manganese texture, wire cut, smooth	$90.55
Cream texture, wire cut, smooth	$92.55
Old colonial sandstock – red	$79.55
Old colonial sandstock – cream	$97.55
Old colonial sandstock – manganese	$96.55
Red Norman textures	$77.55
Cream Norman textures	$97.55
Extruded load bearing commons	$59.55
Seconds	$51.55
Cartage	$11.50 Zoned

Dry press bricks	Kiln prices per 1000
Red face – light and medium	$79.55
Chocolate and Dark Face	$89.55
Cream face	$90.55
Clinkers	$79.55
Commons	$59.55
P.K. callows, D.K callows	$59.55
Cartage	$11.50 Zoned

Source: Archives of Brickworks Limited

WAR IN THE MIDDLE EAST

Dawes' simmering anger and frustration with local issues pale into insignificance when we consider international events. In the early hours of 6 October, Egyptian and Syrian armies attacked Israel, advancing rapidly on two fronts. Recovering from the onslaught, Israel then forced the Syrians and Egyptians back, establishing a salient on the west bank of the Suez Canal. Making good on previous threats, 11 of the oil-producing Arab states announced production cuts in a bid to pressure the West into restraining Israel. The effect was immediate: by the end of October prices had soared by 70 per cent, and the supply of Middle East oil to the USA and Europe was reduced to a trickle. The embargo only ended

The Yom Kippur War of 1973. Jacky 'Yaacov' Costi, a conscript in the Israeli Army, took these photographs during a lull in fighting in the Sinai Desert. Yaacov Costi www.costi.name

in March 1974 but the shock effect threatened to bring the western economies to a standstill. Australia, like everyone else, was obliged to pay higher crude prices on the world's spot markets. The increase in oil prices nudged the Australian economy over the edge and into recession, with the house building market in Sydney and Brisbane bearing the brunt.[38] Heavily reliant on oil to fire the kilns (tunnel, patent and downdraught alike), the use of coal was out of the question following the introduction of the Clean Air Act. The plight of the building industry

38. Annual Report, Brickworks Limited, 1974. The report noted that fuel prices had increased by 150 per cent in less than 12 months. Worse was to come. Two years later the Organisation of Petroleum Export Countries (OPEC) doubled prices overnight. Oil which had been US$1.50 a barrel in 1970 was now on the way (albeit briefly) to a high of US$31 a barrel.

Carl Zeppenfeld inspecting kiln control instrumentation at the Punchbowl plant, 1974

went largely unnoticed despite intense lobbying and requests for Budget measures that could provide some stimulus to the economy.[39] Premier Robin Askin even wrote to the Prime Minister claiming that unless the federal government acted quickly in supplying funds, the housing industry would collapse in New South Wales.

BETTER NEWS

It would be a mistake to stigmatise this period as one of unrelieved gloom. Good things were still happening. In June 1974 Brickworks announced a bonus share issue of $8 million out of profits arising from the revaluation of assets.[40] Albert Burgis, general manager of Punchbowl Brick & Pipe Company was able to commission in September 1974 a new 75-metre tunnel kiln fuelled by naptha which was convertible to light oil. In a development that reflected the times more women were being employed in the brickmaking industry. During March of that year an article in *The Sun* newspaper opened with the line: 'Want a job that gets you into fantastic shape, offers equal pay and a happy working atmosphere? Australia's biggest brick company is employing women as brick stackers and strappers – and after nearly three months there is a mutual feeling of delight.' 'Amen to that', was the response of male employees in what had hitherto been an exclusively masculine occupation.

> *The manager of Austral Bricks' Wallgrove complex, Bill Reilly, said, "In January, 10 women started as a crew at the No. 2 Plant and three weeks ago we started another crew of 12 at the No. 3 Plant. We were advertising for men and getting no results, so we ran a few ads for women and the 'phones are still ringing."*[41]

39. 'Talking Trade,' *Daily Mirror,* 16 October 1974.
40. Ordinary General Meeting, 18 June 1974, Archives of Brickworks Limited.
41. 'We're muscling in on the brickie boys,' *The Sun* 20 March 1974, p. 82.

The writer went on to say how a 'Mrs June Moore, aged 36 and with six school-age children, thought she would die after the first few days, but now she regards stacking a minimum of 40,000 bricks a week as just a job. "My muscles have toned up, I've lost weight and feel absolutely marvellous. The physical side of the work doesn't worry me at all now," she said. The plant foreman, Ron Manning, chuckled. "There's been a slight difference in shape since some of the ladies started work," he said.'[42] The stacking mentioned by June Moore could have been a reference to either a carousel, a rotating chain to which bins were attached, or the kiln cars. The carousel demanded concentration and great physical and mental dexterity. As the bins clanked along the rack, bricks were neatly stacked, bound together by a metal tie, placed on a wooden pallet and then picked up by a forklift truck for storage in the yard. The article in *The Sun* newspaper, however, was two years behind the times. In 1972 Brickworks was one of the first to employ women at its Rochedale factory. *The Australian Financial Review* noted that:

> *It's* [the Rochedale plant] *going better than any other brickworks in Brisbane. Before long women will be working in every industry. With increasing automation the work is less heavy, and it will attract more women employees. In the USA women employees in brickyards are common.*[43]

Female employment in the clay industry may have raised eyebrows in Australia, but in Europe it was commonplace and had been so ever since the end of the Second World War. Germany, in particular, was heavily reliant on female labour due to the death of so many adult males during the 1939–1945 conflict. The economic policies of Chancellor Konrad Adenaur of West Germany transformed the country's economy from 1948 and was heralded as an 'Economic Miracle' *(Wirtschaftswunder)*. In the face of huge demand for the rebuilding programme, the German brick industry installed the latest in tunnel kilns. These were far less reliant on manual labour, although setting, stacking and unloading the kiln cars remained labour intensive until automation arrived during the early 1960s.

Women or no women, the work was physically demanding. Kevin William 'Bill' Larfield, who retired as manager of No. 3 Plant at Wallgrove in 2006, recalls that the majority of workers at the plant were farmers, many of them vegetable gardeners originally from southern Italy. Among their number was a fair smattering of tomato farmers whose first love was the land. Paid work at the brick factory put bread on the table, which was a necessary fallback between harvests. Yet a ripening tomato waits for no man, with the result that labour simply disappeared from the yards during picking times, making it difficult to operate a brick plant in the heart of market garden territory.

THE DISMISSAL

It was Whitlam's misfortune to have been left a poor legacy of high unemployment and inflation by the previous government. Unfortunately Labor's policies required him to deal with an economy in recession, curb inflation and open up spending on much-needed reform initiatives, all at the same time. By the middle of 1974 labour disputes and militant unionism were on the increase. The government was becoming deeply unpopular. Dawes continued his attacks on the miserable state of affairs created by constant industrial stoppages, which he partly blamed on government policies.

> *Mr J. B. Donovan, Chief Economist of W.D. Scott & Co. Pty Ltd in a recent address referred to the 'English disease' of senseless, incompetent, and ultimately self-defeating income demands by trade union. This disease should not be allowed to spread in Australia. The only answer appears to be to give the members of the unions the democratic right to determine major decisions by secret ballots, and not allow decisions to be made at minority meetings swayed by emotional speeches and radical demagogues.*[44]

42. Ibid.
43. 'Australia's biggest brickworks is also a major long term property investor,' The *Australian Financial Review*, 5 September 1973, p. 3.
44. Annual Report, Brickworks Limited, 1975, p. 3.

Mr David Smith, official secretary to the Governor-General, reads the proclamation dissolving both Houses of Parliament on the steps of Parliament House, Canberra, on November 11, 1975. Behind him is Mr Whitlam. Australian National Library

In October 1975 the new leader of the opposition, Malcolm Fraser, blocked supply in the Senate. For three weeks there was deadlock following Whitlam's refusal to call an election. Ultimately the Governor General, Sir John Kerr, ended the constitutional crisis and dismissed the Prime Minister from office. The sudden departure of Gough Whitlam on 11 November at 11.00am – Remembrance Day – came as a shock to many people. Malcolm Fraser was appointed caretaker Prime Minister, and in the ensuing election the Liberal-Country Party was voted into office by a huge majority.

RATIONALISATION

The impact of the recession of the mid 1970s was timely in that it brought forward the closure of uneconomic yards. The brickmaking industry was fast being concentrated into the hands of several large companies. In the case of Brickworks absorption and rationalisation meant the closure of many of the older yards and the sale of land. The former Burwood yard at Cheltenham Road is a case in point. The exact date when this yard went out of production is difficult to ascertain, however the annual report for 1974 notes that three brickyards were no longer in production, and that five more were on offer as real estate. Attempts to develop the 14-acre site were rejected by Burwood Municipal Council in September 1974.The architect's vision for the disused pit was for terraces that would descend in a series of steps to the bottom of the pit 30 metres below. Each terrace would provide residential accommodation and associated commercial premises. It was either too visionary or unworkable because the idea was killed off in the planning stage. On the North Shore, the Thornleigh yard ceased production in 1975, although it continued to sell its brick stockpile until August 1978 when the yard closed for good. Working machinery was dismantled and used as spare parts; the machine shed, kilns and office stood vacant for another 10 years. Treasure hunters and scavengers picked over the carcass. Large portions of the land were sold to Hornsby Council which later developed a sports complex on part of the site.

One person who was not there to witness the passing of these yards was Ray Dawes who announced his resignation from the board of Brickworks and other companies within the group on 29 March 1976. Having spent over 30 years in the industry it was time to move on and make a life outside the close confines of his father's creation. Like many others Ray Dawes could see tentative signs that the recession was gradually beginning to lift. In New South Wales and Queensland the market trend for Brickworks was upward, with consolidated net operating profit increasing by 11 per cent for the year to 30 June 1976.[45] Notwithstanding, the Wallgrove and Rochedale plants were still operating below capacity with mountains of bricks stacked in the yards. Throughout 1977 stocks continued to remain high and there were moves to curtail production and lay off staff. It was a time for belt-tightening all round. Arthur Dawes, a director and General Manager of the Wallgrove plants, wrote to brick carters. Ross Bell was one who received the letter on 9 March 1977.

> *To help hold and improve our present level of sales, the management wishes to bring to each brick carrier's notice the very important part he plays in presenting the company's image to our customers.*[46]

Arthur Dawes' letter was a reminder of the importance of maintaining customer loyalty in a market that was highly competitive. In another initiative designed to improve production and cut costs the Wallgrove plants were converted from oil to natural gas. The gas was piped via a branch pipeline constructed by Brickworks and connected to the main line operated by The Australian Gas Light Company Limited. The gas originated from the Moomba field in South Australia.

Meanwhile, during May 1977 Arthur Dawes offered the Euston Road site (the City yard) to Sydney City Council for use as a refuse disposal depot, but this was turned down. In Sutherland shire the Heathcote yard was closed at the end of 1978. A dry press operation since its establishment in 1948, Heathcote had supplied basic cream bricks but was forced out of production due to the high price of oil. Like so many others the abandoned pit was used to dispose of waste, in this case truckloads of sewage spoil processed by the Metropolitan Water and Drainage Board in the course of its operations. Closure of the Heathcote pit obliged the company to source more of its clay and shale for the Brookvale operations from Duffys Forest, as well as Canoelands near Wiseman's Ferry. The lean years quickened the pace of yard closures, although by the middle of 1978 the recession had reached its low point. There was, however, a financial cost and the annual report for 1978 records losses due to the closure of four dry press plants, and the scrapping of the kilns, buildings and machinery. The report stated that 'these old dry press brick plants were established over 60 years ago and have reached the end of their economic life. Plant and kilns have no residual value as the method of operation is virtually obsolete.'[47]

The Bedford Brick Works opposite St Peters railway station was one of the plants mentioned. In February the yard burned the last of its famous bricks, although in an interview with the local newspaper, Bruce Parks, Austral Bricks' valuer and real estate consultant, assured readers that the works were good for another 40 years' production.[48] Parks' statement would have created a degree of anxiety for South Sydney Council, which was considering a development proposal for the site. With the yard winding down, the majority of local residents were looking forward to a life without black smoke, brick

45. Reported in *The Australian Financial Review*, 22 September 1976.

46. Letter from Arthur Dawes to brick carters, 9 March 1977. Private papers of Ross Bell.

47. Annual Report, Brickworks Limited, 1978.

48. Reported in the local newspaper, *Guardian*, 22 February 1978.

dust and blasting. The proposal was accepted in August 1978 although it never went ahead. Instead, the land was acquired as public space by the state government, yet it was several years before plans for a new recreation area – Sydney Park – began to take shape. The resumption was part of a wider plan announced in July 1978 to 'green' Sydney. Numerous industrial sites, many of them former brickworks, were purchased in the municipalities of Burwood, Ashfield, Drummoyne, Leichhardt, Canterbury and Hurstville. Their rebirth as city parks continue to remind us of the city's industrial past and the importance of the brick industry. Until work started on Sydney Park the downdraught kilns at the Bedford Brick Works found a less elevated role as the residence for a succession of homeless people.

Closure of Sydney brickyards 1967–1985

1967	• Symons Bros, Fairfield
1970	• Enfield Brick, Greenacre • Albion Brickworks, St Peters • Auburn Brick, Tile, & Pottery Co
1973	• Bankstown Brickworks
1974	• Sutherland yard of Punchbowl Brick & Pipe Co Pty Ltd.
1975	• National yard of Brickworks Limited, Thornleigh.
1978	• Bedford yard (Newtown/St Peters) of The Austral Brick Co. Pty Ltd • Punchbowl yard of Punchbowl Brick & Pipe Co Pty Ltd. End of brickmaking. The 70m brick chimneystack was demolished on 16 September 1980, and the pit filled to make way for expansion. • Heathcote yard of Punchbowl Brick & Pipe Co Pty Ltd. • South Ashfield (Ashbury) and Burwood pits sold to SPEC in June 1978 for $1.25million.[49]
May 1983	• St Peters yard of The Austral Brick Co. Pty Ltd • Brookvale yard of Brickworks Limited – downdraught kilns closed.
1985	• Merrylands Steam Brickworks (acquired by Monier in 1979) • New South Wales Brick Co. Ltd

49. A 1982 report by Hughes, Truman and Ludlow, commissioned by the Department of Environment and Planning (who had acquired the land for open space), examined the old Burwood works. Their report stated that a square-sided stack, a square-ended Hoffman kiln, four downdraft kilns (two free-standing, and two enclosed in a steel-framed building), a timber framed brick making factory and brick retaining walls, were all still standing on the site.

FROM QUARTER ACRE BLOCK TO HOME UNIT

Throughout the 1970s Sydney continued to expand in leaps and bounds towards the western horizon. Field and woodland levelled for housing plots were re-made in brick, tile and asphalt. Commerce, government and property developers, hungry for office space, whittled away even more of the historic inner city. Among the defenders of architectural and aesthetic integrity, as well as an increasing number of horrified bystanders, there was consternation over Sydney's disappearing architectural inheritance.

Whether domestic, retail or commercial, it is difficult to escape the conclusion that much of what was designed and built during this period was bland, utilitarian, and downright ugly. You do not need to be a heritage specialist to appreciate how much was being lost. For example, villas and mansions, already disfigured by subdivision and conversion into hostels, were razed to make way for home units, continuing a practice that had been developing momentum since the end of the Second World War. In architectural terms, 'design' often meant little more than the retention of stone walls and other vestigial remains of good taste embraced by previous generations. Extruded texture brick, especially light brown and cream, became the scapegoat for all that was considered to be in poor taste. The real culprit was not the choice of building materials, but poor design, usually the result of cost cutting on the part of the developer. A minority of architect-designed flats made a positive statement, but most of them were simply the result of a production-line system. Building plans simply rolled off the conveyor to accommodate Sydney's ever-growing population. Greed, opportunism, and indifferently applied planning regulations also worked in favour of the rush to develop medium-density housing. Arguably, much of the development was hasty, and a by-product of rapid growth in Australia's population driven, in part, by the relaxation of strict immigration laws and the abandonment of the White Australia Policy. These measures encouraged Asian immigration from China, Vietnam, the Philippines, and Hong Kong. This was good news for the

Brick shapes made at the Punchbowl yard which closed its brick operations in 1978

industry, which found its products in demand. Migrants from south-east Asia chose to rent home units because they were affordable, clean, modern and often a huge improvement on accommodation in their home country. The old argument over brick versus stone or fibro was no longer relevant.

The combination of ongoing profitability and an asset revaluation enabled Brickworks in October 1979 to announce the capitalisation of nearly $1.6 million and its distribution to shareholders as fully paid bonus stocks. The ties between the two companies were also drawn tighter between 1977 and 1979. During this period Soul Patts' holding in Brickworks increased from 28.6 per cent to 33.2 per cent, thence to 38 per cent. At the same time, the housing market in New South Wales continued to boom, in contrast to Victoria. In March 1980 Brickworks reported a doubling of profit with a 39 per cent increase in sales in New South Wales, and, to a lesser extent, in Queensland. The result overshadowed performances from brickmakers in Victoria, including Brick and Pipe Industries and Clifton Brick.[50]

50. Earnings rose from $1.69 million to $3.5 million in the December half-year. *The Age*, 25 March 1980.

SYDNEY'S DISAPPEARING BRICKYARDS. AERIAL VIEWS OF THE LOCATION OF BRICKWORKS IN SYDNEY, 1943 AND 2008

Scanned 1943 aerial photography prints courtesy of Roads & Traffic Authority, New South Wales. For comparison with 2007 views use the SIX viewer at www.maps.nsw.gov.au 2008 images courtesy Google Earth.

Brookvale yard, Warringah Road, Brookvale, closed 1998. See Google Earth coordinates 33° 45' S 151° 16'E

Ashfield Brickworks (left) and South Ashfield Brickworks (right). Holden Street runs from top to bottom on the right of the photo. Milton/King Street runs from top left to bottom centre. Trevenar Street (formerly Goodlet) bisects the picture from top right to bottom left. See Google Earth coordinates 33° 52' S 151° 07'E

National yard of Brickworks Limited, Pennant Hills Road, Thornleigh. See Google Earth coordinates 33° 43' S. 151° 05' E

Top left: Burwood Brickworks. Bottom left: Croydon Steam Brickworks. Right: Excelsior Brickworks. Croydon Road runs diagonally from top right to centre left of the photo. See Google Earth coordinates 33° 52' S 151° 06'E

District of St Peters-Marrickville showing four brickyards. See Google Earth coordinates 33° 55' S 151° 09'E

Dwellings completed in Sydney region 1957–1996

Town houses, flats

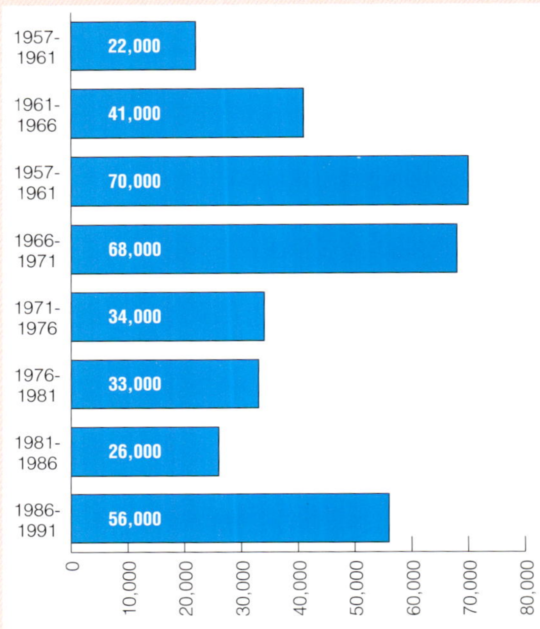

Detached houses

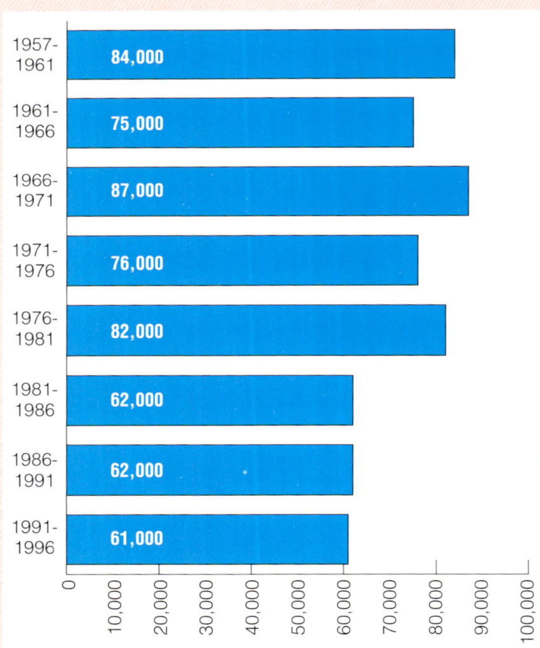

Sources: Commonwealth census; M. Daley et al., Monitoring Sydney, Australian Housing and Urban Research Institute, Melbourne, 1997; P. Spearritt and C. Demarco, Planning Sydney's Future, Allen & Unwin, Sydney 1998

END OF AN ERA

Time was running out for William King Dawes who turned 80 in December 1978. For much of the decade he had been dogged by various health problems including chronic alcoholism. Estrangement and divorce from Emily Louisa, his wife of 50 years, created disharmony within the family. The marriage had never been easy, but the final break devastated Louisa who was a gentle and considerate woman. As for the business of bricks, Dawes' gradual disengagement was plain for all to see, although he remained chairman of the board until the very end.

In poor health, he was painfully aware that Arthur and William Frederick would never be able to take the company forward. Despite their respective skills neither of the sons still working in the business had the 'right stuff'. As the decade drew to a close the spectre of the corporate raider returned. This time it was the turn of Soul Patts to be spooked by nominee buyers from Melbourne. In December 1979 Millner turned to Dawes who wasted no time buying Soul Pattinson shares at up to $10 per share.[51] In fact, the approach of Christmas was a depressing time for Dawes. Arthur was being treated for cancer and within six months, at the relatively young age of 61, he was dead. As joint general manager of Brickworks with his brother, William Frederick, he had been associated with the company since the 1950s. His role was acknowledged by the Board which met on 23 June 1980:

> *It was largely due to his efforts, the three Wallgrove plants, and the Rochedale extruded plants were established, and over the years have contributed substantially to the prosperity of the Group. His wise counsel and experience will be sadly missed.*[52]

The death of Arthur Dawes was a serious blow, leaving the company with no obvious successor to Dawes senior. Messages of condolences flooded in

51. 'Soul Patts sends for Dr Dawes', *The Australian Financial Review*, 20 December 1979.

52. Board minutes, Brickworks Limited, 23 June 1980.

Occupied dwelling: A generalised comparison

Country	Year	Owner %	Renter %
Australia	1947	52.6	43.4
	1954	62.9	34.1
	1961	69.9	27.4
	1966	72.5	25.1
	1971	68.7	26.4
	1976	66.7	25.2
	1980	71.6	24.9
England and Wales	1971	50.1	49.7
Canada	1975	62.0	38.0
France	1975	47.1	42.3
New Zealand	1976	69.5	27.0
Sweden	1975	38.9	56.0
Switzerland	1970	28.5	68.4
USA	1977	64.8	35.2
West Germany	1972	33.5	66.5

Source: Figures cited in 'Australia ahead of most of the world in housing', The Age, 17 September 1981

for a man who was widely respected.[53] With Ray pursuing his own business interests, Dawes senior had long known that William Frederick was not the man to step into the shoes of his late brother. Instead, Tony Padden, who had managed No. 1 and No. 2 plants was appointed general manager of extruded operations at Wallgrove. Padden, an industrial chemist by training, had joined the company in about 1971 and won rapid promotion following his discovery of cream-burning shale on the company's extensive landholdings. At Wallgrove he worked alongside the new office manager, Dennis Downie, who had been employed as a floor sweeper and general labourer in the early 1960s. Until recently Downie had worked at the St Peters yards with Bill Swan whose retirement (and Arthur's untimely death) presented a further opportunity for Downie to advance himself.

By now, however, the factories were seriously run down, the damage caused by years of neglect. No. 1 Plant was fully operational, although No. 2 Plant had been closed for some time due to the recession and lack of demand. No. 3 Plant was decommissioned in 1980 for major refurbishment. The dry press yards were in a similar predicament. In a belated attempt to address long-standing maintenance issues, and lack of investment, the ailing Dawes agreed to convert kilns at the dry press yards of Eastwood, St Peters and Alexandria (City yard) to burn natural gas. This may have been somewhat redundant in the case of Alexandria since the state government was considering resumption. Laurence Taylor wrote to Bill Dawes on 20 February 1981, outlining the difficulties if the company held on to its land at the City yard.

> *If the Department's offer [Department of Environment and Planning] is not accepted it would be open to the Department to resume the area. Bruce Parks is of the opinion that it would be difficult to establish before a Court on a compensation claim a figure as high as the offer made by the Department.*[54]

Head Office of Brickworks Limited, Cheltenham Road, Burwood, New South Wales

In a move that reflected the westward shift of Sydney's geographical centre of population, the company's head office continued its march. Its progression formed a wide arc that so far had taken in St Peters, Cheltenham Road near the Burwood brickyards, and Punchbowl in 1979. Today, it is located at Horsley Park.

Goaded into action, Dawes instructed Bruce Parks to negotiate the sale of the Alexander property for the sum of \$3.6 million with separate lease back arrangements for the engineering workshop and plant repair workshop areas for a period of 10–12 years. A final offer of \$4.4 million from the Department of Environment and Planning was accepted at the board meeting of 23 February 1981. At this point, although for not much longer, the plant was still operating at a reduced level. Taylor also noted a sharp fall in the company's market share from 40 per cent to 23 per cent in New South Wales and from 17 per cent to 16 per cent in Brisbane. However unpalatable, the truth was that Brickworks

53. These included messages from Pullman Swindell in Melbourne and Pittsburgh, Doug Lanceley and Directors of the North Sydney Brick & Tile Company, Margaret and Preston Steele of North Carolina (J C Steele), the Midland Brick Company, the Brick Carriers Association, the brick carriers from Wallgrove, A. E. Baker and Company, Clifton Brick Holdings, Jack Petersen, Lion Tile, Tucknott's Brick Company, Merrylands, Clark Brick Limited and PGH Ceramics at Doonside.

54. Letter from Laurence Taylor to W.K. Dawes, 20 February 1981, Archives of Brickworks Limited.

Right: William King Dawes, 1898–1981

Oil on board, by N. David, 1974

Above: Arthur Dawes, 1924–1980

had fallen asleep on the job. Laurence Taylor's letter pulled no punches.

> *In addition we do not appear to have the range of bricks that our competitors have. Our competitors have embarked, and apparently with great success, on programmes of expansion – particularly in the case of Boral and Acmil, and also Zacuba. If we sit by and do nothing it would appear that they will be continuing to increase their share of the market in both States at our expense. We have substantial liquid funds available, the value of which is being eroded by inflation. It seems essential that we should be seeking to increase our productive capacity and to employ our liquid funds for this purpose. Any further delay will, no doubt, result in increased costs of the necessary equipment to enable us to increase production.*[55]

55. Ibid

Stung by the implied criticism, Dawes approved the upgrade of the Rochedale plant in April 1981 and the installation of the latest Swindell Dressler machinery. The refurbished plant would cost $8.7 million. The decision was to be his last.

DEATH OF W.K. DAWES

The end for Dawes came on 10 August 1981 when he died at the age of 83. A board meeting hastily convened for the following day elected Millner as Dawes' successor. Hugh Dixson was appointed to fill the vacant seat. Dawes was laid to rest in Sydney's Rookwood cemetery following a funeral service at St Peters Church, Cook's River. Apart from a short obituary in the industry magazine and *The Australian Financial Review* his death attracted little attention from the media, which was hardly surprising. Years before Dawes had given up on racing horses, which must have disappointed gossip columnists looking to fill the social pages. Having erected a two-metre brick wall at the front of his Vaucluse home in Sydney to deter snoopers and unwanted visitors, Dawes simply wanted to be left alone in his final years. The industry journal, *Australian National Clay*, concluded:

> *His memorials will be the group's works at Wallgrove, New South Wales and Rochedale, Queensland, two of the finest brick plants in the world, operating ten tunnel kilns between them.*
>
> *It was a source of pride to Mr Dawes that under his management the group produced a profit every year of operation. One of the older style of industrialists, he eschewed the formality of his office suite and was more likely to found running a keen eye over a manufacturing operation or a kiln, ensuring that everything was done yesterday.*
>
> *Though a forceful character, he was unassuming. His friends included a Governor-General, a Chief Justice and Cabinet Ministers, yet he had little concern for social prominence. Mr Dawes encyclopaedic knowledge of the industry and his genius for figures were coupled with the gift of inspiring loyalty. There will be many long-time employees who will mourn his passing more as friends than as servants of the company.*[56]

The achievements of William King Dawes far outweigh his idiosyncrasies, faults and indiscretions. Those who knew him – admirers and detractors alike – regard him as having been a most remarkable leader. His energy and single-mindedness propelled the brick industry – and Brickworks Limited and Austral Bricks in particular – to great prosperity as well as the peak of technical innovation. The tunnel kiln operations had been his crowning achievement and one that enabled Brickworks to dominate the industry. Yet, with the opening of the last of the extrusion plants at Wallgrove Road in 1974 his work was effectively over.

Ambition, a vision for technological change, force of personality, and a ruthless determination to succeed ensured a place at the top for Dawes. He was a giant among men and deserves to be remembered as one of Australia's greatest, but least known industrialists. He belonged, however, to another age where the attitudes of late Victorian society shaped his personality and outlook. Perhaps the enigma of William King Dawes and his life and times can best be summed up in the words of the English novelist, L.P. Hartley. 'The past is a foreign country; they do things differently there'.[57]

56. *Australian National Clay*, Vol. II, No. 16 September–October 1981, p. 3.

57. L. P. Hartley, *The Go-Between*, Penguin, London, 1953.

By the million. Brick panorama at the Wollert plant of Austral Bricks, Victoria

Mass production 18

Tunnel vision – Technical edge – Kiln cars – Swindell-Dressler – Stages in the brickmaking process – Stockpiling the clay – Primary and secondary crushing – Pug milling and extruding – Cutting, setting and drying – Firing and dehacking – Storage and despatch

TUNNEL VISION

Brickmasters such as William King Dawes and his contemporaries envisioned modern plants that could mass produce bricks using production line methods. Such a thing was only possible by adopting expensive new technology in the form of tunnel kilns which operated on an entirely different basis from the century-old patent and downdraught kilns. We have already seen how experiments in the use of primitive tunnel kilns to fire ceramic ware and other clay products were being conducted in Australia as early as the 1880s (Chapter 5). Whilst Austral Bricks was not the first Sydney brickmaker to construct a tunnel kiln plant for the manufacture of extruded bricks, it quickly established itself as the largest and most efficient user of the technology.

For the benefit of the layperson, a tunnel kiln is a structure that allows the continuous exchange of heat between gases and clay in a controlled environment. For the technically minded a kiln is a counter current heat exchanger in which hot gases are directed from the exit end to the entrance. At the same time kiln cars loaded with packs of dry clay bricks carry the bricks in the opposite direction, from entrance to exit. There are several parts to a tunnel kiln – the foundation, the walls, roof, firing system and kiln cars which are the most costly component and extremely expensive to maintain.

TECHNICAL EDGE

Any discussion of tunnel kilns inevitably leads to a comparison between kilns designed and manufactured in Europe and the United States. Tunnel kilns in the USA developed from a knowledge of side-fired downdraught kilns and were designed around a neutral draught in the fire zone. Neutral draught was due to air movement travelling upwards from the bottom of the kiln. Since there was minimum leakage of air into the kiln at the fire zone, little emphasis was placed on kiln car seals. Fuel was also relatively cheap, which made heat loss unimportant. Initially, kilns in the USA began as short and narrow, principally because of the under-fired system where the setting was continuous and the amount of heat drawn forward minimal.

On the other hand, European tunnel kilns evolved from Hoffman kilns and their derivatives, patent kilns, which relied on a high draught to bring heat forward from the cooling zone into the pre-heat zone. This reduced the amount of fuel required for combustion, although Hoffman kilns were still expensive to fire. Accordingly, European designed tunnel kilns were top-fired and required good seals due to the high negative pressures developed as a result of strong draughts. They were also long and narrow and designed to allow for the heat exchange effect to occur from exhaust gas to ware.

By the late 1950s in Europe and the USA longer kilns were being built to accommodate greater capacity. However, this required more powerful fans to pull the air forward, which only increased the leakage of hot gases. It also led to a steep increase in the number of under-fired bricks being produced which prompted designers to engineer tunnel kilns that were short, wide and low. The impact on kiln cars was profound; cars became more complex, especially with the need for a wider deck. A greater number of wheels was needed to support the load and inter-car seals became a major source of leakage. Furthermore, due to poor insulation, heat under the deck became more difficult to remove. Today, most kilns are fired from the side of the kiln in the pre-heating zones and from the to in the higher temperature zones.[1]

1. Information in this introductory section is an edited version of an article by brickmaster, William Emmett, 'After some history, thoughts on kilns, refractories, cars, decks', Australian Clay Journal & Ceramic News, Vol 9, No. 4 July/August 1998, pp. 5–7.

KILN CARS

Jamie Pearson, Manager, No. 3 Plant, 2007

Lining the deck of the kiln car is a layer of refractory brick manufactured to withstand successive firings with minimal cracking and spalling. They must also endure the wear and tear of unloading and be as light as possible to reduce fuel usage and heat losses from the kiln. Furthermore, they should allow heat to penetrate under, and into, the bottom layer of brick on the deck.

For its part, the steel-framed car deck must be able to retain its integrity and remain level through successive firings. This will reduce the loss of product through cracking caused by bridging over uneven deck blocks. The car deck should seal on all sides and be well insulated so that under-car temperatures stay within 90-120C. Gaps can open up between the refractories lining the decks due to the ingress of brick chips or sand into cracks formed where different blocks of the deck meet. Hence the need for expansion joints on the car deck. People walking on the deck or packs falling from machines also has an impact on the integrity of the refractories. Furthermore, car decks can 'grow' due to poorly fired refractories and movement of the deck caused by setting and dehacking machines. Progressive shrinkage of set product can also drag the blocks out position. A powerful vacuum can be used to clean the decks after each passage. Sweeping the decks is better than nothing, but this can force brick chips into the blocks and aggravate the problem. Where sand is the main culprit, the only remedy is to rebuild the top structure of the car deck every one or two years.

SWINDELL-DRESSLER

Austral Bricks' relationship with the US kiln manufacturer, Swindell Dressler, commenced in the late 1950s. During the 1960s Samuel J. Tricase, a kiln engineer with the Swindell Dressler Company, was a regular visitor to Australia and the plants of Austral Brick on Wallgrove Road, Horsley Park. In March 1965 Tricase supervised the modernisation of No. 1 Plant which involved the installation of an additional production line, a continuous dryer and a Dressler tunnel kiln. The plant was shut down in 1980 and was idle for many years until it was re-opened in June 1990 following a complete rebuild. In April 1968 a second factory (No. 2 Plant) was opened for business and extended in 1971. In March 1972 a third facility (No. 3 Plant) was officially opened.

Cathy Inglis, Group Technical, Research and Engineering Manager, 2007

Aerial views of the operations of Austral Bricks, Horsley Park, 2002

Modern brickmaking plants are organised for efficiency and safety. Stockpiles of clay and shale excavated from the pit can be seen in each photograph. Burnt bricks are neatly stacked in the yard ready for delivery. No 1 and 2 Plant produce mainly face bricks, whereas No. 3 Plant is set up for the production of common or engineering brick. An aerial photograph of No. 3 Plant can be found on page 303.

STAGES IN THE MODERN BRICKMAKING PROCESS

This numbered illustration shows the different stages in production, from the pit to packaging of the finished bricks.

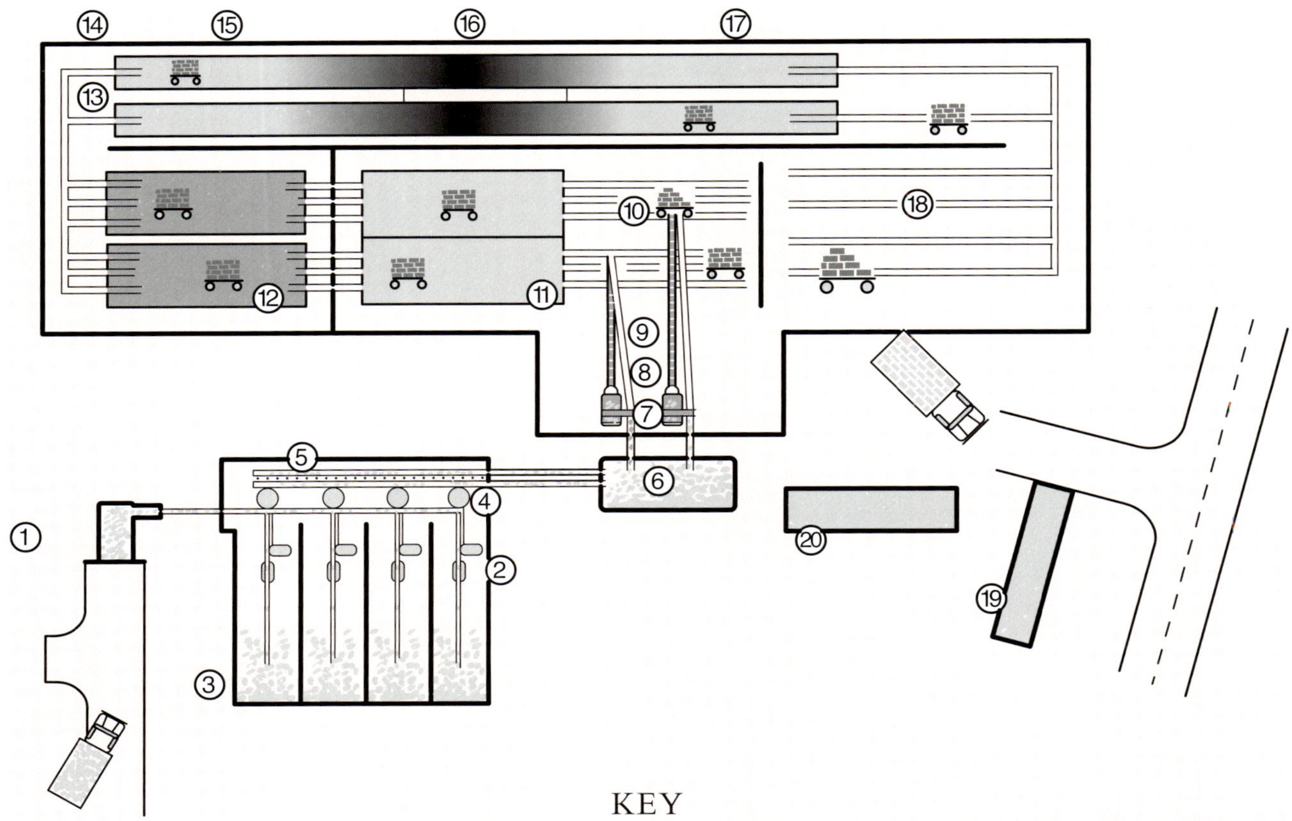

KEY

1. Primary crusher receives raw material.
2-3. Clay storage areas
4. Pan mills
5-6. Conveyor belt leading to surge bin (6)
7. Treatment in the pug mill
8. Extrusion into columns of clay
9. Clay cut into lengths and then wire cut
10. Setting the bricks
11. Kiln cars entering a heated holding room
12. Tunnel dryer
13-15. Entering the tunnel kiln and pre-heat zone
16. Firing zone
17. Cooling zone
18. Dehacking area
19-20. Storage

STOCKPILING THE CLAY

Clay and shale excavated from the pit is deposited in layers to form a premixed stockpile of approximately 50,000 tonnes of material. Each mound measures approximately four metres high, 12 metres wide at the crown, 20 metres wide at the base and 50-70 metres long. Dirt is left to weather ('sour') in the paddock for about six months. During this time the shale breaks down by the action of rain and sunshine. At the 3-month mark another mound is started to ensure a continual flow of raw material. The paver plant at Punchbowl uses approximately 100 tonnes of dirt each day, which is trucked in from Wallgrove Road. Plants No.1, No. 2 and No. 3 at Horsley Park each use approximately 800,000 tonnes a year.

PRIMARY CRUSHING

A 988 H Caterpillar front-end loader removes material from the premixed stockpiles and transfers it to a dump truck which tips the dirt into a hopper for primary crushing

Loading position above crusher hopper

Empty crusher hopper with moving slat floor

The crushed material is transported by conveyor to a clay storage area

Clay is evenly distributed into piles in storage bays to ensure consistency of mix

Feed hoppers (usually three or four) located in the pan room are filled with the blended dirt

Dirt is gravity fed on to a slat conveyor which discharges into the Incla Pan Mill. Secondary crushing in the pan room

SECONDARY CRUSHING IN THE PAN ROOM

Incla grate-type discharge dry pans have a capacity of 40–50 tonnes per hour. Clay discharged through the pan grates is picked up by a bucket elevator and dumped on a horizontal belt conveyor fitted with three adjustable ploughs. Material is evenly distributed into heated, vibrating Leahy screens, passing through the 8-mesh screen. It then falls to a common conveyor and is carried to surge bins. Tailings return to the Incla pan. Up to this point the material has been kept dry

A vibrating screen obtains the correct particle size and distribution

Steve Wall (centre), an experienced and highly knowledgeable brickmaker, was the plant manager at Bowral Bricks in the Southern Highlands before joining Austral Bricks in 2001

PUG MILLING

Once the material has been screened it is sent to the surge bins and fed to the pug mill where it is tempered to make a homogeneous plastic mass ready for shaping. The pug mill consists of a chamber housing a shaft with knives rigidly attached. Water is added at the drive-end of the pug mill tub to the required ratio of 14 per cent moisture. Blades attached to the shaft mix the clay and water into a consistent putty-like mass. The material moves towards the discharge end where a tapered auger forces the wet clay through a rotating shredder into the de-airing chamber. The wet clay falls into the main auger in the extruder barrel where air is removed by a vacuum pump to eliminate any laminations in the final brick.

EXTRUDING

Clay is forced out by powerful motors (up to 500hp) through a Steele 90 Extruder in a continuous, solid column. The installation of a bridge and core buttons at the die produces the familiar shape of a perforated brick. The column is textured as it leaves the shaper cap of the extruder. Different types of shaper cap can be fitted, depending on whether a smooth-face brick is required. At this point two of the brick dimensions are sized: length and width. The column of clay can also be coated with special finishes to achieve the desired colour effect during firing.

In the stiff extrusion process moisture content is approximately 14 per cent, whereas for soft extruded bricks, moisture content is about 18 per cent to achieve plasticity. The paving tiles produced at the Punchbowl yard use the soft extrusion method.

Top view of the Steel 90 extruder

Edge treatment of stiff extruded column

Trimming of extruded column to give cut finish

During the 1960s and 1970s a combination of the Steele 50-F Extruder and Sealing Pug Mill was standard issue for stiff extrusion plants in Australia.

CUTTING

The column of clay is cut into slug lengths which are conveyed to the setter. The extruded column is pushed through a wire cutting bow giving the third dimension of height. In 1965 each of the three production lines at No. 1 Plant had a capacity of 12,000 bricks per hour.

Bottom left:

Reel cutter and texture machine at the Golden Grove plant of Hallett Brick in South Australia, circa 1970. Reel cutting machines were an integral part of manual brick setting operations and are no longer in use. Extruded clay passes through a texturing machine and is sliced into brick shapes by the wheel-shaped reel cutter. The reel cutter was almost certainly manufactured by Ron Hamden, the only Australian business to manufacture extrusion clay equipment. The business finally closed in the late 1970s

SETTING AND DRYING

Air bags on the setting head line up with the separated bricks ready for pick up

The setting head picks up a full table of bricks and places them into position on the kiln car, which can be seen exiting the setting area

The loaded cars are transferred to controlled humidity tunnel dryers

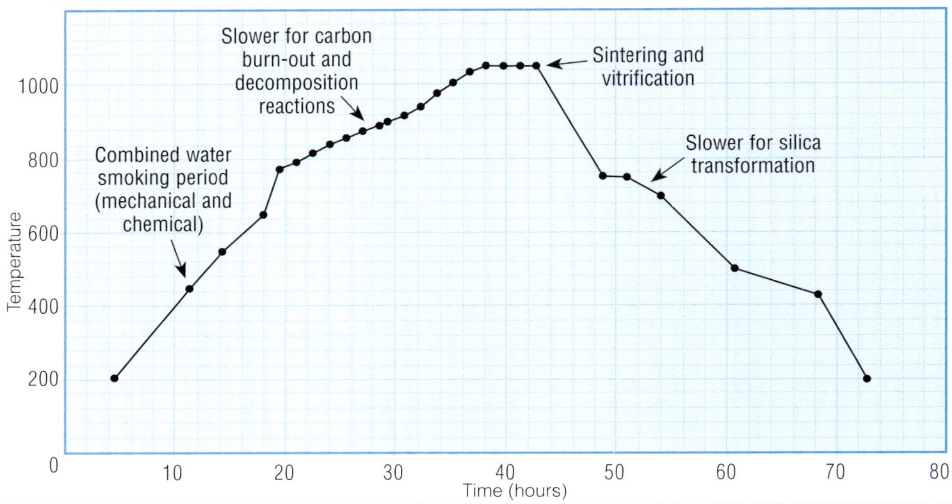

FIRING THE BRICKS

Loaded cars enter the tunnel kilns by the vestibules then travel through the pre-heat zones into the furnace zone where the temperature reaches 1,100°C approx. Firing brings out the colours in the bricks. After moving through the cooling zones they emerge with their load of fired bricks. This can be anything from 24 hours to several days.

Bricks falling from kiln cars during the firing process are a particular hazard since they can block the passage of cars through the fire and are difficult to remove. The kilns are shut down for cleaning and maintenance over the Christmas period of each year during which time brick debris is retrieved. Some plants may go for as long as five years without stopping.

Top: A 1972-vintage photograph showing kiln cars set by hand emerging from a tunnel kiln at No. 2 Plant

Centre: Graph showing kiln temperatures over a 72-hour period

Left: The interior of a tunnel kiln during annual maintenance. Temporary lighting has been installed

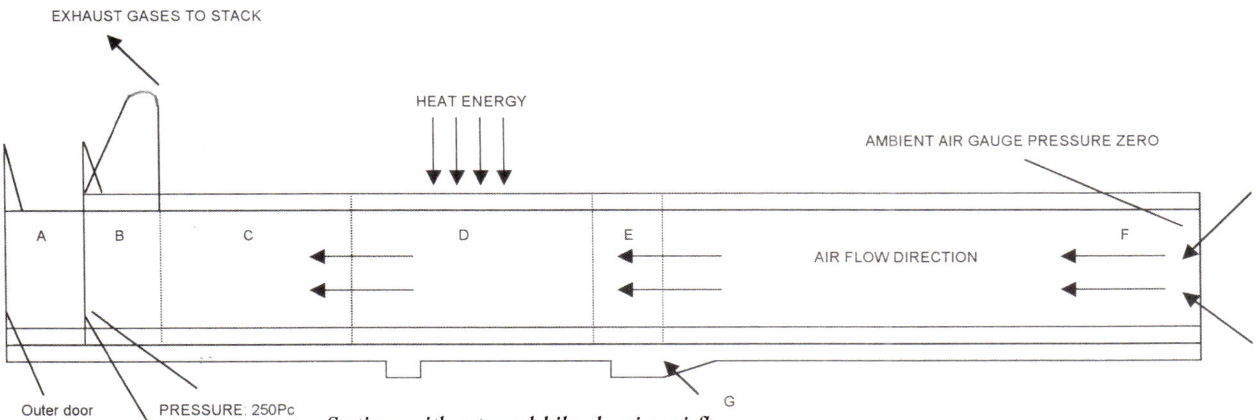

Sections with a tunnel kiln showing airflow

Key: A – Kiln Vestibule B – Kiln Exhaust C – Kiln Preheat D – Firing Zones E – Rapid Cooling F – Heat Recovery G – Undercar Cooling

DEHACKING

When sufficiently cool the burnt or finished bricks are unloaded, packaged and stored in the yard ready for delivery. Proper sorting (dehacking) and packaging of the brick after burning is extremely important. This photograph, taken in 1972 at No. 2 Plant, shows a 'carousel' which allowed blending of bricks of different colours and tones. White texture bricks are being hand set at a pace which required the operator to keep up with the machine. The work was extremely arduous. Note the operator at centre right who is wearing protective leather 'cots' referred to in Chapter 14

Today the work is performed by automatic dehacking machines which have significantly reduced the incidence of broken, twisted and otherwise mechanically defective bricks. Brick colour and range is carefully monitored to assure a quality product. The finished product is packaged and banded into cubes of approximately 500 bricks, inventoried by lots and distributed to customers.

STORAGE AND DESPATCH

Bricks being unloaded at the end of the production line. The packaged bricks are then graded and stacked in the yard by forklift according to product type

WOLLERT 2008

The latest production techniques are similar to those used in the 1960s and 1970s, although touch-screen computers, robotics and factory design at the newly opened plant of Austral Bricks at Wollert in Victoria provide a much cleaner and more efficient working environment. Using advanced technology state-of-the-art plants such as Wollert produce brick and clay products at greatly reduced costs per units.

***Main picture:* Brick factory, steel chimney. Afternoon sky over Wollert, Victoria**

***Top:* Night Vigil**

***Above:* Eyes and ears**

The chimneys of the Bedford Brick Works on the north west corner of Sydney Park. Photography by Josh Hill

Fightback 19

Audit – Albert Burgis – Knockdown, rebuild – Cultural shift – Recruiting drive – Boardroom – Payoff – Rochedale revival – Industry and regional trends – An unusual turn of events – There and back again – Upgrades – Generational change

AUDIT

In the months following the death of his father, William Frederick Dawes gave little thought to anything else but the future. His immediate concern as a director of the company and general manager of dry press operations at Brickworks Limited was: 'who would come out on top?' In fact, he may even have had an inkling that the family's long reign over The Austral Brick Company and Brickworks was about to end.

Elected unanimously as chairman of the board, Jim Millner drew up a report on the group's overall position, calling on the company auditor, Geoffrey Travis, for assistance. Whilst they made rather depressing reading, the findings came as no surprise. By early 1980 the factories at Wallgrove Road were already in an advanced state of disrepair. Maintenance had been cut to the bare minimum in a bid to save money. No. 1 Plant was little better than scrap and remained closed for nearly a decade. No. 2 and No. 3 plants spluttered along. In fact, conditions were so bad at No. 2 that most workers could hardly wait to get out. Only the depressed state of the economy and gloomy job prospects stemmed the rate of desertion. In all of the yards the maintenance of kiln cars in the Swindell Dressler tunnel kilns had been sorely neglected. Each of these cars had a complex top that required regular maintenance. As for the kilns and dryers: they were old, leaked gases and were totally run down. The wastage was enormous since the defective kilns damaged the bricks during the firing cycle, leaving them either crazed or fused to the bottom of the kiln car.

As conditions deteriorated the exodus of skilled workers continued with the result that wages doubled in a bid to retain and attract staff. Whether or not poor maintenance was specifically due to lack of resolute action on the part of the joint general managers, the late Arthur Dawes and his brother William Frederick, is debatable. Regardless of who was to blame, Millner was determined to reverse the downward spiral. In doing so he clashed with the last member of the Dawes family to hold a position in the business. The showdown was inevitable. In fact, it is difficult to avoid the conclusion that Millner had scant regard for the abilities of young Bill. His father, on the other hand, had been cut from a different cloth. The annual report for 1981 included a short reflection on the life of W.K. Dawes.

> *It was with deep regret that the Board recorded the death of their late Chairman, Mr W.K. Dawes. Mr Dawes was associated with Brickworks Limited since its inception, and was Managing Director for many years. Due to his ability, foresight and drive Brickworks Limited became one of the largest brick manufacturing organisations in Australia, and the company's present financial status and profitability substantially resulted from his efforts.*

With the economy entering a severe downturn, relations between Millner and William Frederick Dawes worsened to the point where Dawes absented himself on several occasions from directors' meetings. This may have had something to do with his plans to instal a mud brick plant at Rochedale, which required his presence at the Queensland factory. Unfortunately his attempt to take the company forward with a new product failed. According to Bill Dennis, 'we were working together on a proposal for a new plant. I recall there was some sort of upheaval. Matters came to a head, planning stopped and he left the company soon afterwards.'[1] Dawes announced his retirement (effectively, his resignation) as general manager by letter which was read at a board meeting in April 1982. The departure was timed to coincide with the end of the financial year. A week or so after the meeting, Millner spoke to financial journalists.

1. Interview with Bill Dennis, Swindell Dressler, Dandenong, 2 March 2005

The second fire at the Brick Manufacturers' Association of New South Wales was recorded by amateur photographer, Paul Costantouras, who happened to be passing at the time. Photography by Paul Costantouras

> *Mr William Dawes will be retiring as general manager of Brickworks Limited later this month. His retirement will mark the end of the Dawes family's direct control of the company's management. Mr Dawes will be replaced by Mr A.W. Burgis as group general manager and Mr D. Downie as assistant group general manager. Mr J. Willis has also announced his retirement as the company's secretary. He will be replaced by Mr F. Elsworthy.*[2]

Having effectively relinquished the position of general manager, the final break for the 64-year old William Dawes came on 13 September 1982 when he resigned from the board. Angry at what he perceived to have been shabby treatment from colleagues, Dawes sold his shares and had nothing further to do with the company. He died in 2001.

Dynasties come and go, and the disappearance of the Dawes' name from history's page attracted little attention. The domination of a large-scale industrial enterprise by a single family is not unusual, but in this case the succession collapsed following the death of W.K. Dawes. Some would go as far as to say that the succession was fated to end following the merger with Soul Patts in 1969. The Millner family had an uncanny knack for making good investments, and entered the world of Brickworks at the precise moment when Dawes' powers were in decline.

Another departure of a distinctly different kind was that of Darcy Dugan, the career criminal, who made a last tilt at fame and fortune during the winter months of 1982. As plans were being made

2. *The Australian Financial Review*, 2 June 1982. Frank Elsworthy joined Punchbowl Brick and Pipe Co. Ltd in 1967 as the company accountant.

to transfer ownership of the City yard on Euston Road to the City of Sydney for waste disposal, Dugan snatched the payroll from the site office. The City yard appears to have been popular with Dugan. Another, final, prison term put paid to his career.

The retirement of William Frederick Dawes was eclipsed by an event that, quite literally, put the past almost completely beyond recall. In the early hours of Sunday 16 May 1982 a fire broke out in a milk bar situated in a Burwood (Sydney) arcade of shops and business premises. The circumstances were suspicious. Seven days later a second fire broke out in the fire-damaged premises and quickly spread throughout the arcade to adjacent offices, one of them being the New South Wales Brick Manufacturers' Association. Arson – an insurance job – was the most likely cause. Tom Edwards, secretary of the association, and Robert 'Bob' Rushton, the industrial officer, were able to salvage records from filing cabinets that lay beneath the blackened, sodden mass of fallen beams.[3] The New South Wales branch of the Federated Brick, Tile and Pottery Industrial Union offered to replace documents with copies from their own records. Yet the real damage lay in the destruction of photographs of every member of the original Brickmasters' Association and of the original Council of Brick Manufacturers which dated from 1934. The association was also the only known source of a complete set of its journals. Although arson was never proved the finger was pointed at one of Sydney's flamboyant underworld identities who died recently.

ALBERT BURGIS

Millner's audit confirmed what everyone working in the plants had known for some years; that the Wallgrove operations were in a parlous state. Not so at the yard of Punchbowl Brick and Pipe where its general manager, Albert Burgis, ran a highly successful business making salt glazed clay pipes and bricks. Born in 1923 at Glen Innes, New South Wales, Burgis was an accountant by training and commenced working at a small company, Richardson Holdings, after the end of the Second World War. Burgis joined Reg Allott's Punchbowl firm in 1959 which was acquired by Brickworks Limited in 1961. Under Burgis the clay pipe business in the suburb of Punchbowl became highly profitable, and the leading manufacturer in Australia, although competition with the firm of PGH was intense. On the advice of Ray Dawes Millner appointed Burgis as group general manager of Brickworks. Albert Burgis took up his appointment on 27 June 1982 and the following year was offered a directorship. He relinquished his seat on the board in May 1988, although he continued as a non-executive director, retiring in November 2000. A year or so before his death in 2006 Burgis recalled:

> *If it hadn't been for Millner then I would not have been given the job of running Brickworks Limited. If I went to him with a problem he would give me a free hand. Millner stood up for me during board meetings and I'll always appreciate that.*[4]

Burgis's elevation came amidst the steady disintegration of the federal coalition under Prime Minister Malcolm Fraser who fought off a leadership challenge from his rival, Andrew Peacock. The country, especially in rural areas, was also in the grip of a devastating drought. By strange coincidence the election of Robert 'Bob' Hawke in March 1983 as Labor prime minister was followed by drought-breaking rain several months later. As the former leader of the Australian Council of Trade Unions (ACTU), Hawke pledged to end the industrial relations turmoil that the coalition government of Malcolm Fraser had failed to curb. Years of political and industrial trouble had taken their toll on the Australian economy. The home building industry was hardest hit due to steeply rising energy costs, cartage rates, labour costs and workers' compensation premiums. Millner put on a brave, public face. In an interview with *The Australian Financial Review* he commented:

3. Bob Rushton joined the New South Wales branch of the association in 1972 as an industrial officer. By this date the association had 25–30 members, which reflected the declining number of manufacturers in the state. Tom Edwards died in office in 1983 and was succeeded by Rushton who retired in 2006.

4. Interview with Albert Burgis, Sydney, 15 November 2004.

Brickworks had carried out a major rationalisation of employees and plant in the year. The reduced profits are also symptomatic of the dramatic reduction of building approvals in New South Wales in the year.

He also stated that

only two of the seven kilns at Brickworks' biggest plant, in the Sydney suburb of Wallgrove, were operating, but idle kilns could be re-fired if demand returned to the building industry.[5]

Taking advantage of the recession Burgis ordered the closure of Austral's remaining plant at St Peters, which burned its last bricks on 31 May 1983. On the same day the last batch of face bricks was fired in the dome kiln at Brookvale, although production of commons continued. Unfortunately, resentment over the loss of jobs and the closure of yards simmered, and erupted in strikes at the Wallgrove Road plants in early 1984.

Helping to shore up the bottom line during 1982 and 1983 was the sale of investments, as well as land from the closed yards at Heathcote, Bayview and Thornleigh.[6] Land development was now commonplace for all brick companies; in fact it was highly lucrative and took advantage of the historic necessity for brick manufacturers to hold substantial reserves of acreage. The case of Norwest Business Park, the brainchild of Doug Lanceley, is instructive. Lanceley, the son of E.D. Lanceley was managing director of North Sydney Brick and Tile. In the early 1950s the company purchased a large tract of land, close to the intersection of Burns Road and Old Windsor Road at Parklea. Lanceley and his other directors soon found the cost of transporting the clay and shale to St Leonards prohibitive and moved the brick plant to the quarries. Having acquired far more land than was needed for brickmaking Lanceley decided to create a business park modelled on examples that he had seen in England and the United States. Unfortunately development costs

Albert Burgis, 1962 and 1985

blew out, which was a serious financial drain. Ultimately the company was taken over by the New Zealand businessman, Sir Ron Brierley, who later sold out to Malaysian interests. The century-old history of North Shore Brick and Tile ended with a whimper.

KNOCKDOWN, REBUILD

Burgis now faced the daunting task of rebuilding the Austral brand. By the end of 1982 No. 1 Plant finally expired. No. 2 Plant continued to operate, but for how long? It was apparent that without millions of dollars of capital investment the business would continue to slide. Without any hesitation the board ordered the phased rebuilding of the Horsley Park plants beginning with No. 3 Plant. In January 1983 Burgis, Millner and Dixson left on a fact-finding tour of European manufacturers of brickmaking equipment. Like Dawes, 20 years before, the three men took with them a shopping list. A visit to the French firm of CERIC (Centre d'Etudes et de Realisations Industrielles et Commercials) in Paris convinced them that European manufacturers led the world in the design of casing-type kilns and drying equipment. Having obtained what they were after the party returned to Sydney several weeks later, and on 3 February 1983 the board approved the contract for the supply of a kiln by CERIC. In March contracts for the supply of brickmaking equipment were also let with a Spanish company, Agemac. The new CERIC kiln was side-fired by

5. 'Brickworks Earnings Fall 38.1 per cent to $5.64', *The Australian Financial Review*, 27 September 1983.
6. A net capital gain of $979,000 resulted from the sale of land at Heathcote, Bayview and Thornleigh (Annual Report, Brickworks Limited, 1982). For the financial year ending June 1982, the sale of investments realised an additional $500,000.

gas using 100 EHB jet burners spread across 22 separately controlled firing zones. All functions were coordinated by a Merlin Gerin PB600 programmable control equipped with computer screens. The CERIC kilns were designed for a round-the-clock 365 day operation at a maximum firing temperature of 1150°C.

Work began immediately at the plant which had been originally designed as a large production facility for common extruded bricks. One of the two Swindell Dressler kilns, barely 10 years old, was demolished to make way for the 157 metre-long CERIC kiln which could accommodate 55 kiln cars. Regarded as one of the most modern and efficient brick factories in the world, No. 3 Plant reopened in 1984 with a new, fully automated production line, including automated de-hacking units. Burgis also installed a dry press section that was intended to produce a superior common brick for the commercial market. Unfortunately, the venture failed.[7] Modernisation came at a price. More than $11 million in capital expenditure was needed to turn the tide at No. 3 Plant. Most of it was raised through short-term loans and by taking advantage of tax relief allowable on depreciation. Production requirements for many years to come were able to be met by continued expansion. For example, in 1986 a German packaging machine was installed at No.3 Plant, bringing capacity to two million bricks per week. Of these, approximately 75 per cent were extruded bricks and 25 per cent were extruded clay pavers. Meanwhile, the dry press yards at Eastwood and Brookvale continued to produce bricks, demand for which showed no sign of abating.

The year 1983 ended without fanfare. With the closure of the Austral yard at St Peters, it was announced that from 5 December 1983 the company's registered office would relocate from King and Cowper streets to the Punchbowl yard on Belmore Road. Brickworks Limited would also close its Burwood office and join its subsidiary in Punchbowl. Still at St Peters, the resumed brickpit in the area bounded by Campbell, Barwon Park, Mitchell and Euston Roads was coming to the end of its life as a garbage disposal venue.[8] In January 1984 Sydney City Council acquired the excavated sections of the shale quarries of the City yard as an extension of the existing landfill depot from its owner, the Department of Environment and Planning. The firing of bricks continued by agreement but ceased altogether early in 1985. This brought to an end over 100 years of brickmaking in the St Peters–Alexandria district. The near coincidence of the completion of the waste filling operation with Australia's Bicentenary in 1988 prompted the government to develop the site as parkland, or Sydney Park as it became known. All along the intention had been to recycle this land for recreational use.

CULTURAL SHIFT

Regardless of the enormity of the task Albert Burgis pressed on, although he soon realised that strong and competent leadership among the depleted ranks of senior managers at Wallgrove was in short supply. Anxious to place the day-to-day running of operations in the hands of an experienced brickmaker and manager, Burgis went outside the company. Few people at the time realised how significant the appointment of Alan Bentley as general manager of Austral's New South Wales operations in November 1984 would be to the company and its fortunes.[9] Bentley's move to Austral coincided with the departure of Tony Padden, general manager of extrusion operations at the Wallgrove Road plants. Padden's tenure had not been a happy experience all round. In November 1984 he joined the old-established firm of North Sydney Brick and Tile in a senior role.

Driving along the Great Western Highway to Horsley Park on the first day of work, Bentley must have had ample time to reflect on the present,

7. The Spengler press had a tendency to blunt the sharp edge (arriss) on each brick, rendering them unusable as face bricks. There were also difficulties in finding sufficient local skilled labour to work the presses.

8. Since the mid 1950s, the City of Sydney Council had conducted a sanitary landfill refuse disposal depot in the area, successfully reclaiming the disused brickpits known as St Peters, Vulcan, Bakewell, Carrington and Warren. All of these, with the exception of St Peters and Vulcan had been purchased from The Austral Brick Co. Pty Ltd

9. At that time the Eastwood and Brookvale factories in Sydney, as well as the Brisbane operations, were run as part of Brickworks Limited and were not under Bentley's control. He assumed control of these plants when he was appointed CEO of Brickworks in May 1988.

Alan Bentley, circa 1987

and also the past. Less than a week before he had been general manager of PGH's brick operations, working out of the Schofields plant which was near Blacktown. Alan Bentley's working career commenced on 9 September 1957 when he joined the firm of Stewarts & Lloyd Australia Pty Ltd (now part of OneSteel Ltd) at Mayfield, on the outskirts of Newcastle as an apprentice fitter and machinist.

> *When I finished my time as a toolmaker I moved on to an engineering firm for a while as works foreman. Then in 1966 Stewarts and Lloyd invited me to become their technical representative which involved selling tubes and valves to power stations, and the mines in the Hunter valley.*[10]

In a career move Bentley joined PGH in 1970 as sales manager at the company's East Maitland brick plant. He was promoted to general manager for northern New South Wales in 1974, and in 1978 he moved to PGH's Eureka tile plants at Doonside.

10. Interview with Alan Bentley, Sydney, 4 November 2004.

In 1980 Bentley went back into the brick business at PGH until he was approached by a corporate headhunter, Geoff Morgan, in 1984 to work at Austral Bricks.[11]

For Burgis, the preceding two years had been an ordeal and he was heartily relieved to have found a highly experienced and charismatic general manager. Bentley was given a free hand to set things straight. Leading from the front he began to instil a new sense of pride among the company's employees. Determined to turn the tide on poor work habits, Bentley launched a blitz at the company's plants. Dubbed 'Mr Clean', he was fanatically tidy and demanded that brick machinery, vehicles and the yards be kept spotless, or as near to, considering the circumstances. This didn't always work out as planned. Human nature being what it is, people generally jib at authority. If workers continued to drop their empty coke cans and cigarette butts, they were careful to put the place in order before an inspection visit. Anxious to avoid Bentley's fiery blasts, supervisors now made sure that rubbish bins, stiff brooms and shovels were always within reach. Pilferage, a regrettable practice that is an almost universal problem in the workplace, was stopped in its tracks by the simple expedient of on-the-spot sackings. For those with something to hide, Alan Bentley's habit of turning up unannounced was disconcerting. Thieves took the hint. Yet, if Bentley drove his employees hard it was because he, too, was reading from the same song sheet – diligence, honesty and a tidy work bench. Such habits would have been reinforced during his early years at Stewarts and Lloyds. As a machinist the need for precision and a clean working environment nurtured habits that, once learned, were never forgotten. There was, however, another side to Alan Bentley which bears comparison to the late William King Dawes. At a time when brickmaking was still the exclusive domain of men, Bentley, like Dawes, ruled with an iron fist. Yet the industry was changing profoundly, and relied less on 'brawn' although some of the

11. PGH operations in New South Wales comprised brick plants at East Maitland, Horsley Park and Enfield. There were also two pipe plants at Doonside which manufactured terracotta pipes, clay pavers, whiteware and extruded quarry tiles. Eureka also operated another plant in Victoria at Ballarat. An additional brick and paver plant was located at Schofields.

work was still intensely physical, especially strapping and packing. Only recently with the introduction of automation, robotics and plasma screens have brick plants ceased to be rough and tumble. Like Dawes, Bentley also displayed a more private and compassionate side and could show great concern for the welfare and health of staff.

RECRUITING DRIVE

Unlike Dawes, however, Bentley was a teacher. In his previous incarnation at PGH, he had been instrumental in developing the skills – and careers – of some outstanding people. In fact, PGH produced a rich crop of talented men who followed Bentley to Horsley Park. Bentley's active recruitment of former colleagues to senior positions was significant in that a new culture began to emerge which owed nothing to the Austral way of doing things and all its associations with W.K. Dawes, which in some respects was just as well. The appointment of Bentley gave Burgis access to a different breed of men who were reliable, loyal and technically qualified. Lindsay Partridge, the current managing director of Brickworks Limited, was one such newcomer who had worked at both the PGH plant at East Maitland, and its whiteware factory at Doonside until his move to Austral Bricks in late 1985. A university-trained ceramics engineer, Partridge was mentored by Alan Bentley who appointed him operations manager at Horsley Park in 1987.

One other PGH employee who followed the yellow brick road to Horsley Park was Bill Larfield, a plant operator and former union representative at PGH Doonside. Initially, he was reluctant to apply for a job.

> *In March 1986 I telephoned Lindsay Partridge and enquired about vacant positions. Having been a union delegate when Alan Bentley was at Doonside, I was concerned with how he would receive me. Lindsay got back to me and asked when could I start. Alan Bentley gave it the go-ahead.*[12]

12. Interview with Bill Larfield, No. 3 Plant, Horsley Park, 4 July 2005. At about the same time, Lindsay Partridge recruited another skilled brickmaker, Jamie Pearson.

Larfield had gained a reputation at Doonside as a tough union negotiator, but clearly had a great many other skills to offer. By the time Alan Bentley left to join Austral Bricks, Larfield had worked his way through the ranks and was manager of the Doonside plant in all but name. A personable fellow like many other PGH recruits, he got together with other colleagues, most notably Joe Davies, to form a social club. Davies, another ex-PGH brickmaker, was also instrumental in establishing the National Association of Testing Authorities (NATA) laboratory at Austral's Horsley Park operations in 1988. While the main event was a children's Christmas party, the party for the employees was for bravehearts – something you only went to if you liked to socialise and conclude the evening with a fistfight. For many organisations the 'works' Christmas party' remains a long established tradition and a way of saying 'thank you' to employees. It was certainly the case that in a less sophisticated age children never tired of endless rounds of musical chairs, pin the tail on the donkey, races, blind man's buff and visits from Santa. And, of course, a table groaning under the weight of tasty tucker.

Meanwhile, the closure of No. 1 Plant in 1980 left no margin for error. Every effort was made to maintain production, especially at No. 2 Plant where Bill Larfield stepped in for a short while as manager during May 1987. The arrival six weeks later of Mark Chellew, another experienced brickmaster, was more or less coincidental with the elevation of Lindsay Partridge to the position of operations manager. After just 12 months Chellew succeeded Partridge who had been appointed general manager. Such was the demand for bricks that three, round–the–clock shifts comprising six crews were introduced to cope with demand. Teams of men working on the carousels graded and packed the bricks by hand. The relentless pace, moderated only by good wages, led to an alarmingly high rate of attrition. Indeed, the work was so physical that only the most resolute survived. Recalls Larfield, 'for a short period I had to let a man go every two weeks or so, although things settled down after about a year.'[13] The wear and tear

13. Ibid.

Bill Larfield

David Fitzharris

on equipment was such that every week the kiln cars would need to be stripped down and rebuilt with new skirts and wheels. Larfield and Partridge could often be seen using heavy crowbars to prise clumps of vitrified brick that had fallen from the side of the cars during firing. So successful was Larfield at No. 2 Plant that he was promoted in 1990, prior to its much-needed refit. In the space of 12 months he had transformed it into a clean, dependable, highly profitable operation.

Bentley recruited and promoted other resourceful young managers including Alex Payne, who became company secretary in August 1987 following the retirement of Frank Elsworthy who had been with the company for 20 years. Like his colleagues Alex Payne understood the need to get on the front foot:

> *The industry knew that Austral Bricks had huge capacity and modern plants, especially at No. 3 Plant, but it wasn't performing and it lacked direction on both the production and sales side. Bentley started a new era in brickmaking by getting the plants going, especially No. 3 Plant. He brought the Austral business, including the Rochedale operations, back to where they should have been. Bentley reversed falling sales, and revitalised marketing by introducing new products and radically improving the quality of our bricks.*[14]

For all of these newcomers, Austral Bricks provided a stimulating environment in which they could grow professionally and rise up the ladder. One such was the Englishman, David Fitzharris, who entered the workforce doing something he loved best – playing soccer. After leaving technical college Fitzharris joined the northern English premier league side of Bolton Wanderers in 1980 where he remained as a professional footballer until 1983. Buffeted by the cold winds gusting from the surrounding hill country, Fitzharris must have concluded that Australia was an inviting prospect – the place to be. Certainly, it was far away from the industrial confrontations being fought out between trade unions and the Tory government of Margaret Thatcher. At the end of 1983 he packed his bags, moved to Sydney, and continued playing soccer for the Penrith club of the National Soccer League (NSL). He returned home for a while to start a new career in sales although it was not long before he was back in Sydney. In 1987 he joined Austral Bricks and rose steadily through the ranks to become Group General Manager Sales and Marketing. Fitzharris brought enthusiasm

14. Interview with Alex Payne, Horsley Park, 9 August 2007.

Valet. Jim Liepins, engineer with Brickworks Limited for over 30 years, on his retirement. Front row: Jim Liepins, Ray Dawes, Arthur Sherman. Back row: (Unknown – thought to be one of the office staff at Eastwood yard), Bill McGregor, chief engineer of Brickworks Limited, Ron Hagstrom, Harold Williams, the yard foreman at Eastwood

and youthful vitality to a portfolio that was sorely in need of direction. Casting an eye over the challenging sales landscape, he worked assiduously, building relationships where previously none had existed. Over time, his efforts paid off handsomely, especially with major trade customers (builders and developers) who source building products from the company. Open to new ideas, and intensely loyal, Fitzharris's impact on the fortunes of the company has been profound.

BOARDROOM

Another new face, this one in the boardroom, was that of Major General C.M.I. 'Sandy' Pearson who was invited by the chairman, Jim Millner, to fill a vacancy left by the sudden death of Jack Taylor in 1983. Pearson was a highly decorated soldier, having earned the MC in New Guinea, the DSO in Vietnam and an OBE for his work in Indonesia. The former commander of Australian forces in Vietnam from 1968–1969, Pearson retired from the Australian Army in 1975, and for eight years was chief executive officer of the Royal Agricultural Society of New South Wales. He was awarded the AO for services to the nation.

Sandy Pearson was a dignified, likeable man who had known Millner during their schooldays together at Newington College, Sydney. His military training had stood him in good stead. In fact, Pearson's ability to absorb and understand detail was impressive, although he, like many others before, gleaned little from the minute book.

> *I knew nothing about brickmaking, so the first thing I did was read the minutes of meetings from the beginnings of Brickworks Limited. They [the minutes] were pretty unforthcoming.*[15]

Board minutes rarely offer much insight into the workings of a company, although for Pearson references to landholdings kindled an interest.[16]

15. Interview with Sandy Pearson, Northbridge, Sydney, 10 September 2005.

16. Like the carpet in Millner's office, board reports were also remarkable for their thinness. Millner never bothered with a chairman's address and only disclosed the minimum required by the Australian Stock Exchange.

Sir Joh Bjelke Petersen, Premier of Queensland, 1968–1987

Home approvals, Australia, 1987–1988

# building approvals	1987-1988 $6.53 billion
Brick houses	95,591
Double brick	22,425
Brick veneer	73, 166
Total approvals (incl brick, stone, concrete)	117,691

Home approvals, selected states, 1986–1988

	1986-1987	1987-1988
Victoria		
Brick veneer	18,756	18,632
Double brick	363	470
Total brick	19,119	19,102
Queensland		
Brick veneer		19,993
Double brick	1,230	1,884
Total brick	13,996	21,877
Western Australia		
Double brick	11,598	14,937

Source: Australian Bureau of Statistics

Frank Elsworthy was also curious and urged Pearson to find out more. Several visits to the Land Titles office of the New South Wales Department of Lands enabled him to compile a precise inventory of property, a seemingly routine task, but one which allowed the company to put its property portfolio in order. Pearson and Burgis would drive out to Duffys Forest and Menai where they inspected large tracts of land held as clay and shale reserves. Having become an expert in the subject, Pearson found himself at the centre of negotiations over the acquisition and disposal of land, especially the sale of St Peters site to the Walker Corporation owned by Lang Walker, the Sydney property developer.

PAYOFF

At a time when the brick industry was still struggling with a severe downturn, the decision to upgrade the Horsley Park plant and commit substantial capital was a bold, yet necessary, initiative. With the re-opening of No. 3 Plant Austral Bricks could now produce the widest range of sand-stocks, clinkers and commons to meet the needs of all sectors of the market at prices that catered to budget-conscious customers. Taking advantage of its improved market position the company also revitalised customer services, adopting new marketing measures to alert brick consumers of product benefits. By the end of 1986 Albert Burgis was pleased with results all round. 'We have spent more than $30 million during the past couple of years and now have the most modern brickworks in Australia, producing the best bricks at the cheapest price,' he told journalists.[17] The financial commentators were also impressed with the turnaround, having softened their views of the defensive share swap placement in 1969 made as a protection against takeovers.

> *The relationship today is even stronger and the shares of both companies sell at a substantial premium based on their asset strength and longer-term hopes that the position could change one day … the real reason why Brickworks' shares – $24.50 on Friday – were commanding a premium of 18.8 times the June half*

17. *The Australian Financial Review*, 30 September 1986.

Export drive – 1985. Trench digger preparing the site for clay sewer pipes. The Punchbowl plant supplied thousands of clay pipes such as these to Singapore from 1981

annualised results, is the underlying value of the company's property and investment portfolio.[18]

ROCHEDALE REVIVAL

The makeover at the Wallgrove yards of Brickworks Limited was part of a wider strategy to put the house in order. Several months before Bentley's arrival Burgis and the board made a decision to upgrade the brick plant at Rochedale, which had last been modernised in 1967. With the Queensland housing market on a bull run, and driven by an influx of retirees and young families from the southern states, the plant was in danger of falling behind its competitors. Plans were drawn up for technologically advanced equipment to be placed within the existing factory. First, however, a huge stockpile of bricks was put aside so that production could be wound back during the construction period, which was estimated to take at least 14 months. On 14 December 1983 Burgis signed a contract with CERIC for the supply, erection and commissioning of a casing type kiln and 151 kiln cars. The project called for demolition of two of the three existing Swindell Dressler tunnel kilns and their associated tunnel dryers, and replacement with one wide kiln and two dryers comprising two tunnels. Modification of the existing buildings commenced in November 1984 with work continuing over the Christmas shutdown period. With limited production capacity restored, work on the new plant went ahead over the next 15 months.

The testing of kiln fans and burner equipment began on 1 March 1986 and the first car of fired brick emerged from the kiln later that month. The new facility, which cost $16 million, gave the plant a capacity of 75 million bricks per year and was opened by the Premier of Queensland, the late Sir Joh Bjelke-Petersen. A further investment in 1993 of $20million brought production capacity up to 120 million units per year. The upgrade included the installation of a Swindell Dressler kiln and dryers, and dedicated paver extrusion and setting equipment. Carl Zeppenfeld returned to Australia for one last time to assist with this project, as well as the construction of the Riverview plant for Brick & Pipe, which was later taken over by Austral.

INDUSTRY AND REGIONAL TRENDS

Fortunately for Brickworks Limited the process of renewal came at a time when the home building industry was enjoying a boom after several years in the doldrums. In the financial year 1987–1988 the value of all building approvals rose by 38.1 per cent over the previous year. Similarly, the number of brick homes approved during the financial year ending June 1988 was up 28.7 per cent.

By the middle of the 1980s, however, issues that had been developing over the previous decades were consolidating into longer-term trends. For

18. 'Generosity cracks the cross-holdings', *The Australian Financial Review*, 8 December 1986.

Training co-ordinator Tony Fitzpatrick (left) with David O'Connell at Brickworks' training school, Horsley Park

example, the industry was no closer to remedying the shortage of bricklayers (especially in the eastern states) that had become apparent in the early 1960s. Despite various attempts to recruit apprentices the situation was becoming critical. The situation had reached crisis point in New South Wales. Even in Western Australia the shortage was being felt to the extent that bricklayers were being recruited from the UK during the building boom of the 70s and 80s. The turnaround for Western Australia eventually came in 1987 when the state's brick manufacturers agreed to sponsor bricklayer training through a levy of 0.5 per cent on turnover. The scheme continues today in a modified form, although the shortfall persists. In that same year Western Australia was believed to have had one of the highest standards of bricklaying in the world.[19] The dire shortage appears to have stemmed partly from training provided by the country's technical and further education system (TAFE), which favoured a three-year programme. Private training institutions considered this to be outmoded and were able to turn out bricklayers in 12 months. The problem was nationwide, although in Western Australia matters came to a head when the local Construction Mining and Energy Union tried to close private bricklaying schools.[20] The attempt failed but it did force the brick industry and TAFE to look at ways of reforming a curriculum that was increasingly out of touch with reality. In truth, it did not take three years to train a skilled bricklayer.

BRICK CAPITAL OF AUSTRALIA

Regional differences in the brickmaking industry were also asserting themselves in a variety of ways. Full double brick homes had always been the norm in Western Australia since the late nineteenth century. Government legislation in the early days of settlement, and the shortage of timber suitable for building, help to explain the situation in that state. The result was that generations of licensed builders were trained in bricklaying, and with the benefit of strong demand for brick homes builders were able to maintain the tradition. This was not the case on

19. *Australian National Clay*, Vol.17, No.5, September–October 1987, p. 1.

20. *Australian National Clay*, Vol.18, No.4, July–August 1988, p. 13.

the eastern seaboard, especially in New South Wales, where carpentry had been the preferred trade of many builders. Carpenters tended to favour easier – and faster – construction methods and promoted the benefits of brick veneer homes with their reliance on wood-framed walls. The trouble for manufacturers was that brick veneer homes used only 8,000 to 9,000 bricks. The drift away from building double brick was also accompanied by the replacement of brick footings by the floating concrete slab during the 1970s. This alone put a huge hole in brick sales.

The divergence between east and west was reflected in the statistics. In 1987 the brick industry in Western Australia was selling 420 bricks per head of population, which was more than anywhere else in the world. This was three and four times the number being sold in New South Wales. The tally for the US was 35 (and declining) and 50 per head for the United Kingdom. According to Geoffrey Noble, an industry leader with many years' experience, Western Australia was jokingly referred to as the 'brick capital of the world', and retains the position even today.[21]

Another development across the continent was the closure of yards, confirming a tidal drift that can be traced as far back as the late 1880s. The latest casualty was the once mighty firm of Goodlet and Smith which was producing 10 million dry pressed bricks a year at its Merrylands site as late as December 1987. By this time the company was a shadow of its former self with a market share of only 1 per cent of the total market, and just 8 per cent of the dry press market.[22] By the end of 1989 the plant was closed. The site lay idle for many years until it was transformed into a heritage display centre. Hooker Real Estate, the property division of The Hooker Corporation, developed a large portion of the adjoining land for apartments.[23] It was also the end of the road for the State Brick Works at Homebush Bay and its sister plant at Blacktown. On 2 June 1988 the government announced that it would cease manufacture of bricks and close the Homebush Bay yard. Plans to sell the Blacktown yard as a going concern came to nothing and the site was sold to a developer of residential property.

ROOFING TRENDS

The roofing tile business, a close relative of brickmaking, was also experiencing dramatic change. Brickworks Limited was anxious to participate in this market, although its attempt to take control of the American-owned Monier Ltd, the major player in the tile sector, was thwarted. The market for terracotta roof tiles found itself under attack during the 1980s, especially in Queensland, where factories closed and demand slumped. Most plants were small and inefficient, besides which terracotta was losing its traditional hold due to the fact that it was expensive and more time consuming to fit. On the other hand, concrete tiles were cheaper, stronger, survived better in a coastal salt water environment and were easier to make. They were less prone to buckle during manufacture, and unlike terracotta, were pressed on a metal plate which went into the kiln along with the tile during curing.

The transition to cheaper and more convenient concrete tile turned the tables on the sombre, darker hues of traditional terracotta. Taking advantage of the Mediterranean climate and lifestyle, concrete roofing tile manufacturers introduced strong colours. Two-tone tiles were popular, especially those with the outward-facing sides coloured red and black. In response, terracotta manufacturers also adopted two-tone colour. Blending techniques improved, solving earlier problems of colour consistency. Companies such as CSR Limited and Monier Roof Tiles experimented with pastel 'mushroom' which was a pink beige coloured tile. Others followed suit, with Pioneer Concrete producing its own version in coral pink. The 'sunshine state' of Queensland blazed the colour trail, whereas southern markets played catch-up. The demand for a wider choice of shades in roofing and floor tiles was a trend that soon

21. *Australian National Clay*, Vol.17, No.5, September–October 1987, p. 3.
22. Bricks were made in a low thermal mass side-draught kiln built by a New South Wales manufacture, Ceramic and Kiln Industries Pty Ltd, of Wilberforce.
23. In 1955 Newbold Silica and Firebrick Co. Pty Ltd, a company owned by Newbold General Refractories Limited, acquired the issued capital of Goodlet and Smith Ltd. In March 1960 it also acquired the Clyde Brick Co. of Sydney. In 1974 Newbold was taken over by Manufacturing Resources of Australia and its assets broken up and sold off. The Hooker Corporation closed the Merrylands plant and developed its extensive landholdings for residential apartments.

became an established fact of life. Stylish products now came at affordable prices. Pressed, corrugated sheet steel roofing also made an appearance and was in demand for its qualities as a low maintenance, long lasting, waterproof product.

THE LOOK

Popular tastes had been evolving since the end of the Second World War in response to advances in manufacturing technology as much as the desire to look and feel 'modern'. Evolution in the way Australians perceived themselves, and how they lived, took a quantum leap during the 1980s. Builders and developers listened to their customers, pooling market intelligence with manufacturers. Product innovation, inspired by a variety of factors, advanced rapidly. One notable success which continues today was Austral Bricks' 'Through Wall Brick' (TW), which was designed to replace a twin leaf cavity wall with a single leaf. Today, intense competition based on choice, appearance and price continues to drive innovation and adaptation. For purists, the last word belongs to clay and its qualities, for despite the growing popularity of steel roofing and steel wall cladding, the thermal performance of clay tile and brick has been proved to be superior.[24]

AN UNUSUAL TURN OF EVENTS

Unfortunately, the dark side was never far away. To the dismay of just about everyone the prosperity bubble burst on Monday 19 October 1987 when stock markets around the world came crashing down in a wave of panic selling. So-called Black Monday was the second largest one-day percentage decline in stock market history. By the end of October, markets in Hong Kong had fallen 45.8 per cent, Australia 41.8 per cent, the United Kingdom 26.4 per cent, the United States 22.68 per cent, and Canada 22.5 per cent. Since no major news or events had occurred in the lead up, the collapse seems to have come from nowhere.

The events of late October failed to dampen the enthusiasm of Jim Millner and the board for a partnership with the firm of CSR Ltd which was anxious to become a brick manufacturer. Over the years it had grown from being a sugar refiner into a building products company. On 7 December 1987 the two companies announced that they would form a joint venture to operate and develop brick and pipe making operations in New South Wales and Queensland. At this time Brickworks was the major player in the metropolitan areas of the aforementioned states, holding about 40 per cent of the Sydney market and 23 per cent in Brisbane. Undoubtedly, CSR was positioning itelf to capitalise on the expected growth of Sydney and Brisbane.

Millner recommended that 'the sale to CSR will bring together the production skills of Brickworks and the marketing skills of CSR, benefiting the shareholders of both companies.' Ian Burgess, Chief Executive Officer elect of CSR Ltd, spoke of the agreement in glowing terms.

> *We are entering into a partnership with the best and lowest cost producer of clay bricks and pipes on the east coast of Australia.*[25]

A general meeting to consider the sale of 50 per cent of Austral Brick to CSR was held on Wednesday 23 December 1987 at Goldfields House, Sydney Cove. Shareholders voted to accept. Under the agreement which became effective on 1 February 1988, CSR paid $75 million to acquire a half share in Austral Bricks which would carry on the business as a joint venture. Initially everything went reasonably smoothly, although CSR disposed of land at Duffys Forest, developing it for a golf course. Jamie Pearson, the current manager of Austral's No. 3 Plant recalls that the sale of the company's last deposits of white clay threatened the production of the popular white and cream bricks at the Brookvale yard. Excavators and trucks worked round the clock to extract huge quantities of dirt before the sale was finalised.[26]

24. A construction materials testing laboratory accredited by NATA is located at No. 3 Plant at Horsley Park. The laboratory plays an important role in research into physical attributes of clay bricks.

25. Press release, 'CSR Limited and Brickworks Limited', 7 Dec 1997, Archives of Brickworks Limited.

26. Jamie Pearson joined Austral Bricks in 1986 and has occupied a number of positions in senior management positions. Interview with Jamie Pearson, Horsley Park, 10 September 2005.

Several months later, on 21 May 1988 Albert Burgis retired as managing director. His legacy was enormous. At a time of great uncertainty and instability Burgis had inspired confidence. He revived the fortunes of Austral Bricks by investing millions of dollars in plant refurbishments. Bentley was the obvious successor; the board admired his skills and dynamism. Jim Millner certainly liked him for his aggressive ambition and determination to use Austral's production capacity to dominate the New South Wales and Queensland markets, undercutting competitors such as PGH. An inspiring leader, Bentley led his team through upheavals in the industry, including price-cutting wars and a damaging strike in 1990 by brick carters.

WHY NOW?

Back to our story. What was the motivation behind the joint venture with CSR? There are a number of possible explanations. Millner may have seen the partnership as an opportunity to bring about much needed rationalisation in New South Wales and Queensland, which were suffering from overcapacity.[27] Despite extensive competition and price cutting Brickworks Limited had been immensely profitable, although earnings performance during this period began to slip. Perhaps Jim Millner believed this latest joint venture was no different to the 1969 share swap – and that only good things would come of the union. For years ACI had been wooing Millner to try to persuade him to sell but with no success. To the bean counters at Soul Patts the CSR offer was inviting, to say the least.

CSR had its own reasons, although these will never be known. Quite possibly, it regarded the joint venture as a springboard for expansion, initially at the expense of Brick & Pipe. Had CSR been able to purchase this company outright, and acquire a greater share in Brickworks Limited, then it would have been able to form the largest brick operation in the eastern states.

27. Victoria had fared better; industry observers believed that the state's producers were receiving \$100 to \$150 more per 1,000 bricks than their counterparts north of the border.

We're as Australian as a magpie up an old gum tree
As solid as two mates drinkin' billy tea
We're Austral Bricks, bricks you can depend on.
We're as loyal as a cattle dog and his mate
As faithfully as two friends yarnin' at the gate

Refrain

Austral Bricks, bricks you can depend on
Austral Bricks, building up Australia
Austral Bricks building it good and strong
Austral Bricks, building up Australia
Austral Bricks, bricks you can depend on

Confidently Australian, and drawing on references to loyalty and the land, this jingle was written for a radio advertisement in 1988 and reflected the spirit of the times. It was similar to many other advertisements of the period.

THERE AND BACK AGAIN

Alas, the joint venture was not destined to last. The contract between the two parties contained the dramatically-titled, 'shotgun clause' which, at first glance looks somewhat contradictory. The clause required that should CSR wish to acquire Austral outright then its offer price would need to be unacceptable to Millner and the board, ie on the low side. Rejection of the offer would set the market price and pave the way for the sale. Thus, CSR's offer hinged on the company setting a price per share that Brickworks Limited would

Brickworks Limited, consolidated profits 1981–1998

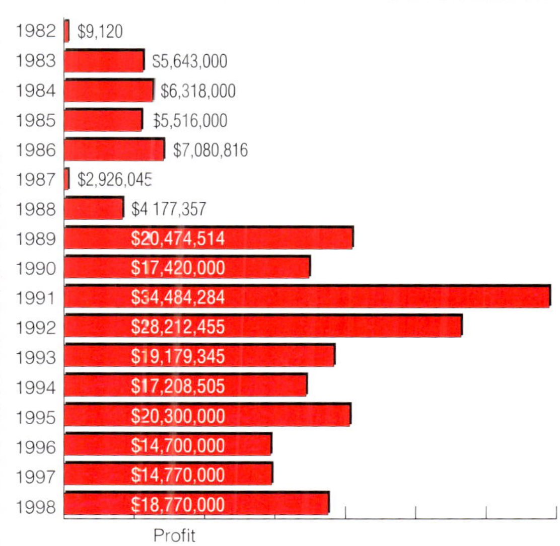

reject as unacceptable. On the other hand, if Millner accepted the offer then Brickworks had the right to buy CSR's 50 per cent shareholding at the offer price. Legal experts acting for both sides during the original negotiations had inserted the clause, which, according to Alex Payne, had been initiated by CSR and negotiated by Millner's advisers. 'Jim Millner was not mindful of it until the fight started. No one really thought he would buy back Austral. He simply outfoxed CSR.' [28]

Like steam in a pressure cooker tension between the two parties began to build. There were disagreements over pricing policies, the refurbishment of No. 1 Plant and the use of certain technologies. CSR-appointed managers to the joint venture clashed with Alan Bentley who had the full support of Millner. Tensions spilled over into executive meetings in a series of acrimonious exchanges between the two sides. Events unfolding in another brickmaking sideshow, as well as Millner's mounting unease over CSR's ambitions, may have contributed to worsening relations. In 1988 Monier PGH Ltd was offered to the Brickworks–CSR joint venture but this was rejected by Brickworks on the grounds that CSR had overvalued the assets. Holdings in the Monier group consisted of the PGH clay brick, paver and pipe business, the Zacuba brick and the Monier roof tile businesses in Australia and New Zealand.[29] Rebuffed by Brickworks, CSR went on to form a joint venture with the UK firm of Redland Plc to purchase Monier PGH Ltd.

The scene was set for a showdown. When CSR Ltd offered to sell its share of Austral to Brickworks it fully expected Millner to take the cash and walk away. Millner accepted the offer then turned the tables on CSR by exercising Brickworks' right to acquire CSR's half share of Austral Bricks, enabling it to buy back the family farm. Indeed, had CSR offered another $10 million more, then the bid may have succeeded, consigning Brickworks to oblivion with Soul Patts not far behind. Happily, it was not meant to be. CSR miscalculated, lining itself up in the cross-hairs of Millner's hunting piece. Besides, he loved the brick business and had done so ever since his first visit with Dawes to Horsley Park. It was a near miss. Millner must surely have been relieved. Perhaps the spectral figure of Dawes had reached out from the grave and tapped him on the shoulder – reminding him that it had always been the strategy of Brickworks to hold land in reserve for brickmaking. The land, however, had become extremely valuable and its loss would have been a huge blow to the company.

It was typical of Millner's conservative philosophy that Brickworks used its substantial cash holdings and a one for one rights issue at $20 a share to finance the transaction. This deft move ensured the company's balance sheet continued to have no gearing. Reunited with its parent, Austral Bricks continued to thrive, outshining its main rivals, Boral and CSR Ltd. 'Why is Brickworks apparently doing so much better than Boral and CSR?', asked the *Business Review Weekly* in June 1992. Millner offered a simple explanation:

Women employed on the floor tile production line at the Punchbowl yard, circa 1995. Brickworks Limited

28. Interview with Alex Payne, Horsley Park, 5 November 2007.
29. In 1995, CSR acquired Redland's 49 per cent of Monier PGH Holdings, so that all the joint venture's brick and paver plants, as well as roof tile operations, became fully CSR owned.

Michael Flack

Rod Clark

Our plant is modern. We shut down some of the plant types our competitors are using 15 years ago. Some are 30 years out of date.[30]

Asked if he could confirm whether or not Brickworks Limited had a cost advantage of 10–15 per cent, Millner said that it was even better but refused to elaborate. Since the early 1980s Austral has remained at a competitive advantage by modernising plants to ensure efficiency and reduce production costs.

Remarkably, within 18 months following the unwinding of the joint venture cash holdings had been restored, standing at $41 million in June 1991. At this date Soul Patts held 49.9 per cent of Brickworks, which in turn held 44 per cent of Soul Patts. The main shareholders, the Pattinson, Dixson and Millner families, held about 10 per cent of Soul, amounting to control of Brickworks in conjunction with the crossholding. Soul Patts – and by extension, Brickworks – remained one of the last family controlled listed companies, with some of the shareholdings in the hands of third generation family members. Historically the management of family companies tends to weaken with successive generations. James Millner proved an exception.

30. 'Brickworks builds on lead', *Business Review Weekly*, 12 June 1992.

UPGRADES AND EXPERTS

Bentley relished a challenge, and was delighted when the board approved plans for the complete automation of No. 1 Plant at a cost of $42 million. Supplied by Keller of West Germany, one of the world's most experienced and respected brick plant builders, it became the third automated brick making facility at Wallgrove. So advanced was the computerised operation, that it enabled Austral to produce almost every type of brick in large quantities, and at literally the press of a button. The plant concentrated on producing the Governors (Australian Colonial Sandstocks) and Rebel ranges. The existing plants, and the newly re-opened No. 1 Plant, increased the production capacity of the Wallgrove plants to five million bricks a week. In 1992 it was the turn of No. 2 and No. 3 plants for major refits to the kilns and dryers. Austral and Swindell Dressler worked together with Austral carrying out most of the work in No. 2 Plant and Swindell Dressler carrying out the majority of work in No. 3 Plant.

Following in the footsteps of a previous generation of engineers and plant managers, Bill Larfield, manager of the plant, and Tony Watling, the operations manager of Horsley Park, were sent to the United States on a tour of brickworks. Returning home, one of the kilns was taken out of commission, allowing

Peter Mahony

production to continue using the second kiln. Day and afternoon shifts were organised and with careful management Larfield managed to produce with one kiln what had previously been done with two. 'It was the best $4 million I ever spent', said Alan Bentley.[31] For some, however, the upgrade to No. 2 Plant, spelled the end of their career with Austral. With the number of kilns reduced to one, which was still able to handle production demands, almost half the men employed as kiln burners were paid off, their jobs having been made redundant. In January 1996 Larfield was appointed to manage No. 3 Plant which had been experiencing production difficulties, especially with its clay and shale mix. Before long it was running efficiently although his insistence on things being done his way proved too difficult for some of the factory supervisors, who left. Ultimately the plant became highly profitable, an achievement due, for the most part, to Bill Larfield's ability to run machines at peak efficiency with a minimal labour force.

Meanwhile, in 1992 a new plant to manufacture high quality soft extrusion pavers and cobble tiles was commissioned on the Punchbowl site.[32] This was in addition to the existing production facility which made salt-glazed clay pipes. In a series of name changes reflecting new realities, Punchbowl Pipes Pty Ltd was renamed Austral Pipes Australia Pty Ltd and later still as The Austral Tile Company Pty Ltd. The cobble tiles part of the operation was known as The Austral Tile Company. Overseeing the changeover from pipes to tiles at Punchbowl was Michael Flack who had joined Brickworks Limited in 1991 as technical manager at Punchbowl. A typically phlegmatic Yorkshireman from the Pennine district in the north of England, Flack continued to make pipes alongside the pavers and tiles. Flack is acknowledged to have produced some of the best unglazed floor tiles available in Australia. During 1995 he was despatched to Italy where he researched tilemaking methods and consumer trends. It was during this overseas trip that Flack and another senior colleague, Rod Clark, were involved in a spectacular crash whilst driving through Spain. Both men survived without serious injury, crawling out from underneath the car, which was a total write-off. In 1997 Brickworks Limited ceased to manufacture clay pipes, although to this day it continues to lease part of the yard at No. 1 Plant to the UK firm, Hepworth Building Products. This company imports clay pipes for local distribution and also happens to be the original employer of Michael Flack.

Michael Flack's move to Eastwood in 1994 to oversee the dry press yard was highly significant, for it was here that the team comprising Flack, Arthur Sherman and Brian Binskin achieved outstanding results from the gas-powered patent and dome kilns. Flack advanced the art and craft of brickmaking at Eastwood with the stockpiling of specific types of clay and shale in layers. Stockpiling had long been routine practice at the Wallgrove Road plants where dirt was left to 'sour' in the open. At Eastwood it was a radical departure from the usual practice of hauling skiploads of dirt up to the mixing shed. Stockpiling was a prudent measure since an increasing amount of material was being trucked in from outside due to the depletion of shale reserves in the quarry. With a more homogenous mix, Flack re-configured burning in the patent kiln to improve the quality of dry press bricks. It was left to Brian Binskin, the kiln

31. Interview with Bill Larfield, No. 3 Plant, Horsley Park, 4 July 2005.

32. The brick plant and chimneys at Punchbowl were demolished in 1978 and the adjoining pit filled. Today the site is used as a container depot.

boss, to convince old hands among the burners to learn new ways. It was an uphill struggle.

Rod Clark, who exited the hire car in Spain by unconventional means, was another veteran brickmaker. Clark's career in the clay industry had commenced in 1967 at the Greenacre plant of Enfield Brick (originally Eagle Brick) where he worked as a maintenance fitter. The buyout by PGH in the following year saw him move to the firm's newly built sandstock brick manufacturing plant at Schofields in western Sydney where Clark met the young Lindsay Partridge. Moving through a series of promotional roles, Clark spent 18 years with PGH, leaving in 1984 after being hired to run a privately owned brick plant, North Brick, in Darwin. Clark joined Brickworks Limited in 1994 with responsibility for raw materials procurement and union matters at the Wallgrove plants, an unusual combination. Twomonths later he was put in charge of the Rochedale operations. For 10 years he was general manager of Queensland operations, ending his career as general manager for national exports. By the time Clark retired in 2006 he had worked in the industry for over 40 years.

The recruitment in June 1997 of Peter Mahony as general manager in charge of manufacturing continued the decade-long practice of attracting outstanding talent. When Mahony joined Brickworks he had already been in the industry for more than three decades, his most recent employer being Boral at Badgery's Creek. Mahony was originally from Victoria and had risen through the ranks at Brick & Pipe Industries Limited.[33] Moving to New South Wales he had worked alongside Alan Bentley, Michael Flack, John Giffen, Alex Payne and Lindsay Partridge, in the Hunter Valley town of Maitland. At Horsley Park his brief was to focus on the problematic No. 1 Plant and its manufacturing glitches. Mahony was someone who thought carefully before speaking his mind and was dogged when it came to the work. He was patient, cool under pressure and had a peculiar knack for being able to solve complex, multiple problems at the one time without losing his temper. Time and again he would return to a plant to fix a problem, or a series of inter-related problems that were seemingly intractable.[34] As a teacher Mahony had few peers and passed on a wealth of knowledge, broadening the technical understanding of many employees.

INNOVATION

By the mid 1990s, whether it was bricks, roofing tiles, pavers or floor tiles, the 'look' had become a critical factor to sales. Customers wanted building products that reflected not only their lifestyle, but also their aspirations. Indeed, the demand for aesthetically attractive building products has become a permanent feature of market demand. Unsurprisingly, the strapline to Pioneer Concrete's market slogan in the new millenia for its range of concrete roof tiles was: 'we've got the look'. Trying to second-guess market directions, however, is always difficult. In the face of intensive competition and a narrowing technological advantage product quality and customer service assumed even greater significance.

The importance of teamwork and sales success was stressed by Alan Bentley, who introduced new initiatives designed to help individuals work together more effectively. One of many successful programmes was 'Pathways to Excellence', which was as far away a person could get from the stereotype of a brick salesperson. The programme was designed with sales, administration and production staff in mind and also involved all levels of management. At the same time product innovation continued. One notable success was the introduction in 1995 of the 'Maxi Brick'. The equivalent of two and a half standard bricks, the Maxi was designed as a lightweight, load-bearing unit with superior fire resistance. The following year two new lines, Pastel bricks and ClayLink retaining wall blocks, were introduced. The latter was developed by Rod Clark during his time at Rochedale. Unfortunately, the idea of linked clay brick walling did not take off and the product

33. A reference to Brick & Pipe owned Hallet Industries, the South Australian brick manufacturer. See Appendix 5 for more details. Peter Mahony joined PGH in the early 1980s and moved to the company's Bowral plant as general manager in about 1985. In the course of his career Mahony has worked for all three of the major manufacturers, Boral, Austral Bricks and CSR-owned PGH.

34. Interview with Lindsay Partridge, 31 July 2006, Horsley Park, Sydney.

The road to St Peters Brickworks, 2005. Oil on canvas by artist: Jane Bennett

was subsequently withdrawn. Undeterred, Clark continued to tweak the design. Following repeated requests for retaining wall blocks it was reintroduced at the Riverview plant in 2001. More recently it was further modified to sit within kilns, which had been increased in height.

GENERATIONAL CHANGE

By 1996 the times certainly were changing. Thirteen years of federal Labor government came to an end with the defeat of Paul Keating in the federal election of March 1996. Voter rejection severely traumatised the Labor Party and its supporters who had progressively lost the confidence of the Australian people. Under the leadership of John Howard, the Liberal-National Coalition government began to steer a course that was avowedly pro-business. Following the introduction of enterprise bargaining even greater emphasis was placed on the need to engage more effectively with the workforce. At each of the plants consultative committees were formed representing all areas and shifts, and provided valuable feedback on ways to improve efficiency and reduce costs.

As the year 1997 gave way to 1998 Brickworks experienced its own change in leadership. By this time Alan Bentley had led the company for more than 10 years, during which the revival in the fortunes of Austral Bricks had been remarkable. The firm had regained leadership of the industry with efficient, modern plants that were the envy of its competitors. Credit for this belongs to Bentley and his close-knit team, especially Lindsay Partridge who had grown in stature as a leader. In the years since joining Austral Bricks Partridge had mastered the intricacies of running an operation that was highly technical in nature, but which also required an understanding of market trends. An astute businessman who knew where he wanted to take the company, Partridge was ready for the top job. On Bentley's advice the board appointed Lindsay Partridge as CEO in 1998. The following year Partridge was given a seat on the board and confirmed as the company's managing director.[35] At more or less the same time, Jim Millner stepped down as chairman of Brickworks Limited. After a run of 30 years he relinquished the position to his nephew, Rob Millner, who was also chairman of Soul Patts. Where Partridge was about to take Brickworks was an open secret among close colleagues such as Millner. In getting there his actions transformed – and continue to transform – the company.

35. Partridge's elevation to the board continued a long tradition. Bentley was made a director in February 1995 and continues to sit on the board as a non-executive director.

East meets West 20

Learning curve – North of the border – Sydney Olympics – GST boom time – Optimism – Trends and transformations – Bowral Bricks – Industry downturn – Exports – Eureka Tiles – Land And Property – Where there's muck – East meets west – Everywhere – Wollert – Bottom of the graph – Must haves – Something unique

Brick delivery trucks are owned by contractors as well as by Austral Bricks. This photograph was taken on the company's property at Horsley Park, New South Wales, 2004

LEARNING CURVE

What goes on over the garden fence has always held a special fascination for young children who like to peep through gaps in the lattice, or poke sticks over the top. As a child Lindsay Partridge was no exception. An inquisitive fellow, he soon discovered that the next door neighbour was a potter. Objects carefully packed into the kiln intrigued the young Partridge. What came out of it was equally fascinating. Colours, tints, hues, shapes – all of these captured the boy's imagination, opening the door to a hobby that became a lifelong career. Drawn into a world of trial and error, repetition and problem solving, Lindsay Partridge was learning valuable lessons. His play was his work. He played with clay and different coloured glaze, encouraged first by the neighbour and then by his father who set him up with his first kiln when he was a teenager. By the time Partridge commenced an undergraduate course in ceramic engineering at The University of New South Wales, he had acquired a level of understanding that was unusual for a person of his age. Shortly after, he won a cadetship with PGH

Lindsay Partridge

Industries, which at that time was owned by Acmil Industries Limited. Before he graduated in 1978 spare days, weekends and holidays were spent at the company's various brick, pipe and tile factories, including the Ballarat tile plant.

Having recognised the potential of their charge, PGH appointed 23-year old Partridge in late 1979 as unit manager of its Maitland factory which soon gained a reputation for being the firm's most efficient plant. This was followed in mid-1980 by a posting to San Jose, California where he worked at Stonelight Tile, one of Acmil's North American subsidiaries. With the US economy in recession and the building trade flat, Partridge returned to Australia in 1982 and was appointed by PGH Eureka Tiles to manage its national operations. Uncertainties created by changes in management and ownership prompted him to follow Alan Bentley to Austral Bricks in 1985. Thereafter, a series of promotion in quick succession marked a rapid career progression. From being works manager at No. 2 Plant, he was appointed operations manager of the Horsley Park plants in 1987, becoming general manager for New South Wales in 1989. Almost a decade spent as understudy to Alan Bentley helped to prepare Partridge for the top job. Rob Millner, who became chairman of the board of Brickworks in 1999 upon the retirement of his uncle, Jim Millner, was the first to congratulate him. Millner was impressed with his credentials and entrepreneurial flair, especially his intimate knowledge of the business:

> *Lindsay Partridge was involved at every level. He understood that each plant had a different way of making bricks and that each state produced a different product.* [1]

Partridge knew that the strength of the company's management team lay in the breadth and depth of experience of individual members. Michael Flack had the advantage of following the career of Lindsay Partridge from the time both men worked at the Doonside plant of PGH during the late 1970s. Peter Mahony was another 'elder statesman' of the ceramics industry and an early mentor who encouraged Partridge. Both Flack and Mahony were consummate professionals whose advice was sought – and heeded. The aforementioned were typical of men who had risen through the ranks of building-related industries. They gained their reputation and the respect of others simply by getting their hands dirty. The business of brickmaking is 'hands on' with the result that experience at plant level is a necessary prerequisite for anyone aspiring to reach the top. Brick plants are also great levellers; people with tickets on themselves don't usually survive for long.

NORTH OF THE BORDER

Understandably, Partridge was impatient to stamp his own style of leadership on the company. Wasting no time he took Brickworks and Austral in a radically

1. Interview with Robert Millner, Sydney, 4 July 2007.

new direction. Growth by a series of strategic acquisitions transformed the company into an organisation that now bridges the island continent of Australia. How Lindsay Partridge brought east and west together and the ongoing consequences of this manoeuvre occupies much of this chapter. First, there was work to be done north of the border. Starting in September 1999 the board approved the purchase of a brick manufacturing plant at Riverview, west of Brisbane, from the firm of Nubrick, a subsidiary of Pioneer. The plant's advanced, fully automated kilns were designed to carry smaller and lower loads allowing for fast drying and had a production capacity of 40 million bricks per year. For a total purchase price of about $13.5 million for the factory and its inventory, the cost was well below what it would have taken to construct a similar facility from scratch. Doug Wilmot, the current Group General Manager Manufacturing and Major Projects for Brickworks Limited, installed the plant during his time at Nubrik. He was also responsible for the construction and commissioning of Nubrik plants in South Australia, Canberra, Tasmania and Victoria. [2]

SYDNEY OLYMPICS

Meanwhile, Sydney was preparing to host the eagerly awaited 2000 Olympic Games. The announcement on 23 September 1993 of Sydney's winning bid thrilled the nation. Referred to as the Millenium

In July 1987 Pioneer Concrete Services Limited took over the Queensland operations of Besser Queensland which made concrete building blocks. Pioneer renamed the Besser operations, Pioneer Building Products. Pioneer then bought Nubrik in Victoria. Another Pioneer company occupied a site at Dinmore, Queensland. Pioneer used vacant land at this location to build a brick factory. It sold its brickmaking operations to Austral Bricks in 1999. Shortly after, in December 1999, Pioneer Building sold Nubrik and its roof tile business to Bristile Limited, which was based in West Australia. Brickworks Limited acquired Bristile in 2003.[3]

2. Earlier in his career Doug Wilmot worked under Jack Peterson, an influential figure in the brickmaking industry who was also managing director of Clifton Brick. This firm was later acquired by Nubrik, part of the Bristile group, which was taken over by Austral Bricks in 2003. Peterson was the driving force behind the construction of Nubrik's Riverview plant in 1994 and worked closely with Doug Wilmot during installation. See Appendix 4 for more details.

3. In January 1989 Pioneer Concrete Services Ltd changed its name to Pioneer International Limited. It was delisted from the ASX in December 1989 and taken over in May 2000 by Hanson Australia Proprietary Limited, a subsidiary of Hanson PLC.

Doug Wilmot, brickmaster and electrical engineer whose 37-year career in the clay products industry culminated in the commissioning of Austral's Wollert plant in Victoria, 2007

games it was a major coup for the city and a huge fillip to the building industry which was beginning to emerge from years of recession. Plans were drawn up to transform the enormous expanse of land at Homebush, the former home of the State Brick Works and the State Abbatoir. Actual building work did not commence for several more years, but when it did the task proved to be Herculean. The acquisition of the Riverview plant the previous year was timely. The yard's entire stockpile of 12 million bricks, and its current production, were trucked to Sydney in an immense shuttle operation that lasted months. In fact, every single brick was sold as part of the Olympics boom.[4]

BRICKIES

As one might expect, the Olympic preparations acted as a magnet for labour, in much the same way as the minerals boom in Western Australia or, for that matter, the gold rushes of the nineteenth century. As the clock ticked towards the new millennium the shortage of skilled tradespeople worsened with the result that the cost of residential and commercial

4. Interview with Lindsay Partridge, Sydney, 21 December 2007.

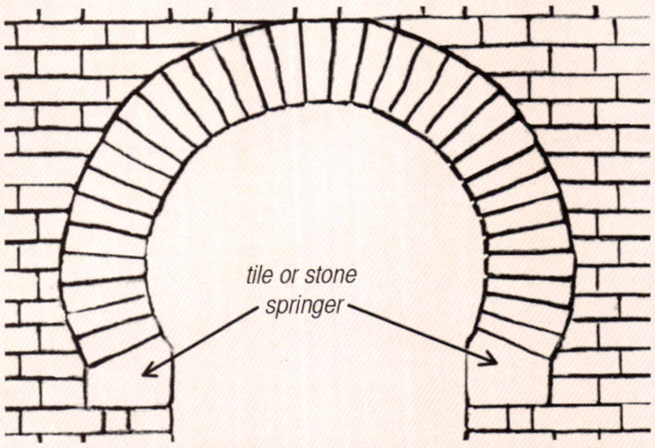

Horse shoe arch. The origins of this arch was probably oriental. The crown of the arch often pointed instead of circular, and in such a case it was known as a 'Moorish' arch

Elevation of arch showing orders of brickwork

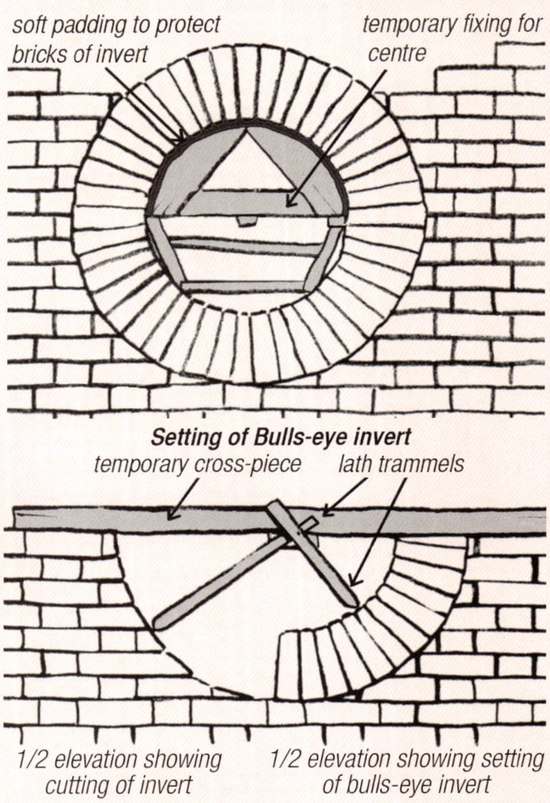

Circular Bull's Eye. The setting out and cutting of this bull's eye differs slightly from the ordinary arch in that the lower half or invert does not require temporary supports

developments began to spiral upwards. Finding a bricklayer for a home renovation was on a par with the quest for the Holy Grail. For reasons that are self-evident, bricklaying is labour intensive, having changed little apart from the replacement of the hod with an electric hoist, and the introduction of the onsite portable cement mixer. It is also a craft practised increasingly by an ageing workforce, many of them in their late forties. At Austral Lindsay Partridge offered financial support to older employees to enter the company's mature age apprenticeship scheme. This was both a pragmatic and progressive move and also a recognition of the experience, work habits and attitudes of men and women over the age of 40. The bricklaying school established under Partridge in the late 1980s eventually grew into The Australian Brick and Blocklaying Training Foundation Limited (ABBTF). By 2005 Austral 's Graduate program for those with a degree was also bearing fruit.[5]

At the height of the pre-Olympic building boom, the Australian Competition and Consumer Commission (ACCC) approved a levy of $1 per 1,000 bricks sold in order to part fund the training of bricklaying apprenticeships. Administered on behalf of the industry by ABBTF the levy was matched by contributions from manufacturers.[6] New South Wales, Queensland and the Australian Capital Territory (ACT) were quick to introduce the scheme which had an immediate effect on the number of apprentices in training. In July 2003 Victoria set up its own training foundation.[7] South Australia and Western Australia followed suit in 2007. By 2004 plans were also well advanced to form a National Roof Tiles Association to take responsibility for instigating a roof tiler training program.

GST BOOM TIME

Escalating labour costs were further complicated by the debate on a Goods and Services Tax (GST)

5. Interview with Doug Wilmot, Horsley Park, 27 June 2005.
6. In 2007 the levy stood at $2 per thousand on standard bricks. The average house in Western Australia has approximately 20,000 bricks.
7. This organisation goes by the name of the Victorian Brick and Blocklaying Training Foundation Ltd (VBBTF).

which was considered by many to have been as dead as the Dodo. In the run-up to the 1998 election John Howard, the prime minister, announced that if re-elected his Coalition government would replace all existing sales and wholesale taxes with a single tax which would apply to all goods and services. Having narrowly won the election the Howard Government enacted the necessary legislation that came into force on 1 July 2000. The ensuing rush to beat the GST sparked a home building and DIY boom during which record volumes of bricks, pavers and tiles were sold. With its brick and tile factories running at capacity Austral cancelled the usual Christmas maintenance shutdown. The downside to this brick bonanza was that stocks were run down to critically low levels, with sales exceeding manufacturing capacity. Builders, developers and bricklayers had a field day, but those waiting on their new home were caught between the pincers of the GST and the Olympic inspired building boom. Escalating costs and rapidly inflating land and house prices across the sprawling metropolis of Sydney cut deep into the hip pocket, although some relief for cash-strapped house hunters was earned in early 2001 with the introduction of a $14,000 First Home Buyers Grant by the Federal Government.[8]

OPTIMISM

The arrival of the new millennium was more a cause for relief than celebration, especially for those dependent on silicon chips to run cars, buildings and assorted electronic equipment. As the clock struck midnight on 31 December 1999 the sky did not fall in and lifts continued to function. IT consultants and doomsayers who had made a fortune on the back of real and imagined problems quietly left the building (by the lift) having enjoyed seven years of plenty.

With money in their pockets delivered by rises in real wages consumers had a field day – and not just in Sydney. Spending as if there was no tomorrow, shoppers, mortgagees and lifestylers embarked on a credit binge that promised to leave a nasty hangover. The rate of personal debt in Australia soared, but so

8. The subsequent reduction of the grant to $10,000 on 30 June 2002 led to a drop in housing sales.

FINISHES TO TOPS OF WALLS

Stone coping

Saddle back coping from bricks flat

Tile-creasing and bricks on edge

Tile band and bricks on edge

Saddle back coping purpose-made engineering brick

Saddle back coping from bricks on edge

Architect-designed home showing facing wall of Governors bricks

did living standards and employment opportunities in the boom economy.

Riding a wave of new investment and soaring world demand for minerals, the Australian economy grew strongly. The housing market in particular enjoyed a long boom that only began to run out of steam in 2004. Just as it is today, the building of private housing in the early years of the new century was extremely important. In 2003 it represented about 2.6 per cent of Gross Domestic Product (GDP), compared to the agriculture sector that made up about 2.2 per cent.[9] Numerous factors are at work in the housing sector, including land and house prices, rental vacancies and levels of home ownership, housing debt and building activity. Each part of the building lifecycle shows periods of expansion and contraction, although movements in the dwelling cycle are usually much greater than most other components in the economy.

If the buoyant Australian economy gave rise to a feeling of general optimism, then the escalating share price of Brickworks made people like the New Zealander Sir Ron Brierley sit up and take notice. On 6 April 2000 Brickworks received an unsolicited (and unwelcome) offer from Brierley's Guinness Peat Group (GPG). After having received no reply to questions put to GPG by the Brickworks board, Soul Patts advised the financial press that it did not intend to accept the offer. The retired Jim Millner was incensed. On 13 June 2000, with just over a 10 per cent interest in Brickworks, GPG extended its offer by three months. Frustrated by the lack of response, Brierley stood up in the November annual general meeting and challenged the board. Not one to hold back Jim Millner let fly, adding a touch of drama to the proceedings. Brierley's attempt to gain control of Brickworks was never going to succeed, especially when the Foreign Investment Review Board (FIRB) ruled that the existing media interests in Soul Patts and the current bid would put any prospective deal in breach of legislation governing the ownership of Australian companies by foreign nationals or entities.

9. Australian Bureau of Statistics, Australian Economic Indicators, 1350.0 December 2003, p. 5.

Margaret 'Maggie' O'Leary

What goes on at the first point of contact with visitors and callers is important for any organisation. 'Maggie O'Leary – professional, diplomatic and a wealth of knowledge – has been with Austral for over 20 years as front of house receptionist

TRENDS AND TRANSFORMATIONS

Necessity being the mother of invention the trade shortage, especially in bricklayers, helped to drive innovation in construction materials and methods. One example was Slick Bricks® which was introduced by Austral in 1999. Jim Millner is credited with the idea after concluding that bricklayers were holding the industry to ransom with their exorbitant rates. The system used precision, diamond ground clay bricks with an interlocking key that eliminated the vertical joint. Mortar was replaced by a form of glue that was intended to make bricklaying easier and quicker, and walls stronger. In theory this would result in productivity gains, improved on-site quality and cost savings. Unfortunately, substantial competition from panel type products made progress slow.[10] 'Panelisation', a streamlined building system that instals pre-fabricated homes on site has profound implications for the Australian building industry. Such advances are an extension of the move towards off site assembly of building components, including frames, trusses and windows using a range of materials from timber and ply through to cement and steel. Prefabrication in a controlled factory environment has long been applied to commercial constructions and it was only a matter of time before the same methods began to make inroads into the residential building sector. The increasing popularity of 'applied finishes' (render), in lieu of face bricks has been another trend. Whether built in brick or blocks, walls rendered in concrete and painted in pastels have captured people's imagination – at least for a while. Increasingly, building companies are becoming aware of the reduced life expectancy and maintenance costs associated with these more expensive finishes.

At the heart of any discussion about future trends must be the role and relevance of brick in modern dwellings. Less than 10-20 per cent of a brick home is structural; the balance is cladding, which has everything to do with colour and fashion. In short, buildings have increasingly become a statement about the way we see ourselves – a matter of style. This, perhaps, is what drives consumer sentiment; playing on our desire to look and feel good about ourselves. Marketing appeals to the emotions as much as it relies on our notion of affordability.

Meanwhile, another Austral innovation, Mighty Bricks™, was launched for medium height buildings of up to eight storeys. In 2001 Austral's e-brick™ system was introduced, allowing customers to place orders, view account details and organise deliveries via the internet. At the same time traditional markets in paving bricks were also changing from the standard brick size to large format pavers. Austral released new paving ranges produced at the Punchbowl plant in New South Wales. It also relaunched the Claylink retaining wall system in an updated format.

10. C. Inglis, 'Slick Bricks – An Innovative New Masonry System', *Journal Of The Australasian Ceramic Society*, Volume 39, Number 1, (2003).

Composite design using face bricks produced by Bowral Bricks

THE CUSTOMER IS KING

With the dawn of a new century product development and innovation across the brick industry found themselves up against fragmenting popular taste. Actually, the process had been underway for at least a decade. The notion that 'one size fits all' applied to an earlier age of uniformity where manufacturing facilities produced a limited range of basic shapes and patterns. This had as much to do with cars as it did with bricks, whiteware or any other type of factory-made consumer good. During times when demand outstripped supply, the customer's personal preference came second place to getting the finished goods out the door, lending weight to the expression: 'you can have any colour as long as it's black'. The assertion of the individual over the collective marks a more recent trend in developed economies, not only in manufacturing but also across the political and social spectrum. For example in Australia the major political parties have seen their electoral advantage eroded by single interest groups. Customers walk out of stores if they can't buy at prices they are prepared to pay, in the process displacing traditional notions of value. How low can you really go? Markets of all types now differentiate premium products from standard offerings which are priced accordingly. Clearly, price, customer expectations and the desire for seemingly infinite choice are complicating factors for industries trying to come to grips with the fickleness of the market. Brick manufacturers have worked hard to produce clay products that serve aesthetic (facing brick) as well as practical purposes (engineering or 'commons').[11] In this regard, Austral remains an industry leader although such a position means that the company must remain ahead of the game through constant product innovation.

BOWRAL BRICKS

One traditional brick product that remains in high demand from architects is dry press face brick which is manufactured in traditional sizes as well as shapes

11. Traditional engineering brick, a bluish-black in colour, was highly dense and could withstand intense pressure and extremely heavy loads. Great Britain pioneered its use in railway tunnels and sidings, prior to the introduction of concrete construction.

Bob Edwards, a long-time employee at Bowral Bricks, NSW, is typical of men whose working life in brickyards has seen the transformation of the industry

known as 'specials'. In 2001 Austral's Eastwood yard, the company's only producer of dry press face bricks, was close to retirement. For many residents the end could not come soon enough. The noise of blasting and the nuisance of smoke were too much for groups who busily lobbied the local council to have the yard shut down. There was no point denying the obvious; brickyards located in densely populated areas were increasingly expensive to run in terms of meeting environmental requirements. The throwaway comment decades earlier by William King Dawes that one day land would be worth more than the brickmaking operations was amply demonstrated in the case of Eastwood. Austral capitalised by laying plans to remediate the site for use as residential building plots. Yet, with Brookvale closed, where would the company source its dry press bricks? The answer lay far to the south west in the Southern Highlands. In February 2001 Austral Bricks purchased Bowral Brickworks Pty Ltd from MBS Securities Pty Ltd of Western Australia, a family firm owned by Ric New.

The plant was conveniently located a short distance from a major highway. In addition to the 27-hectare factory site, there were two clay and shale properties totalling approximately 130 hectares with reserves in excess of 40 years, and an approved landfill site at Bowral. Since acquisition, an adjoining property of 1.14 hectares has been purchased allowing for plant expansion and additional storage. Originally, Bowral's patent kilns were designed to burn commons only, but modifications allowed the plant to manufacture face bricks in a 50-50 ratio. At the time of purchase Bowral Bricks was in need of a major injection of capital which the new owner would in due course make.

Established by the Baker family in the early decades of the twentieth century (see Appendix 1), Bowral Bricks enjoyed a well-deserved reputation. The premium face brick and shaped bricks produced by the company were esteemed by architects and builders who were anxious to create inspiring buildings and also to match modern renovations with century-old dry press bricks.[12] Besides having the good sense to 'stick to its knitting', the company had a sales advantage. Under its longstanding sales manager, John Brady, Bowral continued to win most of the tenders supplied to potential clients.

The purchase of Bowral returned Austral Bricks to its previous position as the leading manufacturer of architectural bricks. The reader may recall that in 1925 Austral acquired a substantial stake in Central Brick and Tile, a leading manufacturer of face, commons and shaped bricks. The Central yard was an immediate neighbour of Austral at St Peters on the Princes Highway and was purchased outright by Brickworks in the late 1930s. Unfortunately, the factory was destroyed by fire in April 1951 and never made another brick, regardless of its shape. The purchase of Reg Allot's Punchbowl Brick and Tile Co. in 1961 restored the company's reputation as a producer, although not for long. For some reason Dawes chose not to continue the tradition.

INDUSTRY DOWNTURN

In view of the fact that Eastwood's days were numbered, the purchase of Bowral Bricks in 2001 was a necessary and logical step in the rationalisation of the company's landholdings and production facilities. Unfortunately, it was overshadowed by a temporary slump in house building activity. The GST induced boom and preparation for the Sydney Olympics were the main and well-publicised causes of this unprecedented draw forward and subsequent collapse in demand. In fact, the housing industry was suffering its worst downturn in thirty years with commencements falling 34 per cent to 114,000 nationally in 2000-01.[13] Brick and paver sales fell by a third, although unit selling prices continued to increase beyond unit production costs, cushioning the impact of the slump. This was achieved despite aggressive discounting by competitors and numerous alternative products seeking to gain a foothold in traditional markets. Falling brick sales were further impacted by the spectacular multi-billion collapse of HIH Insurance through corporate fraud and ineptitude at senior levels of the company. Some developers with long-term associations with Austral were left uninsured by the fiasco and were unable to commence new work. It was some time before the insurance industry recovered and underwriters could recommend cover for those trapped beneath the wreckage of HIH.

This did not appear to affect the Punchbowl tile plant which was undergoing a $21.5 million refit. Work progressed rapidly once government and local council approval had been secured with the plant scheduled to commence operation in January 2002. Next in line for modernisation was Riverview which replaced the existing palletised brick packaging system with the company's first robot operated dehacker. Robot technology was installed at Bowral Bricks several years later.

EXPORTS

The appointment of Lindsay Partridge also coincided with the growth in export markets. Australia's export record had been woeful compared with that of other developed economies, a situation that played on the mind of Federal Treasurer, Paul Keating. In an interview on the John Laws radio program on 14 May 1986 Keating referred to the size of the foreign debt relative to GDP, stating that Australia was in danger of becoming a 'banana republic'. By 2000 the country had done much to live down its poor export performance. In the case of Punchbowl Pipes a limited market in clay pipes had been established by Albert Burgis which continued to grow under Alan Bentley. During the late 1990s Asian and New Zealand markets were explored, the drive to sell clay products overseas being part of a wider initiative by sales manager David Fitzharris and his colleagues to establish new markets. The company's representative criss-crossed Taiwan, South Korea, Hong Kong, China and Japan, making contact with new customers. Japan became a major buyer of pavers and brick facings that were being produced and exported from the Queensland operations. For a short time Japan's DIY market experienced boom conditions.

12. The shale at Bowral is very hard and remains hard when it is ground and burned, although exposure to the elements at the bottom of the pit makes it easier to crush.
13. Annual Report, Brickworks Limited, 2001.

Clay tiles – Terraçade – are used extensively as a design feature on the exterior wall of the Adelaide Oval. TerraçadeXP® first gained a foothold in Victoria and was quickly embraced by architects for its design clean lines, open jointed, long and rectangular pattern. The system is used as external cladding for prestige, branded buildings and is easy to install, easily cleaned and colour fast

The new terracotta floor tile plant at Punchbowl, with its high level of efficiency, also allowed the company to compete effectively overseas. When it was finally commissioned in early 2002 unit production cost was substantially below the imported price of equivalent products and below all local competitors. Moreover, quality was well above imported Italian or Spanish tiles and other locally produced terracotta (quarry) tiles. In an effort to sustain its profitable export market Austral continued to develop new products such as a range of pavers, Silk Road, which was introduced in 2003 specifically for Hong Kong. These pavers were developed at the firm's Rochedale plant in Queensland.

Export success also attracted attention from those who understood the importance of international trade. For example in October 2000 Austral's export record was recognised when the Japan Chamber of Commerce conferred its 'Large Advanced Manufacturer' on the company at a ceremony in Brisbane. Austral was also a nominated finalist in this category for the 2002 Premier of Queensland's Export Awards. Export growth continued to rise, especially in New Zealand where new distribution arrangements were struck with Carter Holt. This was timely in view of a consumer backlash against light-weight polystyrene based walling systems which were notable failures. In fact, growth in New Zealand was unprecedented with with demand outstripping supply. The company's annual report for 2004 could not resist making the point that:

> *many New Zealand bricklayers and consumers sensibly realised there was no substitute for face bricks, to avoid the expense of re-applying fading and flaking applied finishes. New Zealand builders and bricklayers have very quickly appreciated the benefits of Slick Bricks™ and are showing increasing interest and demand for this leading product.*[14]

Instrumental in this export surge was Rod Clark, a veteran brickmaster and senior general manager, who was appointed export manager for the group. Clark looked even further afield, establishing a new market in Dubai.

14. Annual Report, Brickworks Limited. 2004.

Exports, however, tended to fluctuate in response to local demand for clay products and movements in the value of the Australian dollar on international currency markets. Unfortunately, the export push began to run out of steam during 2005 when the strength of the Australian dollar coupled with higher transport costs caused by rising oil prices and shipping shortages, began to have a significant impact on margins. New Zealand softened as the housing construction cycle retreated from its peak and volumes to Japan also declined. By 2006 and into 2007 aggressive Asian competitors were making the company's products less attractive resulting in lower exports to the region. An exception to this trend was Hong Kong where Brickworks had opened an office in 2003. Here, sales improved on the back of the government's recommenced urban renewal program. Nearby Macau, a honey pot gambling and holiday destination for China's newly-rich and fast-growing middle class, also provided growth on back of the casino and gaming parlour building boom.

EUREKA TILES

Just over a year after the purchase of Bowral Bricks Austral Bricks purchased Eureka Tiles Pty Ltd in May 2002 through its subsidiary, The Austral Tile Company. Eureka Tiles was based in Ballarat where its main tile factory comprised two production lines – a traditional tunnel kiln and a modern fast fire roller hearth kiln that could produce a range of large format slip resistant glazed tiles. Located nearby was the plant that manufactures bathroom fittings. The combination of Eureka Tiles and Austral Tiles, trading as Eureka Tiles Australia, formed Australia's largest floor tile producer with capacity approaching two million square metres per annum.

A comprehensive sales and marketing drive reinforced growth during this period. Product development continued strongly with the release of a 40 x 40cm unglazed tiles called the Boulevard range. Terraçade®, a terracotta cladding used as a curtain wall in buildings up to 40 stories high, met with immediate success in both the domestic and overseas markets. A new staining system was also launched, winning praise from builders, architects and designers for its technical excellence. However, the construction of a 488-metre long state-of-the-art tile factory at Rutherford, near Maitland, in New South Wales proved something of a dampener. Its owner, National Ceramic Industries, a majority owned South African company trading, was determined to peg back the competitive advantage of Eureka Tiles Australia. The plant opened in March 2004 and was a formidable, low cost producer. This prompted Partridge and the board of Brickworks to shift from domestically produced to imported tiles. Accordingly, the Fast Fire facility in Ballarat was decommissioned during July 2007 as part of the rationalisation program.

LAND AND PROPERTY

Let us take a moment and retrace our steps back to the end of 2001 then Sydney's building industry was beginning to claw its way out of the post-Olympic trough. During 2002 the New South Wales and Queensland markets were performing well, although by mid-year the shortage of land in the greater Sydney area was heading toward crisis point. Writing in the *Sydney Morning Herald*, Anne Davies, the paper's editor for urban affairs, reported on what was already common knowledge: that a land shortage in Sydney would drive up house prices and make the city even more unaffordable. Her source, the New South Wales Urban Taskforce, urged the State Labor Government to reconsider its population projections for Sydney which were too conservative.[15] The warnings went largely unheeded and the government of Bob Carr vowed that land would be developed only if public transport was in place. Yet, without funding to allow a rail spur to be built to the Bringelly release area where 30,000 new homes were planned, any further development was stymied. The release of building plots on the fringe of Sydney fell even further behind schedule, with 'brownfield' sites in the middle and inner ring were nearly exhausted. Add to this the infrastructure levies imposed by the State and local governments, and it is easy to see why the housing situation continued

15. Anne Davies, 'Land shortage raises new alarm over house prices', Sydney Morning Herald, 2 August 2002.

Eastwood Brickworks in the process of being demolished, 2005. Artist: Jane Bennett, oil on canvas

to make headlines in Sydney's newspapers. The hardest hit sector of the community was working class families who either continued to rent or saddled themselves with dangerously high levels of mortgage debt. In the event of a property downturn they would be at risk of defaulting and having the property repossessed. Or equally distressing, see the amount they owed exceed the value of the house.

Meanwhile, with a portfolio of surplus land running into hundreds of hectares, land and property management for Austral continued to grow in importance as a source of revenue, as well as a contributor to profits. We have already seen in previous chapters how the various Austral yards at St Peters were progressively sold to Sydney Council for waste disposal. Had compulsory acquisition not been the trigger for disposal – and supposing that Austral had retained ownership – then the land would now be worth a huge sum of money. Exploitation of land and property assets assumed greater importance with the creation of a new senior position, that of general manager, property and development. Megan Kublins (née Tamsett) took this role in 2001 and under her direction the portfolio has prospered. Anything residential is rezoned and sold. Land is also acquired, not only for its value as clay reserves, but also for commercial development.

BROOKVALE

Over at Brookvale on Sydney's North Shore, the redevelopment of the company's brickyard had commenced as early as 1990, a full eight years before the closure of the plant. Having identified the land as being suitable for medium density housing Warringah Council commissioned a heritage study in 1991. The ensuing Brickworks Conservation Plan more or less rejected the adaptive re-use of the old kilns and buildings and advocated preservation of the site. Who would pick up the bill for turning the site into a museum was not clear. Brickworks challenged the findings and in September 1999 submitted a development application for the demolition of all existing buildings and structures. Finally, in 2002 the company entered an agreement to sell the site to a third party rather than continue as the developer. The deal allowed Austral to withdraw completely from any further involvement that was prudent since there was a high risk where heritage issues were paramount.

Brookvale was just one in a long line of divestments and developments with the last decade witnessing an acceleration in activity. Surplus land at the Punchbowl site was also being prepared for sale in 2001 as the new tile factory took shape. Plans for its subdivision passed through the local council enabling the company to lease the land, some of it for the long term storage of sea shipping containers. In the same year the company lodged a development application to subdivide into commercial lots the 65 hectares of land it owned at Mamre Road, St Marys.

EASTWOOD

Development of the 14.7 hectare Eastwood brick factory site was another ambitious project. In September 2001, six months after the purchase of Bowral Bricks, a plan was put in motion for a future residential development on the site. The following year a master plan was drawn up subject to development approval by Parramatta Council, which allowed filling of the enormous pit to commence. In April 2003 the plan was approved, although objections from some residents over truck movements severely restricted the work for a while. Just like the Brookvale experience, the redevelopment of Eastwood was no easy task. Eventually, the land was sold in September 2007 to the building firm of A.V. Jennings for $70 million, which under zoning requirements allowed the construction of 255 free standing homes, townhouses and apartments. The development company also drew up plans for a heritage precinct that would include a sympathetic interpretation of the operations of The Great Northern Brick Company Ltd established in 1912.

THE VINEYARD

At Horsley Park even bigger things were afoot. During 2002 planning commenced on turning a major quarry comprising 150 hectares of land adjacent to No.3 Plant into industrial property. Referred to by the unusual name of State Employment Planning Precinct (SEPP 59) the immediate surroundings still possessed remnants of grape vines indicating previous use as part of the Penfold Estate. Later, the site was referred to by the more inviting name of 'The Vineyard'. Development consent was granted by the New South Wales Minister for Planning in March 2003.

Work on developing another lot comprising 333 hectares to the west of No. 3 Plant also commenced in March 2003. Approval was given for approximately 60 hectares of the Bedford property to be used for an extractive industry, increasing clay reserves at Horsley Park by an additional 20 years' supply. The remainder was prepared for eventual lease to commercial tenants. In 2004 the New South Wales Roads and Traffic Authority (RTA) resumed large portions of land for the construction of the Western Sydney Orbital (M7) that by-passed Horsley Park. The sale realised a considerable sum in view of the underlying clay reserves.

In 2004 Macquarie Goodman (now Goodman), an industry property expert, was selected as the partner to develop The Vineyard which was finally renamed the M7 Business Hub to reflect both its strategic location and future use. With easy access to the M4 motorway and prime frontage to the M7 motorway, this large tract of industrial land was a rarity in the Sydney market. In 2006 the relationship between Macquarie Goodman and Brickworks deepened when the two organisations created a property trust, the Brickworks Macquarie Goodman Wholesale Trust. One property sold into the trust was the Mamre Road site from which the Interlink Distribution Centre was created in 2006.

Two other properties acquired by Brickworks following the purchase of Bristile Limited were also disposed of at this time. In South Australia, surplus land at Golden Grove outside Adelaide, was sold in early 2005. In Victoria rationalisation of Nubrik's property portfolio commenced in late 2004 with the closure of the Scoresby and Hallam sites in Melbourne. This went ahead following Board approval for the new Austral factory at Wollert. Zoned for industrial development, the Hallam site was recently sold along with Cranbourne in Victoria and Muchea in Western Australia. At the time of writing Scoresby site is out to tender. By March 2007 the company's land and property portfolio comprised 4,570 hectares. Of this figure 900 hectares is considered surplus and available for development and or sale. Needless to say the profit realised from the sale of land has been substantial.

WHERE THERE'S MUCK

The exploitation of landholdings was not confined to the sale of residential plots and development of sites for commercial use. In September 2001 contracts were signed with Collex, a wholly owned subsidiary of Vivendi Universal of France, to commence a waste recycling and landfill operation on the site at No. 1 Plant, Wallgrove Road, Horsley Park. Purchase of the land by Collex was given the go-ahead by the Foreign Investment Review Board and an Environmental Protection Agency (EPA) licence was duly granted. Landfill operations commenced in late July 2002 and the volume of waste that could be tipped was increased following modifications to the site. In 2005 an additional waste facility was opened at the Bowral site. During 2004 Brickworks applied to have land at Cardup in Western Australia rezoned for landfill as well as at Bolinda Road in Campbellfield, Victoria. For as long as Sydney continues to exist, landfill sites such as those owned by Brickworks will be needed – and profitable.

EAST MEETS WEST

If the introduction of tunnel kilns in the 1960s and the share swap with Soul Patts were landmark events for Brickworks, then the purchase of Bristile Limited in 2003 was in a category of its own. Manoeuvring by Brickworks commenced on 17 August 2001 when it spent $71.7 million on acquiring 29 million shares (a 19.6 per cent stake) in Bristile Limited from its owner, Futuris Limited. Bristile was the second largest clay brick and paver manufacturer in Australia with major market shares in Western Australia, Victoria, South Australia and Tasmania. The company owned a stable of illustrious brand names such as Hallett Bricks and Nubrik. It was also Australia's third largest manufacturer of clay and concrete roof tiles and masonry products. The August beachhead was the first in a series of adroit moves by Partridge that ultimately led to outright acquisition by the end of 2003, establishing Brickworks Limited – and the name of Austral Bricks – as a national force. During the late 1960s and 1970s W.K. Dawes may have toyed with the idea of creating a nationwide manufacturing presence, but without doubt he would have faced insurmountable difficulties. The owners of Midland Bricks in Western Australia, Ric New and his brother, Gerry, would have fiercely resisted any attempt to encroach on their territory.[16] Besides, Ric New was a personal acquaintance of Dawes, and for many years there had been a gentleman's agreement not to poach each other's business. Likewise, the firm of Nubrik in Victoria would have been no easy pushover.

In early 2003 the localised, state-based nature of the industry was about to change. On 4 March The Austral Bricks Company Pty Ltd launched a full takeover offer in an attempt to rationalise the Australian brick industry and in the process form the country's largest and most profitable clay brick producer. Had the bid failed then Austral would most certainly have been relegated to a small regional player.

The fight for control of Bristile Limited was not one-sided. Its managing director, David Gilham, forced a second, substantial lift in the takeover offer which valued Bristile at $582 million, a massive $109 million, or 23 per cent more than the $473 million value of the first bid of $3.15 a share made in March 2003.[17] Having secured the best possible terms, Bristile's directors recommended acceptance of the unconditional offer.

To pay for the acquisition the company borrowed money from its banker, National Australia Bank, although it immediately set about clearing the debt by a capital raising of up to $250 million. Existing investors in Brickworks and Soul Patts were invited to take up the offer of a reset preference share (referred to as PAVERS). This is a cross between a corporate bond and a share, sometimes referred to as a 'hybrid security'. A listed investment company, Brickworks Investment Company Limited (BICL), was also created to acquire the share portfolio of Brickworks. The successful float of BICL, combined with the issue of PAVERS, enabled Brickworks to

16. Richard 'Ric' New died in 1989, aged 74. A visionary and aggressive champion of free enterprise, New's influence on the building industry in Western Australia was immense. It is possible that most people in that state have lived in a home, flat or apartment built from bricks made by Midland Brick. In 1990, the Boral Group purchased the company.

17. Ian Porter, 'Bristile cements a higher bid price', The Age, 1 July 2003.

Above and facing page: The Wollert plant of Austral Bricks, 2008

reduce its debt to between $300 and $350 million. Overseeing the entire process was chief financial officer, Alex Payne, and Iain Thompson, the newly-appointed company secretary.

As for Nubrik, following the acquisition of Bristile, it ceased to exist as an active brand and commenced trading under the national brand of Austral Bricks. Austral inherited six of Nubrik's sites across Victoria, located at Scoresby, Craigieburn, Summerhill, Deer Park, Geelong and Dandenong, employing nearly 300 people. Most of these, with the exception of Craigieburn and Summerhill were decommissioned and the land sold. Nubrik's product lines continued but were supplemented by those from the existing Austral range. Meanwhile, in South Australia, the company picked up from Bristile the former plant of Hallett Bricks at Golden Grove, outside Adelaide. The Hallett plant comprised two Ceric kilns, although only one was operating at the time of acquisition. The yard also enjoyed the increasingly rare advantage of having its own clay and shale pit on site. In Western Australia, home to Bristile, Austral acquired a number of plants, including the roofing tile factory at Caversham, and another brickmaking plant at Malaga. The Bellevue factory, formerly owned by Prestige Brick, was yet another addition and continues to produce clay brick blocks in competition with PGH. In fact, so-called 'blocks' – oversize bricks – dominate the market in Western Australia. The two original plants in the state at Armadale and Cardup were also signed over to Austral. Like the Hallett plant, Cardup was self sufficient in dirt. Finally, the small plant at Waterloo with its updraught kiln was retained. In due course plans were put in train to develop the site as a modern, fully automated brick factory. Unlike its main competitor, Midland Brick, Austral's Cardup facility is well situated with substantial reserves of clay on site. The final piece of the jigsaw acquired from Bristile was at Longford, near Launceston in Tasmania.

Major acquisitions,
Brickworks Limited, 1997–2008

Date acquired	Entity purchased
September 1999	Brick manufacturing plant at Riverview, Brisbane (purchased from Pioneer Bricks, subsidiary of Nubrik)
February 2001	Bowral Brickworks Pty Ltd (from MBS Securities Pty Ltd)
May 2002	Eureka Tiles Pty Ltd
June 2003	Bristile Limited
June 2005	Auswest Timbers Pty Ltd Terra Timbers Hardwoods Australia
March 2007	GB Masonry (Gympie) Caloundra Blocks (Caloundra) Whitsunday Concrete and Block (Proserpine) Ayr Masonry (Ayr)

MORE PLEASE

Since the Bristile purchase there have been more acquisitions, although not in bricks. The ACCC may well veto any further expansion in this area on the basis that Brickworks Limited would be in a position to exercise a monopoly. In June 2005 Brickworks further broadened its base in building materials with the takeover of Auswest Timbers, Terra Timbers and Hardwoods Australia. Auswest's plants manufacture heavy structural timber including floor joists, bearers and roof timbers and decking.[18] In the same year Partridge moved to secure brick distribution operations in New Zealand's north and south islands, at Tauranga and Christchurch respectively. By March 2007 the company was supplying bricks direct to Auckland through a national distribution network. All the current signs point to continued growth with Australia's neighbour across the Tasman Sea.

The next foray was into Queensland where GB Masonry at Gympie, Caloundra Blocks, Whitsunday Concrete and Block at Proserpine and Ayr Masonry joined Austral at a total cost of $41 million.[19] This move has radically changed the shape of the Queensland construction industry and delivered to Austral a total manufacturing capacity of about 230,000 tonnes in premium and grey block masonry products.[20]

EVERYWHERE

Before the acquisition of Bristile, Brickworks had been an operation confined to New South Wales and Queensland. Under Austral the combined company became the largest brick and paver manufacturer in Australia, with operations in every major market. The purchase not only enabled Brickworks to spread its geographical base, but also gave it a commanding presence in the roofing tile market. The upgrade of the old Hacienda-style plates at the roofing plant at Wacol in Queensland and similar upgrades at the Dandenong tile plant ensured that Bristile roofing tiles were consistent in thickness, strength and appearance. Bristile's terracotta roof tile business became one of the dominant players in the market under Austral.

A change in ownership also meant the fusion of two cultures, which according to Rob Millner were very similar. 'When you have mergers and acquisitions there's usually a lot of trouble with cultures and egos, which has not been the case with Bristile. They had much the same inheritance and the same competitors, Boral and CSR. We now have the best people in the clay products industry.'[21]

18. Auswest's managing director, Gary Addison, and another director, Geoff Squires, stayed on following the acquisition.

19. Annual Report, Brickworks Limited, 2006.

20. Grey blocks are currently sold under the Austral Masonry™ brand, whilst premium products are sold under the GB Masonry™ brand.

21. Op. cit., Rob Millner.

Product development, 1986–2008

1986	The Manor House and Federation Range gained their appearance from being flashed and rumbled. They are now sold to home renovators at premium prices.
1991	Governors Range
1996	The year 1996 saw the introduction of the South Pacific Range (Coral Sea) of bricks as the craze for Federation style bricks began to abate. Surface coatings and slurries were sprayed over base colours to achieve pink, red or brown bricks. Eventually this gave way to a demand for monotone products such as the Whitsunday Range (2002). Commencing in 1996, Lindsay Partridge developed the Module Maxi and Modular Standard range of brick which were rolled out over a number of years.
1999	Slick Bricks® were introduced at the same time as the Millenium Range. Vogue was a slurry product with rough edges and some indentations pressed during extrusion.
2000	Mighty Bricks™
2001	e-brick™ ordering system and Claylink make their appearance.
2002	Whitsunday Range
2003	Silk Road pavers
2004	Boulevard range and Terraçade®
2005	Introduction of the Elements™ Series, starting with Blue Onyx, a semi-glazed brick and expanding into Zinc, Graphite, Garnet and the Riviera series
2006	Riviera Freedom®, ColourBrick®
2007	Urbans and Blackstone

WOLLERT

The purchase of Bristile also delivered huge land and property holdings to Brickworks which took some time to digest. Rationalisation of the Victorian operations during 2004 went hand in hand with planning for a new plant at Wollert, north of Melbourne. Officially opened in October 2007 as a replacement for the ageing Scoresby brick plant, Wollert uses highly advanced technologies to reduce fuel consumption and cut emissions of carbon dioxide. The plant was designed in-house and the kiln and dryer were built in Australia to a design patented by the French firm of Ceric, an established supplier.

As Wollert began to take shape Riverview in Queensland (2005) and Golden Grove in South Australia (2006) underwent major upgrades. At the last named plant two more Fanuc robots were added (making a total of six), with each capable of lifting over half a tonne of product in one movement. Golden Grove was also fitted with the latest technology strapping system. In Tasmania new dehacking equipment was installed on the packaging line at the Longford plant in order to improve the handling of bricks from the kiln. Finally, upgrades to the plant in Bowral were completed, with commissioning of the robotic unloading machine virtually eliminating manual handling from the production process. By early 2007 Brickworks Limited was close to achieving its goal of eliminating manual handling from the production process at all of its brick plants.

BOTTOM OF THE GRAPH

Commencing in 1997 Australia enjoyed one of the longest housing booms in recent history. This was briefly interrupted by the introduction of the GST in 2000 before peaking in December 2003 at 46,497

Brickworks Limited, consolidated profit, 1999–2007

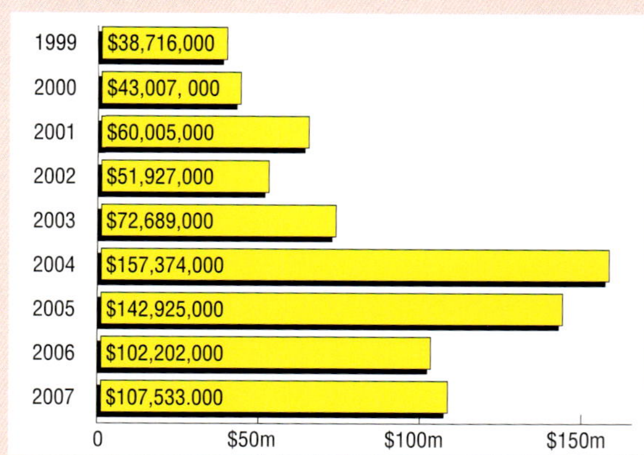

commencements for the quarter. From this point onwards the market went into steep decline – the worst on record. For much of the period, from 2004 to the present, the housing sector has found itself in a precarious state. In parts of Sydney homeowners found themselves with negative equity in their properties. Unlike Sydney, regional New South Wales and Queensland never experienced the same dramatic hikes in real estate prices. By comparison, they had a relatively short way to retreat from their own highs.

Like its southern neighbour, Queensland was one of the first to experience a downturn from early 2004, although 12 months later the strong local economy and the influx of inter-state migrants saw the trend bottom. Notwithstanding, commencements were still bumping along until the end of 2006. In Victoria approvals peaked in late 2004 as builders attempted to beat the introduction of a new 'five star' energy rating system due to come into effect on 1 July 2005. Thereafter the market flattened. Of greater concern was the Victorian government's '2030' planning policy which, despite being a long view, had the immediate and unwanted effect of creating an artificial contraction of land supply. This threatened to put home ownership beyond the reach of many homebuyers. Meanwhile, Tasmania and South Australia slowed from their peaks although they have been less affected than other states.

There were, however, a few bright spots in this sorry state of affairs. During 2005 and 2006 factories were kept running at maximum capacity. Automation and efficiency improvements helped to reduce costs so that brick and paver volumes and prices actually increased. Margins improved steadily despite sporadic discounting by competitors. Western Australia was the standout and proved the value of greater geographic diversity. In sharp contrast to New South Wales, approvals and commencements in Western Australia set new records. When the deluge of orders proved too much for Bristile's plants in Perth, Brickworks turned over one of its Sydney plants to supplying the far west, which although expensive helped maintain production levels in the New South Wales kilns.[22] In the course of 2007 the decline in dwelling commencement finally began to affect Australia's westernmost state. Brickworks' estimated a decline in new starts from 151,000 in 2006-7 to between 148,000 and 150,000 in 2007-8.[23] Given the adverse conditions in New South Wales, a 16 per cent profit rise for 2005-06 and the increased dividend payment was remarkable. The cause was a jump in property profits which provided 36 per cent of group profits, against 42 per cent from building products and 22 per cent from investments.[24] For the year 2006–07 net profit after tax was up 5.2 per cent which defied a 'soft' housing market, a euphemism for a sector of the Australian economy whose growth was flat and falling.

Nonetheless, in early September 2007 the economic forecaster, BIS Shrapnel, declared that housing starts in New South Wales were at their lowest level since 1958.[25] At the time of writing the bottom appears to have been reached, although any upturn will take some time before orders start to pick up substantially. In New South Wales, Brickworks and other major players in the building industry urged the cash-strapped State Government to remove supply restrictions and imposts on land for single dwellings. The response so far has been lukewarm and has done little or nothing to reduce the high price of land which is a major obstacle to home ownership.

MUST HAVES

Thus, by early 2006 the extent of the decline sales throughout Australia's clay products industry was creating a great deal of anxiety among manufacturers and distributors. Was it a short-term event or a long-term trend? We have already seen how consumer tastes change, sometimes at a pace that makes it difficult to accommodate the vagaries of the market. But this is nothing new and only to be expected.

22. Plant locations in Western Australia comprised Armadale, Malaga, Bellevue, Cardup and Waterloo.
23. Annual Report, Brickworks Limited, 2007.
24. *Bulletin with Newsweek*, 12 December 2006, Vol. 124, p. 62-62.
25. Scott Rochfort, 'Housing heading for the wall', Sydney Morning Herald, 28 September 2007.

Western Australia

Pemberton
Hardwood green & dry mill

Cardup
Austral Bricks factory
Brick & paver production

Waterloo
Austral Bricks factory
Brick & paver production

Armadale
Austral Bricks factory
Austral Bricks display
Brick production
Bristile Roof Tile display

Bellevue
Austral Bricks factory
Austral Bricks display
Brick production
Bristile Roof Tile display

Caversham
Bristile Roof Tile factory
Bristile Roof Tile display

Malaga
Austral Bricks factory
Austral Bricks display
Brick and paver production
Bristile Roof Tile display

Busselton
Austral Bricks display
Bristile Roof Tile display

Bunbury
Austral Bricks display
Bristile Roof Tile display

Joondalup
Austral Bricks display
Bristile Roof Tile display

O'Connor
Austral Bricks display
Bristile Roof Tile display

Subiaco
Austral Bricks display
Bristile Roof Tile display

South Australia

Golden Grove
Austral Bricks factory
Austral Bricks display
Brick & paver production
Bristile Roof Tile display

Richmond
Austral Bricks display
Bristile Roof Tile display

- Manufacturing Operation
- Manufacturing Operation & Display Centre
- Company Owned Display Centre

New South Wales

Albion Park
Austral Bricks display

Newcastle
Austral Bricks display

Port Macquarie
Austral Bricks display
Bristile Roof Tile display

Tuggerah
Austral Bricks display
Bristile Roof Tile display

Bowral
Austral Bricks factory
Brick & paver production
Austral Bricks display

Horsley Park
Austral Bricks factory (3)
Austral Bricks display
Brick production
Bristile Roof Tile display

Punchbowl
Eureka Tiles Australia factory
Riviera Paver production
Terracade production
Austral Bricks display

ACT

Fyshwick
Softwood green mill

Fyshwick
Austral Bricks display

New Zealand

Auckland
Eureka Tiles warehouse
Austral Bricks display & distributor

Christchurch
Austral Bricks sales office and display
Austral Bricks warehouse
Bristle Sales office & disp ay
Bristile warehouse

Tauranga
Austral Bricks display & distributor

Wellington
Austral Bricks display & distributor

Queensland

Ayr
Standard masonry production

Proserpine
Standard masonry production

Cairns
Austral Bricks display
Bristile Roof Tile display
Austral Masonry/GB Masonry display

Gold Coast
Austral Bricks display
Bristile Roof Tile display

Toowoomba
Austral Bricks display

Townsville
Austral Bricks display
Bristile Roof Tile display

Gympie
Premium masonry production
Austral Masonry/GB Masonry display
Austral Bricks display
Bristile Roof Tile display

Riverview
Austral Bricks factory
Brick production
Austral Bricks display

Rochedale
Austral Bricks factory
Brick & paver production
Austral Bricks display
Bristile Roof Tile display

Sunshine Coast
Austral Bricks display
Bristile Roof Tile display
Standard masonry production

Wacol
Bristile Roof Tile factory
Bristile Roof Tile display

Tasmania

Burnie
Austral Bricks display

Glenorchie
Austral Bricks display

Launceston
Austral Bricks display

Ulverstone
Austral Bricks display

Longford
Austral Bricks factory
Brick & paver production
Austral Bricks display

Victoria

Ballarat
Eureka Tiles Australia factory (2)
Terracotta floor tile factory (1)
Bathroom fittings production

Brodribb
Hardwood green mill

Bairnsdale
42% owned dry mill

Craigieburn/ Summerhill
Austral Bricks factories (3)
Brick & paver production
Wollert plant (opened 2007)

Deer Park
Austral Bricks display

Newtown
Austral Bricks display

Scoresby
Austral Bricks display

Thomastown
Austral Bricks display

Dandenong
Austral Bricks factory (2)
Besser masonry production
Retaining wall & paver production
Bristile Roof Tile production
Bristile Roof Tile display

The Board of Brickworks Limited, 2008. Back row: T.V. Fairfax, D.N. Gilham, I.H. Thompson (company secretary), A.J. Bentley, A. Payne (chief financial officer). Front row: Hon. R.J. Webster, L.R. Partridge, R.D. Millner, M.J. Millner (Iain Thompson and Alex Payne are seen here in their capacity of senior company officers, and not as members of the board)

One of the lessons to be drawn from history is the cyclical nature of events which can only be read with the long view in mind. The encroachment of panelisation on the brickmakers' turf must be as worrying today as the introduction of fibro was during the 1930s. Yet the industry survived and prospered despite the Great Depression and building restrictions imposed during the Second World War. There is no doubt it will weather the current storm. Despite fibro's cheapness relative to full brick not everyone wanted to live in a particle board house. Likewise, not everyone wants to live in a panel house. And why should panel buildings should exclude the use of brick, especially in exterior feature walls? There is no shortage of aesthetically pleasing face bricks from which to choose. One might be tempted to ask: 'are bricks that much different from any other manufactured products'? The answer is both yes and no. Not all bricks are equal in the eyes of their Maker. Clearly, there are functional (commons) bricks from an engineering and load-bearing point of view. Conversely there are fashion (face) bricks that can lift house and building design from the banal to the sublimely beautiful.

It would be wrong to suppose that manufacturers are simply at the mercy of the market. The relationship between customer and supplier is influenced by many factors. Product development is critical in that it has a strong impact on purchasing decisions, yet this is only part of the story. As Lindsay Partridge observes, today's markets are fragmented.

> *As with every single market, our own have been sliced and diced into smaller niches. Today, we have more competitors in every segment. Being able to have a dominant position in housing is hard to sustain.*[26]

Competition has driven a trend that has been gathering pace for a quarter of a century across the developed economies of the western world. Demand is influenced by price, availability, brand attributes and the consumers desire to own a particular make or model. Perhaps this is why psychologists have been studying consumer behaviour for nearly half a century. Based on empirical research marketing experts create campaigns that tap into our desire to feel good and look good about ourselves. It is no accident that the Apple computer company

26. Interview with Lindsay Partridge, Horsley Park, Sydney, 14 May 2007

THINK BRICK AUSTRALIA

spends millions on developing technology whose external wrappings make them simply irresistible. Must haves that confer status and set us apart from 'the crowd'. In this regard one recent development is worth noting. Shortly before publication of this book, Austral mastered the complex technology of manufacturing a brick with a glazed surface, which gives it the appearance of a ceramic tile. The colours are striking, providing architects, builders and home owners with a new clay product that redefines the accepted notion of 'brick'.

Inevitably, the discussion about what must be done goes beyond what any of the so-called Big Three do as the major producers. If the goal of the industry is to sell more bricks and roofing tiles then it must wage a permanent campaign to promote the benefits of its products and improve market perception. The bid to redefine brick for a modern, contemporary audience and win over hearts and minds is currently underway. The 'Think Brick' initiative, created by the industry's peak body, the Clay Brick and Paver Institute, is working across a number of fronts. The former director, Bob Rushton, and his dynamic successor, Linda Ginger, have brought together architects, research scientists, town planners and industry specialists to develop various aspects of a three-pronged strategy, *Design It, Build It, Live In It*. In particular the environmental advantages of brick is being promoted to redress popular misconceptions. Brick has many qualities among them solidity in construction, security, low maintenance, termite resistance, longevity and thermal mass advantage, ensuring winter warmth and better protection from the summer sun. Work continues on finding ways to attract new entrants to the trade of brickmaking. Training initiatives are welcome, although the warning signs were there for all to see as early as the mid-1960s. The way forward, demands concerted action by all levels of government working closely with the relevant industry groups. The revival in traditional craft trades will take time to achieve but a start has been made, particularly by manufacturers who have a vested interest in attracting new recruits.

SOMETHING UNIQUE

Droughts may be severe, but in due course a correction in the weather pattern reasserts itself and the rains return, sometimes accompanied by floods. In the same way, the business cycle bears comparison to these cyclical expressions of Nature, confirming there is nothing new in this world. With a history that spans more than four millennia, there is no reason why brick should not remain a practical, sustainable and aesthetically attractive building material.

Brickworks Limited under Lindsay Partridge has witnessed enormous efficiency gains and spectacular growth into other industries. An optimist, Partridge has done more than most to turn around public and government perceptions about the place of brick in the built environment. 'We have a good future and lots of options' he confidently asserts. At a time when human society is coming to grips with the impact of global warming, the industry is moving faster than most to become an exemplar of environmental sustainability in terms of water recycling and fuel efficiency.

The story of Austral Bricks and its parent, Brickworks Limited, started in Sydney and, for the most part, remained a local story until the turn of the millennium. Where the two organisations found themselves at this point was due to the decisions made by an earlier generation – from Spackman the founder, to Dawes the builder, and Burgis and Bentley who turned adversity into renewed prosperity. Without their collective brilliance and inventiveness the name Austral would have been nothing more than a footnote to history.

As the flag bearer Austral now finds itself in front of a wider audience, and on a playing field that, one day, may stretch far beyond the physical boundaries of the Australian continent. The architect of expansion and future growth has been Lindsay Partridge whose emotional attachment to Austral and its life and times runs deep. 'There's something unique about this company,' he muses.

Afterword

In this book I have endeavoured to explore how brick has influenced the built environment, especially in Sydney, the birthplace of the modern clay products industry in Australia.

Even today, the city's buildings are mostly brick-built. Wood buildings are few and far between, and so too are dwellings made of stone. From east to west, and north to south the streetscape of our metropolis is one of brick, interspersed with brick apartments. Jet passengers can attest to this. For miles prior to landing, the gradual descent over Australia's cities and major towns reveals a sea of clay roofing tiles and brick houses. A place to live. Home!

However, the face of some inner city suburbs is beginning to change due to the impact of land shortages on the city's fringe. For obvious reasons brick has been displaced as the preferred construction material. Steel and concrete allow multi-storey buildings to be erected that would not be possible using brick. Yet, as a façade and cladding brick is still widely used in landmark buildings.

On the residential front brick still holds sway. Most people prefer to live in a house that offers the solidity of brick, even though the exterior may be rendered prior to painting. Increasingly, modern face bricks in all their various styles, surface textures and colours, continue to re-assert the unique quality of brick as a facing material. In recent years the incorporation of weathered, dry pressed and recycled bricks into architect-design homes and commercial premises are an invitation to 'stop, stare and marvel'. The artistry of skilled bricklaying, architectural design, brick colour and shape is being rediscovered.

This work of history does not pretend to offer an architectural interpretation. It is, for the most part, the story of an industry that has been profoundly influential in so many ways. For centuries to come, clay products, in all their various guises, will continue to shape our built environment and the way we interpret how and where we choose to live.

Published versions of the children's fairy tale, The Three Little Pigs, date back to the late eighteenth century, although the story is thought to be much older. The advantages of building in brick were widely appreciated, especially by the third pig who triumphed over the wolf. Mary Evans Picture Library

1 Appendix

Bowral Brick

NEW SOUTH WALES SOUTHERN HIGHLANDS

Brickmaking in the Bowral district commenced in 1915 when a Matthew Hodgkinson established a brick yard in Rose Street, Bowral. In 1921 the business was offered for sale to brickmakers, Arthur and Charles Baker, who purchased land beside the railway line on Kirkham Road. In September 1922 the first Bowral bricks from the firm of F. Baker & Sons were produced and a Hoffman kiln erected towards the end of 1924. For the rest of the decade the firm continued to prosper, supplying local needs as well as bricks to the fast-growing Canberra region. The Depression years severely affected the industry, including the plant at Bowral. With no prospects of work at the yard, and in the hope of overcoming financial hardship, one of the sons, Ray Baker, took many of his workers to Eden to prospect for gold. Despite a brief period of economic recovery from about 1934 onwards, the Second World War effectively put an end to building work, throwing the industry into a renewed crisis.

With the return of prosperity in the immediate post-war years, and in anticipation of a building resurgence, two downdraught kilns were installed at Bowral. From 1953 to 1972 the company expanded under the leadership of another family member, Ron Baker. During this time new equipment was imported from England and the United States, including a tunnel kiln from Germany which was installed by representatives of its European manufacturers. Meanwhile, the local economy prospered, with the Baker Brothers continuing to provide employment for the menfolk of the town.

CHANGE OF OWNERSHIP

In 1972 Ron Baker sold Bowral Bricks to the building firm of A.V. Jennings who disposed of it to Nefco Nominees in 1984. The new owner refurbished the plant, but geared production towards the lucrative market in architectural face brick rather than common bricks. This decision established a reliable supply source for government departments and architects involved in the rapidly expanding area of heritage work, restoration projects, and upmarket housing.

In a further change of ownership in the early 1990s, the New family of Western Australia gained control of Bowral Bricks through MBS Securities Pty Ltd. Despite it being a time of recession when many brickworks ceased to operate, Bowral Bricks remained a profitable venture, upgrading the dry-pressed tunnel kilns. The initial cost of these upgrades, estimated to run into millions of dollars, was drastically reduced with the purchase of plant and equipment from recently-closed brick plants at Maitland, Homebush Bay, Blacktown and Granville. The upgrade increased plant capacity to three million face bricks a year.

In 1995 the company received development consent from Bowral Council to extend the adjacent shale quarry, securing supplies of clay and shale sufficient to last until 2040. These deposits of Ashfield shale are the same deposits that run under the greater Sydney Basin being used at the Wallgrove Road plants of Austral Bricks at Horsley Park. Moreover, the varying bands of clays encountered in the shale deposits yield bricks that match the distinctive colours of many heritage buildings in the Sydney–Newcastle region. Today, Bowral Bricks is owned by Brickworks Limited, which purchased the company in March 2001.

Construction of the Hoffman kiln at Bowral, circa 1924. Leah Day Collection

Bob Baker (left), owner of Bowral Bricks and Bill Prest, yard manager, pushing a slug of clay through a J.C. Steele reel cutter, circa 1978, probably a trial for a new product. Bill Prest Collection

Appendix 2

Eureka Tiles

Origins of Eureka Tiles – Public company – Depression years – Postwar boom – Hostile takeovers

ORIGINS OF EUREKA TILES

The origins of Eureka Tiles can be traced to a pottery works owned and operated by a German migrant, August E (Otto) Steinkraus, in the city of Ballarat, Victoria during the 1890s. The pottery stood on land adjacent to the Eureka Stockade Reserve. Today, the site is still occupied by the offices and factory of Eureka Tiles.

The fortunes of Steinkraus and his pottery changed when the local architectural firm of Messrs George W. Clegg and William Miller decided to import French terracotta tiles and English slates for use in roof construction. Miller's preference was for a local supplier and in 1910 he approached Steinkraus, persuading the elderly potter to sell his works for conversion into a plant to manufacture Marseilles-type roof tiles. A year later, under the direction of Steinkraus who was hired as manager, the state's first roof tile manufacturing factory commenced operations.

With the business well established, Clegg and Miller decided to expand. On the morning of 12 April 1913 a group of shareholders gathered in the imposing brick edifice of The Royal Craig Hotel in Lydiard Street, Ballarat. A proprietary company trading under the name Eureka Terra Cotta and Tile Co Pty Ltd was duly incorporated. The choice of The Royal Craig for the inaugural meeting reflected the importance of the event. It was here in 1867 that His Royal Highness Prince Alfred, the Duke of Edinburgh, the first member of the Royal Family to visit Ballarat, established his local headquarters during a state visit.

PUBLIC COMPANY

In May 1914 the shareholders met once more and voted to establish a public company, Eureka Terra Cotta and Tile Company of Australia Limited. Matthew Kemp took over management from Steinkraus who was glad to retire knowing that the enterprise he had nurtured for over twenty years was in safe hands.

The driving force behind the public company was William Miller, who became the first company secretary. Before long, the factory was operating on a much larger scale with the addition of six new kilns and an annual output exceeding three million tiles. With the order books full, and eager to install modern and efficient machinery, Miller visited the USA in 1919 on a fact-finding and buying mission. He returned with a wealth of new ideas and products. Through the introduction of tiles, such as English shingle, Spanish Cordova and the 'Hughes Armstrong' tile for which Eureka Tile Co held the patent rights, the company won the respect and support of leading architects who considered it the most enterprising tile manufacturer in Australia.

DEPRESSION YEARS

Despite the economic downturn that lasted from the mid to late 1920s through to the early 1930s, The Eureka Tile Company remained in business, servicing the particular requirements of architectural firms. One such product pioneered during this period was textured 'tapestry bricks'. Developed by industrial chemist, Mr Norm Griffiths, they were particularly successful and used for decorative fireplaces and special features. Orders flooded in from all over the country.

Then, on 6 October 1934, disaster struck when a fire destroyed the factory and its stocks. Fortunately, the kilns and their contents survived but all other work in progress, including 200,000 drying trays made of pine timber, were destroyed. Recovery was swift and within two years a modern steel girder factory had been erected and equipped with the latest tile making equipment imported from overseas. The clay grinders and extruders were of Australian manufacture. The industry faced another crisis at the onset of the Second World War when housing construction ceased and unskilled workers were drafted into the armed services. Eureka contributed to the war effort by supplying acid resistant ceramic rings for vats that were used at the Ballarat Gun Cotton factory.

POSTWAR BOOM

Pent up demand for housing and the influx of overseas migrants created a post-war building boom. Preference was given to the production of roofing tiles. To overcome the shortage of hardwood timber battens used in the construction of a tile roof, the company set up its own sawmill, having obtained an allocation of timber from the Forest Commission. In 1958 the Ballarat factory was extended to house additional tile presses and two large oil-fired downdraught kilns. Eight other downdraught kilns owned by the company were fired with black and brown coal.

Strong demand for tiles in Victoria continued until the end of 1959 when a severe recession took hold in the state. In the face of strong competition from the manufacturers of concrete roofing tiles, the company invited the architect, L.H. Vernon, to develop a popular range of clay building products including sill tiles and fence capping blocks. The new range was a great success.

In 1967 the company was one of the first to purchase a split tile extrusion die from Germany, which was used in conjunction with the recently installed Haëndle De-airing Extruder. This marked Eureka's successful entry into the market for mass-produced quarry tiles.

HOSTILE TAKEOVERS

Unfortunately, an unwelcome takeover bid was in the offing. On 15 November 1971 an unnamed person bought 20 per cent of the company's shares on the Melbourne Stock Exchange. By late August 1972 the board reported that Hebrides Equities Ltd, the company behind the initial purchase, was now the beneficial owner of the majority of shares. When this investment group requested representation on the board, the existing directors refused, prompting the newcomer to unseat the board and seize control.

Within six months the new board of directors commenced the sell-off of the company's assets. On 2 April 1973 the plant and properties were purchased by Sydney-based PGH Industries Ltd, a subsidiary of Felt & Textiles Industries Ltd (F&T). With the sale finalised, the operation began trading under the name of Eureka Ceramics.

In 1974 the board of F&T decided to combine the Ballarat plant with its PGH Quarry Tile and

Bathroom Fittings plant at Doonside, Sydney. This newly merged operations traded as PGH Eureka Ceramics until 1990. In 1976, F&T's parent company changed its name to Acmil Limited – Australian Consolidated Manufacturing Industries Limited – to more accurately describe the company's activities. Unfortunately, Acmil was vulnerable to takeover and in 1982 was acquired by ACI (Australia) Limited.

The next decade was a turbulent time for 'Eureka'. In 1989 ACI Limited was taken over by BTR Nylex Ltd, the Australian subsidiary of UK-based BTR PLC. In a major sell-off BTR retained only the profitable Ballarat and Doonside plants, which continued until 1993 under the name of The Eureka Tile Company. Over the next three years a series of complicated manoeuvres within BTR Nylex saw the company reconstituted yet again. Finally, BTR's English parent purchased the Australian arm in keeping with its global activities. The Ballarat plant, which had absorbed the bathroom fittings components from Doonside in 1992 when the PGH Doonside operation ceased, was put on the market and purchased by a group of six employees. The successful management buy-out defeated rival bids by The Austral Brick Co Pty Ltd, and the UK firm of Johnson Tiles which had operations in Melbourne. Eureka Tiles Pty Ltd once more reverted to being a privately controlled company.

This arrangement lasted until 2002 when Brickworks Limited, the parent company of Austral Bricks, purchased the plant and associated properties. Today, the company operates under the name of Eureka Tiles Australia.

3 Appendix

Bristile Limited

Lance Brisbane – Alfred Wunderlich and H.L. Brisbane – Metropolitan Brick Company Pty Ltd – Creation of Bristile Limited – Brickworks Limited acquires Bristile

In the early years of the state's history the centre of the brickmaking industry in Western Australia was the city of Perth, which lies close to suitable clay deposits. Most of these deposits were, and still are, easily accessible along the banks of the Swan River. Building in brick not only suited popular taste, but also the city's municipal council which was determined to ensure that Perth would become a major centre of trade and commerce. Brick construction conveyed the impression of permanency, solidity and prosperity. From the early 1910s many local municipalities banned unsightly weatherboard and timber construction by declaring 'brick areas'. The rapid growth of Perth's suburbs fuelled the brick industry and sustained the dream of home ownership in the Perth region.

LANCE BRISBANE

The firm of Bristile Limited started life in June 1929 as H.L. Brisbane and Co Ltd, a manufacturer of roofing tiles. Its formation was the work of one man, Lancelot 'Lance' Brisbane, who was born in Victoria in 1893. Shortly after his birth Brisbane's family moved to Perth. After leaving school Brisbane trained as a draftsman. In 1909 he joined the local office of the Sydney building products firm, Wunderlich Ltd, a Sydney manufacturer of roof tiles and pressed ceilings.

Lance Brisbane was among the first to volunteer on the outbreak of war in August 1914. The 21-year old served as a trooper in the famed 10th Light Horse Regiment, and later as an engineer. Returning to Perth in 1919 he rejoined Wunderlich and, following a series of promotions, was sent over to work in the company's Sydney head office. He returned to Perth in 1924. Several years later he decided to strike out on his own, setting up in opposition to Alfred Wunderlich, his former employer and founder of the aforementioned company. The firm of H.L. Brisbane & Co Ltd grew rapidly, taking over the Belmont roofing tile operations of Western Australian Pottery

Company Limited. Brisbane steered the business through the depression years and soon became a powerful player in the Western Australian market for building products. In 1935 the firm listed as a public company on the local stock market.

ALFRED WUNDERLICH AND H.L. BRISBANE

Despite competition between the two firms, Alfred Wunderlich and Lance Brisbane remained on close terms. In 1938 the two men signed an agreement which gave Wunderlich a 38 per cent share in Brisbane's firm, although voting rights were restricted to a 19 per cent shareholding. A new company was established by the name of H.L. Brisbane and Wunderlich Limited.

METROPOLITAN BRICK COMPANY PTY LTD

About this time Metro Brick, as it was commonly referred to, came to play an important part in the fortunes of the aforementioned entity. Metro Brick had grown out of the building and contracting business of Western Australian entrepreneur, Robert Oswald Law who formed the Metropolitan Brick Company Limited in 1922. The firm's main yard was based at Armadale and produced over five million bricks per year. Later in 1922 Law identified good brickmaking clay in the Peninsular at Maylands and acquired a large area of land covering these deposits. It was here that he built a new kiln with a production capacity of seven million wirecut bricks a year. In 1936 a second kiln was built at Maylands which increased output to 14 million bricks per year. Robert Law died in 1947 and was succeeded as General Manager by his son. During 1952 the firm acquired the works of Byford Pressed Bricks Pty Ltd at Cardup and changed its name to Cardup Metro Bricks Pty Ltd. With financial help from the state government a new kiln and machinery were installed, increasing capacity to 11 million bricks. In 1962 Metropolitan Brick Holdings Ltd was incorporated as a public company and took over the assets of Cardup Metro Bricks Pty Ltd and the Metropolitan Brick Company Pty Ltd.

The Caversham brickworks which was opened in April 1953, seen here among the vineyards

During the 1960s, Metro Brick grew rapidly with most of the firm's capital expansion being concentrated at the Cardup works. Maylands and Helena Vale were expected to have limited life spans due to diminishing clay reserves. A slump in the building industry during 1970 and into 1971 saw profits tumble, although it was not long before trade began to pick up.

CREATION OF BRISTILE LIMITED

In 1973 H.L. Brisbane & Co Ltd and Wunderlich Ltd merged with Metropolitan Brick Holdings Ltd, forming a new and more powerful single entity, Bristile Limited. Four years later, in 1977, Bristle exchanged shares with the Perth-based timber milling company of Whittakers Limited.

In 1983 Bristile's new chairman, Terry Hogan, entered private negotiations to buy the Armadale and Waterloo works from Monier Ltd. By this time virtually all of the Western Australian brick industry had been returned to local ownership. For much of the 1980s and 1990s Bristile went from strength to strength, manufacturing and distributing clay bricks and pavers throughout Australia and overseas via its subsidiary, Metro Brick. The recession of the early 1990s slowed building activity which deeply affected Bristile. The company was obliged to accept an injection of equity by Futuris which was the owner of two Western Australian brick companies, Prestige Brick and International Brick and Tile. In 1992 Futuris emerged as the new owner after having acquired 67 per cent of the Bristile Group. Shortly afterwards Bristile, Metro Brick, Prestige Brick and International Brick and Tile were amalgamated under the name, Metro Brick. The plants comprised 14 kilns at five locations: Bellevue, Malaga, Armadale, Cardup and Waterloo, with a total production capacity at that time of 200 million units per year. Determined to expand its interests in the eastern states of Australia, Bristile purchased Pioneer Roof Tiles in 2001.

BRICKWORKS LIMITED ACQUIRES BRISTLE

In the closing months of 2003 Brickworks Limited launched a full takeover bid for Bristile. After an initial negotiation phase the shareholders of Bristile Limited were encouraged by their board to accept the Brickworks offer.

In 2004 Brickworks commenced the process of integrating Pioneer Roof Tiles into the Brickworks group of companies. The various state divisions of Bristile Clay Tiles (WA), Nubrik (Vic) and Pioneer Roof Tiles (QLD & NSW) came together under the new trading name of Bristile Roofing.

4 Appendix

Nubrik

Origins – Clifton Brick Company – Brick and Pipe Industries Limited – Changing fortunes – Pioneer takes charge

ORIGINS

The name Nubrik first appeared in 1948 when it was registered by the New Northcote Brick Company Limited to describe its range of wirecut bricks. New Northcote had a long and distinguished history and was originally established in 1882 as the Northcote Patent Brick Company Limited in the Melbourne suburb of Northcote. Later in the decade the firm changed its name to The New Northcote Brick Company Limited. The latter is mentioned as one of four major industrial companies operating in Melbourne in 1889 and is considered to be one of the founders of the state's brickmaking industry. New Northcote's yard bestrode huge clay reserves, and became a landmark with its tall chimney stacks. In 1959 the company changed its name to Brick Industries Limited.

CLIFTON BRICK COMPANY

In 1891, shortly before the economic depression, The Clifton Brick Company Limited was established in nearby suburb of Preston. As we shall see this privately owned company became intimately associated with New Northcote. An early shareholder emerged by the name of William Angliss, a Melbourne businessman, and the owner of meat processing plants in Footscray. Angliss acquired Clifton in 1906 or possibly earlier and did so in order to secure a reliable supply of common bricks used in the construction of his meatworks. The Footscray works later supplied large quantities of 'bully beef' to feed the Australian Army during World War One. The Angliss connection with Clifton Bricks endured, and was reinforced after 1961 when the company listed on the Australian Stock Exchange, changing its name to Clifton Brick Holdings Ltd. The Angliss family retained a large shareholding and board positions. There followed a period of consolidation within the brick industry with the result that Clifton Bricks and Brick Industries became rivals in their quest to grow by acquisition.

BRICK AND PIPE INDUSTRIES LIMITED

In 1964 Brick Industries Limited adopted the name Brick and Pipe Industries Limited which continued until 1987. Trading as Nubrik, Brick and Pipe took over a number of its competitors, including Commonwealth Pottery, which was later, renamed Viteloy. The company's rise was given additional impetus by the purchase of Hallett Brick Industries Limited in Adelaide in 1969. Key to the success of the rapidly expanding Brick and Pipe was Albert Shergold and his son, John, both of whom were innovators in the industry and outstanding businessmen. Before long, Brick and Pipe's interests extended, among other things, to hotels.

Through its extensive holdings, and under the leaderships of the Shergolds, Brick and Pipe emerged as the largest manufacturer of clay bricks and pavers in Victoria. It also operated a roof tile business manufacturing concrete and clay roof tiles in Melbourne, as well as a concrete masonry business. Major shareholders in the company included the Swire Group, The London Brick Company (later the Hanson Group), and Industrial Equity Ltd, with Sir Ron Brierley as chairman and shareholder.

In August 1987 Brick and Pipe Industries purchased The Clifton Brick Company, which was still 40 per cent owned and controlled by the Angliss family. As mentioned previously, Clifton Brick was a long-established and formidable player in the industry. By the early 1970s Clifton owned five plants, among them Oakleigh (closed in late 1970s), Preston (closed in early 1980s), Brunswick (closed in 1994), and a plant at Colac. Craigieburn, built by the legendary figure of Jack Peterson continues to operate today under its present owner, Austral Brick. Clifton Brick

Holdings Ltd had also built new plants at Longford in Tasmania, Queanbeyan near Canberra, as well as Townsville and Mackay, although about 60 per cent of its business was in Victoria.

CHANGING FORTUNES

Within months of Brick and Pipe's takeover of Clifton Bricks an event occurred that would bring the company's long run of success to an untimely end. In 1988 Abe Goldberg, a Polish migrant with a background in textile manufacturing, managed to persuade the main shareholders of Brick and Pipe Industries to sell their shares. The major shareholders quickly accepted Goldberg's offer, although he was to remain owner of the company for less than seven months. Goldberg knew little or nothing about the brick industry. In fact, his real intentions soon became apparent when it was confirmed that he intended to settle his huge debts by selling the company's extensive property holdings. Fortunately, this never eventuated and Goldberg fled to Poland to escape prosecution for fraud.

PIONEER TAKES CHARGE

Brick & Pipe did not go into liquidation, but as an asset of Goldberg it was sold by the receiver. The brick business was purchased by Pioneer Building Products, a division of Pioneer International. Slowly, the company's fortunes began to turn around under a team of outstanding managers, one of whom was Jack Peterson, the driving force behind the creation in 1994 of a new plant at Riverview in Queensland. At the same time Clifton's historic Brunswick plant was closed due to falling productivity.

In 1996 Pioneer merged all of its building product divisions into a single trading entity called Pioneer Building Products. Two years later, its parent company, Pioneer International, was taken over by the UK company, Hanson, which has a listing on the Australian Stock Exchange. In the meantime Pioneer closed some plants including those at Mackay and Townsville, and also sold off the Riverview plant. Hanson's presence in the UK brick business was sizeable, but its main interest in Pioneer was in that company's worldwide quarrying and concrete business.

Following a period of uncertainty Pioneer's roof tile business was purchased in 2001 by the large Western Australia based building products company, Bristile, but this arrangement was short lived. In 2003 Bristile and all of its subsidiaries was acquired outright by one its major shareholders, Brickworks Limited.

5 Appendix

Hallett Bricks

A South Australian icon – Expansion and modernisation – Hard times – Second World War and after – Merger with Brick and Pipe Industries

A SOUTH AUSTRALIAN ICON

The Hallett story begins in 1879, with the arrival in Port Adelaide of the English migrant and brickmaker, Job Hallett. Like many of his contemporaries Hallett began his venture with a small clay pit and horse drawn pug mill, which a timely inheritance enabled him to purchase in 1886. Having established himself among Adelaide's brickmaking community, Hallett slowly increased the scale of his family-owned operations. By maintaining high standards of workmanship and lowering his prices, he survived the 1890s recession which decimated the brick industry in Australia. With a reputation for producing superior quality bricks, Hallett won a valuable two-year contract in 1900 to supply bricks for the construction of the School of Mines building. From this time onwards the name of Hallett became synonymous with brickmaking in South Australia.

EXPANSION AND MODERNISATION

In 1904 Job Hallett formed a partnership with his first son, Thomas, and the rapidly expanding business was named J. Hallett and Son Ltd. Between 1900 and 1910 they acquired and operated six other brickyards in the plains on the western outskirts of Adelaide. Over an 18-month period (1908–1909) the two men doubled production to almost half a million bricks per week. Unable to recruit local skilled brickmakers, Job Hallett offered free passage, accommodation and a secure job to experienced artisans from his home town of Bridgewater in Somerset. Skilled men such as these enabled him to turn out consistently better bricks than many of the other yards. In fact, Hallett was acclaimed as a leading maker of sandstocks.

Rising labour costs and increased competition prompted him to begin the mechanisation of his works in 1909. Growing prosperity across the state in the pre-war years further enabled Hallett to continue modernising the plants. At the same time he was quick to spot a bargain and in 1910 acquired the Federal Brick and Tile Company following its liquidation. In 1912 a 20-compartment Hoffman kiln of 300,000 bricks capacity was built at the Federal brickyard. The following year a second kiln was installed at the Allenby Gardens works. Electric power was introduced at the plant in 1925. Meanwhile, at Hallett's smaller yards, hand-moulding and the use of horse powered machinery continued into the 1930s. The company responded to the

Flash floods wreak havoc

Between 1917 and 1923, three flash floods inundated the brickyard districts in the low-lying western plains region. On one occasion 1.5 million green bricks were lost, with total losses amounting to £5,000, which was a small fortune in those days. Production was delayed for weeks due to the clean up and draining of the clay pits. The construction of an artificial river channel through these low-lying areas removed the danger of flooding from 1930.

housing trends in the 1920s by creating enamelled face bricks with different colours and textured. Roofing tiles were also manufactured successfully from about 1926.

HARD TIMES

At various times during the 1920s three other sons of Job Hallett joined the business which was renamed Hallett and Sons Ltd in 1927. In 1928 the firm opened a brickworks at Poonindie. By this time, a slowly contracting economy heralded the beginnings of the Great Depression. Brick production fell and weekly output from the combined Hallett works declined to approximately 200,000 bricks, which was half its normal capacity. At the peak of the depression, Hallett employees were given one day's work each week to allow them to pay their rent and food costs. Only the five men in the family worked full-time at the yards.

In 1929 the Halletts, in association with owners of the larger brickyards, formed a syndicate known as The Brick Manufacturers Limited. This company set out to purchase ailing and closed brickyards around Adelaide, Port Pirie and the township of Crystal Brook. One of these was the large-scale operations of the Metropolitan Brick Company. By acquiring the firm, the Hallett family became the largest brick manufacturer in the state and remained so for several decades. By 1934 Hallett Brick had expanded into a range of clay products which included red double-pressed, wirecut or sandstock bricks, salt and enamelled glazed bricks, moulded bricks of various shapes, and terracotta paving tiles of various patterns and colours including Marseilles, Roman Cordova and Shingle tiles.

SECOND WORLD WAR AND AFTER

Job Hallett continued as head of the company until 1936 when he retired after relinquishing control to his sons. When he died in October 1940, the business had grown to become the largest group of works in South Australia, employing more than 170 men. Labour shortages and the lack of building led to the closure of many brickyards shortly after the outbreak of war in 1939, although essential war work continued in the building of munitions factories which provided work for selected brickyards, including Hallett Brick.

In the immediate post-war years the smaller, less profitable yards at Brompton, Thebarton and Poonindie in country South Australia were sold. To encourage firms to modernise the brick industry's ageing plants the state government offered financial incentives. Hallett Brick installed automatic coal-fired stokers into the Hoffman kilns on the understanding that they would repay government loans if the system were successful, as it proved to be.

In 1950, Thomas Hallett became a joint director of the Gilburn Brick Company Ltd - a new company formed by an amalgamation of the Adelaide Brick Company's Burnside plant with the Magill plant, owned by Brick Manufacturers Limited. Gilburn operated until 1982 but was forced to close due to encroaching housing, pollution and the increasing value of the land which was redeveloped for residential dwellings.

With alluvial clay resources in the metropolitan area running out, Hallett purchased reserves at Golden Grove, 16 kilometres north-east of the city. The company also began to mine the strong yellow clay from the Noarlunga district and good red clay for roofing tiles from a deposit at Cherry Gardens. Defying industry conservatism, the firm continued to modernise plant, installing the latest Haëndel machinery from Germany to automate kiln firing and brick handling.

MERGER WITH BRICK AND PIPE INDUSTRIES

In June 1952 Hallett Brick and the Victorian company, Brick & Pipe Industries Limited, officially joined forces to establish a new company, Hallett Brick Industries. The original J. Hallett and Son Limited and the Victorian company each took 40 per cent of the shares. The remaining 20 per cent was made available to the public. By October 1963 a new automated plant at Golden Grove had been completed and quickly proved itself to be highly profitable operation.

By 1969 the fortunes of Hallett were in decline when it was fully acquired by Brick & Pipe Industries, becoming a subsidiary in the process. Rationalisation followed during the recession of the 1970s, resulting in the closure of a number of yards, including Golden Grove. In 1980 Hallett took over the financially troubled Southern Brick Company's works. However, depressed conditions in the building industry and unrealistic demands for over award payments forced the plant's closure in June 1980. Manufacturing of Hallett's pressed clay bricks and paving brick operations were transferred to the Golden Grove plant which was reopened in 1981. The company maintained its leadership in brickmaking in the state and by 1982 was supplying almost 60 per cent of the total South Australian market.

In 1988 Brick and Pipe Industries and its various subsidiaries were acquired by Abe Goldberg, an entrepreneur whose credentials and intentions were somewhat dubious. Following a brief but disastrous, period of ownership Brick and Pipe Industries was broken up. The following year Hallett Industries was sold to Pioneer which was acquired by Bristile Limited in 2001. In one final change of ownership, Brickworks Limited acquired Bristile in 2003. Brick and clay products continue to be made at the former Hallet yard at Golden Grove in South Australia. Sadly, the famous Hallett name has passed into history.

Appendix 6

Directors of Brickworks Limited, 1934–2008

1934-1946*
Charlesworth, A.
Butcher, F.
Davison, C.F.
Dawes, W.K.
Green, A.
Lanceley, E.D.
Mosely, B.
Phippard, F.G.
Spackman, E.
West, W.G.
Wilkinson, R.H.

1947-1950
Charlesworth, A.
Davison, C.F.
Dawes, W.K.
Green, A.
Mosely, B.
Wilkinson, R.H.

1951-1959
Charlesworth, A.
Davison, C.F.
Dawes, W.K.
Green, A.
Wilkinson, R.H.

1960
Charlesworth, A.
Davison, C.F.
Dawes, W.K.
Dawes, R.T.
Wilkinson, R.H.

1961
Allott, R.
Charlesworth, A.
Dawes, W.K
Dawes, R.T.
Thompson, L.H.
Wilkinson, R.H.

1962-1964
Allott, R.
Charlesworth, A.
Dawes, W.K.
Dawes, W.F.
Dawes, R.T.
Eddy, S.J.

1965-1969
Dawes, W.K.
Dawes, W.F
Dawes, R.T.
Taylor, L.W.
Travis, G.F.

1970-1976
Dawes, W.K
Dawes, A.M.
Dawes, R.T.
Dawes, W.F.
Millner, J.S.
Taylor, L.W.
Travis, G. F.

1977-1981
Dawes, W.K.
Dawes, A.M.
Dawes, W.F
Millner, J.S.
Taylor, L.W.
Travis, G.G.

1982
Dixson, H.
Millner, J.S.
Taylor, L.W.
Travis, G.G.

1983
Burgis, A. W.
Dixson, H.
Millner, J.S.
Taylor, L.W.
Travis, G.G.

1984-1994
Burgis, A.W.
Dixson, H.
Millner, J.S.
Pearson, C.M.I.
Travis, G.G.

1995-1997
Bentley, A.J.
Burgis, A.W.
Dixson, H.
Millner, J.S.
Pearson, C.M.I.
Travis, G.G.

1998-2000
Bentley, A.J.
Burgis, A.W.
Fairfax, T.V.
Millner, J.S.
Millner, R.D.
Travis, G.G.

2001-2002
Bentley, A. J.
Fairfax, T.V.
Millner, R.D.
Millner, M.J.
Partridge, L.R.
Travis, G.G.
Webster, R.J. (*The Hon.*)

2003-2008
Bentley, A.J.
Fairfax, T.V.
Gilham, D.N.
Millner, R.D.
Millner, M.J.
Partridge, L.R.
Webster, R.J. (*The Hon.*)

* Records are incomplete for this period, but it is certain that those mentioned were directors at various times during this period.

In addition to the abovementioned, the following directors joined the board in late 1989 following the merger with CSR Limited.

Barton, Dr R. (No other directorships)	Retired 19/2/1990
Baker, B. (Also General Manager, CSR Gyprock)	Retired 26/9/1990
Bennett, W. (Also, Director, CSR Limited)	Retired 26/9/1990
MacMillan, W. (No other directorships)	Retired 26/9/1990
Carew, K. (Also General Manager, CSR Construction Material)	Retired 26/9/1990

7 Appendix

Machinery

TYPES OF MACHINERY USED AT THE YARDS OF BRICKWORKS LIMITED — CIRCA 1955

Brick presses		
Foster Platt	English Platt	Foster Long machine
Foster Short machine	Hankinson Top Lift	Foster Boyd Machine
Clayton Goodfellow*	Whittaker Heavy Duty Press	Wallbank Single
Timing Single	3 pressure roller brick machines	American Boyd machine
Mill pans		
Foster 11' pan	Foster 9' 6" pan	Foster 10' pan overhead driven pan
Hankinson 9' pan	Timming & Gardener pan	Rabin Ecker pan
Incla pan, made by a US company. Brickworks Limited continued to import the large capacity Incla pans when Austral's extrusion plant was opened at Wallgrove in 1961.		
Crushers		
Foster Crusher		

TYPES OF MACHINERY USED BY LOCATION, 1955

Yard	Machinery used
Austral, St Peters	• 4 x 9' 6" Foster pan • 4 x English Platt press machines • 1x Boyd machine • 2 x Foster single machines
Bedford, St Peters	• 5 x 9' 6" Foster pans • 5 x English Platt press machines • 1 x Foster Side Lift machine • 1 x single operated brick press
City, Alexandria	• 3 x 11' Foster pans • 5 x English Platt press machines • 1 x Clayton Goodfellow machine
Radford, St Peters	• 3 x 11' Foster pans • 4 x Foster Top Lift machine • 1 x Foster Side Lift machine
Eastwood	• 3 x English Platt press machines • 2 x H.D. Whittaker machines (replaced by Williams singles) • 1 x Hankinson's machine • 1 x 9'6" Foster pans
Brookvale	• 3 x Clayton Goodfellow machines • 2 x Foster Top Lift machines • 3 x 9'6" pans

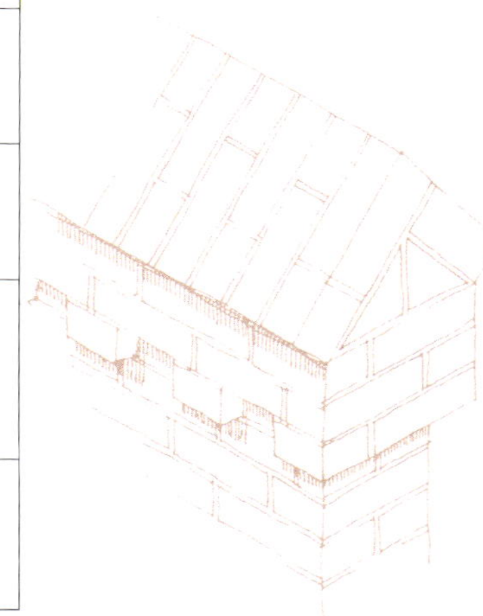

Source: Jim Liepins, Sydney 2007

Appendix 8

Industry bodies

In 1934 the Brick Manufacturers' Association of New South Wales (BMA) established an industry journal, *The Clay Products Journal of Australia*, which changed its name to *Australian National Clay* in 1959. In the same year a select body, The Council of Brick Manufacturers, was created to represent owners of major yards who were also members of the BMA. Some association members, including Speare and Goodlet & Smith, who were also significant producers, refused to join the council. They were suspicious of the motive of W.K. Dawes and other brick manufacturers who wanted to rationalise the industry, possibly at their expense. Soon after its formation, the Brick Council established Brickworks Limited with the express purpose of acquiring the State Brick Works from the government of New South Wales. The council continued to operate throughout the war period, although many of its members progressively sold out to Brickworks Limited, which retained yards with substantial reserves of clay and shale and were profitable. Uneconomic yards were closed. At the same membership of the BMA went into decline so that by 1945 it had virtually ceased to function. Besides, by this time the organisation was not truly independent and represented more the interests of a single company, Brickworks Limited, than the industry itself.

In 1947 brick manufactures in New South Wales formed a new association from the rump of the BMA and also the Council of Brick Manufacturers. Tom Edwards, a former brickmaker, was appointed inaugural secretary and later executive director.

Brick Development and Research Council

In 1969 the Brick Manufacturers' Association established a Brick Development and Research Council (BDRC) with financial support from major brickmakers and the Commonwealth Government. The Council later changed its name to the Brick Development Research Institute (BDRI). This organisation conducted research into a variety of matters, including improved techniques for manufacturing and laying brickwork in highrise buildings. In 1988 the Federal government withdrew funding from the BDRI which amalgamated with the Australian Clay Brick Association to form the Clay Brick and Paver Institute (CBPI).

The Institute operates nationally and is the technical and research arm of the industry body, the Clay Brick Pavers Association (CBPA). It co-ordinates research into the use and serviceability of clay bricks and pavers. It also provides information to the building and landscaping industry through bulletins, manuals, technical notes, techniques sheets and videos for specifying, laying and maintaining clay pavements.

CLAY BRICK AND PAVER ASSOCIATION OF NEW SOUTH WALES

In September 1987 the Association changed its name to the Clay Brick and Paver Association of New South Wales (CBPA, NSW branch), becoming part of a national organisation, the Clay Brick and Paver Association of Australia. As the peak body the CBPA built on its previous function and continued to look after industrial relations from an employer perspective, industry promotion and technical aspects.

Today, CBPA has seven members representing Boral, CSR (PGH) and Brickworks Limited (Austral Bricks) operating in the Sydney metropolitan area, as well as four regional members who are small family companies. These include Lincoln Brickworks at Taree, Glen Innes Brickworks, which still uses the old steam driven equipment. Its management committee comprises three metropolitan manufacturers, the so-called Big Three, who manufacture 99 per cent of clay products in Australia.

The CBPA also owns the Brick Industry Training School, and the Brick Industry Group Training Company which were established to provide basic training and employment of apprentice bricklayers and paviours.

In a more recent change the CBPA now refers to itself as Think Brick.

Source: selections from The Clay Products Journal of Australia, interviews with Bob Rushton, Secretary, Clay Brick and Pavers Association of NSW, 20 October 2004, Sydney

9 Appendix

Inquiries, reports and Royal Commissions into the Sydney brick industry

Date	Inquiry, report or commission
1876	NSW Legislative Assembly Select Committee on the Employment of Children, 1876
1921	C.T. Crockett, Brick Companies under the control of the Metropolitan Brick Co., Report to the Profiteering Prevention Court of NSW, 13 December 1921
1937-38	Royal Commission of Inquiry into charges of fraud and corruption in connection with the sale of industrial undertakings, 1937-38
1939	The Industrial Commission of New South Wales: In respect to the Prices Of Bricks sold by manufacturers operating and selling in the Metropolitan Area and in Suburbs adjacent thereto*
1940	Attorney General versus Brickworks Pty Ltd and others, In the Supreme Court of NSW – Inquiry, commencing 9 October 1940

*The Industrial Commission, established by the Industrial Arbitration (Amendment) Act of 1926, was preceded by the following Courts which passed on all responsibilities to subsequent courts: The Court of Arbitration (1901); The Industrial Court (1908); Court of Industrial Arbitration (1912); The Board of Trade (1918); Industrial Commission (1926).

The Industrial Arbitration Act of 1940 consolidated all previous acts and is the current legislation under which the Commission operates. It has been amended on a number of occasions, with various important changes in the constitution of the Commission and its powers having been made.

Appendix 10

Employment in the Clay Industry
Comparison of figures, 1932–1935

This table shows the effects of the economic depression on employment and earnings in the clay products industry.

	Number of Employees and Contract Carters		Total Earnings for Employees and Contract Carters'	
	Week ending 6 April 1932	Week ending 3 April 1935	Week ending 6 April 1932	Week ending 3 April 1935
Ashfield Brick Co. Ltd., Ashfield	3	47	£6 7s 6d	£219 5s 10d
Austral Brick Co. Ltd., St Peters	65	254	£310 8s 0d	£1,496 6s 1d
Bankstown Brick & Tile Co., Bankstown	2	28	£10 10s 8d	£113 11s 8d
Burwood Brickworks, Burwood	4	44	£10 4s	£192 10s 7d
Butcher Bros & Co., Gore Hill	4	104	£2 13s 4d	£608 7s 4d
Federal Brick Co., Ltd., Hurstville	6	50	£23 0s 6d	£270 12s 6d
Canterbury Brick Co., Ltd., Campsie	5	49	£27 11s	£249 8s 2d
Carrington Steam Brick Co., St Peters	1	45	£5	£181 17s 11d
Central Brick & Tile Co. Ltd., St Peters	86	154	£274	£573 8s
City Brick Co., Ltd., Alexandria	Nil	45	Nil	£185 18s 3d
Clyde Brick Co. Ltd., Auburn	4	91	£17 13s	£505 6s 4d
Great Northern Brick Co. Ltd., Eastwood	Nil	36	Nil	£172 0s 3d
Judd's Hurstville Brick Co., Mortdale	Nil	52	Nil	£246 1s 2d
National Brick Co. Ltd., Thornleigh	6	40	£15 12s 6d	£218 15s
The N.S.W. Brick Co. Ltd., Alexandria	4	43	£21 10s 4d	£207 15s 4d
North Sydney Brick & Tile Co. Ltd., Gore Hill	11	95	£34 1s 1d	£603 14s 1d
Northern Suburbs Brick Co. Ltd., Punchbowl	36	61	£168 10s 6d	£382 14s 6d
Punchbowl Brick Co. Ltd., Punchbowl	7	110	£42 6s 6d	£454 18s 9d
Ryde Brick Works Ltd., Ryde	4	58	£9 10s	£254
St Peters Brick Co. Ltd., St Peters	Nil	36	Nil	£151 17s 9d
Sydney Brickworks, St Peters	1	44	£1	£163 19s 4d
Total	249	1486	£976 18s 11d	£7,542 9s 10d

Source: Figures published in The Clay Products Journal of Australia, 1 June 1935, p. 15

11 Appendix

The earth beneath

The Sydney Basin –The Wianamatta Group – Geological surveys

THE SYDNEY BASIN

For the brickmakers of Sydney, suitable clays and shales lay conveniently close to the surface. Such deposits are found in abundance among the various shale bands of the Wianamatta Group which is part of a much larger saucer-shaped geological feature known as the Sydney Basin. Measuring approximately 350 kilometres in length, the Basin is filled with Permian and Triassic sediments laid down during long periods of subsidence. The maximum thickness is thought to be 5,000 metres, although a complete sequence has not been found at any single location. At one time the total thickness was probably much greater due to erosion of the top deposits of Jurassic sediments.

The geological history revealed in the Sydney Basin confirms the enormous scale of repeated movement in the earth's crust during the Early Permian period approximately 290 million years ago. Over the next 50 million years marine sediments were laid down, which were followed by new forms of sedimentation, fan-shaped alluvial deposits created by streams and rivers. Some of the best preserved Permian sequences containing marine fossils can be found in the Hunter Valley. About the middle of the Triassic Period, some 230 million years ago, the Basin was uplifted to become dry land. From this point onwards erosion has progressively exposed the underlying shale and rock formations.The Cumberland Plain was probably formed about 80 million years ago, associated with the opening of the Tasman Rift off eastern Australia. This disrupted the old stream patterns which largely resulted in the geography that we see today. During the more recent Cenozoic era, humid conditions led to the formation of lateritic soils, rich in iron and aluminium. Since their formation, the rock sequences of the Sydney Basin have undergone only minor folding and faulting.

THE WIANAMATTA GROUP

The Wianamatta Group, also referred to by geologists as the Wianamatta Series, was formed 230 million years ago during the Triassic and Permian periods. It was during this time that the world's first dinosaurs, mammals, ginkgos and pine trees were coming into existence.

Shale formations, up to 300 metres in thickness, predominate within the Group which is strategically located immediately to the west of Sydney. The thickness of the Permian strata, as well as the rapid variations in sedimentary types, make the shale and clay materials suitable for all forms of products used for building purposes, such as bricks, tiles and clay pipes. The Group continues to be the chief source of raw material for brick production in the Sydney Region. Clay minerals comprise from 45 to 60 per cent of shales within the Group. Quartz, the other main constituent, comprises from 25 to 45 per cent.

As a general rule, the shales tend to be horizontally disposed and regularly stratified in fine calcareous lines, extending anywhere between two to 12 metres in depth, and varying in colour from a white to brown, to grey, green, blue or even black hues. Hard shales continue to great depths. Available in infinite grades, qualities and colours, these extensive deposits provided the raw material – clay – which served as the basis of Australia's earliest and most important manufacturing industries.

In the Sydney area, alluvial clays associated with the Georges, Cooks and Duck Rivers consist of material which is generally free of rock fragments. These clays and shales were formed by the weathering of primary rock minerals such as feldspars and micas. The clay minerals are made up of alumina, silica and water, with minor amounts of lime, magnesia, soda and potash. The individual particles are small

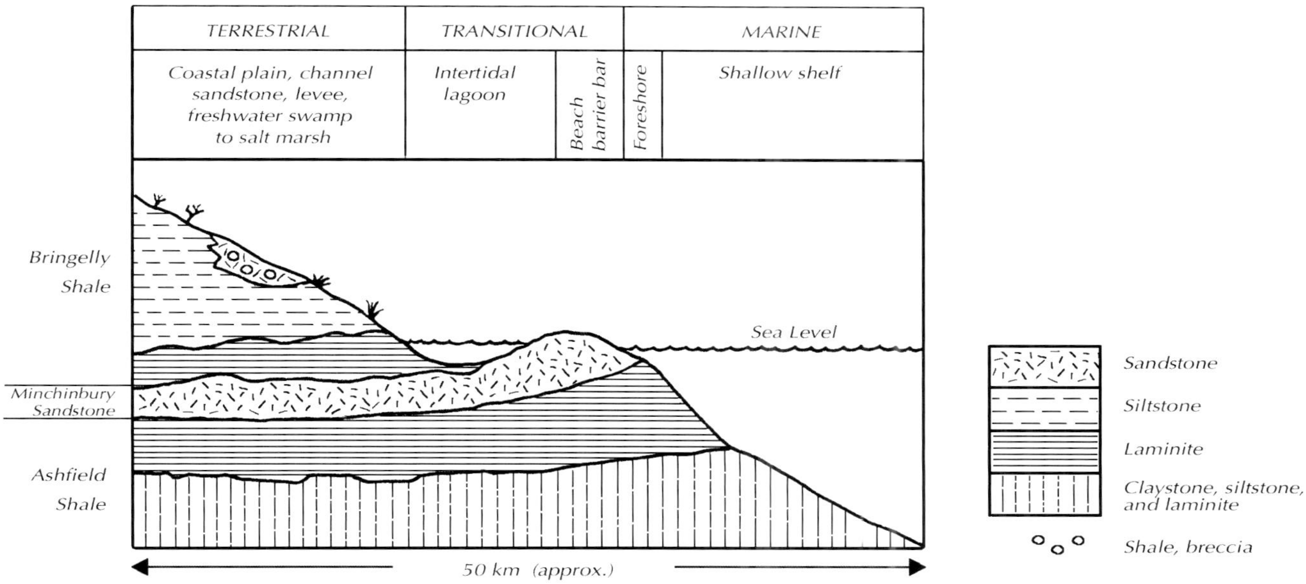

The Wianamatta Group. Vertical sections through the Wianamatta Group show a succession of rock types and sedimentary structures. From the base upwards, Ashfield Shales were deposited as a lake or seafloor sediment in a shallow environment. The Minchinbury Sandstone overlying the Wianamatta Group was deposited as a beach and barrier island complex. Surface deposits of Bringelly Shale form the uppermost layer

(0.1 – 1.0 micrometres diameter) and occur as thin plates. The small size and shape of the clay minerals are the source of the plastic properties on which the workability of the clay depends, and they have important effects on some of the properties of the finished bricks.[1]

Typical of both the Wianamatta Group and the region bounded by the Sydney Basin is a succession of gentle rolling hills which have been extensively cleared for livestock grazing over the past two centuries. Vegetation comprises the eucalypts, grey box and forest red gum. Indeed, eucalypts and the blackthorn bush can be used to accurately predict the presence of the Group and its three main formations – Ashfield Shale, Minchinbury Sandstone and Bringelly Shale. Less diagnostic, however, are the spotted gum and narrow leaved ironbark.

ASHFIELD SHALE

The deposits of Ashfield Shale are located at the base of the Wianamatta Group and comprise dark-grey to black siltstone grading upwards into a laminate, and range in thickness from 45 to 60 metres. The overlying, thin, quartz-lithic Minchinbury Sandstone is less than 4 metres thick. Bands of Ashfield Shale extend from Thornleigh in the north, to Sutherland in the south and west from St Peters to Parramatta and beyond. The dry pressed bricks manufactured from about the middle of the nineteenth century following the introduction of mechanical press machines, utilised these Ashfield Shales.

Plant fossils are rare in the Ashfield Shale, but in contrast, vertebrate and invertebrate fossils are well known, although only from the lower reaches, and chiefly from the St Peters brick pits.The fauna consists of *Unionid pelecypods*, isopods, insects, *labyrinthodonts*, and a diverse fish fauna containing a primitive form of shark, *Pleurocanthus* (Woodward 1908). Evidence of a freshwater lake for the lower Ashfield Shale is indicated by the type of fossils found as well as the presence of well-defined stratigraphic limits of the various fauna.

1. Notes on the Science of Building – NSB 126, July 1973, Department of Work Commonwealth Experimental Building Station.

MINCHINBURY SANDSTONE

Separating the Ashfield Shale from the overlying Bringelly Shale is a thin band (about three metres) of sandstone called the Minchinbury Sandstone.

BRINGELLY SHALE

Immediately above the Minchinbury Sandstone and Ashfield Shales lie the 260-metre thick bands of Bringelly Shale. These surface deposits were laid down in a coastal plain environment and contain a variety of sediments from sandstone to carbonaceous shale and rare thin coal seams. Typically, they were deposited in alluvial and estuarine channels, levees, swamps and coastal lagoons. The Bringelly Shale has a higher plasticity than the Ashfield Shale and is favoured by modern extrusion methods. A range of pleasant bright red colours can be obtained from the Bringelly Shale which may also yield cream-burning horizons if selectively mined.[2]

GEOLOGICAL SURVEYS

Little was known about the geology of the Sydney region in the first hundred years of white settlement in Australia. What may have sparked an interest was the establishment of a pottery works in George Street by Fowlers Pty Ltd in 1837. From this point forward there was an ever-increasing demand for specific types of raw material in the colony of New South Wales.

However, it was not until 1918 later that the first attempt at an appraisal of the State's resources was made in the form of a short report to the Government Geologist. Its author, Matthew Morrison, summarised the scant knowledge and pointed out that ceramic technology in Australia lagged far behind that of Europe and America. Morrison recommended the establishment of a central station for the testing of clays, similar to that in operation at Ottawa and several states of America.[3]

In 1928 Morrison and Harold Raggatt covered a wider field in their 'Mineral Industries of New South Wales'. They included not only clays for ceramic and filler purposes but materials used in the production of structural ware, Portland cement and refractories as well. Again, emphasis was placed on the lack of knowledge of the true nature and properties of the local clay deposits. Although several reports were made on individual deposits it was not until the late 1940s that the next systematic appraisal was made when three Summary Reports were issued by the Bureau of Mineral Resources on the talc, bentonite and fullers earth, and clay deposits. The first work to appear in which modern techniques were employed was by Ferguson and Hosking (1955) who produced a lengthy report on the occurrence, mineralogy and some physical properties of the structural clay and shale deposits of the Sydney Basin.

Further surveys over the last half-century have added greatly to our understanding of the geology and resource potential of the Wianamatta Group and beyond. Notable among them is the work of Chris Herbert (1976 and 1983), and studies by Branagan and Packham published in the year 2000. Both the Department of Mineral Resources and the Geological Survey of New South Wales, which are now part of the NSW Department of Primary Industries, remain at the forefront of geoscience in the State.

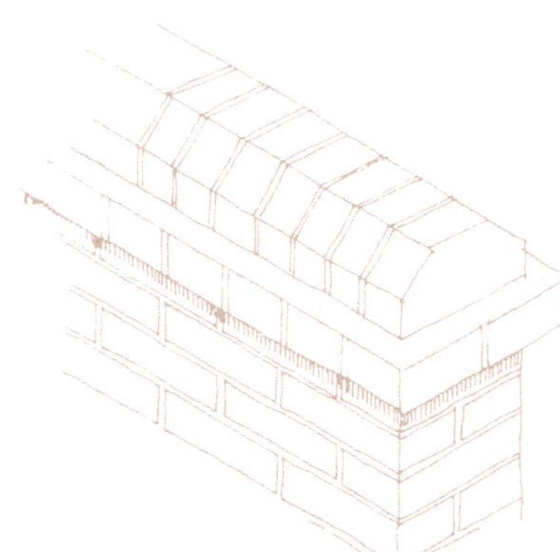

Facing page: Sydney Basin Stratigraphy. Courtesy of Geoscience Australia. From Stephenson and Burch, 2004; after Maung and others, 1997

2. Chris Herbert, *op. cit.*, Abstract.
3. F. C. Loughnan, *The Origin, Mineralogy and Some Physical Properties of the Commercial Clays of New South Wales*, School of Mining, Engineering and Applied Geology, Geological Series No 2, UNSW, 1960, p.1.

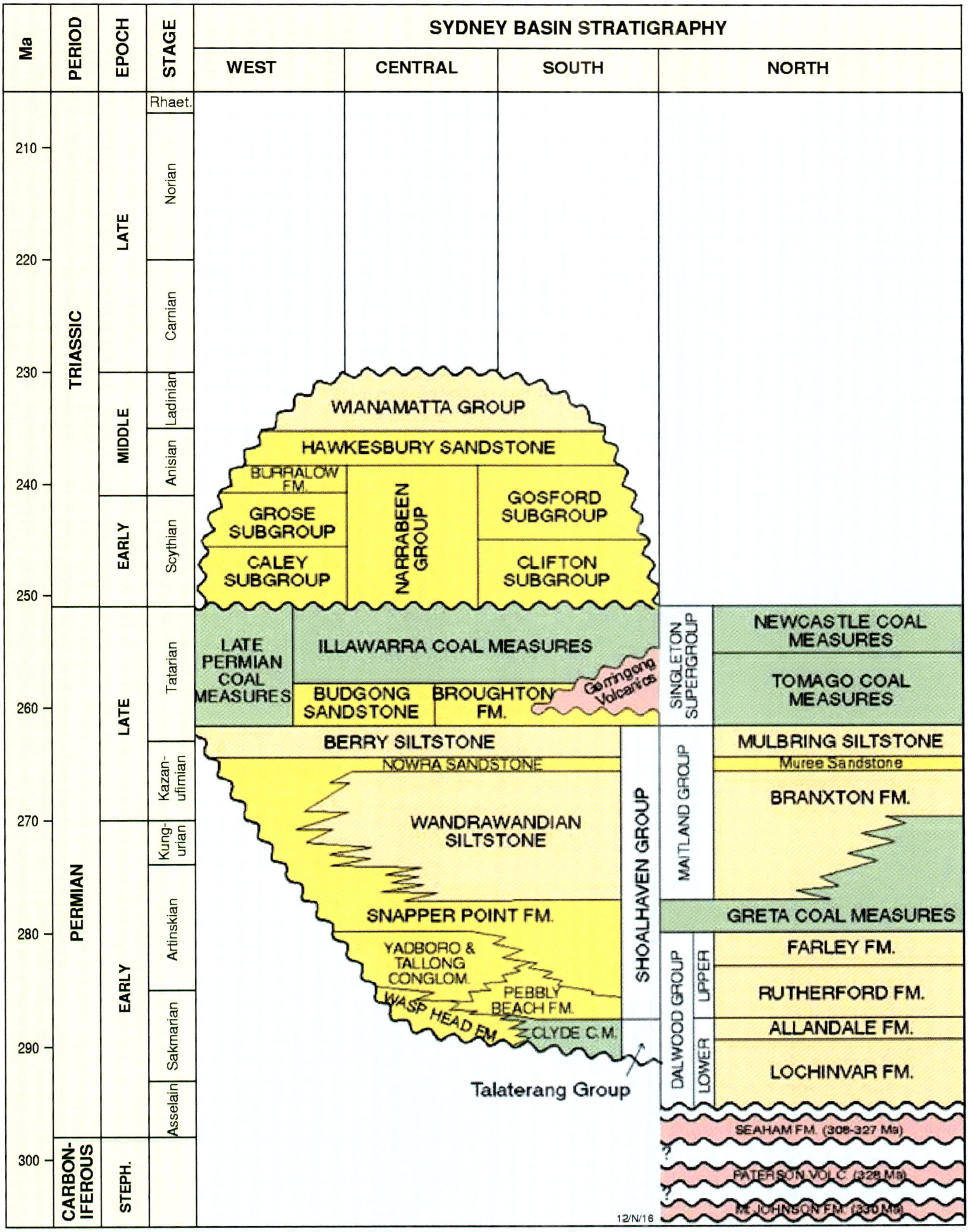

Ma
PERIOD
EPOCH
STAGE
SYDNEY BASIN STRATIGRAPHY
WEST
CENTRAL
SOUTH
NORTH
210
220
230
240
250
260
270
280
290
300
TRIASSIC
PERMIAN
CARBON-IFEROUS
LATE
MIDDLE
EARLY
LATE
EARLY
STEPH.
Rhaet.
Norian
Carnian
Ladinian
Anisian
Scythian
Tatarian
Kazan-ufimian
Kung-urian
Artinskian
Sakmarian
Asselain
WIANAMATTA GROUP
HAWKESBURY SANDSTONE
BURRALOW FM.
GROSE SUBGROUP
CALEY SUBGROUP
NARRABEEN GROUP
GOSFORD SUBGROUP
CLIFTON SUBGROUP
LATE PERMIAN COAL MEASURES
ILLAWARRA COAL MEASURES
BUDGONG SANDSTONE
BROUGHTON FM.
Gerringong Volcanics
SINGLETON SUPERGROUP
NEWCASTLE COAL MEASURES
TOMAGO COAL MEASURES
BERRY SILTSTONE
NOWRA SANDSTONE
WANDRAWANDIAN SILTSTONE
SNAPPER POINT FM.
YADBORO & TALLONG CONGLOM.
PEBBLY BEACH FM.
WASP HEAD FM.
CLYDE C.M.
Talaterang Group
SHOALHAVEN GROUP
MAITLAND GROUP
MULBRING SILTSTONE
Muree Sandstone
BRANXTON FM.
GRETA COAL MEASURES
DALWOOD GROUP
UPPER
LOWER
FARLEY FM.
RUTHERFORD FM.
ALLANDALE FM.
LOCHINVAR FM.
SEAHAM FM. (308-327 Ma)
PATERSON VOLC. (328 Ma)
Mt JOHNSON FM. (330 Ma)
12/N/16

12 Appendix

Brickworks in the Sydney region, including the County of Cumberland, 1955

SOUTH OF SYDNEY HARBOUR

City of Sydney		
Alexandria Brick & Tile Co. Pty. Ltd. (c,n)	Barwon Park Road, Alexandria	B
Bakewell Bros. Ltd.	Mitchell Road, Erskineville	F
Brickworks Ltd., Bedford Brickworks	St Peters Bridge, St Peters	B
Brickworks Ltd., City Yard (c)	Euston Road, Alexandria	B
Brickworks Ltd., Carrington Brick Co. Pty. Ltd. (c,n)	Barwon Park Road, St Peters	B
Commonwealth Ceramics Ltd.	Cr. Gardners & Kent Rds., Mascot	W
The Australian Fireclay Co Pty. Ltd.	Burrows Road, Alexandria	S
Industrial Brick Co. Ltd.	Botany Road, Waterloo	P
Redondo Tile Co. (Aust)	Dowling Street, Waterloo	W
The Morgan Crucible Co. (Aust) Pty. Ltd.	Bourke Road, Alexandria	R
The New South Wales Brick Co. Ltd. (c,n)	Huntley Street, Alexandria	B
Waterloo Firebrick Co. Ltd.	Botany Road, Waterloo	R

Kogarah Municipality		
Judd's Hurstville Brick Co. Pty. Ltd.	Judd Street, Mortdale	B

Sutherland Shire		
F. Mashman & Sons Ltd.	Princes Highway, Sutherland	P
St George Tile Co. Pty. Ltd.	Princes Highway, Sutherland	T
Punchbowl Brick & Pipe Co. Pty. Ltd.	Princes Highway, Sutherland	BD P
Punchbowl Brick & Pipe Co. Pty. Ltd.	Wilson Parade, Heathcote	B R

Strathfield Municipality		
Co-operative Brick & Tile Works Ltd. (c,n)	Water Street, Enfield	B
Enfield Brickworks Pty. Ltd	Juno Parade, Enfield	B
Lion Tile Co. Ltd.	Liverpool Road, Enfield	BT
Western Suburbs Brickworks Pty. Ltd.	Dean Street, Enfield	B

Burwood Municipality		
Brickworks Ltd., Burwood Yard	Cheltenham Road, Burwood	B
Rupert Cook's, Brickworks Ltd.	Mitchell Road, Enfield	B

Randwick Municipality		
S. Dive Pty. Ltd.	Beauchamp Road, Matraville	R

Marrickville Municipality		
Brickworks Ltd., Austral Brick Co. Pty. Ltd.	King Street, St Peters	B
Brickworks Ltd., Austral Brick Co. Pty. Ltd. Ralford Yard	Canal Road, St Peters	B
Brickworks Ltd., Central Brick & Tile Yard (c,n)	Albert Street, St Peters,	T
R. Fowler Ltd.	Fitzroy Street, St Peters	CP PR SW
Speare's Brick & Pipe Works Ltd.	632 King Street, St Peters	BP

Hurstville Municipality		
Fred A. Mashman Pty. Ltd.	Mashman Avenue, Kingsgrove	DE V
J.B. Sutcliffe	150 Kingsgrove Road, Kingsgrove	DE V
The Federal Brick Co. Pty. Ltd. (Pc)	Forest Road, Hurstville	B P

Canterbury Municipality		
Ashfield Brick Co. Ltd.	Milton Street, Ashfield	B
Brickworks Ltd., South Ashfield Yard (c)	Trevenar Street, Ashfield	B
Canterbury District Brick Go. Ltd.	538 Canterbury, Road, Campsie	B
Punchbowl Brick & Pipe Co. Pty. Ltd.	Bond Road, Punchbowl,	BD
Speare's Brick & Pipe Works Ltd	Lyle Street, Kingsgrove Pty. Ltd. (b)	B
The New South Wales Brick Co. Ltd., Excelsior Yard (c)	Penshurst Road, Narwee	B

LEGEND

A Architectural terra cotta
B Bricks – texture, facing, common and ornamental
C Chemical resistant ware (apart from sanitary ware)
D Agricultural pipes or drain tiles
E Earthenware – chimney pots, flower pots and boxes, etc
F Floor tiles, paving tiles and quarries
H Hollow blocks and clay lumber
P Sewer and drainage pipes
R Structural and industrial refractories and firebricks
S Sanitary ware
T Roofing tiles
V Ventilator or air bricks
W Wall tiles

(b) = works not yet in production (c) = works is closed (c,n) = works is closed and not likely to operate again. (Bc, (Pc), (Tc), etc = the brick, pipe, tile plant is closed but not the works as a whole.

NORTH OF SYDNEY HARBOUR

Warringah Shire		
Brickworks Ltd. Brookvale Yard	Federal Parade, Brookvale	B
Warringah Brick & Pipe Works Pty. Ltd. (Tc)	Roseville Road, French's Forest	BP T
Woodall Duckham (Alsia) Pty. Ltd.	Rodborough Road, French's Forest	R

Ryde Municipality		
Marsfield Brick, Tile & Pottery Works, Pty. Ltd.	Quarry Road, Ryde	P
Rymar Brick & Tile Works Ltd.	Victoria Road, Ryde	BT

Willoughby Municipality		
Mashman Bros. Pty. Ltd.	Victoria Street, Chatswood	PS
North Sydney Brick & Tile Co. Ltd.	Herbert Street, St Leonards	B
Northern Suburbs Brick Co. Ltd.	Pacific Highway, Artarmon	B

Hornsby Municipality		
Brickworks Ltd., National Yard	Dartford Road, Thornleigh	B

Baulkham Hills Municipality		
Castle Hill Brick, Tile Pottery	Cranes Lane, Castle Hill Works Pty. Ltd. (Bc)	BC P

OUTER WEST, PARRAMATTA AND THE BLUE MOUNTAINS

Bankstown Municipality		
Bankstown Texture Brick Co.	High Street, Bankstown	B
Bass Hill Tile Co. Pty. Ltd.	Thornton Avenue, Bass Hill	T
Chullora Potteries Pty. Ltd.	Waterloo Road, Chullora	DEV
Earthenware Pipes Pty. Ltd	Canterbury Road, Revesby	DP

City of Parramatta		
Brickworks Ltd., Great Northern Yard	Midson Road, Eastwood	B R
Fifield Magnesite Refractories Pty. Ltd.	North Grand Avenue, Camellia	R
O. Hicks	Bridge Street, Rydalmere	DP
Rigby & McMartin	Woodstock Street, Guildford	DP
Wunderlich Ltd.	Grand Avenue, Camellia	ABP TW

Blacktown Shire		
Cumberland Tile & Brick Pty. Ltd.	Blacktown Road, Blacktown	
Maxwell Porter Sons Ltd.	Railway Crescent, Doonside	
P.G. Clay Industries Pty. Ltd.	Power Street Doonside	
Shannon's Brick Tile & Potteries Pty. Ltd. (Be)	Wentworth Avenue, Wentworthville	
Wilson's Tile Works Pty. Ltd.	Seven Hills Road, Seven Hills	

Fairfield and Liverpool Municipalities		
Clark Kilns (NSW) Pty Ltd		B
Liverpool Tile and Terra Cotta Co. (Bc)		BH
Symons Bros. Pty. Ltd.	Austral Parade, Fairfield	ET

Camden Municipality		
Camden Brick Co.	Narellan	B

Penrith Shire		
Nepean Brick Co. Pty. Ltd.		B

Auburn Municipality		
Clyde Brick Co. Pty. Ltd.	Newton Street, Auburn	B
Mashman Bros. Pty. Ltd.	Parramatta Road, Auburn	P
N.S.W. State Brickworks	Homebush Bay	B
Trenholm Bros	West Street, Auburn	T

Holroyd Municipality		
Clay Products Pty. Ltd.	Pitt Street, Merrylands	T
Ferguson Bros Pty. Ltd	McFarlane Street, Merrylands	P
Goodlet & Smith Ltd.	Crescent Street, Granville	B
Holroyd Potteries A. Harvey	Sherwood Road, Merrylands	S
Langford Pty. Ltd.	Crescent Street, Granville	F
Mosaic Tile & Pottery Co. Pty. Ltd.	Pitt Street, Merrylands	F
The Newman Underwood Tile Co. Pty. Ltd.	Cambridge Street, Merrylands	T
Sun Brick Tile Co. Pty. Ltd.	Clarence Street, Merrylands	T
Tucknott's Tiles Pty. Ltd.	Paton Street, Merrylands	BT
Walker Benson Pty. Ltd.	Paton Street, Merrylands	BP
Walker Benson Pty. Ltd.	Paton Street, Merrylands	P
Walker Benson Pty. Ltd.	Betts Road, Merrylands	T
Walker Benson Pty. Ltd. (c,n)	Merrylands Road, Merrylands	B

City of Blue Mountains		
Katoomba Brickworks (c)	Copeland Street, Penrith	BP

Gosford Municipality		
Gosford Brick, Tile & Pipe Pty. Ltd.	175 Mann Street, Gosford	BD PT

Bowral Municipality		
F. Baker & Sons	Bowral	BR

Source: J.S. Hoskings (ed), Australian Structural Clay Products Works, Division of Building Research, CSIRO, Melbourne, April 1955.

13 Appendix

Brickmakers in the inner city suburbs of Sydney, by name, 1858–2007

A complete and accurate list of brickmakers and brick manufacturing companies operating in the inner suburbs of Sydney is difficult to establish. The names in this appendix were sourced from Sands Directory, GPO telephone directories, trade directories and Telstra White Pages. It is reasonable to assume that many individual or family-based operations were not included in published directories due to the short-lived nature of their business or the advertising fee levied.

Alexandria

Bedford Brickworks (Gentle)

Beulah Brickworks

Bryant, Henry

Carrington Steam Brick Company

City Brick Co Ltd

Dawes, Naasson

Drane, James & Samuel

Edwards, William

Federal Brickworks

Green, Thomas

Griggs and Chappell

Gurney Patent Brick Making Co Ltd

Gurney Steam Brickworks

Halverson and Wright

Harber, Abel and Co

Homobrick Manufacturing Co Ltd

Hunt, Richard

Jubilee Steam Brickworks

Lawler, James

Lloyd, Curlewis and Brown

New South Wales Brick Co Ltd

Newton, John

Patent Plastic Brickworks (Abel Harber)

Squires, James

St Peters' Dry Brick Co (also St Peters Patent)

Tabrett and Draper

Towner, John

Tripp, Charles

Tuck, A. and S.

Tye and Dawes

Vulcan Brickworks (owned by Vickery & Sons)

Waddington H.

Warren Brick Company

White, William

Wilkie, William

Woodley, H. & Co

Wright, William

Young, James

Marrickville

Addison Road Brickworks

Atkinson, D.

Bathgate, Alexander

Beauchamp, George

Bentley, James

Bloomfield, James, Robert and Samuel

Brian Brothers

Challis, Nathan

Clissold, George

Cochrane, George

Cole, James

Collier, James

Cook, Rupert

Curlewis, Frederick

Curren, Edward

Daley, Thomas

Frogmore Brickworks - Despointes, Mitchell

Eagle, James

Frogmore Brickworks (Despointes)

Garing, Henry

Gillies, A.W.

Goold, Mrs.

Griggs & Loveday

Harber, Abel

Hews, William

Hogan, John

Johnston Brothers Patent Steam Brick Works

Keenan, Patrick

King, John & William

Leslie, Robert

Loveday and Griggs

Marshall, John

McCormick, James

Model Brick Works - George Beauchamp

Moncur, William

Moseley, William

Moyes, William

Phoenix Steam Brickworks

Pickvance, Thomas

Porter, Maxwell

Red Cross Steam Brickworks

Sampson & Eagle

Saywell, Thomas

Simpson, John

Standsure Brick Company Limited

Staunton, Thomas

Stonestreet, James

Sydney Brickworks
Toyer Brothers
Toyer, George
Toyer, John
Toyer, W.H.
Williams, John

St Peters

Albion Brickworks
Annabel & Co.
Austral Brick Co Ltd
Ballin, Joseph
Berlin, Charles
Bolton, William
Bradshaw, William
Browning, William G.
Burling, Henry
Cadman, Martin
Cairncross, D.
Campbell, James
Central Brick Co Ltd
Chappell, Henry
Charlesworth & Co
Cheesman, Henry
Collison, James
Cook, James
Dawes, George
Edwards, Henry Jr
Farr, George
Gardiner, John
Garwell, William
Gentle, Josiah
Goodsell, Frederick J.
Griggs, Charles
Hailer, A.
Harber, Emanuel
Hill, John
Lee, Alfred
Louden, John
Martin, Joseph
Moore, John
Ogden, Peter
Osborn, Henry
Playford, William
Riley, John
Russell, James and Frank
Sholea, Stephen
Shortus, William and Henry
Speachley, Philip
Stubbings, Samuel
Tempe Brickworks
Turney, Henry
Walker, Samuel
Willard, Robert
William Collins & Co
Zahra, Henry

Newtown

Anderson, Francis
Baker, Thomas & Hilton (from 1863 then Auburn from 1890)
Barrett, Joseph
Bates, Robert
Birnie, Henry
Burling, John
Eggleton, William
Emery, Simeon
Gibben, Martin
Goodsell & Tye
Goodsell Brothers
Goodsell, William Henry
Goodsell, Alfred
Ives, Frederick
Knight, Henry & James
Mitchell, Frederick
Newtown Brickworks
Plowman, John
Rawlings, James
Speare, Peter
Tattersall, Joseph

Waterloo

Australian Pigment & Firebrick Co Ltd
Bryant, John
Cook, Reuben
Dalton, William
Edwards, Henry
Fullen, John
Gentle Family (also Cooks River Road)
Industrial Brick Co
Lipscombe, George
Paul, William
Pendleton, William
Pritchard, Samuel
Puckeridge, John
Rockley, William
Spitstead (Spilstead), William
Turner, John
Tye Family, Alfred, Frederick, James
Waterloo Firebrick Co Ltd

Enfield

Argyll Brick & Tile Co
Boots, Harry
Card Brick & Tile Co Ltd
Cooks Brickworks
Eagle Brickworks Ltd
Enfield Brick & Pipe Works
Enfield Steam Brickworks
Glazed Brick & Pottery Co Ltd
Kay, William
Lion Tile Co Pty Ltd
O'Hara, James
Pettit, J.
Taylor, John
Tompkins, James
Western Suburbs Brick & Tile Co Ltd

Bibliography

BOOKS

Adams, J.T. *The Complete Concrete, Masonry, and Brick Handbook* Arco, New York, 1979.

Atkinson, W., *View of picturesque cottages, with plans: selected from a collection of drawings taken in different part of England, and intended as hints for the improvements of village scenery* Gregg, Farnborough, England, 1971 edition.

Baker, D.S. *From Yeoman to Brickmaster: A Biography of Frederick Baker, brickmaker 1842–1916 and the history of his family from the end of the Middle Ages until 1940* F. Baker & Sons, Canberra, 1986.

Aldine Centennial History, Vol II, Sydney The Aldine Publishing Co. 1888.

Bender, W. & Händle, F. *Brick and Tile Making: Procedure and Operating Practice in the Heavy Clay Industries* Abacus Press, 1982.

Bertie, C.H. *Story of Sydney* Shakespeare Head Press Sydney, 1933.

Birmingham, J. *Industrial Archaeology in Australia: Rural Industry* Heinemann, Melbourne, 1983.

Blainey, G. *The Steel master: a life of Essington Lewis* Macmillan, Melbourne, 1971.

Bowley, M. *Innovations in Building Materials* Duckworth, London, 1960.

Bowyer, J., *A History of Building* Crosby, London, 1973.

Boyd, R. *Australia's Home: Its Origins, Builders and Occupiers* Melbourne University Press, Melbourne, 1952.

Bridges, P. *Foundations of Identity: Building Early Sydney 1788–1822* Hale & Iremonger, Sydney, 1995.

Broadbent, J. *The Australian Colonial House: Architecture and Society in New South Wales, 1788–1842* Hordern House, Sydney, 1997.

Bures, S. *The House of Wunderlich* Kangaroo Press, Sydney, 1987.

Burton, J.H. *Emigrant's Manual: Australia, New Zealand and America and South Africa: Emigration in its practical application to individuals and communities* William and Robert Chambers, Edinburgh, 1851.

Burwood Council *Brickworks in Burwood* Burwood Council, 2005.

Campbell, J. & Pryce, W. *Brick: A World History* Thames & Hudson, London, 2003.

Carroll, B. *A Very Good Business: One hundred years of James Hardie Industries Limited 1888–1988* James Hardie, Sydney, 1987.

Carroll, B. & Rule, I. *Preston: an Illustrated History* Victoria: City of Preston, 1985.

Clews, F.H. *Heavy Clay Technology* British Ceramic Research Association, London, 1969.

Clifton-Taylor, A. *The Pattern of English Brickwork* Batsford, London, 1962.

Collins, D.	*An Account of the English Colony in New South Wales [Volume 1]: With Remarks on the Dispositions, Customs, Manners etc of the Native Inhabitants of that Country. To Which are Added, Some Particulars of New Zealand: Compiled by Permission, From the Mss. Of Lieutenent-Governor King*	T. Cadell Jun. and W. Davies 1798.
Coghlan, T.A.	*Results of the Census of the Seven Colonies of Australasia*	General Post Office, Sydney, 1894.
Coupe, S. & R.	*Speed the Plough: Ashfield 1788–1988*	Council of the Municipality of Ashfield, Ashfield, 1988.
Cowan, H.J.	*From Wattle and Daub to Concrete and Steel*	Melbourne University Press, Carlton, 1998.
Crowley, F.	*A Documentary History of Australia, Vol. 2 (Colonial History)*	Thomas Nelson, Melbourne 1973–1980.
Cunningham, P.,	*Two Years in New South Wales, facs. edn., Vols 1 & 2*	Libraries Board of South Australia, Adelaide, 1996.
Davison, G., Dingle, T. & O'Hanlon, S. (eds)	*The Cream Brick Frontier: Histories of Australian Suburbia*	Monash Publications, Melbourne, 1995.
Denholm, D.	*The Colonial Australians*	Allen Lane, Ringwood, Victoria, 1979.
Dickens, C.	*The Drunkard's Death, Sketches by Boz*	Macrone, 1836.
Dobson, E.	*"A rudimentary treatise on the manufacture of bricks and tiles" (1850), edited with an introduction, biography, notes, bibliography and index by Francis Celoria Stafford*	George Street Press, 1971.
Dunlop, E.	*Between Two Highways, The Story of Early Croydon*	Self published work, 1969.
Dyster, B.	*Servant and Master: building and running the grand houses of Sydney, 1788-1850*	Sydney, UNSW Press, 1989.
Egan, J.	*Buried Alive, Sydney 1788–92: Eyewitness accounts of the making of a nation*	Allen & Unwin, Sydney, 1999.
Flannery, T.	*The Birth of Sydney*	Text Publishing, Melbourne, 1999.
Fleming, G.	*Merrett, D. & Ville, S., The Big End of Town*	Cambridge University Press, Cambridge, 2004.
Ford, G.	*Australian Pottery: The First 100 Years*	Salt Glaze Press, Wodonga, 1995.
Forster, H.W.	*Preston: Lands and People 1828–1967*	F.W. Cheshire, Melbourne, 1968.
Fowlie, T.	*A History of Granville, 1880–1920*	Granville Historical Society, Granville, 2001.
Fraser, H.	*The Federation House: Australia's Own Style*	New Holland Publishers, Sydney, 2002.

Freeland, J.M.	*Architecture in Australia: A History*	F.W. Cheshire, Melbourne, 1968.
Gemmell, W.	*And So We Graft From Six to Six – The Brick Makers of NSW*	Angus and Robertson, Sydney, 1986.
Gilbert, L.	*William Woolls, 1814–1893: 'a most useful colonial'*	Mulini Press, Canberra, 1985.
Gilbreth, F.B.	*Bricklaying System*	Myron L. Clark, New York, 1909.
Hammond, M.	*Bricks and Brickmaking*	Shire Publications Ltd, Buckinghamshire, 2001.
Harrington, J.	*An archaeological and historical overview of limeburning in Victoria*	Heritage Council Victoria, 2000.
Hartley, L.P.	*The Go-Between*	Penguin, London, 1953.
Hayward, R.	*The Brick Book*	Rigby, Melbourne, 1978.
Herman, M.	*The Early Australian Architects and their Work*	Angus & Robertson, Sydney, 1954.
Hodge, J.C.	*Brickwork for Apprentices*	Edward Arnold, London, 1971.
Houghton, N.	*A Century of Country Clay. Selkirk The First 100 Years 1883–1983*	Selkirk Brick Pty Ltd, Melbourne, 1983.
Hughes, J. & Reynolds, P.	*John Horbury Hunt. Radical Architect, 1838–1904*	Historic Houses Trust, Sydney, 2002.
Hutton, J.	*Building and Construction in Australia*	Cheshire, Melbourne, 1970.
Inglis, J.	*Our Australian Cousins*	Macmillan & Co, London, 1880.
Irving, R., (ed)	*The History and Design of the Australian House*	Oxford University Press, Melbourne, 1986.
Jubilee of the Municipal Council of St Peters, 1912.		
Karskens, G.	*Holroyd: A Social History*	University of New South Wales Press, Sydney, 1991.
Kennedy, M.J.	*Hauling the Loads: A History of Australia's Working Horses and Bullocks*	Melbourne University Publishing, Melbourne, 1992.
Lang, J.D.	*An Historical and Statistical Account of New South Wales both as a Penal Settlement and as a British Colony, Vols 1 & 2*	Cochranes and McCrone, London, 1834.
Leigh, S.T.	*The Handbook to Sydney and suburbs with a plan of the city and map of the roads of the Colony*	S.T. Leigh, Sydney, 1867.
Lemon, A.	*The Northcote Side of the River*	City of Northcote in conjunction with Hargreen Publishing Company, Melbourne, 1983.

Mackaness, G.	*Admiral Arthur Phillip, Founder of New South Wales, 1738–1814*	Angus and Robertson, Sydney, 1937.
Maclehose, J.	*Picture of Sydney and Stranger's Guide for New Scuth Wales for 1838*	Maclehose, Sydney, 1838.
Mallory, K. & Ottar, A. 1973.	*Architecture of Aggression*	Architecture Press,
Mardi, M.	*Bricks and mortar: a celebration of the 50th anniversary of Midland Brick*	City of Swan Libraries, Perth.
Maung, T.U., Alder, D., Shaw, R. & Hawley, S.	*Offshore Sydney Basin*	Bureau of Resource Sciences, Petroleum Prospectivity Bulletin, 1997–1.
McKell, W.	*Five Critical Years, Australian Labor Party*	Sydney, 1946.
Meader, C. & Cashman, R.	*Marrickville: Rural Outpost to Inner City*	Hale & Iremonger, Sydney, 1990.
Mies van der Rohe, L.	*A look at bricks in architecture*	JA, Japan, 2001.
Moore, B.	*From the ground up: Bristle, Whittakers and Metro Brick in Western Australian History*	University of Western Australia Press, Nedlands, 1987.
Morrison, F.W.	*The Aldine Centennial History of New South Wales, Vol. II*	The Aldine Publishing Co, Sydney, 1888.
Nangle, J.	*Australian Building Practice: A Treatise for Australian Students of Building Construction*	G. Robertson & Co., Melbourne, 1900.
Norman, L.G.	*Historical Notes on Newtown*	Council of the City of Sydney, 1963.
Oswald, A.	*Country Houses of Dorset*	Country Life Ltd, London, 1935.
Parkes, H., Sir	*Freehold Homes in a Gold Country: Two Public Addresses on the present condition and natural resources of the Colony of New South Wales,*	delivered at Derby and Birmingham, Hudson Ellis, Birmingham, 1861.
Peek, N. & Pratten, C.	*Working the Clays: The Brickmakers of the Ashfield District*	Ashfield Historical Society, Ashfield, 1996.
Péron, F.	*Description of Sydney Town, the capital of the English Colonies in New South Wales*	The Monthly Magazine, 1 September 1809.
Plumridge, A. & Meulenkamp, W.	*Brickwork: Architecture & Design*	Harry N. Abrams, New York, 1993.
Pyle, K. (ed)	*A Singular Success: Washington H Soul Pattinson 1872–2002*	Focus Publishing, Sydney 2002.
Rashleigh, R.	*Adventures of an Outlaw: The Memoirs of Ralpn Rashleigh. A Penal Exile in Australia 1825–1844*	Jonathan Cape and Harrison Smith, New York, 1929.
Reeves, S.	*Jubilee celebration of the proclamation of the Town of St. Peters: a brief summary of the early history of the town*	Hassell Press, Adelaide, 1933.

Reynolds, P., Muir, L., Hughes, J.,	*John Horbury Hunt. Radical Architect, 1838–1904*	Historic Houses Trust, 2002.
Roe, J., (ed)	*Twentieth Century Sydney. Studies in Urban and Social History*	Hale & Iremonger, Sydney, 1980.
Rosen, S.	*We Never Had a Hotbed of Crime*	Hale & Iremonger, Sydney, 2000.
Ross, S.	*Bricks by the millions: a brief history of Lilydale's brick making industry*	Lilydale and District Historical Society, 1996.
Sawtell, M.	*Job Control: The basis of Future Society,* issued by the Bricklayers' Union of New South Wales	The Worker Print, Sydney, 1926
Scott, G.	*Sydney's Highways of History*	Georgian House, Melbourne, 1958.
Searle, A.B.	*Bricks and Artificial Stones of Non-Plastic Materials*	J. & A. Churchill, London, 1915.
Searle, A.B.	*Modern Brickmaking, 1911–1956*	Ernest Benn, London 1956.
Serventy, V. & Raymond, R. (eds)	*Australia's Wildlife Heritage, Our Plants and Animals*	Paul Hamlyn Pty Ltd, 1973.
Sherington, G.	*Australia's Immigrants*	Allen & Unwin, Sydney, 1990.
Shying, B. Technical	*The Beginning of Brickwork in Australia*	Department of and Further Education, New South Wales, 1976.
Sidney, S. (ed)	*Sydney's Emigrant's Journal and Traveller's Magazine, Second Series, London*	Orr & Co., 1850.
Sidney, S.	*The Three Colonies of Australia: New South Wales, Victoria, South Australia; their pastures, copper mines and Gold Fields*	Ingram Cooke, London, 1852.
Sowden, R.L.	*A Pictorial History of Marrickville*	Marrickville and District Historical Society, 1968.
Spearritt, P.	*Sydney's Century: a history*	SydneyUNSW Press, 2000.
Stapleton, M. & I.	*Australian House Styles*	Flannel Flower Press, Brisbane, 1997.
Budd, D., et al	*Wait here for Trams: Sydney Trams during the Last Decade of Operation*	Nostalgia Press, 1974.
Stephenson, A.E. & Burch, G.J.,	*Preliminary evaluation of the petroleum potential of Australia's central eastern margin*	Geoscience Australia Record, 2004/06.
Tench, W.	*Sydney's First Four Years*	Angus & Robertson, Sydney, 1961.
Toghill, J.	*A Century of Achievement: Stuart Bros 1886–1986*	Jeff Toghill Productions, Mosman, 1986.

Troy, P. (ed)	*European Housing in Australia*	Cambridge: Cambridge University Press, 2000.
Turnbull, L.H.	*Sydney: Biography of a City*	Random House Australia, Sydney, 1999.
Wakely, M.	*Dream Home*	Allen & Unwin, Sydney, 2003.
Wentworth, W.C.	*Statistical, Historical and Political Description of the Colony of New South Wales and its dependent Settlements in Van Diemen's Land, with a particular numeration of the advantages which these colonies offer for emigration, and their superiority in many respects over those possessed by the United States of America*	London, Whittaker, 1819.
Woodforde, J.	*Bricks to Build a House*	Routledge & Kegan Paul Ltd, London, 1976.
Worgan, G.B.	*Journal of a first fleet surgeon*	Library Council of New South Wales in association with the Library of Australian History, 1978.
Yeomans, D.	*Construction Materials since 1900*	Batsford, London, 1997.

REPORTS

Coghlan, T. A.	*General Report on the Eleventh Census of New South Wales, Sydney*	Government Printer, 1894.
Cumberland County Council	*County of Cumberland Planning Scheme Report,*	CCC, Sydney, 1948.
Kass, T. 1995.	*A History of Goodlet and Smith Brickworks*	HLA-Envirosciences Pty Ltd, Sydney,
Pratten, C. & Irving, R.	*Brookvale Brickworks And Surrounds – Draft Land Use Study*	Warringah Municipal Council, Sydney, 1997.
Proudfoot, H.	*The First Government House, Sydney*	Report for the Department of Environment and Planning, March 1983.
Russell, A. & Vogel, R.	*Evaluation of Hoffmann Kiln Technology*	Immediate Technology Consultants, 1999.

THESES

Cousins, D.	*The Melbourne Brick Combine, 1896–1913*	Honours Thesis, Monash University, Melbourne, 2000
Pavlou. O.S.	*The History of Bricks and Brick Making in New South Wales, 1788–1914*	Honours Thesis, Faculty of Architecture, UNSW, Sydney, 1975.
Varman, R.	*Bricks and nails: building materials as criteria for dating in Sydney and environs from 1788; a documentary survey and assessment of dating potential*	PhD thesis, University of Sydney, Sydney, 1993.

PROCEEDINGS

Boden, E.G.	*Tempe Wesleyan Chapel, 1855–1939. 80 Year Romance of a Small Suburban Mission*	The Australian Methodist Historical Society, Journal and Proceedings, Vol VIII – Part 2, January 1940.
Proceedings of the Federated Master Builders Association of Australasia, Vol. II, 1909.		
Smythe, R.B.	*'On the Comparitive Value and Durability of Building Materials in Use in Melbourne'*	Transactions of the Philosophical Society of Victoria, Vol. 1, 1855, Melbourne, James Blundell, 1855.

PAPERS AND ARTICLES

Davison, A.	*'Australian Suburban Imaginaries of Nature: Towards a Prospective History'.*	Australian Humanities Review. Issue 37, 2005.
Dawes, A.	*Early Brickmaking in Australia: Slump of the Nineties and After*	The Clay Products Journal of Australia, 2 December 1935.
D. Hutchinson and P. Lee	*'Disclosure: Imperfect Information and the Development of an Equity Market'*	Paper presented at the Fourth Accounting History International Conference, Braga, 2005.
Loughnan, F.C.	*The Origin, Mineralogy and Some Physical Properties of the Commercial Clays of New South Wales*	School of Mining, Engineering and Applied Geology, Geological Series No. 2, University of 1960.
Ministry of National Development, Division of Industrial Development,	*Brief Review of the Australian Clay Brick and Roofing Tile Industries*	Commonwealth Government of Australia, April 1951.
Spennemann, D.H.R	*Identifying historic brickwork-techniques, styles and conservation problems*	School of Environmental and Information Science, Charles Sturt University, Albury, 1994.
Watson, D.M.S.	*A New Labyrinthodont (Paracyclotosaurus) from the Upper Trias of New South Wales*	Bulletin of the British Museum (Natural History) Geology Vol. 3,7.

ARCHIVES

Austral Brick Company, 1922–1971, Mitchell Library MSS

Brickworks Limited

City of Sydney

Colonial Secretary In-Letters, 4.1741, Bun.12, No.70–97

Eureka Terracotta & Tile Company

Federated Brick, Tile and Pottery Industrial Union

Heritage Council Annual Report, 1982

Historical Records NSW (HRNSW), Vol. IV

Historical Records of Australia, Series 1, Vol.1

Notes on the Science of Building – NSB 126, July 1973, Department of Work Commonwealth Experimental Building Station

ARCHIVES Continued

NSW Royal Commission into allegations of fraud in the sale of state industrial enterprises

Sands Directory

Select Committee on the Employment of Children, Progress Report, New South Wales Legislative Assembly, 1876.

Sydney Park, Plan of Management, Conybeare Morrison & Partners P/L July 1982.

NSW Parliamentary Papers – 1938–1940, Vol. 6, pp. 747–779, Mitchell Library

Manuscript Card and Index Catalogue Mitchell Library

Manuscript Leaf Catalogue No. 1 and No. 2, Mitchell Library

Trade Catalogues, Mitchell Library

NEWSPAPERS AND PERIODICALS

Australian Gold Digger's Magazine

Australian National Clay

Brick Trade Journal

Builder and Contractor's News

Building Development Research Institute

Building: *Official Journal of the Master Builders Association*

Colonial Magazine

Colonial Mining Journal

Daily Telegraph

Illustrated Sydney News

Land Magazine

Maitland Mercury

Southern Courier

Sutherland Shire Voice

Sydney Daily Mirror

Sydney Gazette

Sydney Morning Herald

The Australian Financial Review

The Bulletin

The Clay Products Journal of Australia

The Sydney Telegraph

Glossary

Acid-resistant brick	Brick that is suitable for use in contact with chemicals, usually in conjunction with acid-resistant mortars.
Adobe brick	Large roughly-moulded, sun-dried clay brick that may vary in size.
Angle brick	Any brick shaped to an oblique angle to fit a corner.
Arch brick	Wedge-shaped brick for special use in an arch. Also, an extremely hard-burned brick from the arch of a scove kiln.
Arriss	The sharp edge formed by any two surfaces meeting at an angle.
Art Deco	An artistic movement of the 1920s and 1930s characterised by clean, sharp lines.
Banding	Brickwork, normally of a contrasting colour or texture, one or more courses high.
Batch	The specific production run of bricks or pavers.
BCA	Building Code of Australia.
Bevelled	The edges of the face of some bricks are bevelled or sloped, thus providing a softer decorative effect.
Blending	Mixing two or more brick colours or shades for effect.
Bond	Tying various parts of a masonry wall by lapping units one over another, or by connecting with metal ties. Also patterns formed by exposed faces of units. Can also refer to adhesion between mortar or grout and masonry units or reinforcement.
Bottle kiln	A traditional type of kiln, circular in plan and tapering to the top, and similar in shape to a bottle.
Brick	A solid masonry unit of clay or shale, formed into a rectangular prism while plastic and burned or fired in a kiln.
Brick Combine	A group of brickmakers who form an association to set the price of bricks. Also referred to as the Brick Tryst.
Brick mould	A bottomless rectangular box of timber, slightly deeper than the height of a brick.
Brick sand	Also called 'brickies sand'. Common term for sand used for making mortar. Colour can vary by region and therefore affect the mortar colour achieved.
Brick veneer	Construction where a timber or metal frame is covered with a lining material for internal walls and a single leaf or layer of brickwork forms the exposed exterior of the building.
Brickbats	Broken bricks.
Brickfield Hill	The brickmaking area that comprised the upper part of what used to be known as Paddy's Market bounded by George, Campbell, Elizabeth and Goulburn Streets.
Brickfielders	Blinding storms of red brick dust whipped up by on-shore southerly winds.
Brindles	Bricks that have been unequally fired.
Bristol ware	Glazed stoneware pottery, so called as the glaze was developed in Bristol, England. Normally in the form of utilitarian pottery, e.g.. ginger beer bottles, handled bottles and bung jars. A zinc oxide glaze developed to replace poisonous lead glaze in the 19th century.
Bungalow Style	Houses built with a low-pitched roof, designed intentionally to provide greater protection against the sun through wider eaves.

Burns	An aesthetic treatment during firing to create light to dark blushes. Also known as 'flashing'.
Cadigal	Inhabitants of Sydney's northern shore. People who called themselves Cadigaleans, Cadi being their name for Sydney harbour, and galean meaning 'the people of'.
Calcined magnesite	Magnesium oxide.
Callows	Bricks from the outer face of the clamp that were soft and underburnt, creating the pleasant light-brown or salmon-coloured bricks, termed 'callows' or 'doughboys'.
Ceramics	A term used to describe the whole range of pottery and claywork articles made from potter's earth. Derived from the Greek, Keramos. The correct pronunciation of the first letter is pronounced as 'S'.
Chilean Mill	A machine which reduces raw clay or fired shards to powder.
Chinaman	A central loading platform at the bottom of the pit.
Clamps	One-off kilns.
Clappers	Wooden paddles used to unload bricks from the hack to avoid damage to the arrisses of the soft brick.
Clay	The basic ingredient of all pottery bodies. Hydrated silicate of alumina containing, in addition, varying amounts of free silica, lime, iron and the alkalis of residue of naturally decomposed felspathic rock. Clay is plastic when sufficiently wetted, rigid when dried and vitrified when fired to a sufficiently high temperature.
Clinker	A brick that has intentionally been overburnt in a kiln – giving it a rough, dark appearance. Bricks that were stacked nearer the flues and fuel supplies and so became fused and distorted.
Clinker brick	A very hard-burned brick whose shape is distorted or bloated due to nearly complete vitrification.
Clot	The measure of clay which is forced into a stock mould.
Colonial bond	Three courses of stretchers followed by one course of headers.
Common brick	Moderately priced, well-shaped bricks that are sufficiently strong to support a normal load. May also be used where structure and appearance is not important. More often referred to as building brick, ie any brick not especially treated for texture or color.
Compass bricks	Bricks that taper in thickness and/or width.
Compression tests	Machine-tested bricks to establish the breaking point when pressure is applied.
Crazing	Fine, non-structural cracks on the face of the brick.
Damp course	A course, or layer, of impervious material in a wall or floor to prevent the migration of moisture. Also called 'damp-proof course' or DPC. The Building Code of Australia requires the DPC to visibly extend beyond the mortar.
Dampers	Metal rods that could be raised or lowered through the roof of the patent or Hoffman kiln. Dampers could close or open the flue and thus controlled the rate of burn in a brick kiln.
Downdraught kiln	Designed so that the heat and gases are passed up to the crown of the kiln and

	are then drawn down to escape through flues in the floor before being exhausted. They were also referred to as dome kilns, characterised by a permanent arched brick roof, fire holes and wicket in the walls.
Drawing (a kiln)	Unpacking it after firing.
Dry press brick	Brick formed in moulds under high pressures from relatively dry clay, eg 5 to 7 per cent moisture content.
Earthenware	Pottery which is opaque, porous, made of plastic clay, and fired at about 1000°C. Glazed earthenware bodies commonly used in domestic tableware, such as teapots, jugs, and bowls etc.
Efflorescence	A crystalline deposit (white or sometimes coloured powder) on the surface of a brick, caused by soluble salts in the water used for the brickmaking process. The problem is usually temporary and will disappear in time. In places where the land is waterlogged, salts will tend to rise through the soil. This defect can be avoided by not using the top layer of topsoil and clay in extraction.
Engineering brick	A dense, robust brick, suitable for use in heavy duty structures, ie viaducts. Moulded under high pressure and fired by controlled burning to ensure that bricks conform with defined water absorption and strength requirements. Typically red or blue in colour. Dry pressed engineering bricks were used extensively throughout the 19th and 20th century but were largely made redundant by the introduction of pre-stressed concrete structures.
English bond	Alternate header and stretcher courses.
Eora	Eora who lived in the area between Botany Bay and Broken Bay.
Etruria	Wedgwood's Staffordshire pottery at Etruria, Staffordshire.
Extruded bricks	Bricks formed by squeezing – or extruding – wet clay through a machine.
Face	The long surface of a brick to be exposed in a wall.
Facing brick	Dense and exceptionally strong brick made especially for facing purposes, made of selected clays, or treated, to produce desired colour and texture.
Fire box	The chamber in which fuel is burned in order to heat the kiln.
Fire brick	Brick made of refractory ceramic material which will resist high temperatures.
Fireholes	Small, rectangular holes set into the roof of the chamber in a patent kiln, down which the burner would pour pulverised coal onto the bricks. Fireholes were also called 'eyes'.
Fireman	Person in control of the firing process of a kiln. Also referred to as a burner.
Firing	The process of changing dried mud into an approximation of stone by heating to extreme temperatures in a kiln.
Flashing	A method used during firing to create light to dark blushes or areas of deepened colour.
Flemish bond	Alternate headers and stretchers in each course.
Fletton Process	A new method of brick making developed by a 19th century brickmaker in Peterborough, England, who discovered a deeper shaley clay called Oxford Blue which is high in carbon. Once heated the bricks would virtually fire themselves. This method meant that bricks could be produced more economically.
Flue	Passages through which heat travels in a kiln.

Footings or foundations	The base of a building, usually concrete, designed to transfer loads to the ground.
Freemasonry	The word mason derives from the French maçon – builder of walls or stonecutter. The most commonly recognised Masonic symbol is the square and compass.
Frit	A glass substance melted onto bricks to create glossy spots.
Frog	The hollow indent on the bottom of dry press bricks.
Frog	The indentation in the face of a brick, which makes it both easy to handle and bed into mortar.
Full or double brick	Construction where most of the external and internal walls are made of brick. The external walls are usually composed of two layers of brick, with a 50mm cavity between them.
Garden wall bond	Three stretchers to one header in each course.
Gauged brickwork	Precisely made bricks laid with fine joints.
Glaze	Impervious coating applied to the surface of pottery which becomes a layer of glass fused into the surface during firing.
Green bricks	Newly cast, wet but firm bricks, which are put into a drying shed or hack to harden up before being transferred to the kiln.
Grinding pans	Machines comprising a set of heavy rollers connected to a central shaft and driven by an electric motor. A wide range of crushing mills was manufactured to service the brick industry. These include Incla, etc.
Grog	Previously fired material ground to a powder. The addition of grog to clay reduces shrinkage, facilitates drying and reduces firing, cracking and washing.
Hack	The stacking area for bricks.
Header	A brick laid so that only its end appears on the face of the wall.
Hearting	A red, or sometimes orange, aesthetic burn on the brick face.
Hoffman kiln	A continuous firing kiln that was originally circular in plan and worked on the principle that the fire, unlike in earlier kilns, was never extinguished. The chimneystack of the Hoffman was always in the centre of the kiln. Later variations on the traditional Hoffman were called 'patent kilns' and distinguished by the rectangular form with the chimneystack placed at the end.
Hollow brick	A masonry unit of clay or shale whose net cross-sectional area in any plane parallel to the bearing surface is not less than 60 percent of its gross cross-sectional area measured in the same plane.
Hovel	A rough open-sided timber shelter with a roof supported on wooden posts, and thatched with reeds or covered with bark.
Joint	How the layer of mortar between two bricks is finished.
Kaolin	China clay.
Kiln	A furnace or oven in which pottery is fired.
Kiln furniture	The props and shelves used to support pottery inside a kiln.
Lever press	A hand-operated machine where the lever was depressed forcing clay into the shape of a brick. By repeating the action several times, bricks were re-pressed

	to ensure adequate compaction of the clay, producing a brick with fine, straight edges (arriss).
Limeburning	Primitive method of burning of shells in specially constructed rectangular trenches.
London stock brick	London stock brick is the most commonly used type of building brick used in London. Its distinctive colour and soft appearance comes from the yellow clay the bricks are made from, which is found in Kent.
Manganese	A metal of which the oxides and salts are used as glaze colouring pigments which produce brown and purple colours.
Mortar	A mixture of lime, cement, sand and water used to bond the bricks together.
Mottle	A mixture of different colours within one brick.
Owner builder	A builder who controls the building process of their own home, managing the project and coordinating subcontractors.
Oxford Blues	Bricks made by a firm of Oxford brickmakers in the United Kingdom using clays with a high iron content from the Oxford Clay belt . 'Oxford Blues' were a dark blue colour and were impervious to water. They were widely used as a damp course prior to the use of slate and lead.
Pack	A package or bundle of bricks or pavers prepared for delivery.
Patent kilns	A continuous burning kiln so-called because the design was patented. Patent kilns were based on the original Hoffman kiln.
Paving brick	Vitrified brick especially suitable for use in pavements where resistance to abrasion is important.
Plastic brick	Brick made from clay which has been converted into a highly plastic paste, or in which the plasticity has been developed as fully as possible. All hand-made bricks and tiles are of this kind, but the term is also used in connection with machine-made goods, particularly with loamy clays.
Plasticity	A quality of clay which enables it to be readily shaped into different forms without cracking, crumbling or sagging.
Platt press	Dry press brick machine of the type manufactured by the engineering firm of Platt Brothers.
Polychromatic brickwork	The juxtaposition of different coloured bricks to achieve a decorative effect in buildings.
Pressed bricks	Solid bricks that are made by pressing clay into individual moulds.
Pugged or puddled clay	A mass of clay that has been mixed to a smooth and uniformed state in a pug mill.
Pugmill	A machine which mixes clay to an even consistency, de-airs it and forms it into a plastic state in readiness for the manufacture of brick or pottery objects.
Pyrometer	A scientific instrument used in the measurement of temperature in a kiln.
Rat trap	A construction of headers and stretchers in which the bricks used are laid on edge, as a result a cavity is formed.
Red oxide of iron (Ferric Oxide)	Occurs widely in clays and is widely used as a colouring agent in glazes, producing browns and tans.

Refractory bricks	Specially manufactured bricks made from fire clay and used to line kilns. Raw fireclay is usually found in conjunction with coal deposits and has a different chemical composition to other brick clays and shales. Refractory bricks were also used extensively in iron and steel foundries,
Rockface	A brick face texture where the face of a brick is chipped off.
Rubbers	Large bricks which are then rubbed to size giving a very precise, smooth finish.
Salmon brick	Generic term for under-burned brick which are more porous, slightly larger, and lighter coloured than hard-burned brick. Usually pinkish-orange colour.
Salt glazed	When common salt is thrown in to fireboxes during the final stages of a stoneware firing soda is released. When combined with the silicate within the clay a skin (hard glassy surface) of sodium aluminium silicate forms on the surface of the pottery.
Sand moulding	Traditional brickmakers sprinkled sand on the stock board, mould and the table before kneading a 'clot', a piece of clay into the rough shape of a brick. This produced a brick with a sandy textured face, hence the name sandstock brick.
Sands Directory	An early residential and commercial directory published in Sydney between 1858 and 1931–32.
Scotch kiln	A series of rectangular open-topped chambers about 3.5 metres wide by 3.6 metres high.
Scottish bond	One row of headers to five of stretchers.
Semi plastic bricks	A type of extruded brick.
Setting	The placement of green bricks in kilns in readiness for firing.
Sewer brick	Low absorption, abrasive-resistant brick intended for use in drainage structures.
Skintled	Bricks laid at an angle on the hack. A small space was left between each brick to aid the drying process.
Slurry	A texturing process where a coloured solution is applied and kiln fired into the body of the brick to create a permanent, ceramic finish.
Smoking	The slow preheating of bricks in a kiln to drive out moisture content.
Soft mud brick	Brick produced by moulding relatively wet clay containing 20 to 30 percent moisture. Often a hand process.
Soldier course	A course of bricks laid with the long sides upright.
Specials	Bricks which have to be made for a specific purpose such as forming features or construction details that cannot be built from standard bricks.
Stack bond	Stretchers laid one above the other with the perpendiculars in line
Stiff mud brick	Brick produced by extruding a stiff, but plastic, clay (12 to 15 percent moisture) through a die.
Stiff plastic bricks	A type of extruded brick.
Stock brick	A brick, hand made, using a stock mould. Later came to mean a large number (stock) of bricks all manufactured in the one locality, ie London stock brick.

Stoking	Adding fuel to the fire of a kiln.
Stretcher bond	A brick laid with its long side to the face of the wall.
Stucco	A lime based render applied to the exterior of a building. Often scored to imitate courses of masonry, then called channelled, and sometimes more deeply incised to give the appearance of roughly hewn stone, in which case it is rusticated.
Suffolk Punch	Breed of draught horse named after the English county.
Terrace housing	Row of houses, usually working man's cottage.
Terracotta	Unglazed, low-fired pottery produced from a fine clay, usually of reddish colour. Terracotta ware commonly includes architectural pottery, garden ornaments, flower pots, and butter coolers.
Texture bricks	Extruded brick whose sides pass through a series of rollers that mark the side to produce a textured effect.
Tumbled or rumbled	A term used to describe the distortion of brick edges.
Tunnel kilns	Rectangular-shaped kilns where kiln cars stacked with bricks moved on rails through a series of fire zones.
Updraught (Kiln)	Kilns in which the heat and gases from the fire boxes pass directly through the ware to the chimney or stack.
Vanadium	A form of efflorescence derived from yellow or green salt deposits that is particularly visible on light-coloured bricks.
Weathering	To expose clay to the elements after it has been mined in order to break it up and improve its plasticity.
Weighted Sound Reduction Index (Rw)	A measure of the reduction in sound level passing through a wall.
Wheeler	A barrowman who would load green bricks onto a barrow and then wheel the load to the wickets of the patent or downdraught kiln.
Wianamatta shales	Bands of shale found throughout the Sydney region.
Wicket	The entrance to the kilns chambers which, when sealed prior to burning of the bricks, was used as the wicket in a game of cricket.
Wire cut bricks	Wire cut bricks were extruded bricks made from semi plastic clays i.e. stiffer than that preferred by hand moulders. The process was ideal for the softer, alluvial clays of Victoria.

Index

Page numbers in italics refer to illustrations.

Index

M

N

O

P

T

U

V

W

TELEPHONE: L 1044
BRICK SHEDS
FACING BRICK KILN
KILNS
BRICK
FACING
HARDY
J. GENTLE
OFFICES